AF770619

# Marine Acoustics

## The Physics of Sound in Underwater Environments

**Jens M. Hovem**

**Peninsula Publishing**
**Los Altos Hills, California**

Published in the United States of America by Peninsula Publishing
Los Altos Hills, California
www.peninsulapublishing.com

Library of Congress Control Number: 2012948251

Hovem, Jens M., 1937-
Marine acoustics : the physics of sound in underwater environments / Jens M. Hovem. -- 1st ed.
p. : ill. ; cm
Includes bibliographical references and index.
ISBN: 978-0-932146-65-6
1. Underwater acoustics. 2. Sonar. I. Title.
QC242.2 .H68 2012
620.25

ISBN: 978-0-932146-65-6

Cover illustration courtesy of Kongsberg Maritime ASA

Printed by Lightning Source Inc., La Vergne, TN

Printed in the United States of America

# *Preface*

The objective of this new book, *Marine Acoustics*, is to provide graduate and undergraduate students of sonar and underwater acoustics and their teachers, engineers, and technicians with an unusually *complete* text covering the wide gamut of sonar engineering. The text is supported by an exceptionally strong foundation of underwater acoustic theory and the book provides a wealth of up-to-date references that can assist the reader to further detailed study. The book has resulted from a long number of years teaching underwater acoustics at the Norwegian University of Science and Technology (NTNU).

*Marine Acoustics* is an application-oriented introduction to sonar engineering, the fundamentals of detection and classification of fish, plankton, ships, submarines, autonomous underwater vehicles, torpedoes and mines, and all underwater objects. It addresses acoustic remote sensing of topography, structure and composition of the seabed, as well as mapping oceanographic conditions in the sea; it includes technical subjects such as acoustic transducers, parametric sources, sound in marine sediments, geoacoustic measurements, and, of course, sound propagation models and performance analyses of sonars. Over the years in the writing of this text I have gradually expanded the book by the inclusion of basic wave theory and the addition of advanced topics and applications, now making this underwater acoustics book, I believe, of broader scope than any other book now available. Indeed, the applications I have addressed have much in common with the use of acoustics in ultrasound in medicines, materials testing, and seismic methods of geophysics.

The author is indebted to many institutions and persons for their involvement in transforming this text into a textbook. Most important have been the encouragement and support from the Applied Research Laboratories, The University of Texas at Austin (ARL) with the Executive Director, Dr. Clark Penrod and the critical and constructive reviews from David Knobles, Bob Koch, Evan Westwood, Nick Chotiros, and Preston Wilson. The author also thanks his colleagues at NTNU, in particular Professor Hefeng Dong, for her assistance in checking and correction of the many equations in the text. Finally, but not the least, I am grateful for the continued support from the scientific editor Dr. Ross Chapman, the book editor John Hammett, and the publisher Charles Wiseman.

—Jens M. Hovem

July 2012

# *Contents*

# *Chapter 1*

# *Introduction and overview*

## 1.1 Background

The word *acoustics* originates from the Greek word meaning "to listen." The original meaning concerned only hearing and sound perception. The word has gradually attained an extended meaning and, in addition to its original sense, is now commonly used for almost everything connected with rapidly varying mechanical vibrations, from noise to seismic and sonar systems, to ultrasound in medical diagnosis and materials technology.

An important technical application of acoustics is related to undersea activities, where acoustic waves are used in much the same way that radar and electromagnetic waves are used on land and in the air—for the detection and location of objects and for communications. The reason that acoustic rather than electromagnetic waves are used in seawater is simple: electromagnetic waves are strongly attenuated in salt water and would, therefore, have too short a range to be useful for most applications. Acoustic waves, on the other hand, are attenuated significantly less than electromagnetic waves and can propagate with greater strength to longer ranges.

The purpose of this book is to give an introduction to the use of underwater acoustics for the detection and classification of objects such as fish and plankton, submarines and mines, and for communications. It also serves as a basic introduction to acoustic remote sensing of the seabed's topography, structure, and composition, as well as to mapping oceanographic conditions in the sea. In addition to the basic concepts, the book also covers more technical subjects such as acoustic transducers, sonar systems, and sonar performance analysis. These applications also have a lot in common with the use of acoustics in ultrasound in medicine, materials testing, and the seismic methods of geophysics. This textbook is therefore relevant for most applications of acoustics.

This book resulted from a number of years of teaching underwater acoustics at the Norwegian University of Science and Technology. The initial purpose was to give a technical and application-oriented introduction to sonar and sonar engineering. Gradually the book has been expanded by the inclusion of more basic wave theory and by the addition of more advanced topics and applications; the book now provides a broader scope than any other textbook generally available.

The interests and personal experience of the author naturally influence the selection of themes, and the need to consult other sources must be emphasized. Unfortunately, there are not many books that cover the whole field of underwater acoustics. Most of the literature related to acoustics primarily focuses on electroacoustics and air acoustics, and this focus affects the presentation and the examples found there. Besides, these books often lack sufficient coverage of elastic wave propagation in solid materials.

Some of the best general textbooks on acoustics include Kinsler et al. (2000) and Pierce (1989). The first also has a chapter on underwater acoustics worth reading. The work by Morse and Ingard (1968) is a classic and is often used for lectures in theoretical acoustics. The books by Clay and Medwin (1977) and Medwin and Clay (1998) come closest with regard to the selection of topics in this book. The book by Brekhovskikh and Lysanov (2003) is a more theoretically oriented book on ocean acoustics and is highly regarded. The book by Lurton (2002) is also recommended as a general textbook in underwater acoustics, as is the more specialized book on fishery acoustics by Simmonds and MacLennan (2005). The collection of benchmark papers edited by Albers (1972) contains reprints of some of the most significant work in underwater acoustics up to 1971 and is interesting both from scientific and historic viewpoints.

When it comes to sonar engineering, the book by Urick (1983) is a mixture between a textbook and a reference work, and it is the central source of reference for applied underwater acoustics, as well as being a useful and popular book. Other books on sonar systems include those authored by Burdic (1984) and Waite (2002). Among the more specialized textbooks is that by Jensen et al. (1994). It offers a thorough treatment of modern mathematical-numerical modeling of acoustic and seismic wave propagation. Within seismology there is, of course, a wide range of relevant books on different levels. As a basis and a source for further references, the book by Sheriff and Geldhart (1995) is recommended.

## 1.2 Applications

The military applications of underwater acoustics have always been of great importance; since World War I, they have been especially necessary for the detection and classification of submarines. In such applications, both active and passive systems are used. The active systems are based on the transmission of acoustic signals and echo analysis, while the passive systems receive and analyze the acoustic noise emitted from distant vessels of interest. Another important military application is in the detection of mines, either in the water or buried, or partly buried, in the seabed. For all systems, range and reliability are of major importance, and these factors strongly depend on the prevailing oceanographic conditions with regard to the temperature and saline content of water, as well as the action of the wind and waves. For long-distance sonar applications that use very low frequencies to compensate for seawater absorption, the properties of the seabed are very important. In many cases, conversion to elastic waves and absorption by the seabed are the most important transmission loss mechanisms.

The most relevant civilian applications have traditionally been the ones related to fisheries and to fishermen's use of acoustic methods to search for fish. Equally important, and more demanding, is the scientific use of acoustic methods in the oceans to determine the distribution of fish species and the sizes of the fish, information that is required for the sustainable exploitation of ocean resources.

Offshore activities have benefited from underwater acoustics in several respects. Acoustics is used in reference systems for oil rigs, diving vessels, and other services. In recent years, acoustic systems have been particularly useful for remote sensing of the seabed, both as hydrographic systems for mapping of the seabed's topography and as systems aimed at providing information about the structure, composition, and physical qualities of the seabed. Other civilian applications related to offshore activities are the location and control of oil and gas pipelines, wireless telecommunications with subsea installations, and underwater vehicles. These applications have grown in importance as the deep-water extraction of oil and gas has increased.

The link between acoustics and oceanography makes possible the use of acoustics for the monitoring of oceanographic conditions. Acoustic tomography, based on measurements of travel times between a number of transmitters and receivers, can be used for monitoring the temperature structure in the sea, and in this way can supply information about oceanographic fronts and eddies. Of particular interest is a series of ongoing international projects where sound is transmitted across great distances, up to 20,000 km, with a transmission time of about three and a half hours. By measuring the travel time very accurately over several years, it might be possible to determine long-term variations and subsequently to be able to draw conclusions about possible changes in global temperature (Munk, Worcester, and Wunsch 1995).

Acoustics is also used for medical diagnosis and in materials research in order to detect internal inhomogeneities and objects. Such applications of acoustic waves are commonly known as ultrasound, which in many ways is an unfortunate term. Ultrasound refers to acoustic waves and signals with frequencies above the frequencies of hearing, which for young, healthy humans reaches up to about 20 kHz. However, as there is no fundamental difference between sound above and below this frequency, such a division can seem quite meaningless. We will still use the term *ultrasound* but maintain that in principle there is little difference between, for instance, the application of sound waves at the megahertz range in medicine and the use of acoustics for detection of fish using kilohertz, or the mapping of the seabed with a frequency measured in hertz. Common to all the above applications is the demand for knowledge about how acoustic waves propagate in fluid and solid materials—both by those who work with constructing underwater acoustic systems and those who use them.

## 1.3 Historical background

Underwater acoustics is not a new science and technology, as many may be tempted to believe. A pioneer in so many other areas, Leonardo da Vinci was also the first man to grasp the possibilities of acoustic detection. Around the turn of the $15^{th}$ century, he wrote that "If you stop your ship, put a long pipe down into the water and listen, you are able to hear the noise from distant ships." What is described here is the principle of passive sonar, which currently is used for the detection of submarines at distances of meters to several decades of kilometers and which has an immense strategic and tactical significance.

Da Vinci also understood that sound propagates with a defined speed and that this fact could be used to establish the distance between a sound source and a receiver. This principle, which he described around 1500, was used by Colladon and Sturm in 1826 to measure the sound speed in the fresh water of Lake Geneva (Laskey 1977). An acoustic wave was released in the water simultaneously as a glimpse of light was flashed in the air. By measuring the time difference between the received acoustic and optical signals, Colladon and Sturm could determine the speed as being 1438 m/s at a water temperature of 8°C, a value of surprisingly high accuracy, given the means and methods available then.

This value can be compared to that obtained through the theoretical calculations of Sir Isaac Newton, who in his 1687 book, *Mathematical Principles of Natural Philosophy,* formulated much of the basic knowledge about acoustics, among other things assessing sound speed in gases and fluids. Almost two hundred years later, in 1877, Lord Rayleigh (John W. Strutt) published his two volume *Theory of Sound*, a monumental work that still defends its place as a standard reference.

The first "useful" application of subsea detection with acoustic waves took place in the wake of the *Titanic* disaster in 1912, when R. A. Fessenden demonstrated the detection of an iceberg at a range of 2 km. He used an electromagnetic device for both transmitting and receiving; it was similar to the one used in loudspeakers. The frequency was quite low, about 1 kHz, so direction determination proved to be difficult.

The possibility of applying acoustics for the detection and positioning of subsea objects was limited for a long time by the lack of effective sound sources and detection instruments. During the 19th century, a number of discoveries were made that changed the situation. Magnetostrictive, and later piezoelectric, effects were discovered and became the basis of the electromechanical transducers used today.

During World War I systematic use of underwater acoustics was first made for the passive monitoring of submarines. This developed into active acoustic methods toward the end of the war. Around 1917, the Frenchman Paul Langevin developed a new kind of electromechanical transducer based on piezoelectricity, with which he made the first successful detection of a submarine at a distance of about 8 km. By using the relatively high frequency of 38 kHz, he was able to produce a transducer large enough to generate a powerful acoustic signal in a narrow beam, so that both the distance and direction of the objective could be measured.

In the years between the two world wars, there was relatively little development in the military acoustics field, but it was during these years that the first civilian applications were introduced. The echo sounder was invented and used for measuring water depths and charting of shipping channels. A little later it was discovered that the same principle could be used for the detection of fish. Here Norwegians were among the pioneers, and a fundamental work was O. Sund's article from 1935, "Echo sounding in fishery research." Also of interest is the book *Fiskeletning med sonar,* or in English, *Fish Detection with Sonar* (Midtun 1962), which has contributions from a number of Norwegian pioneers in underwater acoustics.

Developments gained further speed during World War II. For the allied forces to carry out their operations in Europe, it was essential to keep open the supply lines to the USA, which were under continuous threat from German submarines. This threat led to intensified research in order to improve underwater acoustic techniques. Most of the effort was put into the development and use of active sonar for the detection of submarines at close range. The importance of the results achieved in the field of acoustics is comparable to that of achievements in the better-known field of electromagnetic research. The terms *sonar* (sound navigation and ranging) and *radar* (radio detection and ranging) both originated at this time. By the end of the war, the basis for what we today know as underwater acoustics was widely known—at least qualitatively. Many of the discoveries from this period are presented in *Physics of Sound in the Sea* (U.S. Government Printing Office 1969).

After World War II, new threats appeared with the introduction of large submarines and long-range missiles, which required detection at longer distances. Active sonar, which struggled with two-way transmission loss, was no longer sufficient. Thus, much research during the Cold War years was devoted to the development and use of passive acoustic systems for the detection and classification of distant vessels, based on the reception and analysis of the acoustic noise generated by such objects. For this kind of long-range detection, oceanographic conditions are of great importance, which has led to a close link between underwater acoustics and oceanographic research. In recent years, military interests and demands have again changed, and the use of active sonar for the detection of mines in coastal waters has become increasingly relevant.

As the author of this book, I am pleased to say that Norwegian industry and research institutions have a respectable position in underwater acoustics, with the Havforskningsinstituttet (Marine Research Institute) in Bergen being a leading institution in fishery acoustics and the Forsvarets forskningsinstitutt (FFI, the Norwegian Defense Research Establishment) carrying out research in the military field. Simrad, which today is a part of the Kongsberg Group, is an international producer for all kinds of underwater systems and a market leader in several product areas.

The development of underwater acoustics has, as is true for several other fields, come as a result of general technical development, particularly in the areas of electronics and computer science. Improved electronics have made it possible to receive weaker and weaker signals, while advanced computer science has enabled us to implement signal-processing routines that earlier were merely a theoretical possibility. The same applies to mathematical models for wave propagation and reflection in complex geometries and structures. The increasing possibilities of implementing sophisticated mathematical-numerical models for wave phenomena have made mathematical treatment of wave phenomena more important today than ever before.

## 1.4 Contents and organization

Figure 1.1 is a simple sketch illustrating some of the topics that will be treated in this book. It shows a typical echo localization context, in which we use sound to detect and also perhaps to determine the size and other qualities of an object located inside another medium. This situation is relevant for a sonar system meant for the detection of fish or other objects located in the water, or on the seabed, or buried in it.

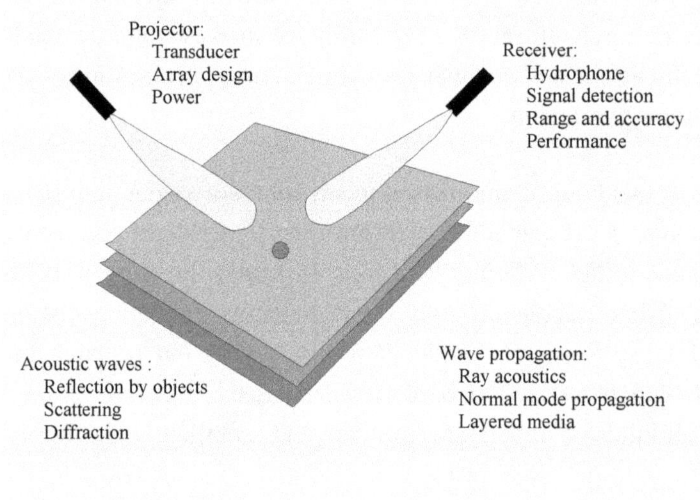


Figure 1.1. Echolocation. The figure schematically depicts the detection of a hidden object by sonar. Some key words relevant to the operation are listed as part of the figure; these terms are treated in this book.

A relatively narrow sound beam is sent from a projector toward the medium, which may consist of several more-or-less distinct layers of materials with different acoustic qualities. Because of these differences, the sound beams are refracted at the layer interfaces, and also partly reflected, before they ensonify the target. Some of the reflected energy in the echo signal from the target is received by the hydrophones and fed to the receiver. From analyzing the echo signal, the receiver must decide whether there is a target present and, if so, classify it with respect to size and composition. In the figure, the projector and hydrophones are separated in space. However, in most cases they are collocated and even use the same transducer.

The problem outlined above is relevant to more than just marine acoustics and echolocation in the sea. In principle, one meets the same situation in a number of other applications, for instance, when using ultrasound for medical diagnosis, for nondestructive testing of materials, or in seismic investigations of the underground in the search for gas and oil reservoirs.

The purpose of this book is to give a comprehensive coverage of all problems related to design and use of such acoustic systems. This necessitates knowledge in a number of fields, most of which will be treated here.

The book's material can be divided into three parts:

- Part I, Chapters 2–7, Basic physical principles
- Part II, Chapters 8–11, Technology and methods
- Part III, Chapters 12–18, Advanced subjects

Part I, composed of chapters 2 through 7, aims to give the basic introduction to underwater acoustics. The acoustic wave equation is derived in chapter 2, and solutions of the wave equations with boundary conditions are presented for plane, cylindrical, and spherical waves. The concept of acoustic impedance is introduced, and the intensity is determined for plane waves and waves radiated from a point source.

The sound speed in the sea varies with temperature, salinity, and pressure, which is a function of depth. The variations are not large, a maximum of approximately 5% around a nominal value of 1500 m/s, but these variations have major importance for sound propagation in the ocean. Chapter 3 contains a brief summary of the oceanographic and acoustical properties of the ocean environment and the seabed.

Chapters 4 and 5 focus on reflection and transmission in the interfaces between media with different acoustic qualities, including the treatment of multiple reflections in layered media.

In chapter 6, the wave equation is solved by means of ray-tracing calculations. This is the simplest technique for sound field calculations, but in most cases the results are valid for active sonar, which usually uses relatively high frequencies. Besides, the results and method are very descriptive, and understanding them is a necessary prerequisite before one can understand the other methods. In shallow waters and low frequencies, sound propagation can more effectively be described by a modal description, where the sea is considered to be a waveguide bounded by its interfaces with the seabed and surface of the sea. This is the theme of chapter 7.

After presenting the basic physical principles in part I (chapters 1 through 7), part II (chapters 8 through 11) aims to describe the applied technology. Again, while this material is relevant for several applications, the emphasis is put on underwater acoustics systems, such as sonar and echo sounder, to provide a basis for the construction and use of sonar for detection and localization in the sea. Chapter 8 treats the electroacoustic transducers used for generating and receiving sound. Acoustic projectors are the focus of chapter 9, and the most important elements of signal processing in sonar systems are treated in chapter 10.

Sonar equations and their different terms are discussed in chapter 11, with several example analyses of realistic sonar systems. The sonar equations are a set of relatively simple relations that make it possible to discuss the performance of various sonar systems and see how changes in the parameters affect signal range and area coverage. This systems-oriented discussion on sonar systems at such an early stage provides the background for the later chapters, where we turn to a more comprehensive analysis of the different problems. Parts I and II are structured to be read in sequence. Still, it is recommended that the student take an early look at chapter 11, the chapter on sonar systems, to increase understanding of the relevance of the more theoretical chapters.

Part III, composed of chapters 12 through 18, is based on the material of part I, and contains a number of new topics as well as extending the discussions of topics that have been touched upon in the earlier chapters.

Chapter 12 gives a short introduction to nonlinear acoustics, with emphasis on parametric sound sources that utilize the nonlinearity of acoustic propagation to generate rather narrow sound beams at low frequency. Such systems have interesting applications, with potential advantages over conventional systems in detection under reverberation-limited conditions, one example being the detection of objects embedded in the seabed.

Chapter 13 treats reflection and scattering from particles and bubbles. Reflection and scattering are important for many marine acoustics applications. For instance, acoustics is used in determining the quantity of plankton, fry, and fish in the sea; the determination is based on measuring the intensity of acoustic echoes. In order to relate the echo intensity to the size of the biomass, it is essential to be familiar with the echo characteristics produced by the various species of marine life. In this chapter we present the basic theory about echo formation of simple objects, such as spheres and gas-filled bubbles. Chapter 14 follows up and deals with reflection, propagation, and diffraction from plane and curved surfaces and in this way provides the basis for understanding what determines the target strength of larger objects. Chapter 14 also treats scattering from uneven and rough surfaces and develops some useful models for the estimation of reverberation levels from the ocean's surface and from uneven seabeds.

Chapter 15 treats acoustic and elastic waves in solid media, where transverse or shear waves may exist in addition to longitudinal waves. This chapter can therefore be considered to be parallel to chapters 2 through 5 but treats wave propagation in solids rather than in seawater. This is essential in the study of sound in marine sediments and rocks. This chapter is relevant not only for seismic applications but also for those used in underwater acoustics, particularly for long-range and low-frequency propagation, where the interaction of longitudinal waves with solid media and their resulting conversion to shear waves may constitute the limiting factor for the performance of an acoustical system.

Chapter 16 deals with the acoustic properties of marine sediments, with the emphasis being on the Biot theory about sound in porous media. This is relevant to a large range of applications, from seabed acoustics to acoustic noise control and sound absorption materials.

Chapter 17 presents the mathematical basis for numerical solutions of the acoustic and elastic wave equations for layered media, describing some of the methods that are the basis for the many numerical wave propagation models existing today. Finally, chapter 18 offers a brief review of the most important methods for marine seismic measurement, with emphasis on methods applicable for investigating layers and structures relatively close to the water-bottom interface.

It is quite possible that some readers will find this textbook too mathematical and would have preferred to see more results and material from experiments at sea. In defense of the theoretical approach of this book, I quote David Weston, one of the best-known underwater acousticians of our time. In an internal report dated 1976, he wrote:*

> It is convenient to base the discussion here on theoretical considerations, since theory tends to be better behaved and more consistent than experiments, and can be used to bring out physical principles.
> However, the comments here have to be, and in this case they are, backed up by the measurements, and for details and for precise levels one must go to measurements.

This textbook aims to cover the theoretical material referred to in the first section of the quotation above. However, backing the theory up with examples from experimental results and real measurements is largely the responsibility of the lecturer, who must draw on her or his own interests and experience.

* Quotation provided courtesy of Ian Roebuck, Winfrith Technology Center, Dorchester, Dorset, DT2 8WX, UK.

# *Chapter 2*

# *Acoustic waves*

## 2.1. Introduction

This chapter introduces the theory of acoustic waves. First we formulate the wave equation for acoustic waves in fluids and then show how to incorporate a source term. We next go on to investigate how the wave equation may be solved in the time and frequency domains and in various coordinate systems. Finally, we introduce the principle of reciprocity, which is of fundamental practical importance for wave phenomena and linear systems. The material contained in this chapter can be found in most textbooks on acoustics. As supplemental reading, the books by Kinsler et al. (2000), Clay and Medwin (1977), Medwin and Clay (1998), Pierce (1998), Frisk (1994), Officer (1958), Morse and Ingard (1968), and Morse and Feshbach (1953) are recommended.

## 2.2. Derivation of the wave equation

Acoustic waves are mechanical vibrations. When an acoustic wave passes through a medium, it causes local density changes that are related to local displacements of mass about the rest positions of the particles in the medium. This displacement leads to the formation of forces that act to restore the density to the equilibrium state and move the particles back to their rest positions. The medium may be a gas, liquid, or solid material. The basic equations of acoustics are obtained by considering the equations for an inviscid and compressible fluid. These equations are expressed with the notation that $p$ is the pressure, $\rho$ is density, $S$ is entropy, and $\mathbf{v}$ is particle velocity. Bold face letters denote vectors throughout the book.

We now formulate more thoroughly the wave equation describing acoustic waves in fluids, using three simple principles:

- The momentum equation, also known as Euler's equation
- The continuity equation, or conservation of mass
- The equation of state: the relationship between changes in pressure and density or volume

We consider a small rectangular volume element as shown in Figure 2.1, with the sides *dx, dy,* and *dz*, volume $V = dxdydz$, and mass $\rho V$ with $\rho$ being the density of the fluid. As a result of the acoustic wave action, mass will alternately flow into and out of the element with displacement velocity, or particle velocity, $\mathbf{v} = [v_x \; v_y \; v_z]$, depending on the position of the element and the time.

The net mass flow out of the element in the direction $x$ is

$$\left\{\left[\rho v_x + \frac{\partial(\rho v_x)}{\partial x}dx\right] - \rho v_x\right\}dydz = \frac{\partial(\rho v_x)}{\partial x}dxdydz \tag{2.1}$$

For the two other directions, we get similar expressions, so that the net mass flow out of the element becomes $\nabla \cdot (\rho \mathbf{v})$, where $\nabla$ is the gradient operator.

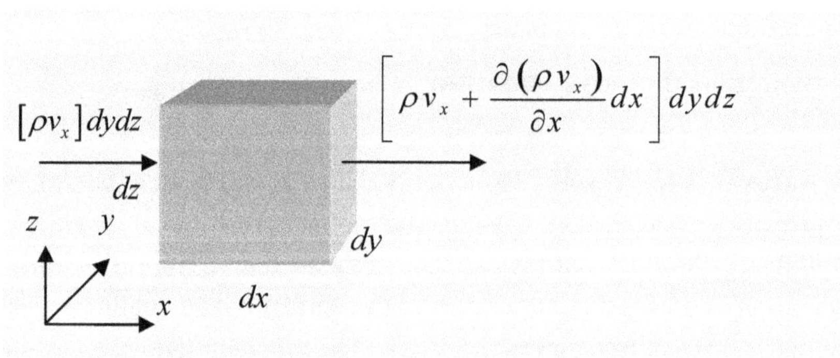


Figure 2.1. The flow of mass into and out of a small volume element. Only the flow in the positive $x$ direction is shown.

***Conservation of mass*** implies that the net changes in the mass, which result from its flow through the element, must be equal to the changes in the density of the mass of the element. This is expressed by the ***continuity equation***

$$\frac{\partial \rho}{\partial t} = -\nabla \cdot (\rho \mathbf{v}) \tag{2.2}$$

The continuity equation, equation (2.2), is nonlinear since it contains the product of density and particle velocity, both being functions of time and position.

***Euler's equation (momentum balance)***

$$\rho\left[\frac{\partial \mathbf{v}}{\partial t} + \mathbf{v} \cdot \nabla \mathbf{v}\right] = -\nabla p \tag{2.3}$$

Equation (2.3) is known as Euler's equation and is an extension of Newton's second law, which states that force equals the product of mass and acceleration. The extension is the second left-hand term in equation (2.3), which represents the change in velocity with position for a given time instant, while the first term describes the change with time at a given position.

An ***equation of state*** is required to give a relationship between a change in density and a change in pressure, taking into consideration the existing thermodynamic conditions. The equation of state may be formulated as pressure as a function of density and entropy $S$ in the form of

$$p = p(\rho, S) \tag{2.4}$$

For a fluid medium it is normally assumed that the passage of an acoustic wave is nearly an adiabatic and reversible process since the thermal conductivity and the temperature gradients of the disturbance associated with the acoustic waves are small enough that no appreciable energy transfer occurs between adjacent fluid elements. This means that the entropy remains constant and that the process is isentropic. With this assumption the equation of state is simply

$$p = p(\rho) \tag{2.5}$$

Equations (2.2), (2.3), and (2.5) are all nonlinear, and in the next section we show how the equations can be linearized and combined to produce the acoustic wave equation.

However, in a real fluid, the dissipation processes like viscosity and thermal conduction act to increase the entropy, as demanded by the second law of thermodynamics. We can say that entropy either remains constant (as for the assumed reversible processes) or increases. The increase in entropy is associated with heating of the fluid as the sound wave passes and a corresponding decrease in the energy of the sound wave. The effect of an increase of entropy leads naturally to the idea of attenuation of the sound.

### *2.2.1 Linearization*

Equations (2.2), (2.3), and (2.5) can be linearized by assuming that each physical quantity is a function of a steady-state, time-independent value and a small fluctuation term

$$\begin{aligned} p &= p_0 + p' \\ \rho &= \rho_0 + \rho' \\ \mathbf{v} &= \mathbf{v}_0 + \mathbf{v}' \end{aligned} \tag{2.6}$$

The variables $p'$, $\rho'$, and $\mathbf{v}'$ are functions of time and spatial positions; the variables $p_0$, $\rho_0$, and $\mathbf{v}_0$ are independent of time, but may be functions of spatial positions. We also assume that there is no mean flow, that is, $\mathbf{v}_0 = 0$. We expand equation (2.5) about the equilibrium values, indicated by the subscript 0, and obtain

$$p = p_0 + \left[\frac{\partial p}{\partial \rho}\right]_0 (\rho - \rho_0) + \frac{1}{2}\left[\frac{\partial^2 p}{\partial \rho^2}\right]_0 (\rho - \rho_0)^2 + \ldots.. \tag{2.7}$$

The first perturbation term of equation (2.7) yields

$$p' = \left[\frac{\partial p}{\partial \rho}\right]_0 \rho' \tag{2.8}$$

As we will see later, the bracketed term in equation (2.8) is identified as the squared value of the sound speed $c_0$,

$$c_0^2 = \left[ \frac{\partial p}{\partial \rho} \right]_0 \tag{2.9}$$

Consequently, equation (2.8) can be written

$$p' = c_0^2 \rho' \tag{2.10}$$

We may also define the term $K_0$ by

$$K_0 = \rho_0 \left( \frac{\partial p}{\partial \rho} \right)_0 \tag{2.11}$$

$K_0$ is called the volume stiffness, volume elasticity, or bulk modulus and is defined by equation (2.11) as the derivative of pressure with respect to density, evaluated at the isentropic conditions. The term $K_0$ has the same dimension as pressure, which is expressed in $N/m^2$ or in pascal (Pa).

Neglecting the higher-order terms in equation (2.7) implies that we have assumed linear elasticity—that is, an assumed linear relationship between the changes in pressure and volume.

Using equation (2.6) we find that Euler's equation [equation (2.3)] reduces to

$$-\nabla p' = (\rho_0 + \rho') \left[ \frac{\partial \mathbf{v}'}{\partial t} + \mathbf{v}' \cdot \nabla \mathbf{v}' \right] \tag{2.12}$$

When

$$(\mathbf{v}' \cdot \nabla) \mathbf{v}' << \frac{\partial \mathbf{v}'}{\partial t} \tag{2.13}$$

and

$$\rho' << \rho_0 \tag{2.14}$$

the first-order term in equation (2.12) yields

$$-\nabla p' = \rho_0 \frac{\partial \mathbf{v}'}{\partial t} \tag{2.15}$$

The continuity equation [equation (2.2)] can be linearized as

$$\frac{\partial \rho'}{\partial t} = -\rho_0 \nabla \cdot \mathbf{v}' \tag{2.16}$$

This assumes that $\rho_0$ is constant in space and therefore can be removed from under the divergence operator. Inserting equation (2.16) into equation (2.10) gives

$$\frac{\partial p'}{\partial t} = -\rho_0 c_0^2 \nabla \cdot \mathbf{v}' \tag{2.17}$$

We now take the partial derivative of equation (2.17) with respect to time and the divergence of equation (2.15) and obtain the acoustic wave equation

$$\nabla^2 p' - \frac{1}{c_0^2}\frac{\partial^2 p'}{\partial t^2} = 0 \tag{2.18}$$

where $\nabla^2$ is the Laplacian operator and $c_0$ is the sound speed at the ambient conditions, which can be defined as

$$c_0 = \sqrt{\frac{K_0}{\rho_0}} \tag{2.19}$$

Thus the sound speed is given by the square root of the ratio between volume stiffness and density. Both volume stiffness and density depend on the properties of the medium, and therefore on ambient conditions such as pressure and temperature that can affect these properties.

The sound pressure $p'$ is given by the difference between the instantaneous pressure and the ambient pressure, $p' = p - p_0$. We define the condensation $s$ as the relative change in density by

$$s = \frac{\rho - \rho_0}{\rho_0} \tag{2.20}$$

The sound pressure can therefore also be expressed as

$$p' = \rho_0 c_0^2 s \tag{2.21}$$

Equation (2.21) shows that the condensation is proportional to the sound pressure. This implies that condensation is also governed by a wave equation,

$$\nabla^2 s - \frac{1}{c_0^2}\frac{\partial^2 s}{\partial t^2} = 0 \tag{2.22}$$

Hereafter, we will omit the primes for the pressure perturbations with the understanding that from now on $p$ is the sound pressure, that is, the fluctuation or change in pressure caused by the acoustic wave.

We can now briefly turn to air acoustics and consider an ideal gas with the equation of state expressed as

$$p = p_0\left(\frac{\rho}{\rho_0}\right)^{\gamma} \tag{2.23}$$

where $\gamma$ is the ratio of the specific heat per unit mass at constant pressure and constant volume, $\gamma = C_p/C_v$. Equation (2.23) also implies that $p \cdot V^{\gamma} = constant$, so that for ideal gases with pressure $p_0$, the volume stiffness is

$$K_0 = \gamma\, p_0 \tag{2.24}$$

and the sound speed is

$$c_0 = \sqrt{\gamma \frac{p_0}{\rho_0}} \tag{2.25}$$

As an example, let the ambient pressure be equal to one atmosphere, $p_0 = 1$ atm $= 1.013 \cdot 10^5$ Pa, the density $\rho_0 = 1.3$ kg/m$^3$, and $\gamma = 1.4$. Equation (2.25) gives the value of the sound speed to be $c_0$ = 331.2 m/s. For an ideal gas, the pressure is proportional to the density and temperature in Kelvin, $T_K$, that is, $p \propto \rho T_K$. Thus, the sound speed $c$ of a gas with any temperature $T$ (°C) is given by

$$c = c_0 \sqrt{1 + \frac{T}{273}} \tag{2.26}$$

For water and other fluids, sound speed must be considered a parameter that has to be measured. In water, the speed may vary from approximately 1450 m/s to 1540 m/s, depending on pressure (depth), temperature, and salt content. (We will treat the variation in sound speed more thoroughly in chapter 3.) The interval according to which the sound speed may vary is no more than about ±6%, though when it comes to sound propagation in seawater, this variation is very important, as we shall see later. In other contexts, however, the correct value of the sound speed is not important, so here we apply the nominal sound speed $c_0$ = 1500 m/s and the nominal density $\rho_0$ = 1000 kg/m$^3$.

### *2.2.2 Velocity and displacement potentials*

Equation (2.18) gives the wave equation for sound pressure. Other variables, such as particle velocity and particle displacement, also must satisfy the same wave equation. For instance, the wave equation for particle velocity **v** is

$$\nabla(\nabla \cdot \mathbf{v}) - \frac{1}{c_0^2}\frac{\partial^2 \mathbf{v}}{\partial t^2} = 0 \tag{2.27}$$

.

In some cases we would prefer to operate with the particle displacement **u** such that the particle velocity is the derivative with respect to time of the particle displacement, that is,

$$\mathbf{v} = \dot{\mathbf{u}} \tag{2.28}$$

Generally, it is more convenient to describe the wave by means of scalar variables, the velocity or displacement potentials. Denoting the velocity potential as $\phi$, the particle velocity is expressed by the gradient of the velocity potential $\phi$ as

$$\mathbf{v} = \nabla \phi \tag{2.29}$$

Similarly, denoting the displacement potential $\xi$, the displacement is expressed as

$$\mathbf{u} = \nabla \xi \tag{2.30}$$

In acoustics, the velocity potential is normally preferred, but in seismic applications with wave propagation in solids, the displacement potential is usually preferred (Aki and Richards 2002). In this book, we mostly use the velocity potential except when discussing waves in solid materials such as rocks and consolidated sediments where we find it more convenient to use particle displacement and displacement potential (chapters 8 and 15).

By combining equation (2.29) with equation (2.27), we find that the velocity potential satisfies the wave equation

$$\nabla^2 \phi - \frac{1}{c_0^2} \frac{\partial^2 \phi}{\partial t^2} = 0 \tag{2.31}$$

It follows that the sound pressure is

$$p = -\rho_0 \frac{\partial \phi}{\partial t} \tag{2.32}$$

Equations (2.29), (2.31), and (2.32) are the ones we apply in our subsequent treatment of wave propagation. Note that the vector operations used are valid in all coordinate systems—that is, for rectangular as well as cylindrical and spherical coordinates.

### *2.2.3 The inhomogeneous wave equation*

The wave equations that were formulated hitherto may be referred to as ***homogeneous wave equations*** because they lack a source term. Introduction of a source terms leads to the ***inhomogeneous wave equation.*** A source can be an external force or an injection of a volume of fluid. Here, we limit ourselves to consider volume sources that are most relevant to acoustics.

A volume source is characterized by its injection of new mass into the medium. The linearized continuity equation, equation (2.16), can then be written

$$\frac{\partial \rho}{\partial t} = -\rho_0 \nabla \cdot \mathbf{v} - \rho_0 f(\mathbf{r},t) \tag{2.33}$$

where $\rho_0 f(\mathbf{r},t)$ represents the rate of mass per unit volume, in kg/m$^3$s, injected by a source located at **r**. Repeating the steps that led to equation (2.18) we obtain the ***inhomogeneous wave equation***, the wave equation with a source term expressed as

$$\nabla^2 p - \frac{1}{c_0^2}\frac{\partial^2 p}{\partial t^2} = \rho_0 \frac{\partial f(\mathbf{r},t)}{\partial t} \tag{2.34}$$

and by the velocity potential as

$$\nabla^2 \phi - \frac{1}{c_0^2}\frac{\partial^2 \phi}{\partial t^2} = -f(\mathbf{r},t) \tag{2.35}$$

Of special interest for future applications is a point source in space given by

$$f(\mathbf{r},t) = \delta(\mathbf{r})q(t) \tag{2.36}$$

where $q(t)$ is the time function of the source excitation and $\delta(\mathbf{r})$ is a three-dimensional spatial delta function with the property

$$\int_V \delta(\mathbf{r})dV = \begin{cases} 1 \text{ when } \mathbf{r}=0 \text{ is inside } V \\ 0 \text{ when } \mathbf{r}=0 \text{ is outside } V \end{cases} \tag{2.37}$$

In spherical coordinates with spherical symmetry the three-dimensional delta function may be written

$$\delta(\mathbf{r}) = \frac{1}{4\pi r^2}\delta(r) \tag{2.38}$$

and in cylindrical coordinates with radial symmetry and a source on the $z$ axis at $z = z_0$,

$$\delta(\mathbf{r}) = \frac{1}{2\pi r}\delta(r)\delta(z - z_0) \tag{2.39}$$

In equations (2.38) and (2.39), $\delta(r)$ is the one-dimensional delta function.

## 2.3. The wave equation in the frequency domain, the Helmholtz equation

We will often treat sinusoidal varying signals and waves, also called harmonic signals and waves. Formally, we move from the time domain to the frequency domain by using the Fourier transformation

$$\Phi(\omega) = \int_{-\infty}^{\infty} \phi(t)\exp(i\omega t)dt \tag{2.40}$$

and back to the time domain by using the inverse transformation

$$\phi(t) = \frac{1}{2\pi}\int_{-\infty}^{\infty} \Phi(\omega)\exp(-i\omega t)d\omega \tag{2.41}$$

Note that this sign convention is usual in the literature on physics but is opposite to the sign convention in the exponents commonly used in signal processing literature. We will generally adopt the sign convention of equations (2.40) and (2.41), but in chapter 8 on acoustical transducers we use the opposite sign convention in order to use material in that chapter from signal processing and electric circuit theory, where a time variation of $\exp(i\omega t)$ is the standard.

Analysis in the frequency domain means that we assume a solution of the wave equation in the form $\Phi(\omega)\exp(-i\omega t)$. After this solution is found, the solution in the time domain, and consequently the time response, is found by using the inverse transformation in equation (2.41). The last step is often not necessary when we treat signals with narrow frequency bands. Active sonar provides examples of narrow bands, which use a bandwidth that rarely exceeds 10% of the center frequency, where a solution at carrier frequency tends to be sufficient. Examples of the opposite case may be found in seismic applications; in these cases the signals typically are transient with relatively broad frequency bands and, therefore, demand a solution for a wide band, followed by a transformation back to the time domain.

The inhomogeneous wave equation for the velocity potential may then be expressed as

$$\left[\nabla^2 + \kappa^2(\mathbf{r})\right]\Phi(\mathbf{r},\omega) = -F(\mathbf{r},\omega) \tag{2.42}$$

where $F(\mathbf{r},\omega)$ is the Fourier transform of $f(\mathbf{r},t)$, defined in equation (2.33), and the wave number $\kappa(\mathbf{r})$ is defined as

$$\kappa(\mathbf{r}) = \frac{\omega}{c_0(\mathbf{r})} \tag{2.43}$$

Equation (2.42) is the wave equation in the frequency domain and is also referred to as the inhomogeneous Helmholtz equation, which is often easier to solve than the corresponding wave equation in the time domain, something we will make extensive use of later. In equation (2.42), we have included a source term and allowed for the possibility that the sound speed is dependent on the coordinates **r**. When there is no danger of misunderstanding, we will remove the argument $\omega$ in the frequency functions, which means that $\Phi(\mathbf{r},\omega) \rightarrow \Phi(\mathbf{r})$.

## 2.4. Absorption

The wave equation was derived under the assumption that there are no losses of energy. While dissipation can be ignored in some cases, there are many situations where absorption is the limiting factor for how far sound can travel. Sound is absorbed in water through three processes: viscous losses, heat conduction losses, and losses associated with intermolecular processes. In chapter 3 we discuss the various contributions and present values for sound absorption in salt and fresh water as functions of frequency.

Here we limit the discussion to show how absorption can be a perturbation in the wave equation that originally was derived neglecting absorption. In the wave equation in the frequency domain, we can account approximately for absorption loss by assuming that the sound speed is complex, adding an imaginary component with value $c_i$ to the sound speed $c_0$. In other words, we make the substitution

$$c_0 \rightarrow c_0\left(1 - i\frac{c_i}{c_0}\right) \tag{2.44}$$

The wave number is now complex. When the loss is relatively small the imaginary component of the sound speed is very much smaller than the real component, that is, $c_i << c_0$ and we can approximate the wave number with

$$\hat{\kappa} = \frac{\omega}{c_0\left(1 - i(c_i / c_0\right)} \approx \frac{\omega}{c_0} + i\omega\frac{c_i}{c_0^2} \tag{2.45}$$

The imaginary component in equation (2.45) is defined as the absorption coefficient $\alpha$

$$\alpha = \omega\frac{c_i}{c_0^2} \tag{2.46}$$

Thus, a plane wave with attenuation is expressed by

$$\Phi(x,\omega) = \Phi_0 \exp(i\hat{\kappa} r) = \Phi_0 \exp(i\hat{\kappa} r - \alpha r) \tag{2.47}$$

This implies that the amplitude of the wave decreases with range according to $\exp(-\alpha r)$, and its intensity by $\exp(-2\alpha r)$. The sign of the imaginary component in equation (2.44) is chosen so that the wave is damped and reduced in amplitude as it propagates. Attenuation, measured in nepers per wavelength, is called the ***log decrement***, $\delta$. The wavelength is $\lambda = c/f$, and therefore the log decrement is defined as

$$\delta = \alpha\lambda = 2\pi\alpha\frac{c_0}{\omega} = 2\pi\frac{c_i}{c_0} \tag{2.48}$$

Alternatively, the attenuation is often given in decibels per wavelength, which is ten times the logarithm to the base 10

$$\delta_{\mathrm{dB}} = -10\log\left[\exp(-2\delta)\right] = -20\log\left[\exp(-\delta)\right] = 8.6859\delta \qquad (2.49)$$

Expressed by the ratio of the imaginary and real parts of the sound speed we obtain from equations (2.48) and (2.49) that

$$\frac{c_i}{c_0} = \frac{\delta_{dB}}{2\pi \cdot 8.6859} = 0.0183\,\delta_{dB} \qquad (2.50)$$

Another often used measure of absorption is dB per meter and per kHz, notation dB/m-kHz.

From equations (2.46) and (2.48), it follows that when the ratio $c_i/c$ is constant the absorption coefficient $\alpha$ is proportional to the frequency and that the attenuation per wavelength is constant. However, this is not the case for pure water, where the damping is proportional to the frequency squared. In saline water, damping is a more complicated function of frequency, as we will see in chapter 3. For acoustic waves in rocks and marine sediments, it is usual to assume that the damping increases proportionally with the frequency; by doing so, the damping per wavelength remains constant and independent of the frequency.

## 2.5. Solution of the homogeneous wave equation

As stated earlier, the wave equation can be expressed in different coordinate systems, provided that we use the correct expressions for the Laplacian operator. We now show what the wave equation will look like in different coordinate systems.

### *2.5.1 Rectangular coordinates (x,y,z)*

The Laplace operator in rectangular coordinates is

$$\nabla^2 = \frac{\partial^2}{\partial x^2} + \frac{\partial^2}{\partial y^2} + \frac{\partial^2}{\partial z^2} \qquad (2.51)$$

It can be shown that the general solution for a plane wave traveling in an arbitrary direction given by **r** is

$$\phi(\mathbf{r},t)=\phi_+\left(t-\frac{|\mathbf{r}|}{c_0}\right)+\phi_-\left(t+\frac{|\mathbf{r}|}{c_0}\right) \tag{2.52}$$

The first part of equation (2.52) represents a wave propagating with speed $c_0$ in the positive **r** direction. The second part of the equation represents a wave in the negative **r** direction. Both waves propagate without any change of amplitude or form, and the wave equation imposes no conditions upon their shapes and forms.

Since the waves are plane, we select the coordinate system so that the direction of the wave concurs with one of the axes, for instance the $x$-axis. In this case, equation (2.52) becomes

$$\phi(x,t)=\phi_+\left(t-\frac{x}{c_0}\right)+\phi_-\left(t+\frac{x}{c_0}\right) \tag{2.53}$$

The sound pressure and particle velocity in the $x$ direction are correspondingly

$$\begin{aligned} p(x,t)&=-\rho_0\frac{\partial\phi(x,t)}{\partial t}=-\rho_0\left[\phi_+'\left(t-\frac{x}{c_0}\right)+\phi_-'\left(t+\frac{x}{c_0}\right)\right]\\ v_x(x,t)&=\nabla_x\left[\phi(x,t)\right]=-\frac{1}{c_0}\left[\phi_+'\left(t-\frac{x}{c_0}\right)-\phi_-'\left(t+\frac{x}{c_0}\right)\right]\end{aligned} \tag{2.54}$$

Here, the $'$ symbol stands for the derivative with regard to time. Notice the minus sign in the expression for the particle velocity for the wave component traveling in the negative direction.

The ratio between sound pressure and particle velocity (which may be complex), called the ***specific acoustic impedance***, $Z$, is equal to the product of the medium's density and its sound speed. For a plane wave propagating in the positive $x$ direction

$$Z=\frac{p(x,t)}{v_x(x,t)}=\rho_0 c_0 \tag{2.55}$$

Plane waves with components propagating in both directions can be written as

$$p(x,t) = p_+\left(t - \frac{x}{c_0}\right) + p_-\left(t + \frac{x}{c_0}\right)$$
$$v_x(x,t) = \frac{p_+}{Z}\left(t - \frac{x}{c_0}\right) - \frac{p_-}{Z}\left(t + \frac{x}{c_0}\right) \quad (2.56)$$

where $p_+$ is the pressure wave propagating in the positive $x$ direction and $p_-$ is the pressure wave in the negative $x$ direction.

In the frequency domain, plane waves propagating in both directions are expressed as

$$\Phi(x,\omega) = A\exp(i\kappa x) + B\exp(-i\kappa x) \quad (2.57)$$

where $\kappa$ is the wave number given by

$$\kappa = \frac{\omega}{c_0} = \frac{2\pi}{\lambda} \quad (2.58)$$

By using equation (2.32), the sound pressure of a plane harmonic wave with a bandwidth of 1 Hz at frequency $\omega$ and propagating in the positive $x$ direction can be written,

$$p(x,t) = p_0 \exp\left[i(\kappa x - \omega t)\right] \quad (2.59)$$

and the particle velocity, with components only in the propagation direction, is

$$v(x,t) = v_0 \exp\left[i(\kappa x - \omega t)\right] \quad (2.60)$$

In equations (2.59) and (2.60), $p_0$ and $v_0$ are the amplitudes of the pressure and the particle velocity, respectively.

Intensity $I$ is defined as the time-averaged energy flux per unit surface normal to the direction of propagation. For plane harmonic waves, and complex notation as in equations (2.59) and (2.60), the time-averaged intensity is given by the real parts of the quantities $p(x,t)$ and $v(x,t)$ as

$$I = \frac{1}{T}\int_0^T \mathrm{Re}\left[p(x,t)\right]\mathrm{Re}\left[v(x,t)\right]dt \tag{2.61}$$

This result can also be expressed as (* denotes complex conjugate)

$$I = \frac{1}{2}\mathrm{Re}[pv^*] \tag{2.62}$$

which, with equations (2.59) and (2.60), gives

$$I = \frac{1}{2}p_0 v_0 \tag{2.63}$$

Using equation (2.55), we find that the intensity of a plane wave in the direction of propagation can be expressed by

$$I = \frac{1}{2}\rho_0 c_0 v_0^{\ 2} = \frac{1}{2}\frac{p_0^{\ 2}}{\rho_0 c_0} \tag{2.64}$$

In equations (2.62), (2.63), (2.64) the intensities are represented by the amplitudes of the particle velocity $v_0$ and the sound pressure $p_0$, therefore the factor is ½. If the root mean squared (rms) of these quantities is used, then the factor ½ disappears. This is common practice, and we use this convention in the following.

### *2.5.2 Cylindrical coordinates (r,φ,z)*

In cylindrical coordinates the axes are $(r,z,\varphi)$, where $r$ is the radial distance, the depth is $z$, and $\varphi$ is the azimuth angle. The Laplace operator in cylindrical coordinates is

$$\nabla^2\phi = \frac{1}{r}\frac{\partial}{\partial r}\left(r\frac{\partial\phi}{\partial r}\right) + \frac{1}{r^2}\frac{\partial^2\phi}{\partial\varphi^2} + \frac{\partial^2\phi}{\partial z^2} \tag{2.65}$$

With azimuthal symmetry, where $\partial\phi/\partial\varphi = 0$, this expression is simplified to become

$$\nabla^2 \phi = \frac{1}{r}\frac{\partial}{\partial r}\left(r\frac{\partial \phi}{\partial r}\right) + \frac{\partial^2 \phi}{\partial z^2} \tag{2.66}$$

Interestingly, a simple form of solution of the cylindrical wave equation in the time domain in the form of equation (2.53) does not exist, but the solution in the frequency domain is straightforward.

The homogenous Helmholtz equation in cylindrical coordinates is

$$\frac{1}{r}\frac{\partial}{\partial r}\left(r\frac{\partial \Phi}{\partial r}\right) + \frac{1}{r^2}\frac{\partial^2 \Phi}{\partial \varphi^2} + \frac{\partial^2 \Phi}{\partial z^2} + \kappa^2 \Phi = 0 \tag{2.67}$$

In underwater acoustics problems we are frequently allowed to assume that there is no dependence on the azimuth angle $\varphi$, in which case equation (2.67) reduces to

$$\frac{1}{r}\frac{\partial}{\partial r}\left(r\frac{\partial \Phi}{\partial r}\right) + \frac{\partial^2 \Phi}{\partial z^2} + \kappa^2 \Phi = 0 \tag{2.68}$$

If, in addition, the sound speed is a function of depth only and not dependent on range we can assume that the solution is a product of two functions that depend only on the depth and radial distance, respectively. This means assuming a solution of the form

$$\Phi(r,z,\omega) = \Phi_r(r,\omega)\Phi_z(z,\omega) \tag{2.69}$$

Inserted in equation (2.68), this gives

$$-\frac{1}{\Phi_r(r,\omega)}\left[\frac{1}{r}\frac{d}{dr}\left(r\frac{d[\Phi_r(r,\omega)]}{dr}\right)\right] = \frac{1}{\Phi_z(z,\omega)}\left[\frac{d^2[\Phi_z(z,\omega)]}{dz^2}\right] + \kappa^2 \tag{2.70}$$

The right-hand side of equation (2.70) depends only on $z$ and the left-hand side depends only on $r$. Both sides have the same constant, which we set equal to $k^2$, and which can be referred to as the separation constant. We will soon discover that $k$ is the horizontal wave number. We therefore get two separate differential equations

***The depth equation***

$$\frac{d^2\Phi_z(z,\omega)}{dz^2} = -\left(\kappa^2 - k^2\right)\Phi_z(z,\omega) \tag{2.71}$$

***The range equation***

$$\frac{1}{r}\frac{d}{dr}\left(r\frac{d\left[\Phi_r(r,\omega)\right]}{dr}\right) = -k^2\Phi_r\left(r,\omega\right) \tag{2.72}$$

The depth equation can be rewritten as

$$\frac{d^2\left[\Phi_z\left(z,\omega\right)\right]}{dz^2} = -\gamma^2\left(z\right)\Phi_z\left(z,\omega\right) \tag{2.73}$$

since $c(z)$ is a function of depth only.

The vertical wave number $\gamma(z)$ in the $z$ direction is defined by equations (2.71) and (2.73) and is given as

$$\gamma^2\left(z\right) = \kappa^2\left(z\right) - k^2 = \left(\frac{\omega}{c_0\left(z\right)}\right)^2 - k^2 \tag{2.74}$$

By introducing the angle $\theta(z)$ with the horizontal plane, we can also write

$$\begin{aligned} \gamma\left(z\right) &= \frac{\omega}{c_0\left(z\right)}\sin\theta\left(z\right) \\ k &= \frac{\omega}{c_0\left(z\right)}\cos\theta\left(z\right) \end{aligned} \tag{2.75}$$

Note that since $k$ is a constant, this leads to Snell's law

$$\frac{\cos\theta\left(z\right)}{c_0\left(z\right)} = \text{constant} \tag{2.76}$$

An increase or decrease in sound speed with depth will therefore lead to a decrease or increase in propagation angle and therefore to refraction. We shall discuss this more in detail in chapters 4 and 6.

This relationship between the horizontal and vertical wave numbers can be depicted geometrically, as in figure 2.2.

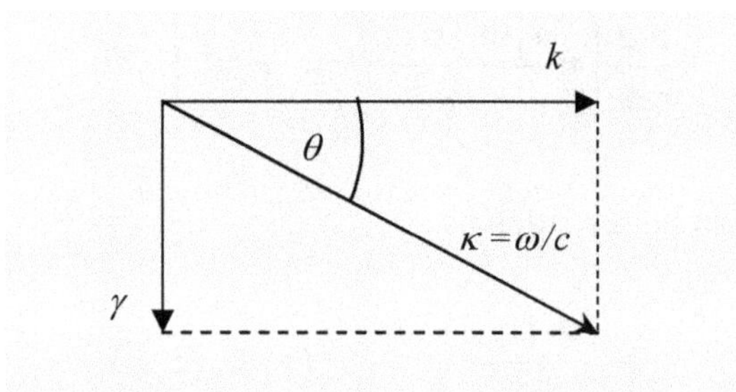


Figure 2.2. Wave number components. The component in the $r$ direction is $k$, while $\gamma$ is the component in the $z$ direction, and $\kappa = \omega / c_0$.

In the special case of the sound speed being constant, $\gamma(z)$ is independent of $z$ and the solution of the depth equation must have the form

$$\Phi_z(z,\omega) = A\exp(i\gamma z) + B\exp(-i\gamma z) \tag{2.77}$$

Equation (2.77) represents two waves, in the positive and negative $z$ directions or in the up and down directions, respectively, with $\gamma$ being the vertical wave number.

The range equation is a standard differential equation, where the solution is a linear combination of the 0-order Bessel functions of the first type $J_0(kr)$ and the second kind $Y_0(kr)$; $Y_0(kr)$ is also called the Neumann function. Thus,

$$\Phi_r(r,\omega) = A_0 J_0(kr) + B_0 Y_0(kr) \tag{2.78}$$

The terms $A_0$ and $B_0$ are constants determined by the relevant boundary conditions. A most useful reference on Bessel functions (and other mathematical functions) is Abramowitz and Stegun's *Handbook of Mathematical Functions* (1965).

Figure 2.3 shows the Bessel and Neumann functions, $J_n(kr)$ and $Y_n(kr)$, respectively, for the orders $n = 0$ and 1 as functions of $kr$.

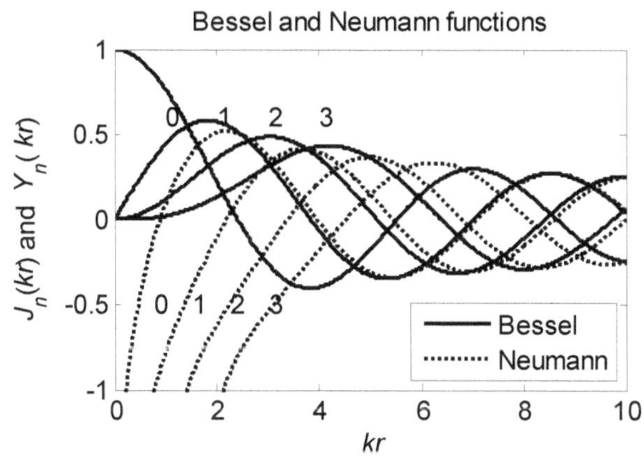


Figure 2.3. The Bessel function $J_n(kr)$ and Neumann function $Y_n(kr)$ as functions of $kr$ for orders $n = 0$ and 1.

The final solution is a product of the solutions of the depth and range equations, that is, the product of equations (2.77) and (2.78). In most cases, a number of the solutions must be discarded for various reasons, as will be demonstrated later. Since $J_0(kr) \to 1$ and $Y_0(kr) \to -\infty$ when $kr \to 0$, as shown in Figure 2.3, it follows either that the part with $Y_0(kr)$ must be left out, or that $B_0 = 0$ when the field includes $r = 0$. Then the solution of the range equation becomes

$$\Phi_r(r,\omega) = A_0 J_0(kr) \tag{2.79}$$

The Bessel functions can be written as the sum of two functions:

$$\begin{aligned} J_0(kr) &= \frac{1}{2}\left\{H_0^{(1)}(kr) + H_0^{(2)}(kr)\right\} \\ Y_0(kr) &= \frac{1}{2i}\left\{H_0^{(1)}(kr) - H_0^{(2)}(kr)\right\} \end{aligned} \tag{2.80}$$

where $H_0^{(1)}(kr)$ and $H_0^{(2)}(kr)$ are the 0-order Hankel functions of the first and second kind.

From equation (2.80), we also find that

$$\begin{aligned} H_0^{(1)}(kr) &= J_0(kr) + iY_0(kr) \\ H_0^{(2)}(kr) &= J_0(kr) - iY_0(kr) \end{aligned} \tag{2.81}$$

For the time convention $\exp(-i\omega t)$, $H_0^{(1)}(kr)$ represents a wave propagating in the positive $r$ direction, that is, outward from the source, while $H_0^{(2)}(kr)$ is a wave moving in toward the source. With the time convention $\exp(i\omega t)$, the opposite is the case.

$J_0(kr)$ may be interpreted as the cylindrical variant of the cosine function, and $H_0^{(1)}(kr)$ and $H_0^{(2)}(kr)$ as the cylindrical variants of the exponential function with positive and negative complex arguments, respectively. This is easily seen when the horizontal distance $r$ is considerably larger than the wavelength, that is, when $kr >> 1$. In this case, we may use the asymptotic approximations

$$\begin{aligned} H_0^{(1)}(kr) &\approx \sqrt{\frac{2}{\pi kr}} \exp\left(ikr - i\frac{\pi}{4}\right) \\ H_0^{(2)}(kr) &\approx \sqrt{\frac{2}{\pi kr}} \exp\left(-ikr + i\frac{\pi}{4}\right) \end{aligned} \tag{2.82}$$

For $J_0(kr)$, the asymptotic approximation is the sum of the two expressions in equation (2.82).

From equation (2.82), we see that for large distances the amplitude decreases according to the square root of the distance. This is as expected for cylindrical propagation and spreading.

### *2.5.3 Spherical coordinates $(r,\varphi,\theta)$*

In the general case, the Laplace operator for spherical coordinates may be written as

$$\nabla^2\phi = \frac{1}{r^2}\frac{\partial}{\partial r}\left(r^2\frac{\partial\phi}{\partial r}\right) + \frac{1}{r^2\sin\theta}\frac{\partial}{\partial\theta}\left(\sin\theta\frac{\partial\phi}{\partial\theta}\right) + \frac{1}{r^2\sin^2\theta}\frac{\partial^2\phi}{\partial\varphi^2} \tag{2.83}$$

In the azimuthally symmetric case, $(\partial\phi / \partial\varphi) = 0$, and equation (2.83) reduces to

$$\nabla^2\phi = \frac{1}{r^2}\frac{\partial}{\partial r}\left(r^2\frac{\partial\phi}{\partial r}\right) + \frac{1}{r^2\sin\theta}\frac{\partial}{\partial\theta}\left(\sin\theta\frac{\partial\phi}{\partial\theta}\right) \tag{2.84}$$

We apply the Laplacian operator of equation (2.84), thereby limiting the validity and the solution to the azimuthal symmetric case where $(\partial\phi / \partial\varphi) = 0$. The homogeneous Helmholtz equation then becomes

$$\left[\frac{1}{r^2}\frac{\partial}{\partial r}\left(r^2\frac{\partial\Phi}{\partial r}\right) + \frac{1}{r^2\sin\theta}\frac{\partial}{\partial\theta}\left(\sin\theta\frac{\partial\Phi}{\partial\theta}\right) + \kappa^2\Phi\right] = 0 \tag{2.85}$$

By assuming the solution to be a product of a radial solution and a polar solution, we find that the total solution can be written for the frequency domain representation

$$\Phi = \sum_{m=0}^{\infty}\left[A_m j_m(\kappa r) + B_m y_m(\kappa r)\right]P_m(\cos\theta) \tag{2.86}$$

$P_m(\mu)$ are Legendre polynomials with the argument $\mu = \cos\theta$, which replaces $\cos(m\theta)$ in the cylindrical case. The Legendre polynomials can be calculated by use of the recursion equation

$$P_m(\mu) = \frac{2m-1}{m}\mu P_{m-1}(\mu) + \frac{1-m}{m}P_{m-2}(\mu) \tag{2.87}$$

with the initial values

$$\begin{aligned} P_0(\mu) &= 1 \\ P_1(\mu) &= \mu \end{aligned} \tag{2.88}$$

Further, $j_m(kr)$ and $y_m(kr)$ are the spherical Bessel and Neumann functions, respectively, which are related to the better-known cylindrical Bessel and Neumann functions $J_m(kr)$ and $Y_m(kr)$ by

$$
\begin{aligned}
j_m(\kappa r) &= \sqrt{\frac{\pi}{2\kappa r}} J_{m+\frac{1}{2}}(\kappa r) \\
y_m(\kappa r) &= \sqrt{\frac{\pi}{2\kappa r}} Y_{m+\frac{1}{2}}(\kappa r)
\end{aligned}
\tag{2.89}
$$

The functions of order $m = 0$ and $m = 1$ are

$$
\begin{aligned}
j_0(\kappa r) &= \frac{\sin(\kappa r)}{\kappa r} & \qquad j_1(\kappa r) &= \frac{\sin(\kappa r)}{(\kappa r)^2} - \frac{\cos(\kappa r)}{\kappa r} \\
y_0(\kappa r) &= -\frac{\cos(\kappa r)}{\kappa r} & \qquad y_1(\kappa r) &= -\frac{\sin(\kappa r)}{\kappa r} - \frac{\cos(\kappa r)}{(\kappa r)^2}
\end{aligned}
\tag{2.90}
$$

The spherical Hankel functions are defined by

$$
\begin{aligned}
h_m^{(1)}(\kappa r) &= j_m(\kappa r) + i y_m(\kappa r) \\
h_m^{(2)}(\kappa r) &= j_m(\kappa r) - i y_m(\kappa r)
\end{aligned}
\tag{2.91}
$$

the functions of order $m = 0$ and $m = 1$ being

$$
\begin{aligned}
h_0^{(1)}(\kappa r) &= \frac{\exp(i\kappa r)}{i\kappa r} \\
h_0^{(2)}(\kappa r) &= \frac{\exp(-i\kappa r)}{-i\kappa r}
\end{aligned}
\tag{2.92}
$$

The general solution is therefore given as

$$
\Phi(r,\omega) = A(k) h_0^{(1)}(\kappa r) + B(k) h_0^{(2)}(\kappa r) \tag{2.93}
$$

We see from equation (2.92) that $h_0^{(1)}(\kappa r)$ represents an outgoing radial wave and $h_0^{(2)}(\kappa r)$ an incoming wave. $A(\kappa)$ and $B(\kappa)$ are, for the moment, unknown coefficients determined by the boundary conditions. We make use of this result in chapter 14 when developing the scattered field of a spherical object.

With spherical symmetry, the wave equation [equation (2.85)] can be written as

$$\frac{1}{r}\frac{\partial^2(r\phi)}{\partial r^2}=\frac{1}{c_0^2}\frac{\partial^2(r\phi)}{\partial t^2} \tag{2.94}$$

In this case there is an explicit solution in the time domain. By inspection, we find that the solution must have the form

$$\phi(r,t)=\frac{1}{r}\phi_+\left(t-\frac{r}{c_0}\right)+\frac{1}{r}\phi_-\left(t+\frac{r}{c_0}\right) \tag{2.95}$$

This also represents any two waves propagating in outgoing or incoming radial directions, though in the spherical case the amplitude decreases proportionally to the distance $r$. This is intuitively correct: the energy spreads across a spherical surface area that increases with the square distance. In the frequency domain this equation becomes

$$\Phi(r,\omega)=A\frac{\exp(i\kappa r)}{r}+B\frac{\exp(-i\kappa r)}{r} \tag{2.96}$$

The first term is again a wave propagating outward from the source and the second term a wave propagating toward the source.

## 2.6. Solution of the inhomogeneous wave equation

In this section we consider the inhomogeneous wave equation but limit the discussion to a point source in an unbounded medium. The case with a source in a bounded medium is treated in section 2.7.

We start with equation (2.42), which for a point source with strength $Q(\omega)$ at position $\mathbf{r}_0$ may be expressed as

$$\nabla^2\Phi(\mathbf{r},\omega)+\kappa^2\Phi(\mathbf{r},\omega)=-Q(\omega)\delta(\mathbf{r}-\mathbf{r}_0) \tag{2.97}$$

The source strength $Q(\omega)$ is the Fourier transform of the source function $q(t)$ introduced by equation (2.36).

The general solution of the homogenous equation under the radiation condition for a source at $\mathbf{r} = 0$ is, according to equation (2.96),

$$\Phi(\mathbf{r},\omega) = \frac{A}{r}\exp(i\kappa r) \tag{2.98}$$

with $A$ given by the boundary condition at the source.

We can prove that equation (2.98) is the solution of equation (2.97), and, at the same time, determines the value of $A$. We first integrate equation (2.97) over a volume $V$ that includes the source position. By using equation (2.37) we thereby obtain

$$\int_V \nabla^2\Phi(\mathbf{r},\omega)\,dV + \int_V \kappa^2\Phi(\mathbf{r},\omega)\,dV = -Q(\omega) \tag{2.99}$$

We then apply Gauss's theorem (see Kinsler et al. [2000], page 105) to transform the first integral of equation (2.99) from a volume integral into a surface integral to yield

$$\int_S \nabla\Phi(\mathbf{r},\omega)\cdot\mathbf{n}\,dS + \int_V \kappa^2\Phi(\mathbf{r},\omega)\,dV = -Q(\omega) \tag{2.100}$$

where $\mathbf{n}$ is the outward unit vector normal to the surface $S$ of the volume $V$.

Substituting equation (2.98) for $\Phi(\mathbf{r},\omega)$ and integrating over a small spherical volume that encloses the source gives the result that $4\pi A = Q(\omega)$, and the solution of equation (2.97) is therefore

$$\Phi(r,\omega) = \frac{Q(\omega)}{4\pi r}\exp(i\kappa r) \tag{2.101}$$

Equation (2.101) gives the field in terms of a distance $r$ from a small volume source at the origin with the source strength $Q(\omega)$. If the source is at position $\mathbf{r}_0$, the field at position $\mathbf{r}$ becomes

$$\Phi(\mathbf{r},\omega) = \frac{Q(\omega)}{4\pi|\mathbf{r}-\mathbf{r}_0|}\exp(i\kappa|\mathbf{r}-\mathbf{r}_0|) \tag{2.102}$$

We illustrate the nature of equation (2.101) [and equation (2.102)] by calculating the sound field as a function of the distance from a point source with the form of a pulsating sphere. This is also an example of how a source can be modeled by introducing dynamically varying boundary conditions.

The sound pressure at any distance can be expressed as

$$p(r,t) = i\omega\rho_0 \frac{Q(\omega)}{4\pi r} \exp\left[i(\kappa r - \omega t)\right] \tag{2.103}$$

The particle velocity, which is generally equal to the gradient of the potential, now has only a radial component and is therefore

$$\begin{aligned} v(r,t) &= \frac{\partial \Phi(r,\omega)}{\partial r} \exp(-i\omega t) = \frac{Q(\omega)}{4\pi} \frac{\partial}{\partial r}\left[\frac{\exp(i\kappa r)}{r}\right] \exp(-i\omega t) \\ &= \frac{Q(\omega)}{4\pi r}(i\kappa)\left[1 - \frac{1}{i\kappa r}\right] \exp\left[i(\kappa r - \omega t)\right] \end{aligned} \tag{2.104}$$

The specific acoustic impedance for a spherical wave is

$$Z = \frac{p(r,t)}{v(r,t)} = \rho_0 c_0 \frac{i\kappa r}{(i\kappa r - 1)} \tag{2.105}$$

The impedance can be split into a real resistive part and an imaginary reactive part:

$$Z = \rho_0 c_0 \frac{(\kappa r)^2}{(\kappa r)^2 + 1} - i\rho_0 c_0 \frac{\kappa r}{(\kappa r)^2 + 1} \tag{2.106}$$

Close to the source, $\kappa r << 1$, the reactive part dominates, whereas at far distances, $\kappa r >> 1$, the resistive part dominates, and the spherical wave impedance becomes equal to plane wave impedance $\rho_0 c_0$.

We consider what happens close to a small sphere pulsating at a frequency $\omega$ with velocity $v(r_0)$ around the mean radius $r_0$. When we assume $\kappa r_0 << 1$ in equation (2.105) we find that the amplitude of the particle velocity, $v_0$, at the surface of the source where $r = r_0$ is $v_0 = -Q(\omega)/(4\pi r^2_0)$. Therefore, we find that the source strength $Q(\omega)$ is

$$Q(\omega) = -4\pi r_0^2 v_0 \tag{2.107}$$

where $Q(\omega)$ represents the rate of volume displacement that the source causes when the surface pulsates with amplitude $u_0$. In general cases, for a source with any given form but with an extent much less than the wavelength, the definition of the source strength is

$$Q(\omega) = -\int_S \mathbf{v}_0 \cdot \mathbf{n} ds \tag{2.108}$$

where $\mathbf{v}_0$ is the particle velocity vector of an element on the surface, $\mathbf{n}$ is the normal to the surface, and $ds$ is a surface element. The integration takes place over the whole surface.

We now go back to equation (2.105), which shows that the acoustic impedance of a spherical wave is a function of its distance from the source. However, we do not have to go far from the source before the wave may be considered locally plane. This is the case when $\kappa r >> 1$, that is, when the distance is considerably greater than the wavelength. Then the last term in equation (2.106) may be ignored and the impedance be given as $Z = \rho_0 c_0$. At a distance $r_0$ satisfying this condition, the intensity consequently becomes

$$I(r_0) = \frac{p_0^2}{\rho_0 c_0} = v_0^2 \rho_0 c_0 \tag{2.109}$$

Note that here we have redefined $p_0$ and $v_0$ as the rms values of the amplitudes of the sound pressure and particle velocity evaluated at $r_0$.

The total acoustic power of the source, $W$, is found by integrating over the surface of the sphere, so that $W$ is defined as

$$W = 4\pi r_0^2 \frac{p_0^2}{\rho_0 c_0} = 4\pi r_0^2 v_0^2 \rho_0 c_0 \tag{2.110}$$

and the intensity $I(r)$ at any distance $r$ is expressed as

$$I(r) = I(r_0)\left(\frac{r_0}{r}\right)^2 \tag{2.111}$$

In practice, the decibel scale is often used, which is ten times the logarithm to the base 10, of a power or intensity ratio

$$10\log I(r) = 10\log I(r_0) - 20\log(r/r_0) \tag{2.112}$$

Instead of specifying strength or power level of an acoustic wave by using the intensity, it is standard practice to characterize the level by the equivalent sound pressure, given that the impedance is $Z = \rho_0 c_0$. The squared sound pressure $p^2$ at a distance $r$ from a sound source radiating an output of $W$ watts under spherical spreading conditions is

$$p^2 = \frac{\rho_0 c_0 W}{4\pi r^2} \tag{2.113}$$

Equation (2.113) is expressed in decibels as

$$20\log p = 10\log\left(\frac{\rho_0 c_0}{4\pi}\right) + 10\log W - 20\log r \tag{2.114}$$

With $c_0$ = 1500 m/s and $\rho_0$ = 1000 kg/m$^3$, the numerical value of the equation's first term is approximately 50.8, so that the sound pressure at a distance $r$ from an omnidirectional source that transmits $W$ watts acoustically into the water with the same intensity in all directions becomes

$$\begin{aligned} 20\log p &= 50.8 + 10\log W - 20\log r \quad && \text{dB re 1 Pa} \\ 20\log p &= 170.8 + 10\log W - 20\log r \quad && \text{dB re 1 } \mu\text{Pa} \\ 20\log p &= 70.8 + 10\log W - 20\log r \quad && \text{dB re 1 } \mu\text{bar} \end{aligned} \tag{2.115}$$

The units are related as follows: 1 $\mu$bar = 1 dyne/cm$^2$ = 0.1 Pa, and 1 $\mu$Pa = $10^{-6}$ Pa.

We illustrate the relevant values by using an example. A very high sound pressure in water is one that equals one atmospheric pressure—that is, $p_0 = 10^5$ Pa. Sound pressure of this magnitude is only to be found in the immediate proximity of a powerful sonar that is operating close to the so-called cavitation limit. With this sound pressure, the amplitude of the particle velocity becomes $v_0 = p_0/(\rho_0\, c_0) = 0.067$ m/s. For a harmonic wave at a frequency of 1 kHz, the particle displacement in this case becomes $v_0 / \omega = \approx 10^{-5}$m. The intensity is $I = 0.5 p_0 v_0 = 0.33 \cdot 10^4$ W/m$^2$, or $I =$ $0.33 \cdot$ W/m$^2$. Note that the particle velocity is considerably smaller than the sound speed and that the particle displacement is very small, even under such high sound pressure. This fact supports the neglect of the product of $v$ terms in linearizing the wave equation [equation (2.12)].

In underwater acoustics, the sound level is normally not defined by intensity but rather by sound pressure, which is measured in dB relative to a reference pressure. In modern literature, the common reference is 1 $\mu$Pa, as used in the second expression of equation (2.115). When using one of the equations (2.115), it is important to remember that the numerical terms apply to sound pressure in water, while for air, the acoustic impedance is about 3600 times lower; furthermore, the reference sound pressure in air is 20 $\mu$Pa. Consequently, the numerical factor becomes approximately 35.5 dB lower in air than that given in equation (2.115).

## 2.7. Solutions of the Helmholtz equation in bounded media

In section 2.6 we found the radiated field of a point source in an unbounded medium. Now we extend the treatment to the more general situation of a source in a bounded medium where we also have to include reflections from boundaries. The situation is depicted in Figure 2.4 where a point source $F(\mathbf{r})$ *is* at position $\mathbf{r}_0$ in a volume of water $V$ that is enclosed by the surface $S$.

The problem is to find the field $\Phi(\mathbf{r},\omega)$ at position $\mathbf{r}$ due to the source $F(\mathbf{r})$ at position $\mathbf{r}_0$, such that $\Phi(\mathbf{r},\omega)$ satisfies the wave equation and the boundary conditions. Intuitively we may think that the solution must consist of two parts, a direct part from the source and the reflected and scattered contributions from the boundaries represented by the surface $S$. In the following we demonstrate that this approach is correct.

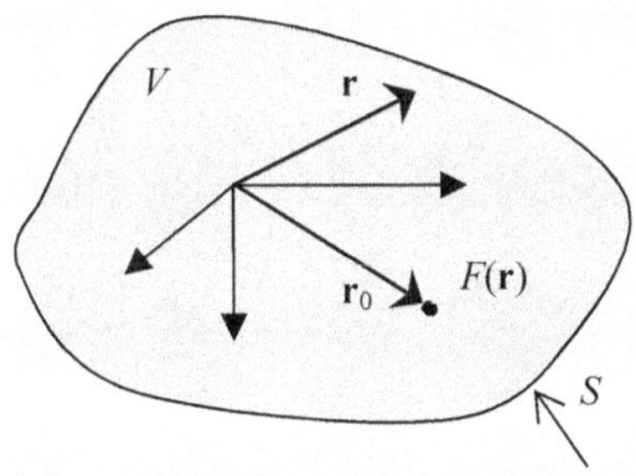


Figure 2.4. Volume $V$ enclosed by the surface $S$ and with a source $F(\mathbf{r})$ at $\mathbf{r}_0$ and a receiver at $\mathbf{r}$.

We start with the inhomogeneous Helmholtz equation given by equation (2.42), which we now write as

$$\left[\nabla^2 + \kappa^2(\mathbf{r})\right]\Phi(\mathbf{r},\omega) = -F(\mathbf{r}) \tag{2.116}$$

In section 2.6, we found the solution of equation (2.116) for a point source at position $\mathbf{r}_0$ of unity strength in an unbounded medium as

$$G_0(\mathbf{r},\mathbf{r}_0) = \frac{1}{4\pi}\frac{\exp(ik|\mathbf{r}-\mathbf{r}_0|)}{|\mathbf{r}-\mathbf{r}_0|} \tag{2.117}$$

The function $G_0(\mathbf{r},\mathbf{r}_0)$ is called the free-field Green's function and it relates the field at the point $\mathbf{r}$ to the radiated field from a point source at $\mathbf{r}_0$ for a unity source strength. From the preceding discussion it is evident that Green's function satisfies the wave equation and the radiation condition. Furthermore we see also that the function is symmetric; we can exchange the positions such that $G_0(\mathbf{r}_0,\mathbf{r}) = G_0(\mathbf{r},\mathbf{r}_0)$, which means that the principle of reciprocity is obeyed.

The general solution $\Phi(\mathbf{r},\omega)$ to the inhomogeneous Helmholtz equation (2.116) can be expressed as the sum of a particular solution and the solution of the homogeneous Helmholtz. The general Green's function, $G(\mathbf{r}_0,\mathbf{r})$, is therefore defined as

$$G(\mathbf{r},\mathbf{r}_0) = G_0(\mathbf{r},\mathbf{r}_0) + H(\mathbf{r}) \tag{2.118}$$

$H(\mathbf{r})$ satisfies the homogenous Helmholtz equation,

$$\left[\nabla^2 + k^2\right]H(\mathbf{r}) = 0 \tag{2.119}$$

The general Green's function satisfies

$$\left[\nabla^2 + \kappa^2(\mathbf{r})\right]G(\mathbf{r},\mathbf{r}_0) = -\delta(\mathbf{r}-\mathbf{r}_0) \tag{2.120}$$

The solution of equation (2.116) may be expressed as

$$\begin{aligned}\Phi(\mathbf{r}) &= \int_{V_0} F(\mathbf{r}_0)G(\mathbf{r},\mathbf{r}_0)dV_0 \\ &+ \frac{1}{4\pi}\int_S \left[G(\mathbf{r},\mathbf{r}_0)\frac{\partial\Phi(\mathbf{r}_0)}{\partial n_0} - \Phi(\mathbf{r}_0)\frac{G(\mathbf{r},\mathbf{r}_0)}{\partial n_0}\right]dS\end{aligned} \tag{2.121}$$

The first part is the direct contribution from the source and involves the integration over a volume $V_0$ that includes the source. The second part is an integration over the surface $S$ and gives the contributions reflected from the boundaries. The derivatives $\partial/\partial n_0$ are taken normal to the surface $S$ in an outward sense.

The proof of equation (2.121) involves some simple, but laborious manipulations that can be found in the books by Frisk (1994), and Jensen et al. (1994). Equation (2.121) is really not the solution, but a reformulation of the problem to a form used as a starting point for other evaluations. We will make use of this in chapter 14 on scattering of surfaces and in chapter 17 where we treat propagation in oceanic wave guides.

## 2.8. Reciprocity

In the previous section we observed that Green's function is symmetric and therefore satisfies the principle of reciprocity. Reciprocity is a fundamental principle related to wave propagation and to linear systems in general. It is therefore of great practical importance. Applied to wave propagation, in brief terms the reciprocity principle enables us to switch the positions of source and receiver and still receive the same acoustic signal.

Figure 2.5 illustrates an experimental situation with transmission of sound between two positions indicated by *A* and *B*. In one experiment sound is transmitted from *A* by a small spherical sound source with strength $Q_A$ and we measure the sound pressure $p_B$ in position *B*. Another experiment uses a source at position *B* with strength $Q_B$, and we receive the pressure $p_A$ at position *A*.

The reciprocity principle states that the ratios of source strengths and received pressures are the same, as expressed by the following relationship,

$$\frac{Q_A}{p_B} = \frac{Q_B}{p_A} \tag{2.122}$$

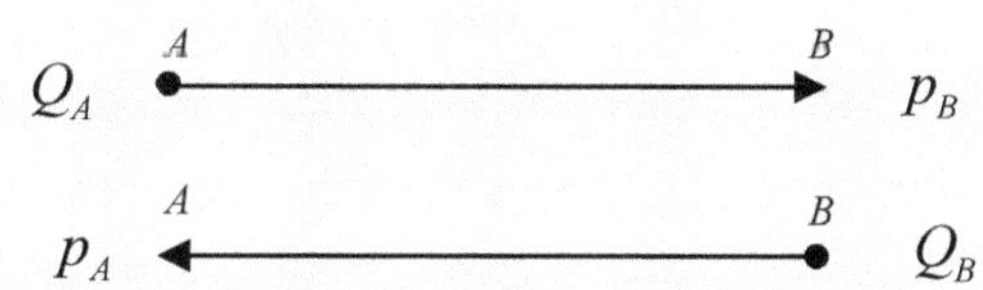


Figure 2.5. Reciprocity geometry. A source at position *A* with the source strength $Q_A$ produces the sound pressure $p_B$ at position *B*. A source at position *B* with source strength $Q_B$ produces the sound pressure $p_A$ at position *A*.

Equation (2.122), which is very general, applies to small nondirectional sources and all transmission relations with or without absorption, as long as the sources have the same frequency. The equation constitutes the reciprocity principle: the pressure at position *B* due to a source at position *A* is equal to the pressure at *A* due to a similar source at position *B*. This result holds also for the case where the medium is composed of several regions. Thus the reciprocity principle is valid in cases where the wave undergoes reflections and refractions on its path from *A* to B (Landow and Lifshitz 1959, and Kinsler et al. 2000).

If we in addition can say something about transmission conditions, the relationship between the source strength and the received sound pressure can be stated more precisely. For instance, with spherical spreading in an unbounded medium, equation (2.103) gives

$$p_B(r) = i\omega\rho_0 \frac{Q_A}{4\pi r} \tag{2.123}$$

In the same way a point source at $B$ with strength $Q_B$ produces at $A$ the sound pressure

$$p_A(r) = i\omega\rho_0 \frac{Q_B}{4\pi r} \tag{2.124}$$

In both cases the ratio of source strength to sound pressure at a distance $r$ from the source is

$$\left|\frac{Q}{p(r)}\right| = \frac{4\pi r}{\omega\rho_0} = 2\frac{r\lambda}{\rho_0 c_0} \tag{2.125}$$

The right-hand side of equation (2.125) is the spherical-wave reciprocity factor $J_s$

$$J_s = \frac{2}{\rho_0 c_0}(r\lambda) \tag{2.126}$$

In similar manners we can derive reciprocity factors for plane waves and for cylindrical waves; see Urick (1983). For a cylindrical transducer with length $L$ the cylindrical-wave reciprocity factor is

$$J_c = \frac{2}{\rho_0 c_0}(\lambda r)^{\frac{1}{2}} L \tag{2.127}$$

and for a planar transducer with surface area $A$, the plane-wave reciprocity factor is

$$J_p = \frac{2}{\rho_0 c_0} A \tag{2.128}$$

Notice that the reciprocity factors do not depend on the properties of the sound transmitter or receiver. The reciprocity factor of equation (2.126) is used in chapter 8 for the assessment of the sensitivity of transmitters and receivers for acoustic transducers.

# Chapter 3

# Oceanographic and physical properties

## 3.1 Introduction

In this chapter we present some values and models for sound speed and absorption in water and other materials of interest in marine acoustics. We also develop a simple equation, known as Wood's equation, for the sound speed of composite materials. We use this equation to calculate the sound speed for a suspension of solid particles and water and to determine the effect on sound speed of free gas or air bubbles in water.

## 3.2 Sound speed and sound speed profiles

In the preceding chapter, we found that the sound speed depends on the volume stiffness, or the bulk modulus, and the density of the medium. In the sea, both depend on ambient conditions, in particular, water temperature, salinity, and the surrounding pressure or depth of the water. Typical values for fresh water are $K = 2.25 \cdot 10^9$ Pa and $\rho = 1000$ kg/m$^3$, which give a sound speed of $c = 1500$ m/s. In the literature there are several formulas for calculating sound speed in water. For our purpose the following simplified formula after Del Grosso (1974) is sufficiently accurate:

$$c = 1448.6 + 4.618T - 0.0523T^2 + 1.25(S - 35) + 0.017D \qquad (3.1)$$

In this equation

$c$ = sound speed (m/s),
$T$ = temperature (°C),
$S$ = salinity (pro mille),
$D$ = depth (m).

Sound speed changes with fluctuations of the temperature, salinity, and depth. In the sea, the temperature and salinity vary according to the oceanographic conditions, and therefore the sound speed varies as well. For normal environmental conditions and at a 10°C water temperature, the gradients are approximately

$$\begin{aligned} \frac{dc}{dT} &= 3.5\,\mathrm{m/s}\ \text{per °C} \\ \frac{dc}{dS} &= 1.25\,\mathrm{m/s}\ \text{per pro mille salt content} \\ \frac{dc}{dD} &= 0.017\,\mathrm{m/s}\ \text{per meter increasing depth} \end{aligned} \tag{3.2}$$

Even if the variation of the sound speed is relatively small, it has a significant effect on sound propagation, and we shall later see many examples of the strong effects the sound speed profile has on acoustic propagation conditions.

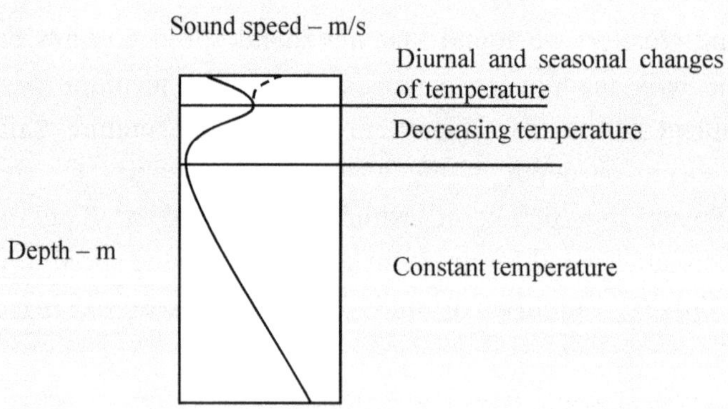


Figure 3.1. Typical sound speed profile for deep water.

We now consider some sound speed profiles typical of summer and winter conditions, as well as for deep and shallow water. A normal sound speed profile for deep water is shown in figure 3.1. The profile is characterized by a number of more or less distinct layers. Near the surface there is a layer where the temperature will be subject to daily or seasonal changes in heating or cooling as well as from the mixing of water masses as a result of ocean wave action. Just beneath this surface layer there may be a seasonally dependent thermocline, in which sound speed decreases with depth because the temperature decreases with depth. In summer, the gradient is often steep as a result of warmer surface water, while in winter the effect is less pronounced. The gradient and the thickness of the layers mentioned above will vary according to geographical position, season, time of day, and meteorological conditions. Beneath these layers is the main thermocline, where temperature decreases with depth with a gradient less affected by surface conditions. At great water depth, the temperature remains essentially constant, but the sound speed begins to increase with depth as a consequence of increasing pressure. This means that there is a depth where sound speed reaches a minimum, creating a sound channel where sound can be focused and concentrated. The axis of this channel will vary with the degree of latitude. In southern waters the axis will normally lie at about 1000 meters depth, while in northern waters it will be shallower and in polar districts close to the surface. The sound speed profile shown in figure 3.1 is typical for the northern Atlantic. In the summer, a relatively pronounced sound channel will be formed at depths of about 50–100 meters, while in winter the sound speed minimum is close to the surface.

Figure 3.2 shows examples of summer and winter profiles in the Norwegian Sea at a location with 300 meters water depth. In winter, the gradient can be positive over the entire water volume, while in summer the gradient is often negative close to the surface but positive at deeper depths with a minimum sound speed relatively close to the surface.

As mentioned earlier, the temperature in the upper layers will strongly depend on the environmental conditions. Especially, wind and waves will sometimes cause a total mixing of the water masses, making the water temperature almost constant in a surface layer. In such a mixed layer with constant temperature, sound speed will have a small positive gradient because of the depth.

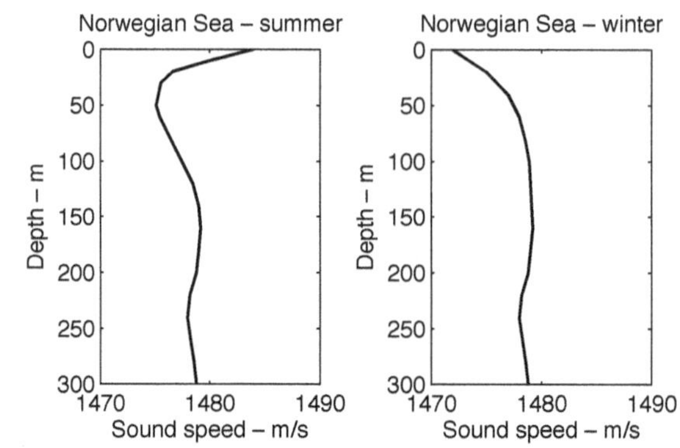


Figure 3.2. Examples of summer (left) and winter (right) sound speed profiles in the Norwegian Sea.

## 3.3 Acoustic absorption

So far in our discussion of the wave equation, we have assumed that the signal has no absorption loss in the medium. Of course this cannot be correct, as there always will be losses caused by viscosity, thermal conductance, and different relaxation phenomena. Ultimately, the acoustic energy is lost to heat due to the interaction of the particles of the medium caused by the wave.

### *3.3.1 Relaxation*

So far we have assumed that, when a medium is subjected to a sudden change of pressure, the change of volume occurs immediately, without any delay in time. This is unrealistic, since all change takes time, however short that time period may be. In order to take the time delay into consideration, we can modify equation (2.21), which gives the relation between the sound pressure $p$ and condensation $s$, the relative change of density, to produce the following differential equation

$$p = K_0 \left( s + \tau \frac{\partial s}{\partial t} \right) \tag{3.3}$$

Here, $\tau$ is the relaxation time, a measure for how rapidly the change occurs. The reciprocal of the relaxation time is called the relaxation frequency $\omega_r$, where $\omega_r = 1/\tau$. There may, of course, be functional forms of change in time other than the one shown in equation (3.3); we will return to this topic later in section 16.4.

In the frequency domain, equation (3.3) is written as

$$P(\omega) = K_0\left(1 - i\omega\tau\right)S(\omega) \tag{3.4}$$

and as

$$S(\omega) = \frac{P(\omega)}{K_0\left(1 - i\omega\tau\right)} \tag{3.5}$$

We have used capital letters for the frequency functions $P(\omega)$ and $S(\omega)$ of the time variation of pressure $p(t)$ and condensation $s(t)$. If, for instance, the volume $V_0$ is suddenly subjected to a jump in pressure of $\Delta p$, then $P(\omega) = \Delta p/(i\omega)$ and

$$S(\omega) = \frac{\Delta p}{K_0}\frac{1}{i\omega\left(1 - i\omega\tau\right)} \tag{3.6}$$

Transformation back to the time domain gives

$$s(t) = \frac{\Delta p}{K_0}\left[1 - \exp(-t/\tau)\right] \tag{3.7}$$

The condensation as we defined in equation (2.10) can also be defined as the relative volume change,

$$s\left(t\right) = \frac{V_0 - V\left(t\right)}{V_0} \tag{3.8}$$

and the size of the volume will therefore change with time for $t > 0$ according to

$$V(t) = V_0\left\{1 - \frac{\Delta p}{K_0}\left[1 - \exp(-t/\tau)\right]\right\} \tag{3.9}$$

Figure 3.3 shows in principle how the volume changes and becomes smaller because of the increase in pressure. Initially at $t$ = 0, the volume change is zero and the medium will be conceived of as infinitely stiff.

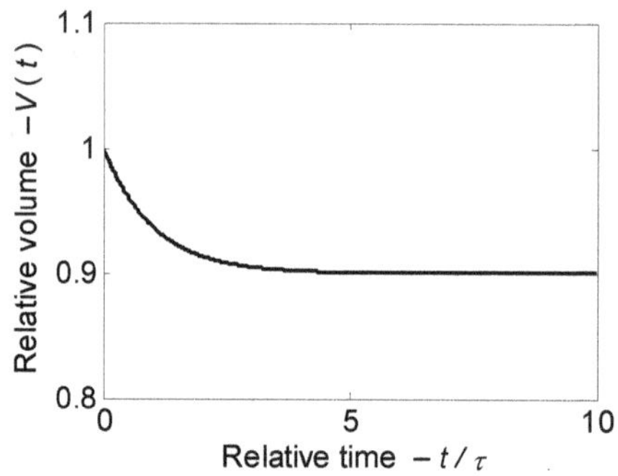


Figure 3.3. Relative volume, plotted as a function of time after the volume has been subjected to a step increase in surrounding pressure.

As time passes, the volume will be compressed and the relative volume change will ultimately become

$$\frac{\Delta V}{V_0} = -\frac{\Delta p}{K_0} \tag{3.10}$$

Therefore, the medium can be considered as having a complex bulk modulus $K$,

$$K = K_0 \left(1 - i\omega\tau\right) \tag{3.11}$$

and the sound speed $c$ becomes complex and equal to

$$c = \sqrt{\frac{K}{\rho_0}} = \sqrt{\frac{K_0\left(1 - i\omega\tau\right)}{\rho_0}} = c_0\sqrt{\left(1 - i\omega\tau\right)} \tag{3.12}$$

with $c_0 = \sqrt{K_0 / \rho_0}$ .

In this medium, a plane harmonic wave will propagate as

$$p(t,x) = A\exp\left[i\left(\kappa x - \omega t\right)\right] \tag{3.13}$$

having a complex wave number $\kappa$ that is defined as

$$\kappa = \left(\kappa_r + i\kappa_i\right) = \frac{\omega}{c_0}\frac{1}{\sqrt{1 - i\omega\tau}} \tag{3.14}$$

The positive sign is chosen for the imaginary component of $\kappa$ in order to have a wave with decreasing amplitude as it propagates. Breaking equation (3.14) into real and imaginary parts (this requires some tedious algebraic manipulations) gives the real part,

$$\kappa_r = \frac{\omega}{c_0}\frac{1}{\sqrt{2}}\sqrt{\frac{\sqrt{1+\left(\omega\tau\right)^2}+1}{1+\left(\omega\tau\right)^2}} \tag{3.15}$$

The phase speed is therefore a function of frequency,

$$c_p = \frac{\omega}{\kappa_r} = c_0\sqrt{2}\sqrt{\frac{1+\left(\omega\tau\right)^2}{\sqrt{1+\left(\omega\tau\right)^2}+1}} \tag{3.16}$$

The imaginary part of $\kappa$ is the absorption coefficient, $\alpha$, which can be expressed as

$$\alpha = \frac{\omega^2\tau}{2c_0^2}\frac{c_p}{1+\left(\omega\tau\right)^2} \tag{3.17}$$

The reciprocal of the relaxation time is called the relaxation (angular) frequency,

$$\omega_r = \frac{1}{\tau} \tag{3.18}$$

Using the relaxation frequency, equations (3.16) and (3.17) can then be expressed as

$$c_p = c_0\sqrt{\frac{2\left(\omega_r^2 + \omega^2\right)}{\omega_r^2 + \omega_r\sqrt{\omega_r^2 + \omega^2}}} \tag{3.19}$$

$$\alpha = \frac{\omega^2 c_p}{2c_0^2} \frac{\omega_r}{\omega_r^2 + \omega^2} \tag{3.20}$$

Figure 3.4 shows in principle how the sound speed and the absorption coefficient vary with the frequency according to equations (3.16) and (3.17).

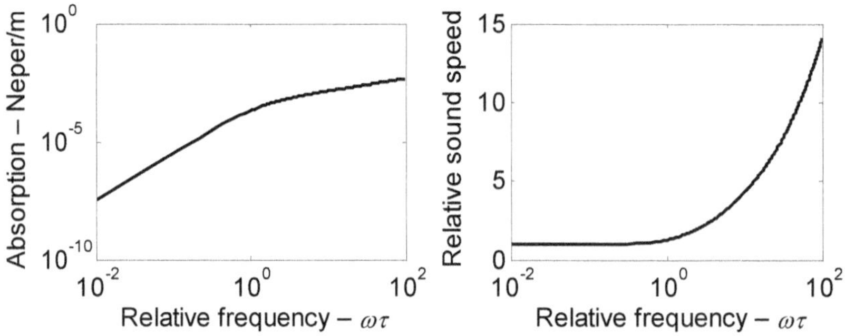


Figure 3.4. Absorption (left) and sound speed (right) in a viscoelastic medium, plotted as a function of the frequency relative to the relaxation frequency.

For frequencies far below the relaxation frequency, that is, where $(\omega/\omega_r) << 1$, we get the limiting values

$$c_p \to c_0$$
$$\alpha \to \frac{\omega^2}{2\omega_r c_0} \tag{3.21}$$

For very high frequencies where $(\omega\tau) >> 1$, the asymptotic limits are

$$c_p \to c_0\sqrt{2\omega / \omega_r}$$
$$\alpha \to \frac{1}{c_0}\sqrt{\frac{\omega\omega_r}{2}} \tag{3.22}$$

Absorption increases proportionally, with the frequency squared for frequencies lower than the relaxation frequency, and with the square root of the frequency at high frequencies. The sound speed is nearly constant at low frequencies, but at very high frequencies it increases with the square root of the frequency. The slightly strange behavior that occurs at very high frequencies can be explained by studying equation (3.22), from which we find that the wave attenuation over a wavelength at very high frequency is

$$\delta = \alpha\lambda = 2\pi \tag{3.23}$$

This is an enormous attenuation, so large that the wave is a diffusion wave rather than a propagating wave.

### *3.3.2 Acoustic absorption in water*

Acoustic absorption, or damping, in pure fresh water is mainly caused by viscosity and leads to the relaxation process described in the previous section. The viscous relaxation time $\tau$ is defined as

$$\tau = \frac{\eta}{K_0} \tag{3.24}$$

The viscosity $\eta$ with the dimension [Pa · s] includes both shear and volume viscosity. For freshwater at a temperature of 14°C, $\eta \approx 4.2 \cdot 10^{-3}$ [Pa · s] and $K_0 \approx 2 \cdot 10^{9}$ [Pa], so that $\tau \approx 2.1 \cdot 10^{-12}$ s. This means that, for all frequencies of practical importance, for which $(\omega\tau) << 1$, the part of the absorption that is caused by viscosity is proportional to the frequency squared. Therefore, for all relevant frequencies, the viscosity gives a contribution to the absorption $\alpha$, which is found by inserting equation (3.24) into equation (3.21), resulting in this definition for $\alpha$:

$$\alpha = \frac{\eta\omega^2}{2\rho_0 c_0^3} \tag{3.25}$$

Using the same numerical values as those cited in the preceding paragraph, we find that the frequency dependence of the sound speed is negligible. Consequently, for all practical purposes the sound speed remains constant and independent of the frequency.

In addition to the viscous absorption, there is also a contribution caused by thermal conductivity when some of the heat caused by the volume changes is diverted and lost from the acoustic wave. The thermal contribution has the same frequency dependence as the viscous contribution, and therefore the two contributions can be added, yielding an absorption coefficient of the form

$$\alpha = A_0 f^2 \tag{3.26}$$

where $A_0$ is a constant to be determined by measurements.

In seawater, there are also other relaxation phenomena that make important contributions to absorption losses. For this reason, the absorption in seawater is considerably greater than in fresh water for frequencies below approximately 1 MHz. This is mainly the result of two relaxation processes that involve boric acid, $B(OH)_3$, and magnesium sulfate, $MgSO_4$, which are present in small amounts in seawater. Each of these processes gives a contribution in the form given in equation (3.17), with a constant sound speed $c = c_0$.

In Francois and Garrison (1982b), the three contributions are combined to form a single equation with both a theoretical basis and satisfactory experimental fit that is widely used by researchers and engineers in the field of underwater sound.

The combined equation is

$$\alpha = \frac{A_1 P_1 f_1 f^2}{f^2 + f_1^2} + \frac{A_2 P_2 f_2 f^2}{f^2 + f_2^2} + A_3 P_3 f^2 \tag{3.27}$$

In equation (3.27) the first term is the contribution of boric acid with $f_1$ being the boric acid relaxation frequency. The second term is the magnesium sulfate contribution with relaxation $f_2$, and the third term represents the viscous absorption loss. The three terms have the forms discussed previously in equations (3.20) and (3.26). The coefficients are given as empirical equations that depend on the environmental conditions in terms of temperature $T$ (°C), salinity $S$ (ppm), depth $D$ (m), and pH value

***Boric acid contribution***

$$
\begin{aligned}
A_1 &= \frac{8.86}{c} \times 10^{(0.78\,pH-5)} \ \text{dB km}^{-1}\text{kHz}^{-1} \\
P_1 &= 1 \\
f_1 &= 2.8(S/35)^{0.5} \times 10^{\left(4-1245/\theta\right)} \ \text{kHz}
\end{aligned}
\qquad (3.28)
$$

***Magnesium sulfate contribution***

$$
\begin{aligned}
A_2 &= 21.44\frac{S}{c}(1+0.025T) \ \text{dB km}^{-1}\text{kHz}^{-1} \\
P_2 &= 1-1.37D\times10^{-4}+6.2D^2\times10^{-9} \\
f_2 &= \frac{8.17\times10^{(8-1990/\theta)}}{1+0.0018(S-35)} \ \text{kHz}
\end{aligned}
\qquad (3.29)
$$

***Fresh-water viscous contribution***

For $T \le 20°\text{C}$

$$A_3 = 4.937\times10^{-4} - 2.59\times10^{-5}T + 9.11\times10^{-7}T^2 - 1.50\times10^{-8}T^3 \ \text{dB km}^{-1}\text{kHz}^{-1}$$

For $T > 20°\text{C}$

$$A_3 = 3.964\times10^{-4} - 1.146\times10^{-5}T + 1.45\times10^{-7}T^2 - 6.5\times10^{-10}T^3 \ \text{dB km}^{-1}\text{kHz}^{-1}$$

$$P_3 = 1-3.83D\times10^{-5}+D^2\times10^{-10}$$

(3.30)

In the above equations $\theta$ is the absolute temperature, $\theta = T + 273$ K, and $c$ is the sound speed

$$c = 1412 + 3.21T + 1.19S + 0.167D \qquad (3.31)$$

Figure 3.5 shows the absorption as a function of frequency for the total (seawater), pure water, and the two relaxation components, calculated for the values of $T$ = 10°C, $D$ = 0, and pH = 7.8. According to equation (3.31) the sound speed in this case is 1485.8 m/s, but equation (3.1) gives 1472 m/s. The difference in the two sound speeds is insignificant for the calculation of absorption.

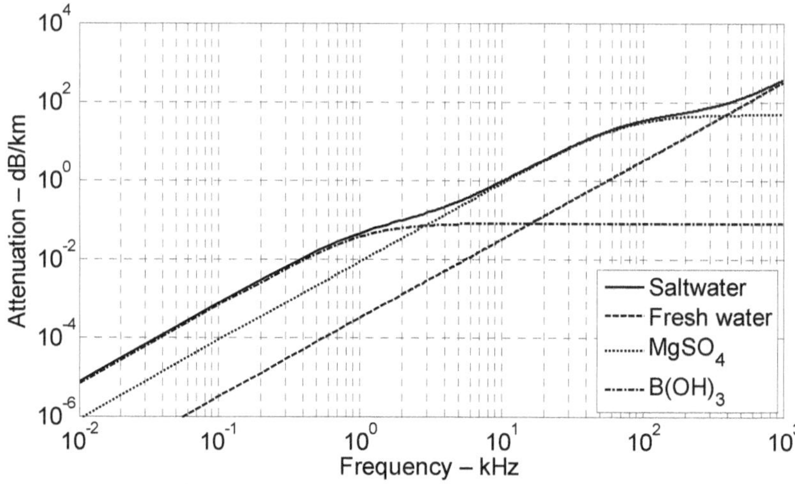


Figure 3.5. Acoustic absorption (dB/km) for fresh water and saltwater, plotted as a function of frequency (kHz) for water temperature of 10°C, atmospheric pressure of one atmosphere (surface), salinity of 35 pro mille, and pH value of 7.8. The legend indicates the various contributions to absorption loss.

In this example the relaxation frequencies are 93.2 kHz for magnesium sulfate and 1.12 kHz for boric acid. These two contributions result in much higher absorption in saltwater than in freshwater for frequencies up to about 1 MHz, as shown in figure 3.5.

For frequencies below approximately 1 kHz, experience shows that the absorption is often greater than the absorption shown in figure 3.5. An empirical expression, based on measurements, is given by Thorp (1965) as

$$\alpha_{dB} = 0.178 \cdot f_{\text{kHz}}^{1.5} \qquad \text{dB/km} \tag{3.32}$$

In equation (3.32) frequency is measured in kHz, and the absorption loss is in decibels per meter. Note that the Thorp formula, equation (3.32), gives higher values than equation (3.20). As an example, for 1 kHz, equation (3.32) gives the value of 0.178 dB/km, as compared with the value of 0.06 dB/km (for salt seawater) in figure 3.5. Thorp's equation is frequently used in numerical models for calculating sound propagation at low frequencies.

## 3.4 Sound speed in marine sediments

The top layer of the seabed often consists of sediments that are relatively loose and less consolidated, such as clay, silt, and sand, or mixtures of these. The sound speed in such sediments is normally a little higher than the sound speed in water, up to 10–15% higher. Particularly in loose masses, the sound speed may sometimes also be less than the sound speed in the water above the sediment. This phenomenon will be explained later in this chapter.

Even in the least consolidated sediments there will be some shear rigidity, which means the material is able to propagate shear waves. However, the shear wave speed is usually very low and rarely higher than 100–200 m/s.

Table 3.1 shows the typical values for sediments, rocks, and other materials. More comprehensive tables can be found in the literature, particularly the work presented by Hamilton and Bachman (1982) and Hamilton (1987).

Table 3.1. Speeds and densities for sediments and other materials.

| Material | Density ($g/cm^3$) | Sound speed (m/s) | |
|---|---|---|---|
| | | Compressional waves | Shear waves |
| Air | 0.0013 | 330 – 360 | |
| Oil | 0.60–0.90 | 1250–1400 | |
| Water | 0.98–1.05 | 1400–1550 | |
| Ice | 0.88–1.07 | 3100–4200 | 1600–2000 |
| Clay | 1.50–2.60 | 1500–2500 | 200–1000 |
| Silt | 1.80–2.20 | 1400–1800 | 150– 450 |
| Sand | 1.60–2.30 | 1500–2000 | 100 – 500 |
| Moraine | 1.50–2.00 | 1500–2700 | 500 –1300 |
| Slate | 2.00–3.20 | 2700–4800 | 1500–2400 |
| Sandstone | 2.10–2.80 | 2100–4500 | 1200–2800 |
| Chalk | 1.60–2.60 | 2100–4200 | 1000–2000 |
| Salt | 2.10–2.40 | 4200–5500 | 2000–3200 |
| Gneiss | 2.40–3.00 | 3500–7500 | 1700–3600 |
| Granite | 2.50–2.90 | 4750–6000 | 2400–3800 |

It should be noted that sound speed in the ocean bottom will normally increase with depth of the bottom because the increase in overburden pressure may result in reduced porosity and stronger interparticle coupling. Theoretical models of idealized materials show that the elastic parameters increase, with the overburden pressure raised to the power 1/3. With unaltered density, this means that the pressure and shear wave speed increase with the overburden pressure to the power 1/6, but the increase is often somewhat larger than this (Hamilton 1987).

## 3.5 Sound speed in composite materials

In this section we develop a very simple model, known as Wood's equation (Wood and Weston 1964), for a medium composed of two or more components. We start by recalling that, in general, the density $\rho$ and the bulk modulus $K$ of the medium give the sound speed $c$ as

$$c = \sqrt{\frac{K}{\rho}} \tag{3.33}$$

For a composite medium, the density and the bulk modulus must be the effective values; the task is now to find these values. To this end, we consider a volume $V$ consisting of a relative amount $\phi$ of fluid and a relative amount $(1 - \phi)$ of solid substance. We call $\phi$ the porosity. If the densities of the fluid and the solid substance are $\rho_f$ and $\rho_s$, respectively, the effective density—the density of the composite material—becomes

$$\rho = \phi\rho_f + (1-\phi)\rho_s \tag{3.34}$$

We find the effective bulk modulus by assuming that a volume $V$ of the mixture is kept in a container where the hydrostatic pressure $p$ is changed by $\Delta p$. The total change of volume is the sum of the volume changes $\Delta V_s$ of the solid substance and the change of volume $\Delta V_f$ for the fluid. The total relative volume change is thus

$$\frac{\Delta V}{V} = \phi\frac{\Delta V_f}{V_f} + (1-\phi)\frac{\Delta V_s}{V_s} \tag{3.35}$$

The effective volume bulk modulus $K_i$ is given by the ratio between the change in pressure and the relative change in volume:

$$K_i = \frac{-\Delta p}{\Delta V_i / V_i} \qquad i = s, f \tag{3.36}$$

Since the pressure is the same on the fluid as on the solid, both phases will be compacted according to their respective stiffness, so that the effective bulk modulus of the mixture becomes

$$\frac{1}{K} = \frac{\phi}{K_f} + \frac{(1-\phi)}{K_s} \tag{3.37}$$

where $K_f$ and $K_s$ are the volume rigidity for the fluid and the solid, respectively.

The sound speed $c$ of the composite medium is then found by inserting equations (3.37) and (3.34) into equation (3.33), which produces

$$c = \sqrt{\frac{K}{\rho}} = \frac{1}{\sqrt{\left[\frac{\phi}{K_f} + \frac{1-\phi}{K_s}\right]\left[\phi\rho_f + (1-\phi)\rho_s\right]}} \tag{3.38}$$

Since the particles of the solid substance are assumed to have no contact with one another and the fluid has no shear rigidity, the shear wave speed is equal to zero.

Equation (3.38), often referred to as Wood's equation (Wood and Weston 1964), predicts that the sound speed of a suspension is determined by the effective compressibility (defined as $K^{-1}$) and density that are given by weighted sums of the compressibility and density of the components, as given by equations (3.34) and (3.37), respectively. An underlying assumption is that there is no relative motion between the phases. Equation (3.38) can easily be extended to media composed of three or more components.

We use equation (3.38) on a mixture of solid particles for which $\rho_s$ = 2650 kg/m$^3$ and $K_s = 3.6 \cdot 10^{10}$ Pa, and water for which $\rho_f$ = 1000 kg/m$^3$ and $K_f = 2.25 \cdot 10^9$ Pa.

Figure 3.6 shows the sound speed, relative to the speed of the water (1500 m/s), as a function of the porosity $\phi$ (that is, as a function of the relative volume of water).

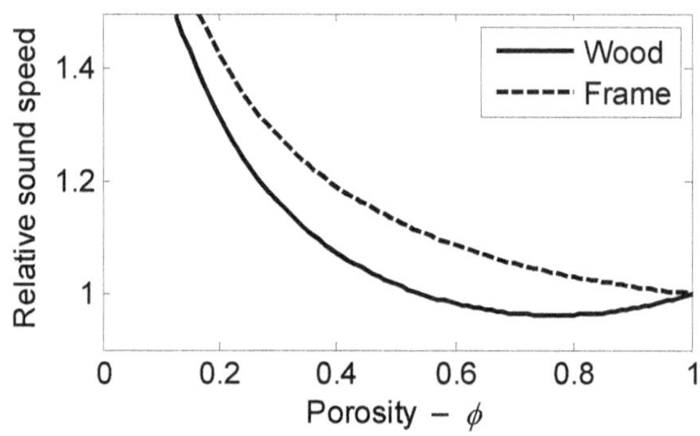


Figure 3.6. Sound speed for a suspension of particles in water, calculated as a function of porosity by Wood's equation. Also shown is the higher sound speed caused by frame rigidity.

It is important to note that use of Wood's equation may result in lower sound speeds for the composite material than for the individual components. When compared to the measured values, Wood's equation often gives values that are too low, particularly for low porosities, because the equation does not take into consideration the connections between the particles therefore ignoring the rigidity of the framework. Consequently, equation (3.38) is valid only for high porosities such as that in fine-grain silts and muds, where the particles are weakly connected to each other. We shall return to this issue in chapter 16. In the first approximation the effect of a framework rigidity is to give an addition $K_r$ to the bulk modulus such that

$$K = K_r + \frac{1}{\dfrac{\phi}{K_f} + \dfrac{(1-\phi)}{K_s}} \tag{3.39}$$

The effect of this addition is to produce a higher sound speed than that found with Wood's equation; this higher sound speed is shown in figure 3.6, where we have assumed that

$$K_r = (1-\phi)K_{r,0} \tag{3.40}$$

In figure 3.6, the value $K_{r,0} = 2 \cdot 10^9$ Pa was chosen to obtain the best fit with measured values and to demonstrate that frame rigidity cannot be ignored.

## 3.6 Sound in water with free air bubbles

Of considerable practical interest, Wood's equation can also be used to determine the sound speed of water that contains free gas or air bubbles. We shall see that the sound speed in water with gas bubbles is considerably lower than in water without gas and is even lower than the sound speed of the gas.

We consider a mixture of water and gas bubbles with the values $\rho_s = 1000$ kg/m$^3$ and $K_s = 2.25 \cdot 10^9$ Pa for the water and $\rho_f = 1.3$ kg/m$^3$ and $K_f = 1.5 \cdot 10^5$ Pa for the gas. Note that we have switched the values used in section 3.5 for the solid, $\rho_s$ and $K_s$, to represent the water's values, and then used $\rho_f$ and $K_f$ to represent the values for air. The values for air produce a sound speed in pure air equal to 340 m/s. The plot in figure 3.7 shows the result of using Wood's equation to obtain the sound speed for this gas-water mixture where the sound speed is a function of the relative volume content of air, even for very low concentrations of air. We see that the sound speed of the mixture is very much lower than the sound speed in pure water.

We can understand why a very little gas produces a strong decrease in sound speed by going back to equations (3.34) and (3.37) for the effective density and bulk modulus, respectively. With air we have that $\rho_f << \rho_s$ and $K_f << K_s$. With very good approximation, the effective values are then $K \approx K_f/\phi$ and $\rho \approx (1-\phi)\rho_s$, when $\phi > 0$. These expressions mean that the sound speed is given as the ratio of the relatively low bulk modulus of the air and the relatively high density of the water; this ratio produces a very low sound speed, as we can see in figure 3.7.

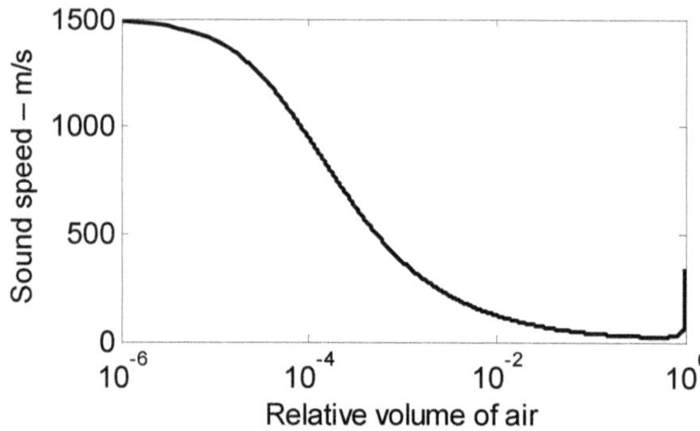


Figure 3.7. Sound speed for water with gas bubbles, calculated by Wood's equation as a function of relative gas content.

Wood's equation to calculate sound speed for water with bubbles is valid for high frequencies and low frequencies, provided they are much lower than the resonance frequencies of the bubbles. For frequencies near the bubble resonance frequency, the sound speed will vary very rapidly with frequency. At much higher frequencies, the sound speed in the mixture again becomes equal to the sound speed in water. This phenomenon will be discussed further in section 13.5.

# *Chapter 4*

# *Reflection and transmission at boundaries between fluid media*

## 4.1 Introduction

In acoustic and seismic wave propagation, an important role is played by reflection and transmission of acoustic waves at the interface between two media with different acoustical properties. In this chapter we consider reflection and transmission of plane harmonic waves at a simple interface between two media with different densities and sound speeds. Whereas this chapter only considers the waves at one simple interface, in chapter 5 we shall also discuss reflection and transmission in media with several layers. In both chapters 4 and 5 we are limiting the discussion to media that can be considered as fluids. In chapter 15 we also discuss interfaces between solids and the conversion of compressional waves to shear waves. In this chapter we also briefly treat reflection from corrugated or rough surfaces; this topic will be discussed in more detail in chapter 14.

## 4.2 Derivation of reflection and transmission coefficients

Consider an interface between two fluid media which we will refer as media 1 and 2, with densities and sound velocities of $\rho_1$, $\rho_2$ and $c_1$, $c_2$, respectively. The geometry is shown in figure 4.1. An incoming plane wave characterized by the velocity potential $\Phi_i$ strikes the interface at the angle $\theta_1$ and produces both a reflected wave with the same angle, the specular angle, and a transmitted or refracted wave with the angle $\theta_2$. Note that these are grazing angles and refer to the horizontal plane.

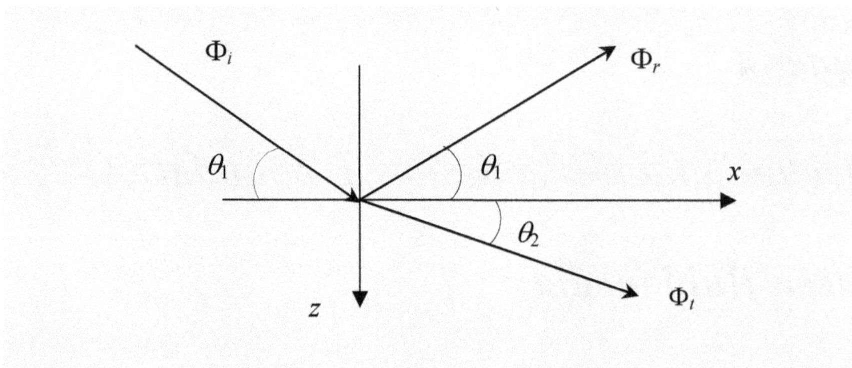


Figure 4.1. Reflected and transmitted waves at an interface between two fluid media: an incident plane wave $\Phi_i$ with the angle $\theta_1$ produces a reflected wave $\Phi_r$ with the same angle and a transmitted wave $\Phi_t$ with the angle $\theta_2$.

We now assume a harmonic incident wave with frequency $\omega$, which results in a reflected and transmitted wave at the same frequency.

The incident, reflected, and transmitted waves can be expressed as follows:

***Incident wave in medium 1***

$$\phi_i(x,z,t) = \Phi_i \exp(i\kappa_1 x\cos\theta_1 + i\kappa_1 z\sin\theta_1)\exp(-i\omega t) \tag{4.1}$$

***Reflected wave in medium 1***

$$\phi_r(x,z,t) = \Phi_r \exp(i\kappa_1 x\cos\theta_1 - i\kappa_1 z\sin\theta_1)\exp(-i\omega t) \tag{4.2}$$

***Transmitted wave in medium 2***

$$\phi_t(x,z,t) = \Phi_t \exp(i\kappa_2 x\cos\theta_2 + i\kappa_2 z\sin\theta_2)\exp(-i\omega t) \tag{4.3}$$

In the equations above, the wave number $\kappa_n$ is defined as $\kappa_n = (\omega/c_n)$, with the horizontal and vertical components $\kappa_n\cos\theta_n$ and $\kappa_n\sin\theta_n$ (where $n = 1, 2$), respectively.

Assuming that the amplitude $\Phi_i$ of the incoming wave is known, we now determine the amplitudes $\Phi_r$ and $\Phi_t$ of the reflected and transmitted waves by applying the boundary conditions that, for physical reasons, the waves must satisfy at the interface.

At an interface between two fluid media there are two conditions that must be met. First, the vertical particle velocity and the pressure must be the same on both sides of the interface. In addition, the horizontal components of the wave numbers have to be equal in both media. This latter condition means that

$$\kappa_1 \cos\theta_1 = \kappa_2 \cos\theta_2 = k \tag{4.4}$$

Since the constant $k$ is actually the horizontal wave number, equation (4.4) gives us Snell's law, which is

$$\frac{\cos\theta_1}{c_1} = \frac{\cos\theta_2}{c_2} \tag{4.5}$$

The vertical wave number components $\gamma_1$ and $\gamma_2$ are defined as

$$\begin{aligned} \gamma_1 &= \frac{\omega}{c_1}\sin\theta_1 = \sqrt{\frac{\omega^2}{c_1^2} - k^2} \\ \gamma_2 &= \frac{\omega}{c_2}\sin\theta_2 = \sqrt{\frac{\omega^2}{c_2^2} - k^2} \end{aligned} \tag{4.6}$$

The wave number components can be graphically represented, as shown in figure 4.2.

Equations (4.1), (4.2), and (4.3) can be rewritten as

$$\begin{aligned} \phi_i(x,z,t)_i &= \Phi_i \exp(ikx + i\gamma_1 z)\exp(-i\omega t) \\ \phi_r(x,z,t) &= \Phi_r \exp(ikx - i\gamma_1 z)\exp(-i\omega t) \\ \phi_t(x,z,t) &= \Phi_t \exp(ikx + i\gamma_2 z)\exp(-i\omega t) \end{aligned} \tag{4.7}$$

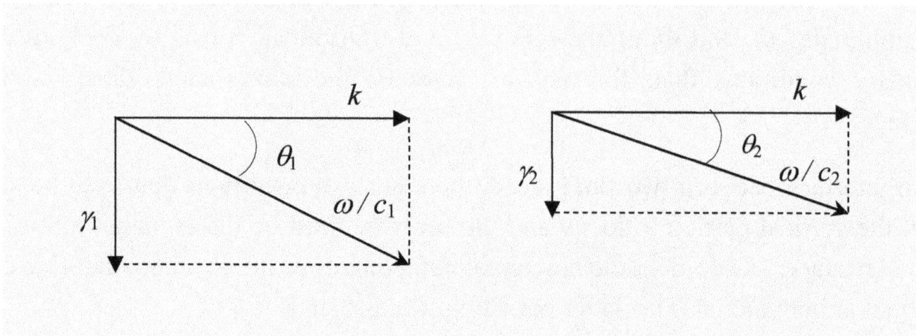


Figure 4.2. The wave number components for the first medium, left, and for the second medium, right. The horizontal wave number component k is the same in both media, while the vertical wave number components γ1 and γ2 are different.

The sound pressure $p$ is defined as $-\rho\partial\phi/\partial t$; therefore, the sound pressure at the interface between the two media, where $z = 0$, is given by the two expressions:

$$\begin{aligned} p_1(x,z,t) &= i\omega\rho_1\left[\Phi_i \exp(i\gamma_1 z) + \Phi_r \exp(-i\gamma_1 z)\right]\exp(ikx)\exp(-i\omega t) \\ p_2(x,z,t) &= i\omega\rho_2\left[\Phi_t \exp(i\gamma_2 z)\right]\exp(ikx)\exp(-i\omega t) \end{aligned} \tag{4.8}$$

The particle velocity is given by $\mathbf{v} = \nabla\phi$, and the vertical components are thus

$$\begin{aligned} v_{z,1}(x,z,t) &= i\gamma_1\left[\Phi_i \exp(i\gamma_1 z) - \Phi_r \exp(-i\gamma_1 z)\right]\exp(ikx)\exp(-i\omega t) \\ v_{z,2}(x,z,t) &= i\gamma_2\left[\Phi_t \exp(i\gamma_2 z)\right]\exp(ikx)\exp(-i\omega t) \end{aligned} \tag{4.9}$$

As earlier stated, the boundary conditions require continuity in pressure and vertical particle velocity at $z = 0$, resulting in two equations

$$\begin{aligned} p_1(x,0) &= p_2(x,0) \\ v_{z,1}(x,0) &= v_{z,2}(x,0) \end{aligned} \tag{4.10}$$

which give us the following relations between the amplitudes:

$$\begin{aligned} i\omega\rho_1\left(\Phi_i + \Phi_r\right) &= i\omega\rho_2\Phi_t \\ \gamma_1\left(\Phi_i - \Phi_r\right) &= \gamma_2\Phi_t \end{aligned} \tag{4.11}$$

When solved with respect to $\Phi_r$ and $\Phi_t$, equation (4.11) gives the reflection and transmission coefficients for the velocity potentials.

$$R_\phi = \frac{\Phi_r}{\Phi_i} = \frac{\rho_2\gamma_1 - \rho_1\gamma_2}{\rho_2\gamma_1 + \rho_1\gamma_2} \tag{4.12}$$

$$T_\phi = \frac{\Phi_t}{\Phi_i} = \frac{2\rho_1\gamma_1}{\rho_2\gamma_1 + \rho_1\gamma_2} \tag{4.13}$$

More commonly, the coefficients are expressed using the angles, densities, and sound speeds for each of the two media:

$$R_\phi = \frac{\rho_2 c_2 \sin\theta_1 - \rho_1 c_1 \sin\theta_2}{\rho_2 c_2 \sin\theta_1 + \rho_1 c_1 \sin\theta_2} \tag{4.14}$$

$$T_\phi = \frac{2\rho_1 c_2 \sin\theta_1}{\rho_2 c_2 \sin\theta_1 + \rho_1 c_1 \sin\theta_2} \tag{4.15}$$

Snell's law, equation (4.5), gives us the angle of the transmitted wave.

The reflection coefficient $R_\phi$ and the transmission coefficient $T_\phi$ which we have found above are valid for the velocity potentials. It is, however, more common to use the coefficients for sound pressures, or vertical particle velocities, so that we define two other sets of reflection and transmission coefficients.

Since the sound pressure is given by the potential $p = i\omega\rho\Phi$, the relationship between the sound pressures in the two media becomes

$$R_p = R_\phi = \frac{\rho_2\gamma_1 - \rho_1\gamma_2}{\rho_2\gamma_1 + \rho_1\gamma_2} \tag{4.16}$$

$$T_p = \frac{\rho_2}{\rho_1} T_\phi = \frac{2\rho_2\gamma_1}{\rho_2\gamma_1 + \rho_1\gamma_2} \tag{4.17}$$

The vertical particle velocity is defined as $v_z = \partial\phi/\partial z$, so that the coefficients of the particle velocity become

$$R_v = -R_\phi = \frac{\rho_1\gamma_2 - \rho_2\gamma_1}{\rho_2\gamma_1 + \rho_1\gamma_2} \tag{4.18}$$

$$T_v = \frac{\gamma_2}{\gamma_1} T_\phi = \frac{2\rho_1\gamma_2}{\rho_2\gamma_1 + \rho_1\gamma_2} \tag{4.19}$$

Alternatively, the coefficients could be expressed by using the acoustic impedances $Z_1$ and $Z_2$, which represent the relations between the sound pressure and the vertical component of the particle velocity for the first and second media, respectively

$$\begin{aligned} Z_1 &= \frac{\omega\rho_1}{\gamma_1} = \frac{\rho_1 c_1}{\sin\theta_1} \\ Z_2 &= \frac{\omega\rho_2}{\gamma_2} = \frac{\rho_2 c_2}{\sin\theta_2} \end{aligned} \tag{4.20}$$

so that

$$\begin{aligned} R_p &= \frac{Z_2 - Z_1}{Z_2 + Z_1} \\ T_p &= \frac{2Z_2}{Z_2 + Z_1} \end{aligned} \tag{4.21}$$

Note that all the expressions developed so far apply with the incident wave in medium 1; the indexes must be interchanged if the incident wave is incident from medium 2.

We now determine the intensities of the reflected and transmitted waves. For a plane harmonic wave, the intensity is given as the real part of the product of the sound pressure and the complex conjugated $z$-component of the particle velocity, equation (2.64). The reflected and transmitted intensities $I_r$ and $I_t$ with respect to the incident intensity $I_i$ are therefore expressed as

$$\frac{I_r}{I_i} = -\mathrm{Re}\left\{R_p R_v^*\right\} = \left|\frac{\rho_2\gamma_1 - \rho_1\gamma_2}{\rho_2\gamma_1 + \rho_1\gamma_2}\right|^2 \tag{4.22}$$

$$\frac{I_t}{I_i} = \mathrm{Re}\left\{T_p T_v^*\right\} = \mathrm{Re}\left\{\frac{4\rho_2\rho_1\gamma_1\gamma_2}{\left|\rho_2\gamma_1 + \rho_1\gamma_2\right|^2}\right\} \tag{4.23}$$

In equation (4.22) the minus sign accounts for the negative impedance of a wave traveling upwards in the negative $z$ direction.

By adding equations (4.22) and (4.23), we can easily see that

$$I_i = I_r + I_t \tag{4.24}$$

Equation (4.24) shows that incident intensity is equal to the sum of the reflected and transmitted intensities, and therefore that conservation of energy is preserved. This does not exclude the possibility of transmission coefficients that exceed unity. This can be seen from equation (4.19) which yields

$$\begin{aligned} T_p &= R_p + 1 \\ T_v &= R_v + 1 \end{aligned} \tag{4.25}$$

Consider as an example that a wave in a low impedance medium is incident on a medium with higher impedance. In this case $R_p > 0$, $T_p > 1$, and $R_v < 0$ and $T_v < 1$. The transmitted pressure is then higher than the incident pressure, and the transmitted particle velocity is lower. The opposite case, with a wave in a high impedance medium that is incident on a low impedance medium, gives $T_v > 1$ and $R_v > 0$.

We comment upon some issues of particular importance. First, consider an interface between water (medium 1) and air (medium 2). At normal incidence, the angle $\theta_1 = \theta_2 = \pi/2$, and we find the ratio of the impedances to be

$$\frac{Z_2}{Z_1} = \frac{\rho_2 c_2}{\rho_1 c_1} \approx \frac{1}{3600} \tag{4.26}$$

Consequently, the reflection and transmission coefficients approximate $R_p \approx -1.0$ and $T_p \approx 0.0$, respectively. This means that only a vanishingly small part of the wave energy is transmitted from air to water or vice versa.

The concepts of a critical angle and total reflection require comment. From equation (4.5) we see that if $c_1 < c_2$ there will be a certain value of the angle $\theta_1$ that will produce $\theta_2 = 0$. This value of the angle $\theta_1$ is called the critical angle $\theta_{crit}$ and is given by

$$\cos\theta_{crit} = \frac{c_1}{c_2} \tag{4.27}$$

For an angle $\theta_1$ that is less than $\theta_{crit}$, the angle $\theta_2$ and the vertical wave number of the transmitted wave in medium 2 become imaginary. This imaginary vertical wave number means that the transmitted sound field will decrease exponentially with depth in medium 2. No acoustic power will be transmitted to medium 2, and we have total reflection. This is seen directly from equation (4.22), which when $\gamma_2$ is imaginary produces $I_r = I_i$. In addition, there is also another important point that needs to be commented upon. From equation (4.16) it follows that, for an incident angle lower than the critical angle, $|R_p| = 1$, but there is a phase term such that the reflection coefficient is

$$R_p = \exp\left[-2i\operatorname{atan}\left(\left|\frac{\rho_1\gamma_2}{\rho_2\gamma_1}\right|\right)\right] \tag{4.28}$$

In consequence of this phase term there is generated a wave traveling along the interface between the two media with the speed of the second medium. This wave and phenomenon are known under different names: head wave, lateral wave, and refracted wave. The wave and phenomenon have both theoretical and practical consequences. We shall not go deeper into the theory but refer to chapter 4 in the book by Brekhovskikh and Lysanov (2003). We shall see some practical consequence of this phenomenon in chapter 6 on ray tracing and in chapter 18 where this wave is used in refraction surveys to determine the sound speed in the seafloor.

The last special case that also ought to be mentioned is intromission, which occurs when the reflection coefficient equals zero and all the sound energy is transmitted across the interface into the next medium. From equation (4.21) we see that the precondition for this case is that $Z_1 = Z_2$. From equation (4.20) and Snell's law, equation (4.5), we find that intromission only takes place when the incident wave arrives at the angle

$$\sin\theta_1 = \sqrt{\frac{1-(c_1/c_2)^2}{1-(\rho_2/\rho_1)^2}} \tag{4.29}$$

This equation shows that intromission can occur only when this equation produces a real solution. This can happen in two ways:

$$\text{when } c_1 > c_2 \text{ and } \rho_1 < \rho_2$$

or

$$\text{when } c_1 < c_2 \text{ and } \rho_1 > \rho_2$$

Both cases are theoretical possibilities, but when the wave makes the transit from water to seabed, the condition will always be $\rho_1 < \rho_2$, so that only the first of these possibilities is relevant. In reality, such conditions are only found in cases with loose and high-porosity sea bottom sediments. See table 3.1 and the discussion in section 3.5.

## 4.3 Reflection from the seabed

Now we will consider examples of the seabed and start with considering a sandy bottom with the following values that may be considered representative:

$$c_1 = 1500\,\text{m/s} \qquad \rho_1 = 1000\,\text{kg/m}^3$$
$$c_2 = 1700\,\text{m/s} \qquad \rho_2 = 2000\,\text{kg/m}^3$$

The values $c_1$ and $\rho_1$ are for the water (medium 1) and $c_2$ and $\rho_2$ are for the bottom (medium 2).

In figure 4.3, the reflection $R_p$ and transmission $T_p$ coefficients are plotted as functions of the incident wave's grazing angle. In figure 4.4, the plots represent the reflected and transmitted intensities for the numerical values listed above. The examples in figures 4.3 and 4.4 are calculated under the assumption that there is no absorption loss in the media. This is unrealistic for reflection at a sea bottom since marine sediments normally will have significant absorption. To study the effects of absorption on the reflection and transmission coefficients we shall now consider a case where medium 2 does have absorption.

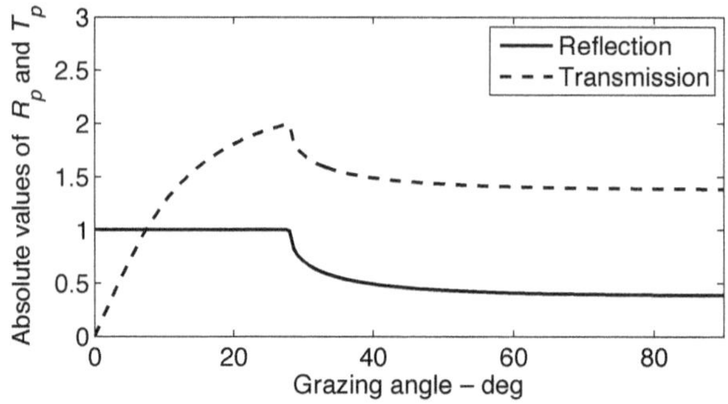


Figure 4.3. Reflection $R_p$ and transmission $T_p$ coefficients for a case where the sound speed for medium 2 is higher than that for medium 1.

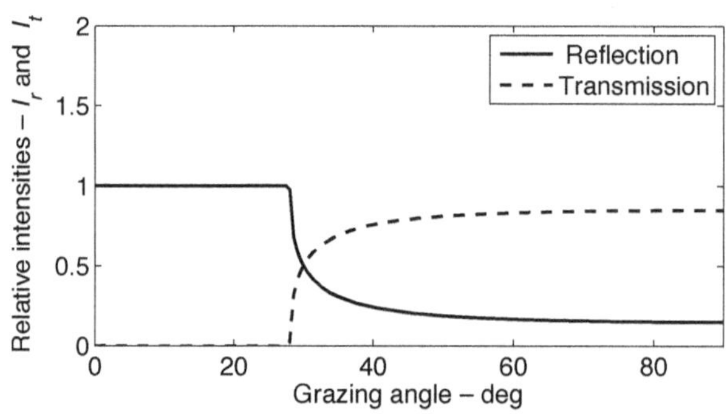


Figure 4.4. Reflected and transmitted intensities for a case where the sound speed for medium 2 is higher than that for medium 1.

We can model absorption by introducing a complex sound speed; see section 2.4 and equation 2.47. If we do this while using the same numerical values as before, we obtain the reflection and transmission coefficients of figures 4.5 and 4.6.

In these figures we have used a value that gives an absorption equal to 1 dB / λ for the imaginary part of the sound speed in medium 2. This is a fairly high, but still realistic, value for absorption in loose bottom sediments. Comparing figures 4.5 and 4.6 with figures 4.3 and 4.4, we see that the effect of absorption can be quite significant on the reflection coefficient at and below the critical angle.

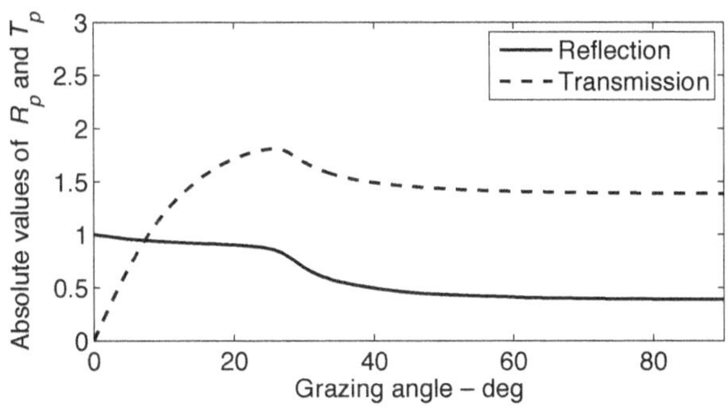


Figure 4.5. Reflection $R_p$ and transmission $T_p$ coefficients when the absorption loss in medium 2 is 1 dB / λ.

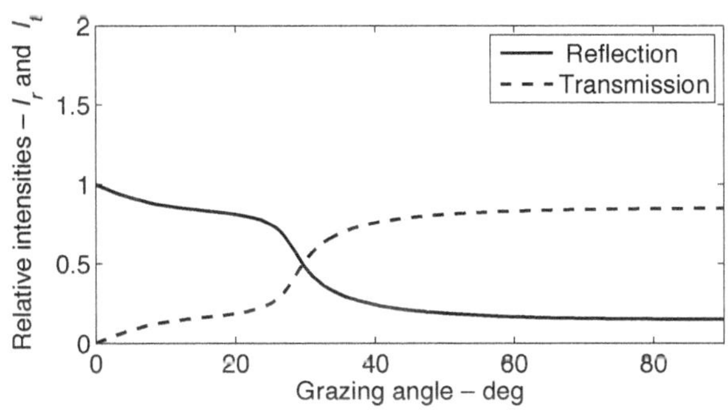


Figure 4.6. Reflected and transmitted intensities when the absorption loss in medium 2 is 1 dB / λ.

The specific acoustic impedance of air is much lower than that of water, and therefore the sound signal will be totally reflected off the surface, at least when the sea state is low. At the sea bottom, the conditions are different and the signal will suffer a reflection loss in decibels, *RL*, of

$$RL = -20\log\left(\left|R_p\right|\right) \tag{4.30}$$

Figure 4.7 shows examples of reflection loss with and without absorption. Without absorption the reflection loss is zero for angles lower than critical and very high for higher angles. With several reflections off the seabed, sound rays hitting the bottom at an angle higher than the critical angle will be severely attenuated and will more likely have several reflections off the seabed in their transit. Possibly only low-grazing angle rays will reach the receiver. However, figure 4.7 also shows that, with absorption, the reflection loss for low grazing angles also can be significant. In this example with the absorption of 1 dB / λ, the reflection loss is approximately 3 dB at angles that are near, but below, the critical angle.

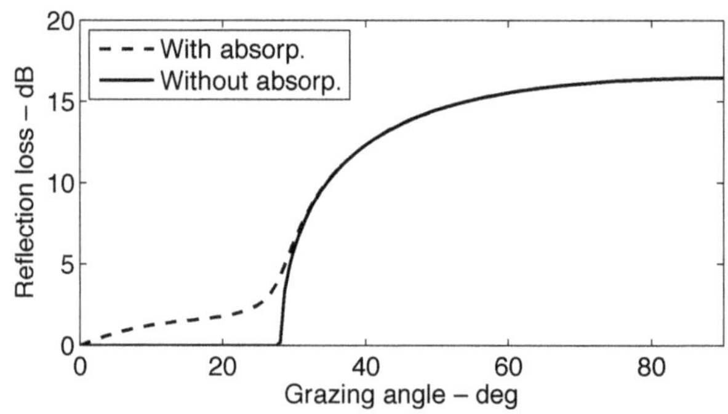


Figure 4.7. Reflection loss as a function of the grazing angle, plotted for both the case with an absorption loss of 1 dB / λ in medium 2 and the case of no absorption.

Even a small reflection loss can have a significant impact on the transmission loss at long distances when sound is reflected from the bottom several times before arriving at the receiver. Thus, bottom absorption is a very important factor in sound propagation in relatively shallow water.

We present an example of intromission, using the following parameters for a muddy seabed:

$$c_1 = 1500\text{m/s} \qquad \rho_1 = 1000\text{kg/m}^3$$
$$c_2 = 1450\text{m/s} \qquad \rho_2 = 1200\text{kg/m}^3$$

Note that here the sound speed at the bottom is lower than in the water—a condition that may be found in high porosity and loss sediments; see also sections 3.4 and 3.5. These values, when used in equation (4.29), produce an intromission angle of $\theta_1$ = 23.5°. In figure 4.8, the bottom refection loss is plotted for this case with three different values of the bottom absorption coefficient: 0.25, 0.5, and 1.0 dB per wave length. The result of figure 4.8 shows that the reflection loss has a maximum value at the intromission angle and that the maximum value is dependent on the bottom absorption coefficient.

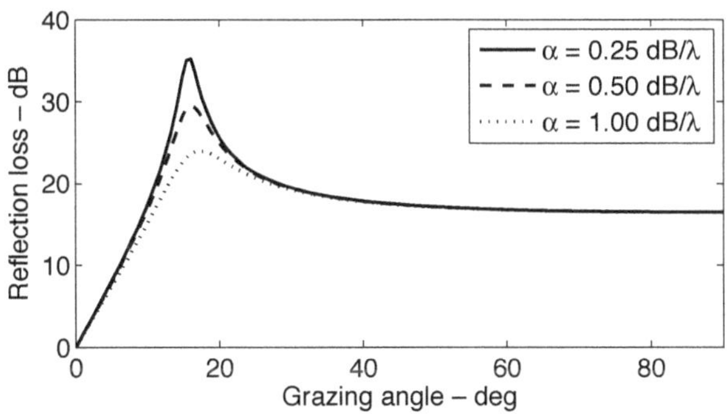


Figure 4.8. Bottom reflection loss for a case where the conditions of intromission are satisfied and plotted for three different values of the bottom absorption coefficient $\alpha$.

## 4.4 Reflection from uneven surfaces

We now consider what happens when the interface is not completely plane but rather uneven or corrugated. When the interface is uneven, we get an additional reflection loss because only parts of the energy are being reflected coherently in a specular direction, with the rest being spread incoherently in other directions. Figure 4.9 contains illustrations of the effects these environments have on scattering.

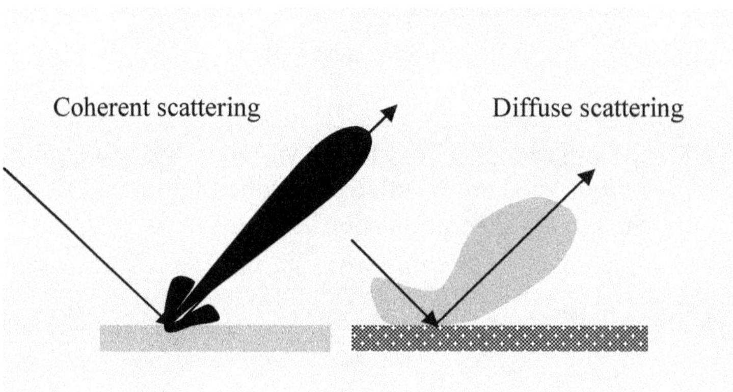


Figure 4.9. Reflection from an even surface mainly gives coherent spreading in the specular direction. An uneven surface gives diffuse or incoherent spreading in many directions.

In order to explain what is meant by ***plane*** and ***uneven***, we will consider a simple theory of reflection from uneven surfaces. Because of local unevenness, a wave hitting the interface may be subjected to an increase or decrease in its length of travel compared to a wave reflected from a plane surface. When the local vertical displacement of the interfaces is $h$, the change in length of travel $l$ between the transmitter and receiver, as shown in figure 4.10, is equal to

$$l = 2h\sin\theta \tag{4.31}$$

When the wave has frequency $f$, the phase contribution $\beta$ is expressed as

$$\beta = \frac{\omega}{c} 2h \sin\theta = 2\pi \frac{2h}{\lambda} \sin\theta \tag{4.32}$$

The sound pressure reflected from one point of the surface, $p_r$, can then be expressed as

$$p_r = p_i \exp\left( -2\pi i \frac{2h}{\lambda} \sin\theta \right) \tag{4.33}$$

where $p_i$ is the reflected sound pressure from an even surface.

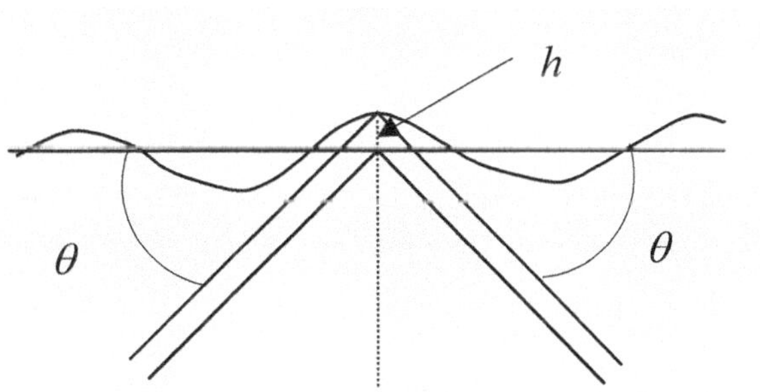


Figure 4.10. A plane wave with grazing angle $\theta$ is striking an uneven surface under two different conditions: one when the surface has a local elevation $h$, the other when the surface elevation is zero. The difference in path length is $l = 2h \sin\theta$.

In the following discussion, certain assumptions with regard to the value and variation of $h$ must be made. A particularly simple and commonly used assumption is that $h$ is a stochastic variable with the probability density $w(h)$. The mean sound pressure in the specular direction then becomes

$$< p_r > = p_i \int w(h) \exp\left( -2\pi i \frac{2h}{\lambda} \sin\theta \right) dh \tag{4.34}$$

where < > means ensemble averaging.

Assuming that the vertical surface displacements have a Gaussian distribution with a zero mean and a standard deviation of $\sigma_h$, which is the root mean square (rms) of the wave surface displacement, then

$$w(h) = \frac{1}{\sqrt{2\pi\sigma_h^2}} \exp\left(-\frac{h^2}{2\sigma_h^2}\right) \tag{4.35}$$

When equation (4.35) is inserted in equation (4.34), we obtain the mean reflected sound pressure

$$< p_r >= p_i \exp\left[-2\left(\frac{2\pi}{\lambda}\sigma_h \sin\theta\right)^2\right] \tag{4.36}$$

This implies that the reflection coefficient for the amplitude is

$$R_{coh} = \exp\left[-2\left(\frac{2\pi}{\lambda}\sigma_h \sin\theta\right)^2\right] \tag{4.37}$$

$R_{coh}$ is called the ***coherent reflection coefficient*** because it only includes that part of the signal being reflected in the specular direction, not the part spread diffusely in other directions. Thus, for an uneven interface between two fluid media, the ***total reflection coefficient*** $R_{tot}$ is expressed as

$$R_{tot} = R_p R_{coh} \tag{4.38}$$

where $R_p$ given by equation (4.16) is the ***reflection coefficient*** caused by the impedance contrast for the flat interface. Consequently, the uneven surface will lead to a coherent reflection loss of $20 \cdot \log R_{coh}$, which is added to the loss due to the impedance contrast of the flat surface. The coherent reflection coefficient and the reflection loss are given by the vertical surface displacements, the acoustic wavelength, and the grazing angle; the coherent reflection loss can be ignored only if $\lambda >> \sigma_h \sin\theta$, that is, in cases with low frequencies and angles close to the horizontal.

Figure 4.11 shows the coherent reflection loss, $20 \cdot \log(R_{coh})$, as a function of $(\sigma_h / \lambda)\sin\theta$, calculated using equation (4.37). This loss is in addition to the reflection loss we calculated earlier. It must be emphasized that these calculations only represent a first-order approach, in which we have assumed an uncorrelated surface by the use of simple Gaussian statistics. We have also ignored a number of important effects, such as shadowing and multiple scattering effects. Some of these issues are treated in chapter 14 on scattering from rough surfaces.

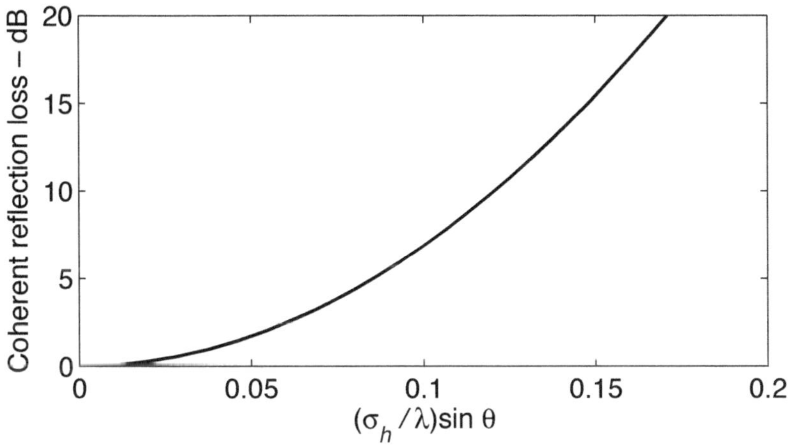


Figure 4.11. When there is reflection from an uneven surface, the signal in the specular direction will be reduced because part of the signal is spread diffusely. Here we have a plot of this loss of reflection, which is a function $(\sigma_h/\lambda)\sin\theta$.

# Chapter 5

# Reflection and transmission in multilayered media

## 5.1 Introduction

We now consider cases where the medium consists of several parallel layers with different thicknesses, densities, and sound speeds. In principle, this is no different from the simple interface studied in chapter 4, insofar that the boundary conditions of continuity in sound pressure and particle velocity must be satisfied for all interfaces. However, when the number of layers exceeds one or two, analytical solutions become too cumbersome, and numerical methods have to be applied. Therefore, much of the intention behind this chapter is to develop effective methods, or algorithms, that will enable us to treat cases with any number of layers in an efficient manner by using computers.

We shall see that such multilayered cases can be treated by use of recursion equations, and we shall develop the relevant expressions to accomplish that. As in many other situations, we find that the frequency domain solution is more convenient, but we shall also demonstrate how the time solution can be calculated using the Fourier transform or, more correctly, the discrete Fourier transform. At the end of this chapter, we present several examples that are of practical interest.

The contents of this chapter have a number of applications. In underwater acoustics and seismic exploration, we frequently deal with layered media, for instance an ocean bottom consisting of several sedimentary layers with different acoustic properties, such as sand, silt, and rocks. In seismic exploration, the task is to reveal the stratification of the subsurface by interpreting reflected acoustic or seismic signals. In underwater acoustics, the sound propagating in the sea is greatly affected by what happens when the sound wave strikes the sea surface or the sea bottom. In fact, at long ranges and low frequencies, the interaction with the bottom is the determining factor for how far the sound will reach.

## 5.2 Derivation of the recursion equations for the multilayered case

Figure 5.1 shows the general situation, with homogeneous layers $n = 0,1,2\ldots, N$ and parameters $\rho_n$, $c_n$ for the densities and the sound speeds, respectively. We assume that the uppermost and the lowest layers are infinitely thick. In the water, an incident wave $D_0$ is reflected and generates a wave train $U_0$ that consists of the wave reflected from the interface between media 0 and 1, as well as all simple and multiple reflections from the layers below. Here $D$ means downward waves and $U$ means upward waves. In the intermediate layers, there will be waves that we can characterize as $D_n$ and $U_n$. In the lowest layer, there is only one downward wave, as no waves are returned from a medium of infinite extent.

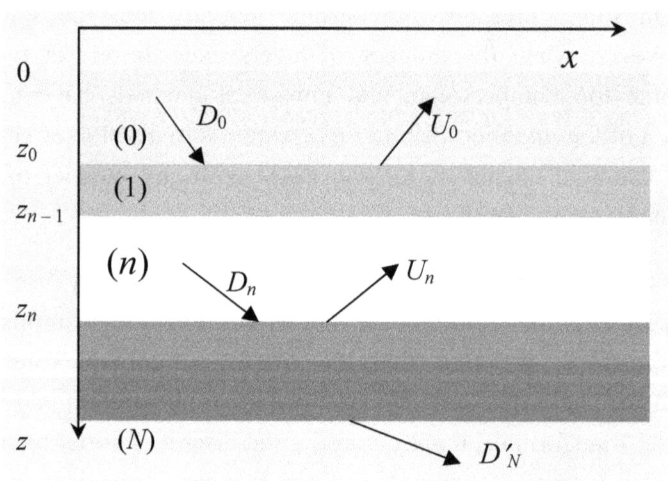


Figure 5.1. A horizontally stratified medium with $N$ layers of fluid. Key: $D_0$, incident wave; $U_0$, reflected wave; $D_n$, downward waves; $U_n$, upward waves; $D'_N$, downward wave in lowest layer.

Figure 5.2 shows the details of the waves that are relevant for a general layer designated as number $n$. Note that waves $D_n$ and $U_n$ refer to the depth $z = z_n$, which is in the bottom of layer $n$, while the waves at the same layer's upper boundary, designated as $z = z_{n-1}$, are $D'_n$ and $U'_n$.

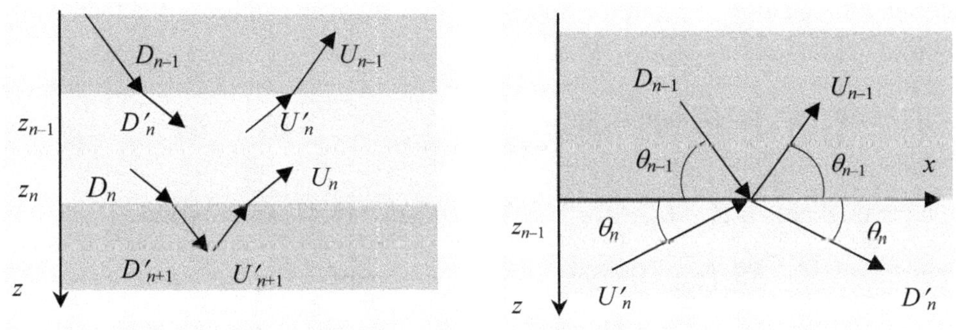


Figure 5.2. The waves relevant for layer number $n$. Left: At the lower interface where $z = z_n$, the waves are $D_n$ and $U_n$. At the upper interface at depth $z = z_{n-1}$, the waves are $D'_n$ and $U'_n$. Right: Details of waves and angles at the upper interface, $z = z_{n-1}$.

We assume harmonic plane waves with frequency $\omega$ and start by writing the solution of the depth equation for layer number $n$ in the form

$$\Phi_n(z) = D_n \exp\left[i\gamma_n(z - z_n)\right] + U_n \exp\left[-i\gamma_n(z - z_n)\right] \tag{5.1}$$

In equation (5.1) $\Phi_n(z)$ is the frequency domain representation of the velocity potential as a function of depth $z$ in layer number $n$. The vertical wave number in the layer, $\gamma_n$, is constant within the layer, and is given by the media parameters and Snell's law as

$$\gamma_n = \frac{\omega}{c_n}\sin\theta_n = \sqrt{\frac{\omega^2}{c_n^{\,2}} - k^2} \tag{5.2}$$

The horizontal wave number $k$ is obtained from the angle of the incoming wave in the upper layer and the sound speed in that layer; $k$ has the same value in all layers.

$$k = \frac{\omega \cos \theta_0}{c_0} = \frac{\omega \cos \theta_n}{c_n} \tag{5.3}$$

According to equation (5.1), at the upper interface in layer number $n$, the total wave field is

$$\Phi_n\left(z_{n-1}\right) = D_n \exp\left[i\gamma_n\left(z_{n-1} - z_n\right)\right] + U_n \exp\left[-i\gamma_n\left(z_{n-1} - z_n\right)\right] \tag{5.4}$$

The layer thickness is $d_n = (z_n - z_{n-1})$, so that equation (5.4) can be written

$$\begin{aligned}\Phi_n\left(z_{n-1}\right) &= D_n \exp\left(-i\gamma_n d_n\right) + U_n \exp\left(i\gamma_n d_n\right) \\ &= D_n' + U_n'\end{aligned} \tag{5.5}$$

Here we have introduced the downward and upward waves $D_n'$ and $U_n'$ at the upper interface, as pictured in figure 5.2.

Equation (5.5) can be rearranged in matrix form to define the relationship between the waves at the top and the bottom in the same layer

$$\begin{pmatrix} D_n' \\ U_n' \end{pmatrix} = \begin{pmatrix} \exp\left(-i\gamma_n d_n\right) & 0 \\ 0 & \exp\left(i\gamma_n d_n\right) \end{pmatrix} \begin{pmatrix} D_n \\ U_n \end{pmatrix} \tag{5.6}$$

Now we can set up two expressions for the sound pressures and vertical components of the particle velocity at the interface between layers $n-1$ and $n$. In layer $n-1$ at the depth $z = z_{n-1}$, we can define the sound pressure $p$ and vertical component of the particle velocity $v_z$ as

$$\begin{pmatrix} p\left(z_{n-1}\right) \\ v_z\left(z_{n-1}\right) \end{pmatrix} = \begin{pmatrix} i\omega\rho_{n-1} & i\omega\rho_{n-1} \\ i\gamma_{n-1} & -i\gamma_{n-1} \end{pmatrix} \begin{pmatrix} D_{n-1} \\ U_{n-1} \end{pmatrix} \tag{5.7}$$

Similarly, in layer $n$ at the same depth, we have

$$\begin{pmatrix} p(z_{n-1}) \\ v_z(z_{n-1}) \end{pmatrix} = \begin{pmatrix} i\omega\rho_n & i\omega\rho_n \\ i\gamma_n & -i\gamma_n \end{pmatrix} \begin{pmatrix} D'_n \\ U'_n \end{pmatrix} \tag{5.8}$$

In equations (5.7) and (5.8), as well as in the following expressions of this type, we have omitted writing the common term $\exp(ikx - i\omega t)$.

The boundary conditions require that the sound pressure and the vertical particle velocity component must be the same on both sides of the interface, and consequently the right-hand side of equations (5.7) and (5.8) must be equal. This gives the following relationship between the waves in medium $n$ and the waves in medium $n-1$:

$$\begin{pmatrix} D_{n-1} \\ U_{n-1} \end{pmatrix} = \begin{pmatrix} i\omega\rho_{n-1} & i\omega\rho_{n-1} \\ i\gamma_{n-1} & -i\gamma_{n-1} \end{pmatrix}^{-1} \begin{pmatrix} i\omega\rho_n & i\omega\rho_n \\ i\gamma_n & -i\gamma_n \end{pmatrix} \begin{pmatrix} D'_n \\ U'_n \end{pmatrix} \tag{5.9}$$

By evaluating the four elements of this matrix, we find that the relationship between the upward and downward waves at the interface between medium $n-1$ and medium $n$ is expressed in matrix form as

$$\begin{pmatrix} D_{n-1} \\ U_{n-1} \end{pmatrix} = \frac{1}{t_{n-1,n}} \begin{pmatrix} 1 & r_{n-1,n} \\ r_{n-1,n} & 1 \end{pmatrix} \begin{pmatrix} D'_n \\ U'_n \end{pmatrix} \tag{5.10}$$

We here have used the reflection and transmission coefficients for the velocity potential as given by equations (4.12) and (4.13), which with the current notation now become

$$r_{n-1,n} = \frac{\rho_n\gamma_{n-1} - \rho_{n-1}\gamma_n}{\rho_n\gamma_{n-1} + \rho_{n-1}\gamma_n} \tag{5.11}$$

$$t_{n-1,n} = \frac{2\rho_{n-1}\gamma_{n-1}}{\rho_n\gamma_{n-1} + \rho_{n-1}\gamma_n} \tag{5.12}$$

By using equation (5.6), we can write a new expression for the upward and downward waves in layer $n$

$$\begin{pmatrix} D_{n-1} \\ U_{n-1} \end{pmatrix} = \frac{1}{t_{n-1,n}} \begin{pmatrix} 1 & r_{n-1,n} \\ r_{n-1,n} & 1 \end{pmatrix} \begin{pmatrix} \exp(-i\gamma_n d_n) & 0 \\ 0 & \exp(i\gamma_n d_n) \end{pmatrix} \begin{pmatrix} D_n \\ U_n \end{pmatrix} \tag{5.13}$$

In a matrix-vector form, this can be written as

$$\begin{pmatrix} D_{n-1} \\ U_{n-1} \end{pmatrix} = \mathbf{S}_{n-1,n} \cdot \mathbf{T}_n \cdot \begin{pmatrix} D_n \\ U_n \end{pmatrix} \tag{5.14}$$

The matrix $\mathbf{S}_{n-1,n}$ gives the reflections and transmissions at the interface $z = z_{n-1}$ between layers $n - 1$ and layer $n$. Often referred to as the ***scattering matrix***, $\mathbf{S}_{n-1,n}$ is given by

$$\mathbf{S}_{n-1,n} = \frac{1}{\mathrm{t}_{n-1,n}} \begin{pmatrix} 1 & r_{n-1,n} \\ r_{n-1,n} & 1 \end{pmatrix} \tag{5.15}$$

The reflection and transmission coefficients are interrelated by

$$\begin{aligned} r_{n-1,n} &= -r_{n,n-1} \\ 1 - r_{n-1,n}^2 &= t_{n,n-1} t_{n-1,n} \end{aligned} \tag{5.16}$$

The ***transport matrix*** for the layer $n$, $\mathbf{T}_n$, is expressed as

$$\mathbf{T}_n = \begin{pmatrix} \exp(-i\gamma_n d_n) & 0 \\ 0 & \exp(i\gamma_n d_n) \end{pmatrix} \tag{5.17}$$

We define the **propagator matrix** $\mathbf{Q}_{n-1,n}$ as the product of $\mathbf{S}_{n-1,n}$ and $\mathbf{T}_n$, so that

$$\begin{pmatrix} D_{n-1} \\ U_{n-1} \end{pmatrix} = \mathbf{Q}_{n-1,n} \begin{pmatrix} D_n \\ U_n \end{pmatrix} \tag{5.18}$$

$$\mathbf{Q}_{n-1,n} = \mathbf{S}_{n-1,n} \cdot \mathbf{T}_n = \frac{1}{t_{n-1,n}} \begin{pmatrix} \exp(-i\gamma_n d_n) & r_{n-1,n} \exp(i\gamma_n d_n) \\ r_{n-1,n} \exp(-i\gamma_n d_n) & \exp(i\gamma_n d_n) \end{pmatrix} \tag{5.19}$$

This propagator matrix $\mathbf{Q}_{n-1,n}$ includes both the transmissions and reflections at the interface between layers $n-1$ and $n$, and the phase shift, or time delay, through layer $n$.

We can also rewrite equation (5.13) as

$$\begin{pmatrix} D_{n-1} \\ U_{n-1} \end{pmatrix} = \frac{1}{t_{n-1,n}} \begin{pmatrix} \exp(-i\gamma_n d_n) & r_{n-1,n}\exp(i\gamma_n d_n) \\ r_{n-1,n}\exp(-i\gamma_n d_n) & \exp(i\gamma_n d_n) \end{pmatrix} \begin{pmatrix} D_n \\ U_n \end{pmatrix} \tag{5.20}$$

Equation (5.20) describes a recursion scheme we can use to calculate the responses for an arbitrary number of layers. The initial condition at the deepest interface is

$$U_N = 0 \tag{5.21}$$

since there is no upward-traveling wave in the lowest layer of infinite thickness.

The method described above, and given in equations (5.18) and (5.19), is commonly known as the ***propagator method.*** In principle, it is simple to implement, treating one layer at a time, which is a clear advantage as it makes fewer demands on computer memory. Before computers reached the current level of sophistication, this advantage was very important and the method consequently was frequently used. A drawback with the method is that it easily becomes numerically unstable. The reason for this can be found in the transport matrix, which contains the exponential functions $\exp(i\gamma_n d_n)$ and $\exp(-i\gamma_n d_n)$. The problem occurs when $\gamma_n$ is imaginary, as is the case when we have an exponentially decreasing or increasing field in a layer. Then one of the exponential functions will be very large and the other correspondingly small. In principle, the large parts are to be canceled by subtraction, but with limited numerical accuracy, the canceling may be incomplete and serious errors may result.

An alternative method that avoids this numerical problem can be found by defining new reflection functions, $R_n = U_n / D_n$ and $R_{n-1} = U_{n-1} / D_{n-1}$. From equation (5.20) we find the following recursive equation

$$R_{n-1} = \frac{r_{n-1,n} + R_n \exp(2i\gamma_n d_n)}{1 + r_{n-1,n} R_n \exp(2i\gamma_n d_n)} \qquad n = 1,\ N-1 \tag{5.22}$$

The starting value of the reflection function at the deepest interface is

$$R_{N-1} = r_{N-1,N} \tag{5.23}$$

where $r_{N-1,N}$ is the simple reflection coefficient between layers $N-1$ and $N$, at the lowest interface with the medium of infinite extent. This method is numerically stable and, in general, the most suitable technique. A drawback to this method is that it is not capable of estimating the waves within the layers; however, for many applications this capability is not of major importance.

The recursive equations we have developed are for particle velocity potentials. Alternatively, we can develop similar equations for sound pressure and for particle velocity. Such equations are sometimes used in the literature, so here we will briefly demonstrate how to develop them. We shall also find the results useful for describing acoustic delay lines, which we will discuss later in chapter 8, "Acoustic transducers."

We consider a particular layer $n$. At depth $z_n$ at the lower interface the downward and upward waves are $D_n$ and $U_n$. At the upper interface the waves are $D'_n$ and $U'_n$. These waves are shown in figure 5.2.

Using equation (5.8) with equation (5.6) gives the pressure and particle velocity at $z = z_{n-1}$ as

$$\begin{pmatrix} p(z_{n-1}) \\ v_z(z_{n-1}) \end{pmatrix} = \begin{pmatrix} i\omega\rho_n & i\omega\rho_n \\ i\gamma_n & -i\gamma_n \end{pmatrix} \begin{pmatrix} \exp(-i\gamma_n d_n) & 0 \\ 0 & \exp(i\gamma_n d_n) \end{pmatrix} \begin{pmatrix} D_n \\ U_n \end{pmatrix} \tag{5.24}$$

At $z = z_n$, we get

$$\begin{pmatrix} p(z_n) \\ v_z(z_n) \end{pmatrix} = \begin{pmatrix} i\omega\rho_n & i\omega\rho_n \\ i\gamma_n & -i\gamma_n \end{pmatrix} \begin{pmatrix} D_n \\ U_n \end{pmatrix} \tag{5.25}$$

By inverting equation (5.25) and using it with equation (5.24), we obtain

$$\begin{pmatrix} p(z_{n-1}) \\ v_z(z_{n-1}) \end{pmatrix} = \begin{pmatrix} \cos(\gamma_n d_n) & -\dfrac{i\omega\rho_n}{\gamma_n}\sin(\gamma_n d_n) \\ \dfrac{\gamma_n}{i\omega\rho_n}\sin(\gamma_n d_n) & \cos(\gamma_n d_n) \end{pmatrix} \begin{pmatrix} p(z_n) \\ v_z(z_n) \end{pmatrix} \tag{5.26}$$

By repeating this operation with layer number $n-1$ and demanding continuity in sound pressure and particle velocity at all interfaces, we again get a set of recursive equations for calculating the waves in all the layers.

### *5.2.1 Acoustic delay line*

We next consider another situation where the result of the last section is very useful: the design and analysis of a common type of piezoelectric transducer, which we will see again in chapter 8.

Figure 5.3 shows a simple rod or plate with length or thickness $l$. For plane waves traveling in the axial direction, equation (5.26) applies, with $\gamma = \kappa = \omega / c$; $p_1$ and $p_2$ are the sound pressures, and $v_1$ and $v_2$ the particle velocities at the end surfaces.

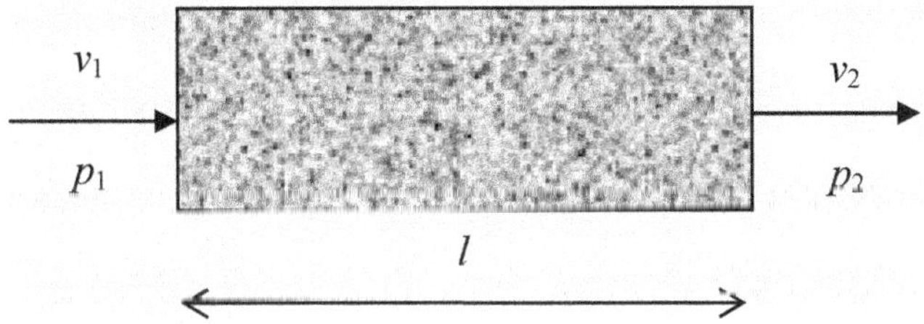


Figure 5.3. Acoustic delay line with the length $l$. Sound pressures at the two ends of the plate are $p_1$ and $p_2$, and the particle velocities are $v_1$ and $v_2$.

Equation (5.26) now becomes

$$\begin{pmatrix} p_1 \\ v_1 \end{pmatrix} = \begin{pmatrix} \cos(\kappa l) & -i\rho c \sin(\kappa l) \\ \dfrac{-i}{\rho c}\sin(\kappa l) & \cos(\kappa l) \end{pmatrix} \begin{pmatrix} p_2 \\ v_2 \end{pmatrix} \tag{5.27}$$

Furthermore, we introduce the specific acoustic impedance $Z_0 = \rho c$, and equation (5.27) can then be rearranged as

$$\begin{aligned} p_1 &= Z_0 \left[ \frac{v_1}{-i\tan(\kappa l)} + \frac{v_2}{i\sin(\kappa l)} \right] \\ p_2 &= Z_0 \left[ \frac{v_1}{-i\sin(\kappa l)} + \frac{v_2}{i\tan(\kappa l)} \right] \end{aligned} \tag{5.28}$$

Equation (5.28) may be represented by the equivalent electric circuit diagram in figure 5.4, where the particle velocities and sound pressures have equivalence to the electric currents and voltages, respectively.

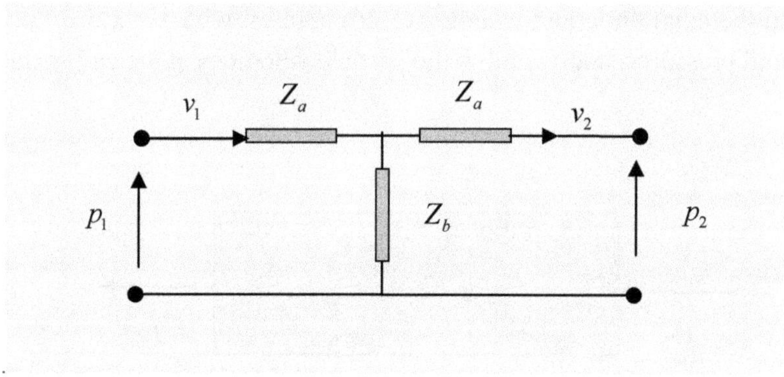


Figure 5.4. The equivalent circuit diagram for a delay line.

By inspecting this circuit diagram, we find that

$$\begin{aligned} p_1 &= (Z_a + Z_b)v_1 - Z_b v_2 \\ p_2 &= Z_b v_1 - (Z_a + Z_b)v_2 \end{aligned} \tag{5.29}$$

We find the impedances $Z_a$ and $Z_b$. Comparing equations (5.28) and (5.29)

$$Z_b = -\frac{Z_0}{i\sin(\kappa l)} \tag{5.30}$$

We then find from our comparison that

$$Z_a + Z_b = -\frac{Z_0}{i\tan(\kappa l)} \tag{5.31}$$

By solving equations (5.30) and (5.31), we find that

$$Z_a = -iZ_0\frac{1-\cos(\kappa l)}{sin(\kappa l)} = iZ_0\tan(\kappa l/2) \tag{5.32}$$

Figure 5.4 and equation (5.29) are both representations of plane waves propagating along the axis of a rod or a plate. These representations are especially appropriate for a number of plates or rods stacked together, where we can just connect the elements as illustrated in figure 5.5, with each element having the impedance $Z_n - \rho_n c_n$, wave number $\kappa_n$, and length $l_n$. In this way, the boundary conditions of continuity of pressure and particle velocities are automatically satisfied.

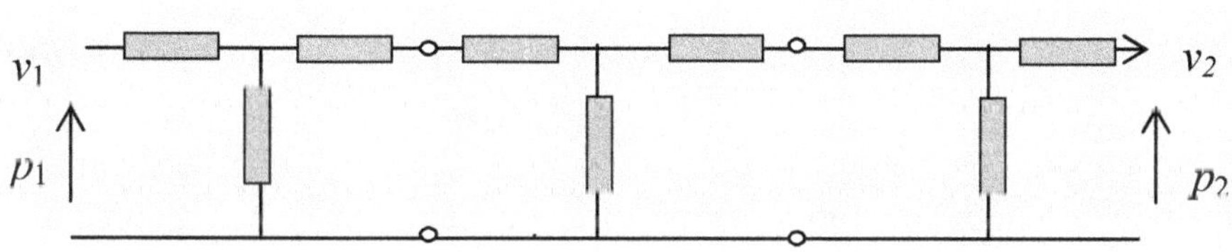


Figure 5.5. Chain coupling of elements, satisfying the boundary conditions at every interface.

At this point, we return to figure 5.4 and the single element. When the line is terminated with the load impedance $Z_2$ on side 2, the input impedance on side 1 is

$$Z_{in} = Z_a + \frac{(Z_a + Z_2)Z_b}{Z_a + Z_b + Z_2} \tag{5.33}$$

and after applying equation (5.32), the input impedance becomes

$$Z_{in} = Z_0 \frac{Z_2 - iZ_0 \tan(\kappa l)}{Z_0 - iZ_2 \tan(\kappa l)} \tag{5.34}$$

We now consider two important special cases. In the first case, the length of the line corresponds to half a wavelength, $l = \lambda / 2$, and $\tan(\kappa l) = 0$ at a particular frequency. Then, according to equation (5.34), the input impedance is

$$Z_{in} = Z_2 \tag{5.35}$$

The impedance is transferred without change by a delay line with a length corresponding to half a wavelength.

The other special case occurs when the length is a quarter of a wavelength, $l = \lambda / 4$ and $\tan(\kappa l) = \infty$, in which we have a so-called quarter-wavelength impedance transformer. From equation (5.34), we find the input impedance

$$Z_{in} = \frac{Z_0^2}{Z_2} \tag{5.36}$$

A delay line with a length equal to a quarter of a wavelength may be regarded as a transformer, which transforms a load impedance $Z_2$ into an input impedance $Z_{in}$, given by equation (5.36). Thus, the quarter-wavelength transformer has the capability to transform a low impedance into a high impedance, and vice versa.

## 5.3 Frequency and time responses

In some applications there may be a need to study the response in the time domain to actually see the reflected and transmitted time signals. We shall see how the time signals can be determined using the discrete Fourier transform.

Let us assume, for instance, that we want to study the reflected time signal of a structure with $N$ layers. The frequency domain solution is found by recursive use of equation (5.22), which results in the reflectivity function $R_0(\omega)$ where we have reintroduced the argument of that equation as a reminder that equation (5.22) is a frequency transfer function.

If the incident signal is a time function $a(t)$ with the associated frequency function $A(\omega)$,

$$A(\omega) = \int_{-\infty}^{\infty} a(t)\exp(i\omega t)\,dt \tag{5.37}$$

the frequency function of the reflected signal is

$$R(\omega) - A(\omega)R_0(\omega) \tag{5.38}$$

The time response thus becomes

$$r(t) = \frac{1}{2\pi}\int_{-\infty}^{\infty} A(\omega)R_0(\omega)\exp(-i\omega t)\,d\omega \tag{5.39}$$

In practice, the Fourier integrals in the equations above will be approximated by a discrete Fourier transform and the calculation will be made using the fast Fourier transform (FFT) algorithm. To accomplish this, we have to discretize both the time function and the frequency function. This means that time and frequency are discretized to

$$\begin{aligned} t &= j\Delta t \\ f &= q\Delta f \\ j,q &= 0,1\cdot\cdot M-1 \end{aligned} \tag{5.40}$$

The inverse product of $\Delta t \cdot \Delta f$ must be an integer, preferentially divisible by two, as in

$$M = \frac{1}{\Delta f \cdot \Delta t} = 2^m \tag{5.41}$$

The sampling interval times and frequencies must be chosen to satisfy the sampling theorem:

$$\begin{aligned} \frac{1}{\Delta t} &> 2 f_{\max} \\ \frac{1}{\Delta f} &> t_{\max} \end{aligned} \tag{5.42}$$

Here, $f_{max}$ is the highest frequency and $t_{max}$ is the total time duration of the signal. If the conditions in equations (5.42) are not fulfilled, aliasing errors may occur in the time response, the frequency response, or both.

We will next apply the results from the preceding section to determine the field in a stratified medium. This operation will be illustrated by the use of examples, which are also selected to show certain structures with practical importance.

### *5.3.1 Example: Reflection and transmission of an infinite plate*

We first consider a plate with thickness $d_1$ and infinite horizontal extent. The plate is embedded, above and below, by two semi-infinite layers, 0 and 2, as illustrated in figure 5.6. With the numbering notation used in section 5.2 and figure 5.1, this is a case with three layers $N = 2$, ($n = 0, 1, 2$). Using equation (5.14) twice, we find that

$$\begin{pmatrix} D_0 \\ U_0 \end{pmatrix} = \mathbf{S}_{0,1} \cdot \mathbf{T}_1 \cdot \mathbf{S}_{1,2} \cdot \mathbf{T}_2 \cdot \begin{pmatrix} D_2 \\ U_2 \end{pmatrix} \tag{5.43}$$

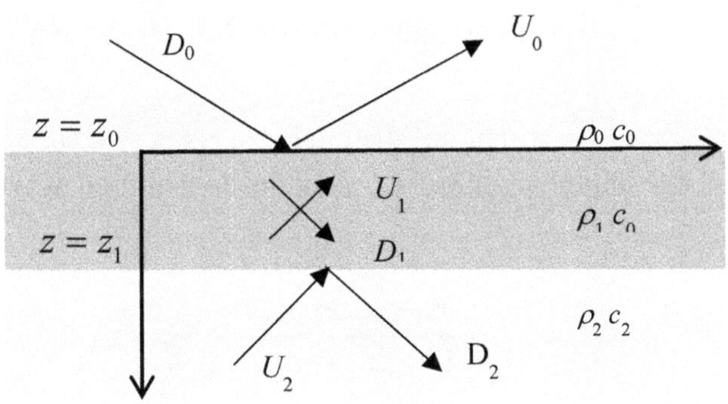


Figure 5.6. The waves for the case where layer 1 lies between two semi-infinite media, layers 0 and 2.

Since, by definition, the thickness of the lowest layer is infinite, we may just as well insert the starting condition for the field at depth $z = z_{N-1}$, so that equation (5.43) can be modified to

$$\begin{pmatrix} D_0 \\ U_0 \end{pmatrix} = \mathbf{S}_{0,1} \cdot \mathbf{T}_1 \cdot \mathbf{S}_{1,2} \cdot \begin{pmatrix} D'_2 \\ U'_2 \end{pmatrix} \tag{5.44}$$

Alternatively, we could insert $d_2 = 0$, which produces $\mathbf{T}_2 = 1$.

Expanded, equation (5.44) is

$$\begin{pmatrix} D_0 \\ U_0 \end{pmatrix} = \frac{1}{t_{0,1}} \begin{pmatrix} 1 & r_{0,1} \\ r_{0,1} & 1 \end{pmatrix} \begin{pmatrix} \exp(-i\gamma_1 d_1) & 0 \\ 0 & \exp(i\gamma_1 d_1) \end{pmatrix} \frac{1}{t_{1,2}} \begin{pmatrix} 1 & r_{1,2} \\ r_{1,2} & 1 \end{pmatrix} \begin{pmatrix} D'_2 \\ U'_2 \end{pmatrix} \tag{5.45}$$

The upward field in the lowest layer must be equal to zero because no wave can be returned from an infinite medium. Therefore, the starting condition for the lowest field is $U'_2 = 0$, and equation (5.45) becomes

$$\begin{pmatrix} D_0 \\ U_0 \end{pmatrix} = \frac{1}{t_{0,1}} \begin{pmatrix} \exp(-i\gamma_1 d_1) & r_{0,1}\exp(i\gamma_1 d_1) \\ r_{0,1}\exp(-i\gamma_1 d_1) & \exp(i\gamma_1 d_1) \end{pmatrix} \frac{1}{t_{1,2}} \begin{pmatrix} D_2' \\ r_{1,2} D_2' \end{pmatrix} \tag{5.46}$$

When the incident wave, $D_0$, which has the angle $\theta_0$, is known, we can determine the reflected wave in the upper medium, $R_0$, and the transmitted wave in the lowest medium, $T_2$, to be

$$R_0 = \frac{U_0}{D_0} = \frac{r_{0,1} + r_{1,2}\exp(2i\gamma_1 d_1)}{1 + r_{0,1}r_{1,2}\exp(2i\gamma_1 d_1)} \tag{5.47}$$

$$T_2 = \frac{D_2'}{D_0} = \frac{t_{0,1}t_{1,2}\exp(i\gamma_1 d_1)}{1 + r_{0,1}r_{1,2}\exp(2i\gamma_1 d_1)} \tag{5.48}$$

Equations (5.47) and (5.48) are recognized as the sums of geometric series, and by expansion we can see how the responses are composed of different contributions. For instance, a series expansion of equation (5.47) gives

$$R_0 = \left[r_{0,1} + r_{1,2}\exp(2i\gamma_1 d_1)\right]\sum_{q=0}^{\infty}\left[(-1)r_{0,1}r_{1,2}\exp(2i\gamma_1 d_1)\right]^q \tag{5.49}$$

The first terms of equation (5.49) can be rewritten as

$$R_0 = r_{0,1} + r_{1,2}\left(1 - r_{0,1}^{\ 2}\right)\exp(2i\gamma_1 d_1) + r_{0,1}r_{1,2}^{\ 2}\left(r_{0,1}^{\ 2} - 1\right)\exp(4i\gamma_1 d_1) + \ldots \tag{5.50}$$

Using equation (5.16), we make the substitution $(1 - r_{0,1}^{\ 2}) = t_{0,1}t_{1,0}$ which produces

$$R_0 = r_{0,1} + r_{1,2}t_{0,1}t_{1,0}\exp(2i\gamma_1 d_1) - t_{0,1}r_{1,2}^{\ 2}r_{1,0}t_{1,0}\exp(4i\gamma_1 d_1) + \ldots \tag{5.51}$$

In the same way, we find that the first terms of equation (5.48) are

$$T_2 = t_{0,1}t_{1,2}\exp(i\gamma_1 d_1)\left[1 - r_{1,0}r_{1,2}\exp(2i\gamma_1 d_1) + (r_{1,0}r_{1,2})^2\exp(4i\gamma_1 d_1) + \ldots\right] \tag{5.52}$$

Consequently, the resultant reflection and transmission responses are the sums of reflections and multiple reflections in all interfaces. The individual terms of the series can be identified, as shown in figure 5.7.

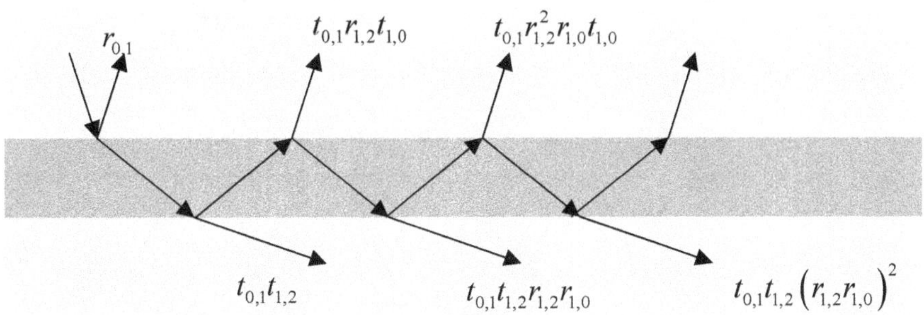


Figure 5.7. The first three contributions to the total reflection from a two-layer structure.

To provide a numerical example, we will use a plate with a thickness of $d_1 = 10$ cm, composed of a material with density $\rho_1 = 1500$ kg/m$^3$ and sound speed $c_1 = 2000$ m/s. On either side we have water, so that $\rho_0 = \rho_2 = 1000$ kg/m$^3$ and $c_0 = c_2 = 1500$ m/s. $R_0$ and $T_2$ are calculated using equations (5.47) and (5.48); in figure 5.8 their absolute values appear as functions of frequency for signals with normal incidence to the plate. We see that both $R_0$ and $T_2$ are periodic functions of frequency, and that $R_0 = 0$ and $|T_2| = 1$ for all frequencies where the thickness is an integer multiple of half-wavelengths. The first zero occurs, therefore, at the frequency of 10 kHz, where the wavelength in the plate equals 20 cm; its thickness is consequently half a wavelength.

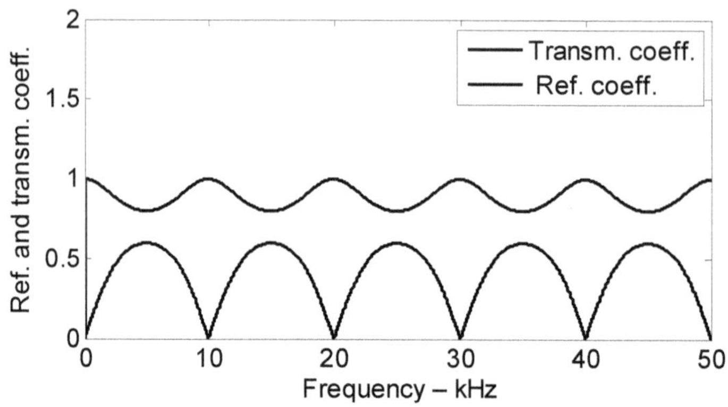


Figure 5.8. Frequency responses at normal incidence of the reflection and the transmission coefficients in a two-layer structure. The parameters are given in the text.

Figure 5.9 is a contour plot of the transmission loss through the plate as a function of frequency and angle of the incoming plane wave. The transmission loss, in dB, for a wave propagating through the structure is defined as

$$TL = -20\log(T_2) \tag{5.53}$$

The plot in figure 5.9 shows that the transmission loss equals zero for a number of frequency bands that vary with the angle. For 90°, the bands with zero or very low transmission loss are the frequencies that make the plate thickness equal to an integer number of half wavelengths, since layers with such thickness are transparent to the incident wave as can be seen from equation (5.34). We also see that the transmission loss becomes very low for a plate with a thickness much smaller than a wavelength.

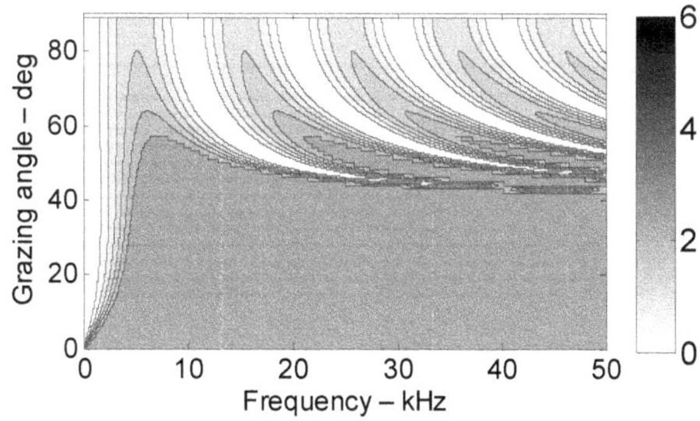


Figure 5.9. Contour plot of the transmission loss in dB as a function of frequency and grazing angle. The transmission loss scale is shown at the left side of the figure.

One application of this information is connected with the construction of a sound window in the form of a plate designed to allow as much sound as possible to pass through. In the example of figure 5.7, for instance, this design requirement is satisfied for all frequencies that are multiples of 10 kHz.

We now consider the reflected and transmitted signals in the time domain. Figure 5.10 shows examples of time responses of a 50 kHz signal calculated as explained earlier in this section. In the example shown in figure 5.10, we have used a relatively short sinusoidal input pulse with only three periods of center frequency at 50 kHz. The sampling frequency is 500 kHz, so that there are ten samples per cycle period. With the FFT length $N = 2^{13} = 8192$, the total time window $\tau_{max} = N\Delta t = 16.384$ ms. The time series have been computed for a number of incident angles at 5° intervals. The 90° angle is normal incidence, and the signal drawn at 100° is the input signal.

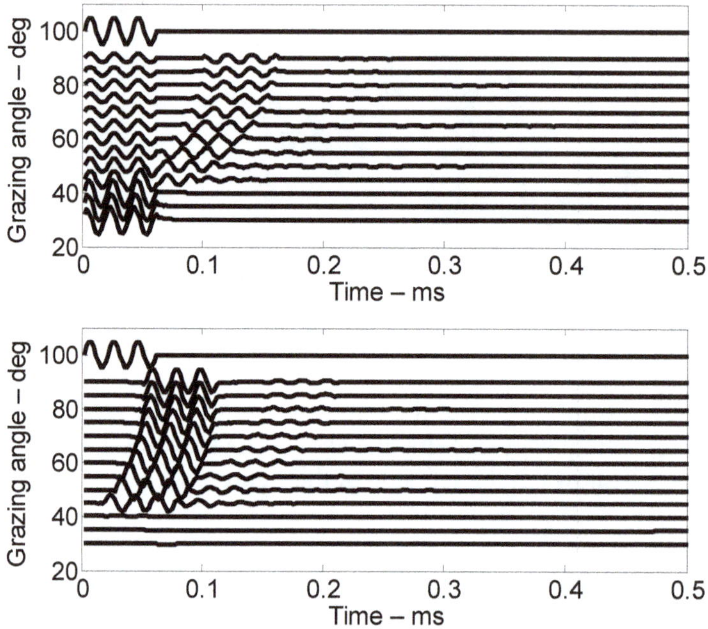


Figure 5.10. Time responses for (top) reflection and (bottom) transmission in a two-layer structure.

The reflected signal in the upper panel of figure 5.10 has many arrivals with arrival times and amplitudes given by the series expansion of equation (5.50) and illustrated in figure 5.7. At normal incidence, the time difference between the multiples is 0.1 ms, which equals twice the time traveled through the plate. With decreasing angle the time difference reduces and becomes zero at critical angle, 41.4°, where total reflection occurs. The first part of the transmitted signal arrives at 0.05 ms, the time to propagate through the layer, but the travel time decreases with decreasing angle, and the transmitted wave vanishes for angles lower than the critical angle, as expected.

We have observed that the transmission time through the layer has a maximum at normal incidence and reduces to zero at critical angle. This may seem strange since one should expect that it will take longer time to propagate through the layer at oblique angles. The reason for this seeming contradiction is that we are considering plane waves, with wave fronts of infinite extent.

Consider figure 5.11 where a plane wave is striking at an angle $\theta_1$. At a specific time, the wave front of the plane incident wave hits the first interface at point A, and then at a later time hits point D. The time difference between the two hit times, $\tau_1$, is given by the distance BD and the sound speed in the upper medium, $c_1$. Simple geometry gives us the expression

$$\tau_1 = \frac{1}{c_1} 2d \frac{\cos\theta_1}{\tan\theta_2} \tag{5.54}$$

The ray path through the layer and back again is the path ADC with the travel time

$$\tau_2 = \frac{1}{c_2} \frac{2d}{\sin\theta_2} \tag{5.55}$$

From simple physical reasoning this is the travel time one should expect but is not what we observe.

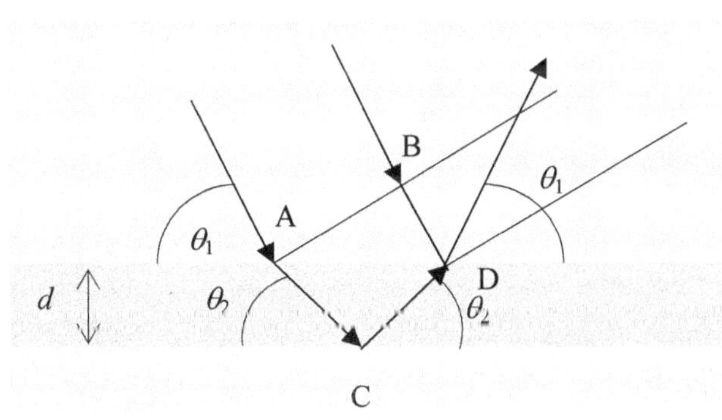


Figure 5.11. A plane wave striking the interface at an angle $\theta_1$. At a specific time, the wave front hits the first interface at point A and also hits point B. Parts of the wave are transmitted into layer 2 with an angle $\theta_2$ and, after reflection at point C, they again hit the first interface, but at point D.

The time difference between the two travel times is

$$\Delta\tau = \tau_2 - \tau_1 = \frac{1}{c_2}\frac{2d}{\sin\theta_2} - \frac{1}{c_1}2d\frac{\cos\theta_1}{\tan\theta_2} = \frac{2d}{c_2}\sin\theta_2 \tag{5.56}$$

Instead, the travel time observed in the simulations shown in figure 5.10 is given by the vertical wave number, and the difference is therefore

$$\tau_2 = \frac{2d\gamma_2}{\omega} = \frac{2d}{c_2}\sin\theta_2 \tag{5.57}$$

The time for transmission twice through the layer with the sound speed $c_2$ and a thickness $d$ along the path ACD is

$$\tau_2 = \frac{1}{c_2}\frac{2d}{\sin\theta_2} \tag{5.58}$$

which is the travel time given by the vertical wave number, equation (5.57).

The vertical wave number travel time therefore represents the difference between the paths ACD and BD in figure 5.11. We should also note that the problem we have discussed here is different from calculation of the reflected response from a point source over a plane-stratified medium.

The field of a point source is a spherical wave and, in order to treat this problem, we must decompose the spherical wave into an infinite number of plane waves representing all directions. Each of the plane waves may then be treated separately with its respective reflection and transmission coefficients, as we have done in this section, before the result is summed (integrated) to give the total response.

### *5.3.2 Transmission and reflection in a periodically multilayered medium*

The following example is intended to demonstrate how the propagator method can be used to analyze sound propagation in media composed of many layers. The example will also give insight in the problem of sound propagation through laminated materials and to determine the effective sound speed of such materials.

Laminated materials are found in natural as well as in synthetic materials. This example is relevant to the issue of sound propagation in finely layered sediments and rocks, but we simplify the problem and consider a periodically laminated material.

Figure 5.12 shows a periodically laminated material consisting of two types of materials, 1 and 2, with sound speeds $c_1$ and $c_2$ and densities $\rho_1$ and $\rho_2$. The bottom half of the figure shows a case with six periods over a given length, while in the top figure there is only one period over the same length. In both cases, the total lengths of the two materials are identical, $d_1$ for material 1 and $d_2$ for material 2, with $d_1 + d_2 = d$. The fractional lengths are $n_1 = d_1 / d$ and $n_2 = d_2 / d$, with $n_1 + n_2 = 1$.

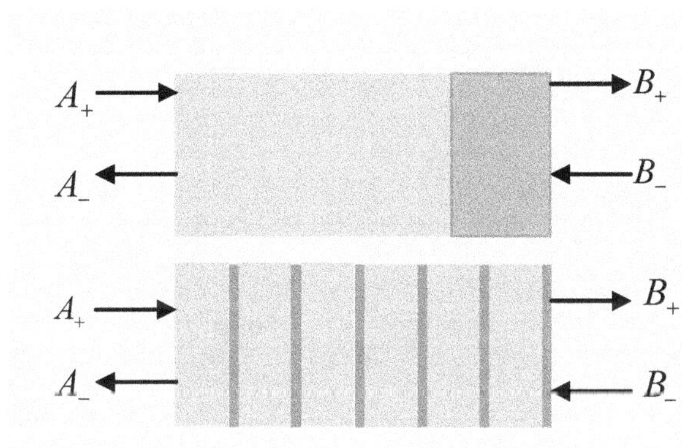


Figure 5.12. Periodic structure of two examples of stratified test pieces consisting of materials 1 and 2. The two materials are periodically divided, with the upper example having one period and the lower example having six periods.

We can imagine a test in which we send a sound pulse through the material and measure the transmission time $\tau$ over the total distance $d = d_1 + d_2$ in order to determine the effective speed: $v_{eff} = d / \tau$. We are interested in finding out if it makes any difference how the material is divided—in other words, whether the travel time is different for stratification with many periods compared to simpler stratification having one or only a few periods.

With waves such as those shown in figure 5.12, a structure with only one period gives

$$\begin{pmatrix} A_+ \\ A_- \end{pmatrix} = \mathbf{Q}_{0,1} \cdot \mathbf{Q}_{1,2} \frac{1}{t_{2,3}} \cdot \begin{pmatrix} 1 & r_{2,3} \\ r_{2,3} & 1 \end{pmatrix} \begin{pmatrix} B_+ \\ B_- \end{pmatrix} \tag{5.59}$$

where $\mathbf{Q}_{0,1}$ and $\mathbf{Q}_{1,2}$ are given by equation (5.19). We have terminated the structure with the materials 0 and 3.

Similarly, a structure with $M$ periods has

$$\begin{pmatrix} A_+ \\ A_- \end{pmatrix} = \frac{1}{t_{0,1}} \begin{pmatrix} 1 & r_{0,1} \\ r_{0,1} & 1 \end{pmatrix} \cdot \left[ \mathbf{Q}_{1,2} \cdot \mathbf{Q}_{2,1} \right]^{M-1} \cdot \mathbf{Q}_{1,2} \frac{1}{t_{2,3}} \cdot \begin{pmatrix} 1 & r_{2,3} \\ r_{2,3} & 1 \end{pmatrix} \begin{pmatrix} B_+ \\ B_- \end{pmatrix} \tag{5.60}$$

with

$$\mathbf{Q}_{1,2} = \frac{1}{t_{1,2}} \begin{pmatrix} \exp\left(-i\gamma_1 \frac{d_1}{M}\right) & r_{1,2} \exp\left(-i\gamma_1 \frac{d_1}{M}\right) \\ r_{1,2} \exp\left(i\gamma_1 \frac{d_1}{M}\right) & \exp\left(i\gamma_1 \frac{d_1}{M}\right) \end{pmatrix} \tag{5.61}$$

$$\mathbf{Q}_{2,1} = \frac{1}{t_{2,1}} \begin{pmatrix} \exp\left(-i\gamma_2 \frac{d_2}{M}\right) & r_{2,1} \exp\left(-i\gamma_2 \frac{d_2}{M}\right) \\ r_{2,1} \exp\left(i\gamma_2 \frac{d_2}{M}\right) & \exp\left(i\gamma_2 \frac{d_2}{M}\right) \end{pmatrix} \tag{5.62}$$

It is relatively simple to develop a computer program that calculates the reflected signal $A_-$ and the transmitted signal $B_+$ for a given sent signal $A_+$ since the last right-hand medium is infinite, so that $B_- = 0$.

As an example, we apply a short transient signal in the shape of a so-called Ricker pulse, whose time function and frequency spectrum appear in figure 5.13.

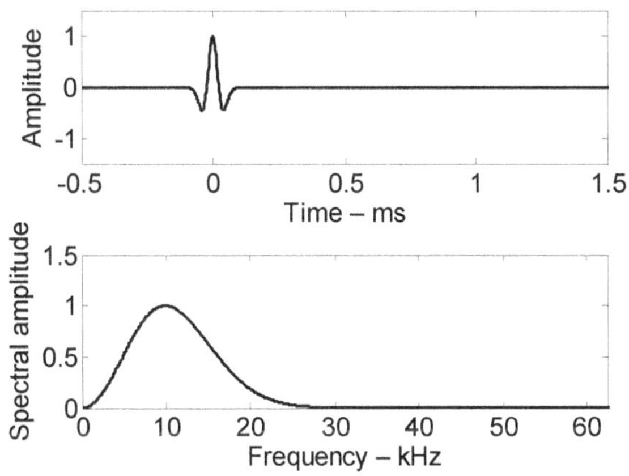


Figure 5.13. The time function (left) and spectrum (right) of a Ricker pulse.

We apply the numerical values given in table 5.1 and calculate the transmitted response for different periodic numbers. Figure 5.14, shows the result for $M = 2^{(m-1)}$ periods in the structure appears, where $m = 1, 2, \ldots 8$. The bottom trace shows the incident Ricker pulse.

Table 5.1

| | |
|---|---|
| $c_1 = 2000$ m/s | $c_2 = 4000$ m/s |
| $\rho_1 = 2000$ kg/m$^3$ | $\rho_2 = 3000$ kg/m$^3$ |
| $d_1 = 0.6$ m | $d_2 = 0.4$ m |

First we consider the signal traveling through the element with one period ($M = 1$), which is most easily interpreted. With the chosen numerical values, the travel time through the first material layer is equal to 0.3 ms; through the second material layer, travel time is 0.1 ms. The first arrival of $M = 1$, as expected, is at 0.4 ms and is a direct transmission through the two layers. The next arrival is at 0.6 ms after a direct transmission through layer 1 and three passes through layer 2. Thereafter, the other multiple reflections follow. To determine the effective sound speed of the composite structure, we may use the measured travel time of the first arrival. In this case, we conclude that the signal speed is 1 / (0.4 ms) = 2500 m/s.

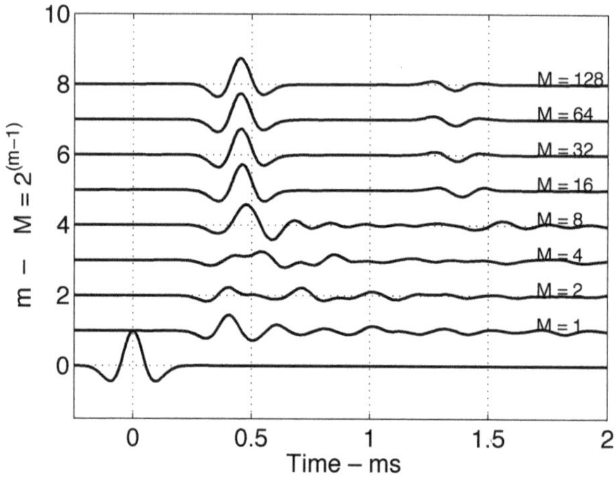


Figure 5.14. Time responses for a transmitted signal traveling through laminated material with a periodic structure. The lower trace is the incident signal.

For $M = 32$ and higher, the response changes character. The direct arrival at 0.4 ms almost disappears, and we instead get one very distinct and isolated response with a travel time of approximately 0.45 ms, and then another response at three times that value, or 1.35 ms. Between the responses there is almost no signal, so the responses for the finely layered structure (with $M = 32$ and higher) behave as if the layer materials were homogeneous. In this case, we would conclude that effective sound speed of the structure is 1 / (0.45 ms) = 2222 m/s.

There are two obvious alternatives for the determination of the effective sound speed. One is based on the average signal speed, calculated on the basis of the total transition time. This gives the sound speed

$$c_t^{-1} = n_1 c_1^{-1} + n_2 c_2^{-1} \tag{5.63}$$

The other possibility is the speed calculated by Wood's equation, developed in section 3.5) and is

$$c_e = \sqrt{\frac{K}{\rho}} \tag{5.64}$$

The average compressibility $K$ and average density to be used in this equation are:

$$\begin{aligned} K^{-1} &= n_1 K_1^{-1} + n_1 K_2^{-1} \\ \rho &= n_1 \rho_1 + n_2 \rho_2 \end{aligned} \tag{5.65}$$

The two materials' bulk moduli are $K_1 = \rho_1 c_1^2$ and $K_2 = \rho_2 c_2^2$, respectively. With the numerical values of this example we obtain two quite different values for the speed:

$$\begin{aligned} c_t &= 2500 \text{ m/s} \\ c_e &= 2236 \text{ m/s} \end{aligned}$$

Since we have observed both of these speeds in figure 5.14 we must conclude that either $c_t$ or $c_e$ may be correct, depending on signal frequency and the thickness of the layers. At high frequencies, where the wavelength is much smaller than the thickness of the layers, the signal detects the discontinuities in the transition between the two materials and the wave is subjected to losses because of its reflection during transitions; the speed measured is the average speed $c_t$. At very low frequencies, where the wave does not have sufficient resolution to detect the layer transitions, the signal propagates without loss, and its effective speed is $c_e$.

The transition between high and low frequencies depends on the values of the sound speeds, the layer densities, and their relative lengths. In the example in figure 5.14, we can say that the transition takes place between $M = 16$ and $M = 32$. The length of a period is thus between $d = 0.0625$ m and $d = 0.03125$ m for the two cases with $M = 16$ and $M = 32$. The frequency spectrum in figure 5.14 shows that, for a central frequency of approximately 10 kHz and a speed of about 2500 m/s, the dominating wavelength becomes approximately 0.25 m. Therefore from this example we can conclude that the transition from low to high frequencies takes place where $\lambda / d \approx 4$. Waves with lower frequencies than those corresponding to this limit will perceive the material as homogeneous and will have a wave speed given by equations (5.64) and (5.65). Waves with higher frequencies will perceive the material as a sequence of different materials, thus generating multiple reflections, which in principle can be analyzed to determine the sound speed of the separate layers. Although this analysis was on periodically layered materials, the main conclusions will, at least qualitatively, be valid for materials with different layer thicknesses and compositions.

There are many important practical consequences to this observation; here we just point to one of these. In petroleum technology the sound speeds of sedimentary rocks are often obtained by measurements on relatively small samples, cores of rock 10 to 20 cm long, using short pulses in a frequency range considerably higher than the frequencies used in seismic exploration. Thus there is a distinct possibility that the measured sound speed values may not be accurate for evaluation and interpretation of the low-frequency seismic measurements in the laboratory which would have been made in situ.

# Chapter 6

# Ray acoustics and ray tracing

## 6.1 Introduction

Ray acoustics and ray-tracing techniques are the most intuitive and often the simplest means for modeling sound propagation in the sea. Ray acoustics is based on the assumption that sound propagates along rays that are normal to wave fronts, surfaces of constant phase of the acoustic waves. When generated from a point source in a medium with constant sound speed, the wave fronts form surfaces that are concentric circles, and the sound follows straight line paths that radiate out from the sound source. If the speed of sound is not constant, the rays follow curved paths rather than straight ones. The computational technique known as ray tracing is a method used to calculate the trajectories of the ray paths of sound from the source.

As will be shown, this approach can be derived from the wave equation when some simplifying assumptions are introduced. The ray theory approach is essentially a high-frequency approximation, applicable to frequencies for which the wavelength is much smaller than the water depth and, in addition, much smaller than the characteristic distance of variation in the sound speed. This method is sufficiently accurate that it is used in applications involving echo sounders, sonar, and communications systems for short and medium-short distances. These devices normally use frequencies that satisfy the high-frequency conditions.

This chapter presents classical ray theory and illustrates the results in a number of examples. We show that ray theory gives a good physical picture of how sound propagates in inhomogeneous media. We also demonstrate that ray theory has limitations and may not be valid for precise predictions of sound levels, especially in situations where refraction effects and focusing of energy are important. There exist corrective measures that can be used to improve classical ray theory, but these are not discussed in detail here. Recommended alternative readings include the books by Officer (1958), Jensen et al. (1994), Frisk (1994), and Kinsler et al. (2000).

First, we derive ray acoustics from the wave equation and examine the requisite assumptions and approximations. Next, we consider the more practically interesting case in which sound speed is only a function of depth, and we develop a simple algorithm for computing the trajectories of all rays emanating from a source. Finally, we look at a number of examples in order to illustrate the effect sound speed profiles can have on sound propagation in the sea and, consequently, the problems one may encounter when using sonar and echo sounders. An advantage of ray theory is that ray characteristics are not functions of frequency and therefore are very efficient for modeling propagation of broadband time signals. We demonstrate this with an example at the end of the chapter.

## 6.2 Theory of ray acoustics

The wave equation for the velocity potential in rectangular coordinates may be written

$$\nabla^2\phi(x,y,z,t) = \frac{1}{c^2(x,y,z)}\frac{\partial^2\phi(x,y,z,t)}{\partial t^2} \tag{6.1}$$

We assume a solution of the form

$$\phi(x,y,z,t) = A(x,y,z)\exp\left\{-i\omega\left[t - \frac{W(x,y,z)}{c_0}\right]\right\} \tag{6.2}$$

where $A(x,y,z)$ is the position-dependent amplitude and $c_0$ is a reference sound speed. The function $W(x,y,z)$ represents the wave fronts because all points in space where $W(x,y,z) - c_0 t$ is a constant have the same phase, as can be seen by examining the exponential in equation (6.2).

Inserting equation (6.2) into the wave equation, equation (6.1), and requiring that the wave equation be satisfied independently for both the real and the imaginary parts results in two equations from which we can determine $W(x,y,z)$ and $A(x,y,z)$:

$$\nabla^2 A - A\frac{\omega^2}{c_0^{\ 2}}\nabla W \cdot \nabla W = -A\frac{\omega^2}{c^2} \tag{6.3}$$

$$2\nabla A \cdot \nabla W + A\nabla^2 W = 0 \tag{6.4}$$

So far, no assumptions have been made, and in principle we can solve these two differential equations, thereby finding a complete solution for wave equations that have three-dimensional variation in the sound speed $c(x,y,z)$. This approach is often referred to as exact ray tracing. In practice, this approach is very difficult because equations (6.3) and (6.4) are nonlinear and coupled. Therefore some simplifying assumptions must be introduced. When we assume that

$$\nabla^2 A << A\frac{\omega^2}{c^2} \tag{6.5}$$

the second term on the left side in equation (6.3) dominates, and the first term can be neglected. This means that the wave front $W$ must satisfy the equation

$$\nabla W \cdot \nabla W = \left(\frac{c_0}{c}\right)^2 \tag{6.6}$$

which can be rewritten as

$$\nabla W^2 = n^2 \tag{6.7}$$

where the index of refraction $n$ is defined as

$$n = \frac{c_0}{c} \tag{6.8}$$

Equation (6.7), the ***eikonal equation***, is a first-order nonlinear differential equation; it gives the geometry for the spatial variation of the wave front $W(x,y,z,t)$. Equation (6.4) the ***transport equation***, gives the sound field's amplitude as a function of spatial position. An alternate and more illustrative derivation of the eikonal equation based on Huygens' principle may be found in the book by Officer (1958).

Two conditions must be satisfied for the assumption in equation (6.5) to be valid. Firstly, the frequency must be sufficiently high that any variation in the sound speed over a distance equal to a wavelength is vanishingly small. Secondly, the spatial variation of the amplitude must be small, which means that equation (6.5) will not be satisfied at the edges of a sound field. A consequence of the second condition is that diffraction effects are not well represented by the ray approximation.

We now continue with the eikonal equation to calculate the trajectories of the ray paths. We first define the unit vector **e** normal to the wave front $W(x,y,z,t)$; from equation (6.7), we find that

$$\nabla W = ne \tag{6.9}$$

Although the solution of this equation defines the ray path direction at any point on the trajectory, it is more convenient to work with the ray path coordinates $\mathbf{s}(x,y,z)$. Consider figure 6.1, which shows the wave front $W(x,y,z,t)$ at several times, and the ray path $\mathbf{s}(x,y,z)$.

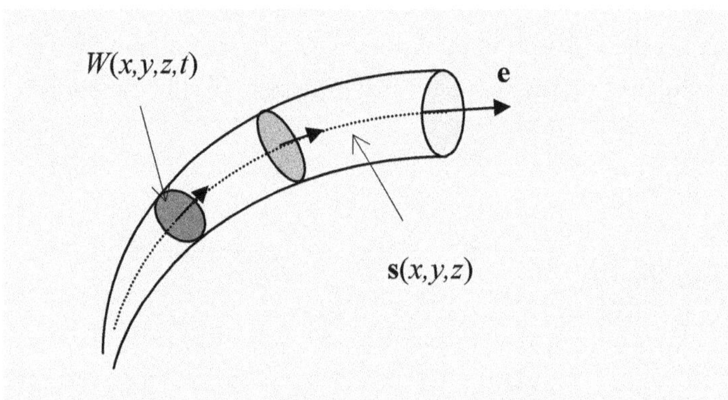


Figure 6.1. Part of the wave front $W(x,y,z,t)$, progressing along the ray path $\mathbf{s}(x,y,z)$ with the local direction **e**.

The ray direction **e** is given by

$$\mathbf{e} = \frac{d\mathbf{s}}{ds} = \mathbf{i}\frac{dx}{ds} + \mathbf{j}\frac{dy}{ds} + \mathbf{k}\frac{dz}{ds} \tag{6.10}$$

where $dx/ds$, $dy/ds$, $dz/ds$ are the direction cosines of the ray and have the property

$$\left(\frac{dx}{ds}\right)^2+\left(\frac{dy}{ds}\right)^2+\left(\frac{dz}{ds}\right)^2=1 \tag{6.11}$$

where $s$ is the distance measured along the ray path.

The direction numbers for the normal to the phase front of $W(x,y,z,t)$ are $\partial W/\partial x$, $\partial W/\partial y$, and $\partial W/\partial z$ for the $x$, $y$, and $z$ directions, respectively. These are proportional to the direction cosines and the proportionality may be expressed by three equations (Officer 1958)

$$\begin{aligned}\frac{dx}{ds}&=a\frac{dW}{dx}\\ \frac{dy}{ds}&=a\frac{dW}{dy}\\ \frac{dz}{ds}&=a\frac{dW}{dz}\end{aligned} \tag{6.12}$$

where $a$ is a constant.

To find the value of $a$ we square and add the members of equation (6.12), use the eikonal equation (6.7) and the properties of the direction cosines in equation (6.11), and find that $a = 1/n$. Equation (6.12) then becomes:

$$\begin{aligned}\frac{dx}{ds}&=\frac{1}{n}\frac{dW}{dx}\\ \frac{dy}{ds}&=\frac{1}{n}\frac{dW}{dy}\\ \frac{dz}{ds}&=\frac{1}{n}\frac{dW}{dz}\end{aligned} \tag{6.13}$$

To find the equation describing the ray path $\mathbf{s}(x,y,z)$, we must eliminate $W(x,y,z,t)$ by taking the differential $d/ds$ along the path. For the $x$-component, this produces

$$\frac{d}{ds}\left(n\frac{dx}{ds}\right)=\frac{d}{ds}\left(\frac{\partial W}{\partial x}\right)=\frac{\partial}{\partial x}\left[\frac{\partial W}{\partial x}\frac{dx}{ds}+\frac{\partial W}{\partial y}\frac{dy}{ds}+\frac{\partial W}{\partial z}\frac{dz}{ds}\right] \tag{6.14}$$

By using equations (6.7) and (6.11), this expression can be written as

$$\frac{d}{ds}\left(n\frac{dx}{ds}\right)=\frac{\partial}{\partial x}\left\{n\left[\left(\frac{dx}{ds}\right)^2+\left(\frac{dy}{ds}\right)^2+\left(\frac{dz}{ds}\right)^2\right]\right\}=\frac{\partial n}{\partial x} \tag{6.15}$$

After similar treatment of the *y*- and *z*-components, we obtain

$$\begin{aligned}\frac{d}{ds}\left(n\frac{dx}{ds}\right)&=\frac{\partial n}{\partial x}\\ \frac{d}{ds}\left(n\frac{dy}{ds}\right)&=\frac{\partial n}{\partial y}\\ \frac{d}{ds}\left(n\frac{dz}{ds}\right)&=\frac{\partial n}{\partial z}\end{aligned} \tag{6.16}$$

In electromagnetics and optics, the speed of light in a vacuum is a fundamental quantity and thus is a natural choice for the reference speed $c_0$. In acoustics, there is no such fundamental reference speed. Therefore equation (6.16) is often expressed by using the sound speed as

$$\begin{aligned}\frac{d}{ds}\left(\frac{1}{c}\frac{dx}{ds}\right)&=-\frac{1}{c^2}\frac{\partial c}{\partial x}\\ \frac{d}{ds}\left(\frac{1}{c}\frac{dy}{ds}\right)&=-\frac{1}{c^2}\frac{\partial c}{\partial y}\\ \frac{d}{ds}\left(\frac{1}{c}\frac{dz}{ds}\right)&=-\frac{1}{c^2}\frac{\partial c}{\partial z}\end{aligned} \tag{6.17}$$

Equations (6.16) and (6.17) are ordinary differential equations that describe how to compute ray trajectories in the general case where the sound speed $c(x,y,z)$ varies with all three spatial coordinates. In practice, this form is rarely used because of its complexity and the lack of detailed knowledge about the sound speed in the horizontal directions.

The next approximation involves assumption of transverse isotropy (TI), which means the sound speed may vary with depth $z$ and range $r$, but not with azimuth. This assumption is also justified physically because the gradients in the ocean variables that affect sound speed (temperature and salinity) are much greater in depth than in $x$ and $y$, as was discussed in section 3.2.

In this case we have cylindrical symmetry around the $z$ axis and we therefore use cylindrical coordinates. Figure 6.2 shows the coordinate system and a small segment of a ray path.

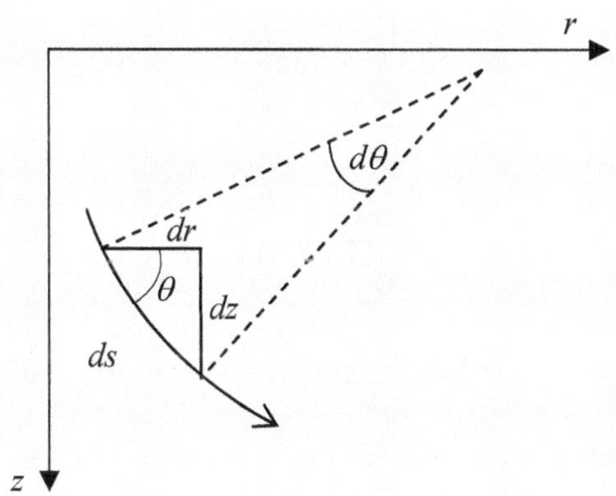


Figure 6.2. A small segment of a ray path in a transversely isotropic medium having the arc length *ds*. The segment has horizontal and vertical components *dz* and *dr*, respectively, and has a grazing angle of $\theta$.

We now introduce cylindrical coordinates and simultaneously introduce two new variables, called the ray parameters, $\xi(s)$ and $\zeta(s)$, defined as

$$\begin{aligned} \xi(s) &= \frac{1}{c}\frac{dr}{ds} = \frac{1}{c}\cos\theta \\ \zeta(s) &= \frac{1}{c}\frac{dz}{ds} = \frac{1}{c}\sin\theta \end{aligned} \tag{6.18}$$

$\xi(s)$ and $\zeta(s)$ are proportional to the horizontal ($k$) and vertical ($\gamma$) wave number components, respectively, that is,

$$\begin{aligned} k &= \omega\xi \\ \gamma &= \omega\zeta \end{aligned} \tag{6.19}$$

By equations (6.17), the wave number components satisfy the differential equations

$$\frac{d\xi}{ds} = -\frac{1}{c^2}\frac{\partial c}{\partial r}$$
$$\frac{d\zeta}{ds} = -\frac{1}{c^2}\frac{\partial c}{\partial z} \tag{6.20}$$

Using equation (6.18) and figure 6.2, we find that the initial values for integrating these equations are

$$\xi = \frac{\cos\theta_0}{c(r_0, z_0)}$$
$$\zeta = \frac{\sin\theta_0}{c(r_0, z_0)} \tag{6.21}$$

where $\theta_0$ is the ray angle and $c(r_0, z_0)$ is the sound speed at the starting point $(r_0, z_0)$.

Equations (6.20) determine the change in ray path direction as the sound speed changes with both range and depth. These equations are therefore the basis for ray tracing in a range-dependent environment, where the sound speed is a function of both range and depth. In principle, equations (6.20) can be integrated when we know the sound speed at all positions in space. The starting values of the integration are given by the initial direction of the ray at the source and the sound speed at the source, as indicated in equation (6.21).

## 6.3 Ray tracing in a horizontally stratified medium

In many situations of practical interest the sound speed profile is well approximated to be a function of depth only. This assumption leads to the so-called range-independent situation, which we consider in the remainder of this chapter. With this assumption the first of equations (6.20) reduces to

$$\frac{d\xi}{ds} = -\frac{1}{c^2}\frac{\partial c}{\partial r} = 0 \tag{6.22}$$

This implies that $\xi$ is constant and independent of $r$. On the basis of equation (6.18) it is clear that $\xi$ is given by

$$\xi = \frac{\cos\theta(z)}{c(z)} = \frac{\cos\theta_0}{c_0} \tag{6.23}$$

which is known as Snell's law and has been derived and discussed in section 4.2.

For a given ray, $\xi$ is a constant for all points on the ray's trajectory. For this reason, $\xi$ is called the ray parameter; we shall make use of this parameter in section 6.5 to describe a simple ray-tracing program. Notice, however, that $\xi$ is not constant when the bottom is sloping since the ray's angle changes by twice the slope angle after it is reflected from the bottom (see section 6.5).

Since the rays propagate along curved paths in accordance with Snell's law, it is quite easy to determine the ray path curvature. The radius of curvature $R$ is defined as the ratio between an increment in the arc length and an increment in the angle

$$R = \frac{ds}{d\theta} \tag{6.24}$$

From figure 6.2 we find that the arc length $ds$ is

$$ds = \frac{dz}{\sin\theta} \tag{6.25}$$

so that the radius of curvature becomes

$$R = \frac{1}{\sin\theta}\frac{dz}{d\theta} \tag{6.26}$$

We take the derivative of equation (6.23) with respect to $z$ and get

$$\frac{d\theta(z)}{dz} = -\frac{\cos\theta(z)}{\sin\theta(z)c(z)}g(z) \tag{6.27}$$

where $g(z)$ is the sound speed gradient

$$g(z) = \frac{dc(z)}{dz} \tag{6.28}$$

The ray's radius of curvature at depth $z$ is, from equation (6.26), therefore

$$R(z) = -\frac{c(z)}{\cos\theta(z)}\frac{1}{g(z)} = -\frac{1}{\xi g(z)} \tag{6.29}$$

where we have used Snell's law and the ray parameter $\xi$, defined in equation (6.23).

Therefore, at any point in space, the ray curvature is given by the ray parameter $\xi$ and the local value of the sound speed gradient $g(z)$. The positive or negative sign of the gradient determines whether the sign of $R$ is negative or positive, and thereby determines if the ray path curves downward or upward.

We now calculate the coordinates of a ray, starting with the angle $\theta_1$ at the point $(r_1, z_1)$, where the sound speed is $c_1$, as shown in figure 6.3.

For the coordinates of the running point at $(r_2, z_2)$ along the ray path, the horizontal distance is

$$\begin{aligned} r_2 - r_1 &= \int_{z_1}^{z_2} \frac{dz}{\tan\theta(z)} = \int_{z_1}^{z_2} \frac{\cos\theta(z)dz}{\sqrt{1-\cos^2\theta(z)}} \\ &= \int_{z_1}^{z_2} \frac{\xi c(z)dz}{\sqrt{1-\xi^2 c^2(z)}} \end{aligned} \tag{6.30}$$

The travel time $\tau$ between the two points is obtained by integrating the quantity $1/c$, the slowness, along the ray path

$$\begin{aligned} \tau_2 - \tau_1 &= \int_{z_1}^{z_2} \frac{ds}{c(s)} = \int_{z_1}^{z_2} \frac{dz}{c(z)\sin\theta(z)} \\ &= \int_{z_1}^{z_2} \frac{dz}{c(z)\sqrt{1-\xi^2 c^2(z)}} \end{aligned} \tag{6.31}$$

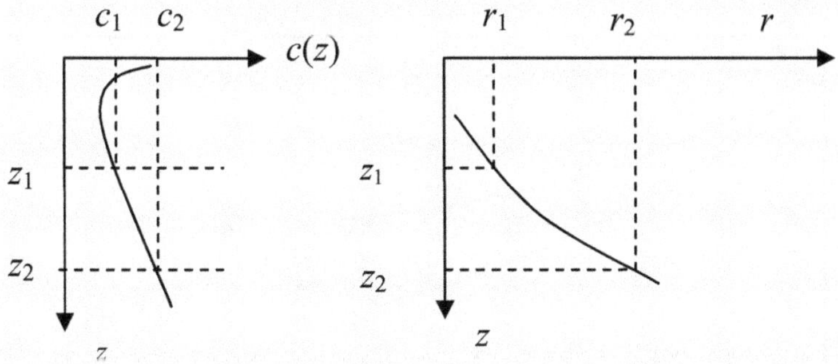


Figure 6.3. Left: The sound speed profile. Right: A portion of a ray traveling from point $(r_1, z_1)$ to $(r_2, z_2)$.

We now return briefly to the transport equation, equation (6.4). This equation describes how the amplitude of the sound field varies in space because of spatial variations of the sound speed. The acoustic intensity can be calculated using the principle that the power within a ray tube remains the same.

Figure 6.4 shows two rays with a vertical angle separation of $d\theta_0$ that define a ray tube centered on the initial angle $\theta_0$. At a reference distance $r_0$ from the source, the intensity is $I_0$. Taking into consideration the cylindrical symmetry about the $z$ axis, the power $\Delta P_0$ within the narrow angle $d\theta_0$ is

$$\Delta P_0 = I_0 2\pi r_0^2 \cos\theta_0 d\theta_0 \tag{6.32}$$

At horizontal distance $r$, the intensity is $I$. In terms of the perpendicular cross section $dL$ of the ray tube, the power is

$$\Delta P = I 2\pi r dL \tag{6.33}$$

Since the power in the ray tube does not change, we may equate equations (6.32) and (6.33) and solve for the ratio of the intensities

$$\frac{I}{I_0} = \frac{r_0^2}{r}\cos\theta_0 \left|\frac{d\theta_0}{dL}\right| \tag{6.34}$$

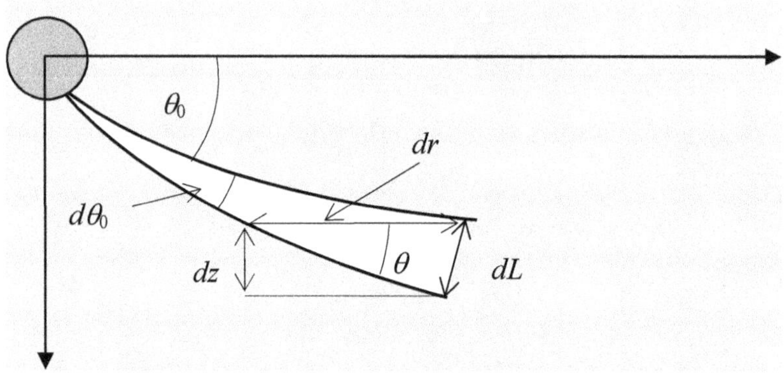


Figure 6.4. The principle of intensity calculations: energy radiated in a narrow tube remains inside the tube; $r_0$ represents a reference distance and $\theta_0$ is the initial ray angle at the source; $d\theta_0$ is the initial angular separation between two rays; *dr* is the incremental range increase; $\theta$ is the angle at the field point; *dz* is the depth differential; and *dL* is the width of the ray tube.

Using equation (6.34) as a basis, several expressions can be derived for the intensity as a function of position. For instance, it may be more convenient to use the vertical height *dz* of the ray separation, which, as can be seen from figure 6.4, is

$$dz = \frac{dL}{\cos\theta} \tag{6.35}$$

Then it follows that

$$\frac{I}{I_0} = \frac{r_0^2}{r}\frac{\cos\theta_0}{\cos\theta}\left|\frac{d\theta_0}{dz}\right| = \frac{r_0^2}{r}\frac{c_0}{c}\left|\frac{d\theta_0}{dz}\right| \tag{6.36}$$

To obtain the last expression in equation (6.36) we have assumed that the ray parameter is constant and we have made use of Snell's law. This requires that the water depth be constant for all ranges. Furthermore, the term $c_0 / c$ is close to unity in water and can be ignored.

As an alternative to the use of equations (6.34) and (6.36), we can also use the horizontal separation *dr* between the pair of rays, which is defined as

$$dr = \left| \frac{dL}{\sin\theta} \right| \tag{6.37}$$

which results in

$$\frac{I}{I_0} = \frac{r_0^2}{r} \left| \frac{\cos\theta_0}{\sin\theta} \frac{d\theta_0}{dr} \right| = \left( \frac{r_0^2}{r} \right) \left( \frac{c_0}{c} \right) \left| \frac{\cos\theta}{\sin\theta} \frac{d\theta_0}{dr} \right| \tag{6.38}$$

Note that $dr$ is defined by the intersection of the ray tube with a horizontal depth line, as shown in figure 6.4

In equations (6.36) and (6.38) we have introduced absolute values in order to avoid problems with regard to the signs of the derivatives and of $\sin\theta$.

With reference to the distance $r_0$, the transmission loss $TL$ is expressed as

$$TL = -10\log(I / I_0) \tag{6.39}$$

The ray amplitude is defined as

$$A_r = \sqrt{\frac{I}{I_0}} = \sqrt{10^{-\left(\frac{TL}{10}\right)}} \tag{6.40}$$

By inserting equation (6.36) into (6.39) the transmission loss can be defined as

$$TL = 10\log\left(r / r_0^2\right) + 10\log\left| \frac{dz}{d\theta_0} \right| + 10\log\left( \frac{c}{c_0} \right) \tag{6.41}$$

As indicated above, the last term of equation (6.41) can usually be ignored since the sound speeds are nearly equal.

Note that in this treatment the transmission loss only includes the geometric spreading loss; the absorption loss must be included separately. We see that the geometric transmission loss in equation (6.41) consists of two parts. The first term represents the horizontal spreading of the ray tube and results in a cylindrical spreading loss. The second and third terms represent the vertical spreading of the ray tube and are influenced by the depth gradient of the sound speed.

In the special case of constant sound speed, $z = r\tan\theta$, and

$$\frac{d\theta}{dz} = \frac{\cos^2\theta}{r} \tag{6.42}$$

Equation (6.36) then becomes

$$\frac{I}{I_0} = \left(\frac{r_0}{r}\right)^2 \cos^2\theta = \left(\frac{r^2{}_0}{r^2 + z^2}\right) \tag{6.43}$$

Equation (6.43) therefore defines a spherical propagation loss, as one would expect for a point source in a medium with constant sound speed. The same result may be obtained from equation (6.38).

We note that equation (6.38) predicts infinite intensity under either of two conditions: when $\theta = 0$ or when $dr / d\theta_0 = 0$. The first condition signifies a turning point where the ray path becomes horizontal; the second condition occurs at points where an infinitesimal increase in the initial angle of the ray produces no change in the horizontal range traversed by the ray. The locus of all such points in space is called a caustic. In both cases there is focusing of energy by refraction and classical ray theory incorrectly predicts infinite intensity. Caustics and turning points will be discussed further in section 6.7.

## 6.4 Ray tracing in a layer with constant sound speed gradient

Knowing the sound speed profile and using the expressions for the ray path geometry, travel time, and intensity, we are able to use the ray theory approximation to model sound propagation in the sea. This is only practically possible with the help of computers, and there are many programs for this application. We will describe a simple ray-tracing program in the next section. Before doing that, it is instructive to consider some particularly simple examples that can be solved without using numerical methods.

In general, the water column may be divided into many thin layers, and the sound speed profile in each layer may be approximated by a linear function. We shall use this approach in section 6.5, but now we will study in more detail the ray paths where a larger section of the sound speed profile can be considered linear with depth. This approximation may be too coarse for most cases, but nevertheless sheds light on what can happen in more complicated cases.

The sound speed profile is assumed to be

$$c(z) = c_1 + g(z - z_1) \tag{6.44}$$

Here, $c_1$ is the speed at depth $z_1$, and the sound speed gradient $g$ can be positive or negative. Depending on the starting angle, we can have four different types of curvature, as depicted in figure 6.5.

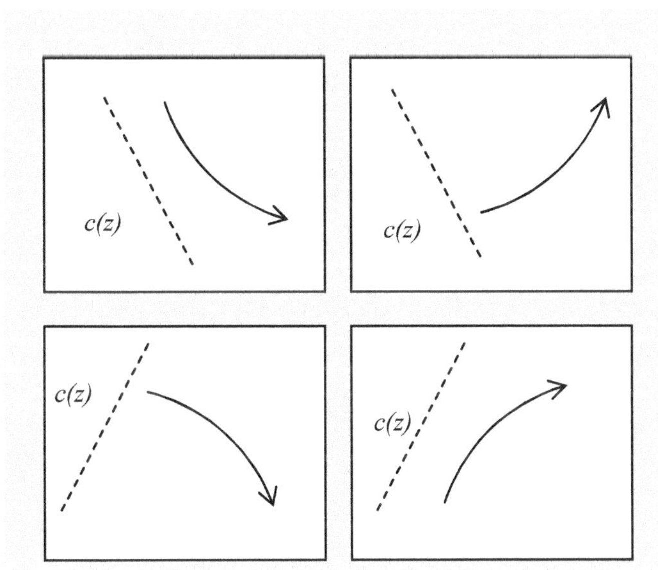


Figure 6.5. The four types of ray curvature that can occur, depending on the combination of the gradient of the sound speed profile and the initial angle of the ray.

We now determine the path of a ray that starts at the point $(r_1, z_1)$ with angle $\theta_1$ and reaches the point $(r_2, z_2)$ with angle $\theta_2$. Because the angles are related by Snell's law

$$\frac{\cos\theta_2}{c_2} = \frac{\cos\theta_1}{c_1} \tag{6.45}$$

we found in section 6.3 that a sound path in a layer with a constant gradient is a circular arc with radius of curvature $R$ given by equation (6.29).

Figure 6.6 shows the case under consideration, where the sound speed increases with depth. In the following discussion, it is more convenient to treat the radius of curvature $R$ as positive, regardless of the curvature direction, and to account for the curvature direction directly from the geometry.

Consequently, we define $R$ as

$$R = \frac{1}{|g|}\frac{c(z)}{\cos\theta(z)} = \frac{1}{\xi|g|} \tag{6.46}$$

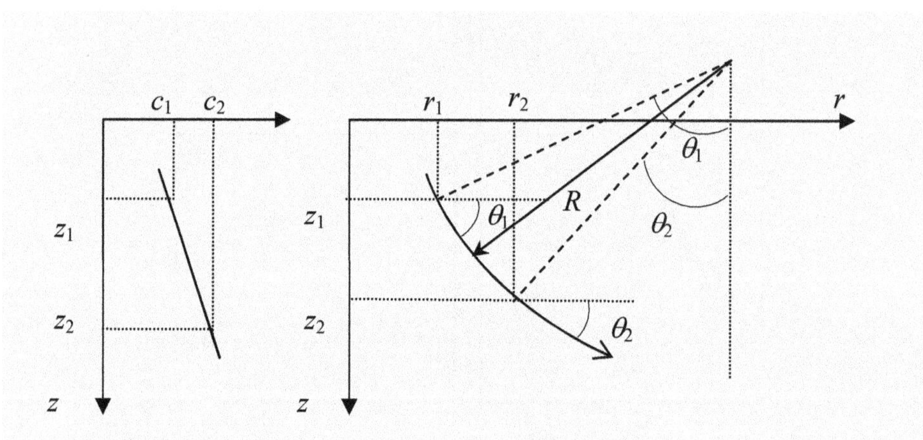


Figure 6.6. Ray path calculations for a case with a constant sound speed gradient. Left: Sound speed profile. Right: Section of the ray path.

After studying figure 6.6, we can write the following equations

$$\begin{aligned} z_2 - z_1 &= R(\cos\theta_2 - \cos\theta_1) \\ r_2 - r_1 &= R(\sin\theta_1 - \sin\theta_2) \end{aligned} \tag{6.47}$$

The travel time $\tau$ between these two points is found by integration over the ray path

$$\tau_2 - \tau_1 = \int_{z_1}^{z_2} \frac{ds}{c(z)} = \int_{\theta_1}^{\theta_2} \frac{R\,d\theta}{c(z)} = -\frac{1}{g}\int_{\theta_1}^{\theta_2} \frac{d\theta}{\cos\theta(z)} \tag{6.48}$$

Using the standard formula for this integral, we obtain

$$\tau_2 - \tau_1 = \left| \frac{1}{g} \ln\left\{ \frac{1+\sin\theta_1}{\cos\theta_1} \frac{\cos\theta_2}{1+\sin\theta_2} \right\} \right| \tag{6.49}$$

Using the ray parameter of equation (6.47) the expressions for the increase in range and travel time can also be written

$$\begin{aligned} r_2 - r_1 &= \frac{1}{\xi |g|}\left[ \sqrt{1-\xi^2 c^2(z_1)} - \sqrt{1-\xi^2 c^2(z_2)} \right] \\ \tau_2 - \tau_1 &= \frac{1}{|g|} \ln\left( \frac{c(z_2)}{c(z_1)} \frac{1+\sqrt{1-\xi^2 c^2(z_1)}}{1+\sqrt{1-\xi^2 c^2(z_2)}} \right) \end{aligned} \tag{6.50}$$

### *6.4.1 Sound speed profile with negative gradient*

We consider now the case where there is a layer near the surface with a constant negative sound speed gradient, a situation that may occur in the summer as a consequence of warmer surface water. The negative gradient causes the rays to bend downward, so that a ray that initially heads upwards is bent downward by the negative gradient (figure 6.7). As a result of this bending, a shadow zone is created in which no sound arrives directly from the source. The ray just touching the sea surface defines the boundary of the shadow zone and is therefore called ***the limiting ray*** (see the thick dashed ray in figure 6.7).

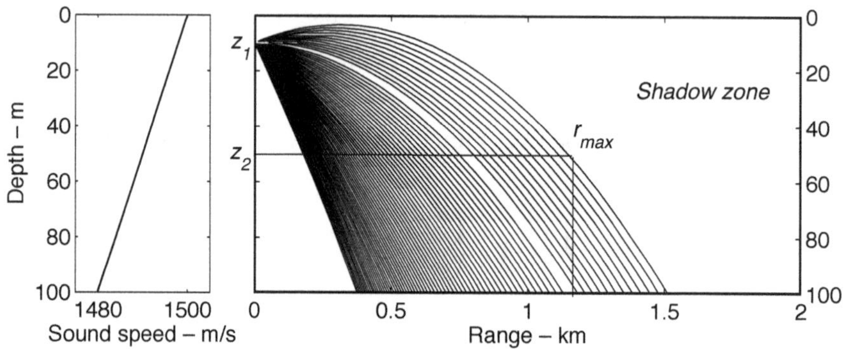


Figure 6.7. With a negative sound speed gradient, rays are bent downward, creating a shadow zone where sound cannot reach. In the example shown above, the sound speed gradient is 0.2 s$^{-1}$ and the source is at depth $z_1$ = 10 m. The thick dashed line is the limiting ray. For a target at depth $z_2$, the maximum range is $r_{max}$.

To solve for the angle and range of the limiting ray to determine the maximum range $r_{max}$ we use the same geometry as in figure 6.6, but flip it vertically so that the gradient is negative. With the source at $z_1$ and the surface is at $z_2 = 0$ the limiting ray must leave the source at angle $\theta_1$ such that

$$-z_1 = R(1 - \cos\theta_1) \tag{6.51}$$

where we have used the first of equations (6.47), and from equation (6.46),

$$R = \frac{c_0}{|g|} \tag{6.52}$$

Substituting equation (6.52) into equation (6.51) and solving for $\cos\theta_1$, we obtain

$$\cos\theta_1 = 1 + \frac{z_1}{R} = 1 - \frac{z_1}{c_0}|g| \tag{6.53}$$

Using the second of equations (6.47), the horizontal distance from the source to the tangent point at the sea surface may be expressed as

$$r_1 = R\sin\theta_1 = R\sqrt{1-\cos^2\theta_1} = R\sqrt{1-\left(1+\frac{z_1}{R}\right)^2} = \sqrt{z_1\left(2R-z_1\right)} \tag{6.54}$$

Since $R >> z_1$, equation (6.54) may be approximated as

$$r_1 \approx \sqrt{2Rz_1} = \sqrt{2z_1\frac{c_0}{|g|}} \tag{6.55}$$

To find the maximum target range $r_{max}$ for a source at $z_1$ and a target at $z_2$ (see figure 6.7), we apply equation (6.55) twice and obtain

$$r_{max} \approx \sqrt{2\frac{c_0}{|g|}}\left[\sqrt{z_1}+\sqrt{z_2}\right] \tag{6.56}$$

As an example, consider the case of a hull-mounted sonar at a depth of 10 m, a submarine at a depth of 50 m, a surface sound speed of $c_0$ = 1500 m/s, and a near-surface gradient of $g$ = –0.2 s. The resulting $r_{max}$ = 1.25 km, which illustrates the difficulty in detecting a shallow target from a hull-mounted surface ship sonar.

### *6.4.2 Sound channel with positive and negative sound speed gradient*

A sound channel is created when there is a layer with a negative gradient above a layer with a positive gradient. We will analyze the simplest case, in which the sound source is located at the depth of the sound speed minimum and the gradients on both sides are equal. From the preceding discussion, it is clear that the sound paths consists of circular arcs with alternating positive and negative curvatures around the sound axis as shown in figure 6.8. In this case the sound speed varies with depth as

$$c(z) = c_0 + g\left|z - z_0\right| \tag{6.57}$$

The sound speed has a minimum at $z = z_0$, the depth of the sound channel axis.

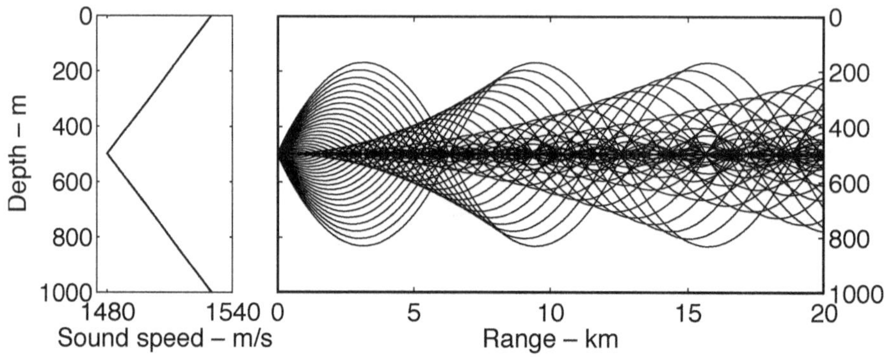


Figure 6.8. A minimum in the sound speed (left) becomes the axis for a channel (right) where the sound will be concentrated. In the example shown above the sound speed minimum is 1480 m/s at 500 m and the sound speed gradient is 0.1 $s^{-1}$. The rays have initial angles between –12° and 12° with 0.5° increments.

Using equation (6.46), the ray with initial angle $\theta_0$ has a radius of curvature

$$R = \frac{c_0}{g \cos \theta_0} \tag{6.58}$$

Using the second of equations (6.47), the ray becomes horizontal at the range $r = R \sin\theta_0$. The travel time of the ray from the source to the turning point is, using equations (6.47) with $\theta_1 = \theta_0$ and $\theta_2 = 0$

$$\tau = \frac{1}{g} \ln \frac{1 + \sin \theta_0}{\cos \theta_0} \tag{6.59}$$

and the travel time when the ray crosses the axis is twice this value.

When we use the fact that $\sin\theta_0 = r / 2R$, we can express the travel time of the ray as it crosses the axis as

$$\tau = \frac{2}{g}\ln\frac{1+\dfrac{r}{2R}}{\left(1-\left(\dfrac{r}{2R}\right)^2\right)^{1/2}} = \frac{1}{g}\ln\frac{\left(1+\dfrac{r}{2R}\right)^2}{\left(1-\left(\dfrac{r}{2R}\right)^2\right)} = \frac{1}{g}\ln\frac{1+\dfrac{r}{2R}}{1-\dfrac{r}{2R}} \tag{6.60}$$

A series expansion of equation (6.60), valid when $R >> r$, gives us approximately

$$\tau = \frac{r}{gR} = \frac{r\cos\theta_0}{c_0} \tag{6.61}$$

It is interesting to examine the travel time to a point $r$ on the axis by studying it as a function of the initial angle of the ray. As the initial angle decreases, the deviation of the ray path from the sound channel diminishes, the path alternates increasingly faster around the axis, $R$ is reduced, and the travel time increases. In the limiting case with the parameters $\theta_0 = 0$ the travel time approaches its maximum value, $\tau_{max}$, which is

$$\tau_{\max} = \frac{r}{c_0} \tag{6.62}$$

This result demonstrates that the rays having the shallowest angles are the slowest and that the last arrival is the direct ray that travels along the sound channel axis. The rays at steeper angles have greater path lengths, but pass through water that has higher sound speed. The net effect is that the signals traveling along steep-angle rays arrive earlier than those along shallow-angle rays. Sound transmitted within a small angular interval around the sound axis remains within a limited depth interval around the axis.

### 6.4.3 Sound speed profile with positive gradient

We now look at an example where the sound speed at the surface is $c_0$ and increases linearly with depth with gradient $g$ in water of depth $D$ as

$$c(z) = c_0 + gz \tag{6.63}$$

Figure 6.9 shows a ray that starts with the initial angle $\theta_0$ and the ray parameter

$$\xi = \frac{\cos\theta_0}{c(0)} \tag{6.64}$$

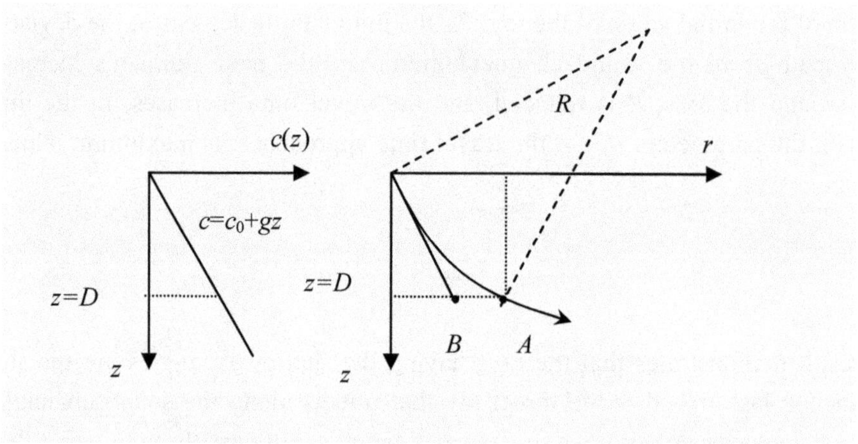


Figure 6.9. Left: Sound speed profile with a constant positive gradient. Right: The circular ray path from $z = 0$ to $z = D$ with radius of curvature $R$. $A$ and $B$ represent the actual and the incorrect target locations, respectively.

Using the first of equations (6.50) with $z_1 = 0$ and $z_2 = D$, we find that the ray intersects the sea floor at a range of

$$r_1 = \frac{1}{\xi|g|}\left[\sqrt{1-\xi^2c^2(0)} - \sqrt{1-\xi^2c^2(D)}\right] \tag{6.65}$$

At the depth $D$ the sound speed is $c(D) = c_0 + gD$.

With equations (6.64) and (6.65) we can calculate the ray paths and the range to the points where the rays hit the bottom for any value of the initial angle.

Figure 6.10 shows the result of such computations, using the values $D = 500$ m, $c_0 = 1475$ m/s, and $g = 0.015$ s$^{-1}$. This gradient is approximately the constant-temperature pressure gradient as described in section 3.2.

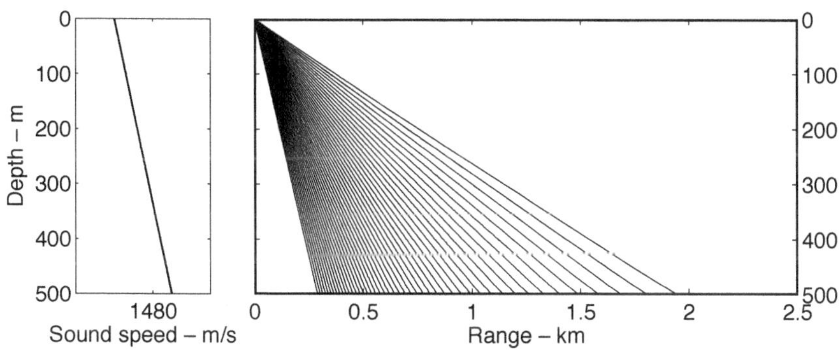


Figure 6.10. A fan of rays in a layer with a small positive gradient. The rays have initial angles from 15° to 60° in 1° increments.

The curving of the ray paths may not appear very important in figure 6.10, but a closer look shows that the assumption of straight-line propagation may result in significant errors. For example, the ray with a 15° initial angle hits the bottom at 1942 m, whereas a straight-line ray strikes at 1866 m. The difference, 76 m, is quite significant and needs to be corrected in many applications, for instance when a multibeam echo sounder is used for seabed mapping.

## 6.5 A recipe for a simple ray-tracing program

This section describes a simple ray-tracing program, but we note that the approach used is only one of many possibilities. The book by Jensen et al. (1994), chapter 3. discusses the particularities of the various approaches. The approach used for the algorithm described here is based on the results of the previous section, where we developed the equations for horizontal distances and travel times of rays for profiles in which the sound speed was a linear function of depth.

We divide the whole water column into a large number of layers, each with the same thickness $\Delta z$. Within each layer, the sound speed profile is approximated as linear so that, in the layer $z_i < z < z_{i+1}$, the sound speed is taken to be

$$c(z) = c_i + g_i (z - z_i) \tag{6.66}$$

where $c_i$ is the speed at depth $z_i$, and $g_i$ is the sound speed gradient in the layer. From equations (6.50) the range and travel time increments in a layer are given by

$$r_{i+1} - r_i = \frac{1}{\xi g_i}\left[\sqrt{1-\xi^2 c^2(z_i)} - \sqrt{1-\xi^2 c^2(z_{i+1})}\right] \tag{6.67}$$

and

$$\tau_{i+1} - \tau_i = \frac{1}{|g_i|}\ln\left(\frac{c(z_{i+1})}{c(z_i)}\frac{1+\sqrt{1-\xi^2 c^2(z_i)}}{1+\sqrt{1-\xi^2 c^2(z_{i+1})}}\right) \tag{6.68}$$

When $\xi^2 c^2 (z_{i+1}) \geq 1$, the ray path turns at a depth between $z_i$ and $z_{i+1}$, and equations (6.67) and (6.68) must be replaced by the following expressions:

$$r_{i+1} - r_i = \frac{2}{\xi g_i}\sqrt{1-\xi^2 c^2(z_i)} \tag{6.69}$$

and

$$\tau_{i+1} - \tau_i = \frac{2}{|g_i|}\ln\left(\frac{1+\sqrt{1-\xi^2 c^2(z_i)}}{\xi c(z_i)}\right) \tag{6.70}$$

The algorithm makes repeated use of the previous five equations, tracking a ray upward or downward until it reflects from an interface or turns within a layer.

In this manner, one can calculate the trajectories and the travel times for any ray's path to the desired range. By applying equation (6.38), the ray transmission loss can also be determined. The simplicity of this method lies in the approximation of the sound speed profiles with straight-line segments and the ray path's subsequent decomposition into circular segments. The method's accuracy is determined by how well the linear fit matches the actual profile. In practice, the sound speed profile is often given as measured sound speeds at relatively few depth points. It is therefore advisable to use an interpolation scheme that is consistent with the usual behavior of the sound speed profile to increase the number of depth points to an acceptable density.

The algorithm also can be used when the bathymetry is range dependent. However, in that case the ray parameter $\xi$ is no longer constant, but changes when the ray strikes the bottom; the amount of change depends on the local slope angle. The change in ray direction is illustrated in figure 6.11, where it is easy to see that, after reflection, the angle $\theta_{in}$ of an incoming ray is increased by twice the bottom inclination angle $\alpha$.

$$\theta_{out} = \theta_{in} + 2\alpha \tag{6.71}$$

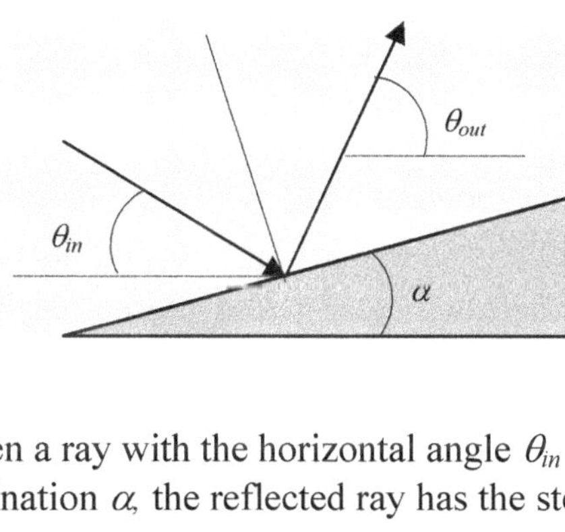


Figure 6.11. When a ray with the horizontal angle $\theta_{in}$ strikes a plane with inclination $\alpha$, the reflected ray has the steeper angle $\theta_{out}$.

Consequently, after the ray is reflected, its ray parameter must change from $\xi_{in}$ to $\xi_{out}$, which is expressed as

$$\begin{aligned} \xi_{out} &= \frac{\cos(\theta_{out})}{c} = \frac{\cos(\theta_{in} + 2\alpha)}{c} \\ &= \xi_{in}\cos(2\alpha) - \frac{\sqrt{1-\xi_{in}^2 c^2}}{c}\sin(2\alpha) \end{aligned} \tag{6.72}$$

The next section contains a number of examples where the described algorithm has been applied to sound speed profiles of general interest.

## 6.6 Ray-tracing examples

In chapter 3 we discussed how the sound speed typically varies in the sea; now we use the computer program described in section 6.5 to demonstrate the effect of the sound speed profile on sound propagation.

### *6.6.1 Deep-water propagation*

As discussed in chapter 3, the sound speed profile in deep water is dominated by two factors: (1) the decreasing temperature with depth, and (2) the increasing pressure with depth. Generally, the result is a profile that has a minimum at a certain depth. The most commonly used parameterization of a deep-water sound speed profile is the Munk profile (Munk, Worcester, and Wunsch 1995), given by the formula

$$
\begin{aligned}
c(z) &= c_0 \left\{1 + \varepsilon \left[x - 1 + \exp(-x)\right]\right\} \\
x &= 2\frac{z - z_0}{z_0}
\end{aligned}
\tag{6.73}
$$

where $z_0$ is the depth of the sound channel axis, $c_0 = c(z_0)$, and $\varepsilon$ is a dimensionless constant with the value of $\varepsilon = 0.00737$. Figure 6.12 illustrates ray paths for three different source depths when the sound speed profile is given by equation (6.73) with $c_0$ = 1500 m/s and $z_0$ = 1000 m.

In the situation depicted at the top, the transmitter is at 200 m, and the sound paths are bent downward because of the upper negative gradient. At deeper depths, the ray gradient is positive and the sound paths bend upward, almost reaching the surface at about 45 km. Then the rays are bent down again, and this cycle repeats itself with a new zone at the surface at a range of about 90 km. The zones where the sound rays are concentrated are referred to as convergence zones and are characterized by particularly high sound levels and the possibility of good sonar contact at considerable distances. In the middle part of the figure, the transmitter is on the axis of the sound channel at 1000 m. A beam of outgoing rays with initial angles of ±10° oscillates around the sound axis and hits neither the surface nor the seabed.

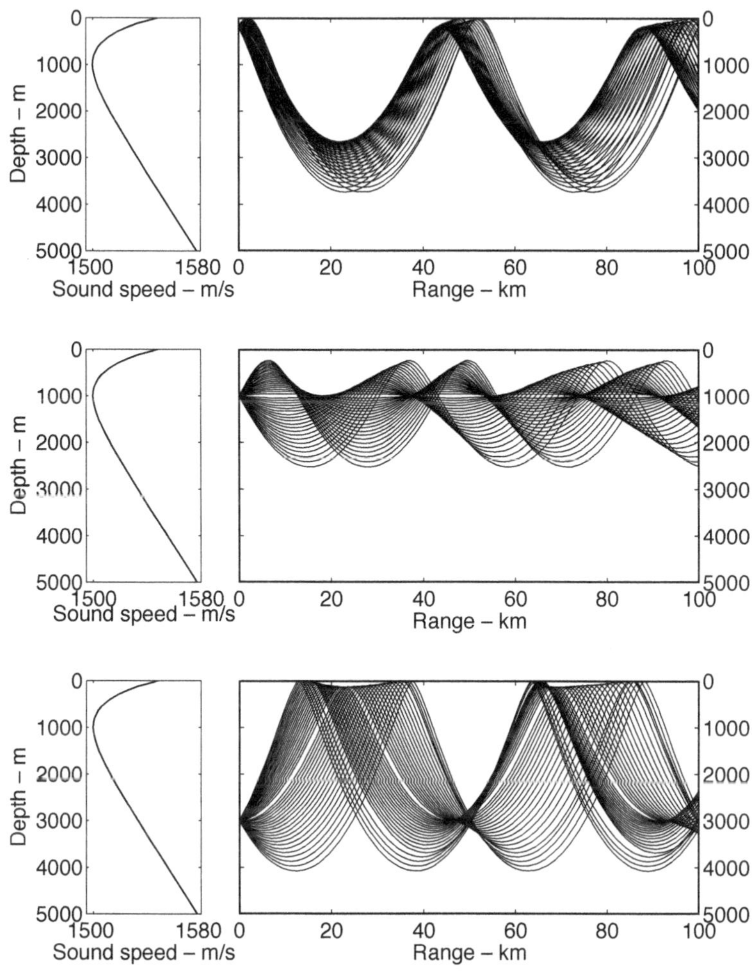


Figure 6.12. Examples of ray traces for deep water. Source depth (Sd.): (top) 200 m; (center) 1000 m; (bottom) 3000 m. Initial angles at the source from −10° to +10°.

Consequently, the transmission loss is quite low because the channeling results in cylindrical spreading. In the bottom part of figure 6.12 the transmitter is located at 3000 m, which is far below the sound axis. The rays are bent upward, which results in insonification of a wide range interval centered at 25 km.

### *6.6.2 Ray paths in the Norwegian Sea: summer and winter conditions*

Figures 6.13 and 6.14 show ray paths in a shallow water environment in the Norwegian Sea during typical summer and winter conditions. The sound speed profiles, shown at the left sides of the figures, are the same as those shown in figure 3.2; the source depths are 5 m.

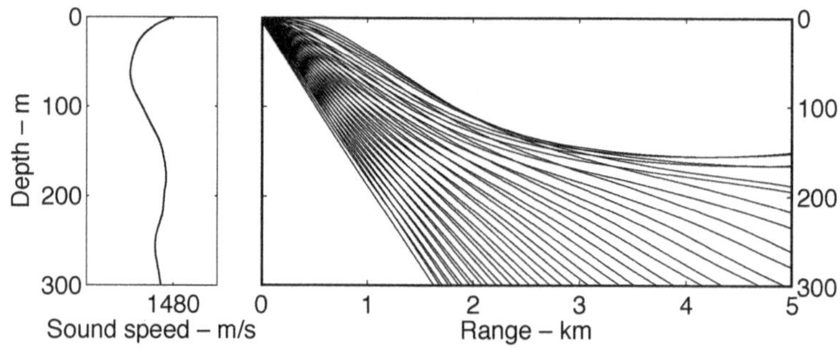


Figure 6.13. Ray traces for a source depth of 5 m in the Norwegian Sea under typical summer conditions.

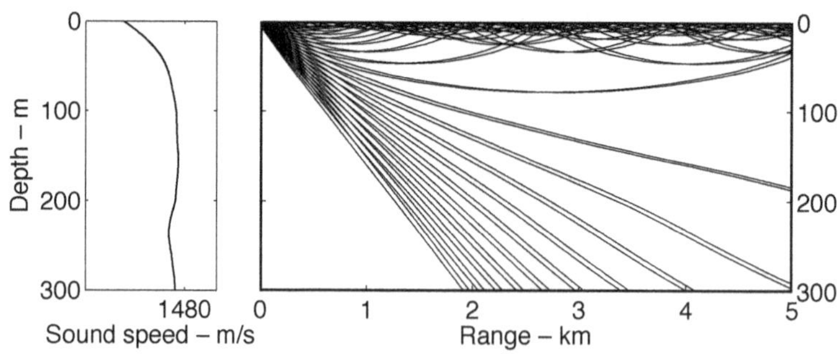


Figure 6.14. Ray traces for a source depth of 5 m in the Norwegian Sea under typical winter conditions.

For the summer profile in figure 6.13 we notice that with the source at 5 m—which roughly corresponds to the depth of a hull-mounted sonar—the direct path's range to shallow targets is very limited. Note that we have chosen not to show rays after they have hit the bottom.

In the winter, as shown in figure 6.14, propagation conditions are quite different. For shallow source depths, the sound rays have a marked upward curvature, with possibilities of good sonar coverage for shallow targets

### *6.6.3 Mixed-layer conditions*

As mentioned in section 3.2, the temperature of the upper layers strongly depends on atmospheric and oceanographic conditions. Wind and waves may sometimes cause a total mixing of the water masses, creating nearly constant temperature and salinity. In such a mixed layer, the sound speed will have a small positive gradient because of the depth effect. Figure 6.15 shows an example where the sound speed gradient in the upper 50 m mixed layer is positive and equal to 0.017 $s^{-1}$ because of the depth effect. Below 50 m the gradient is negative and assumed to be 0.02 $s^{-1}$. The sound speed profile (in m and m/s) is therefore

$$\begin{aligned} c(z) &= 1475 + 0.017(z-50) \text{ for } z < 50 \\ c(z) &= 1475 - 0.020(z-50) \text{ for } z \geq 50 \end{aligned} \tag{6.74}$$

Figure 6.15 shows that the positive gradient results in upward refraction within the mixed layer, but that the negative gradient below the layer causes rays to refract downward and hit the ocean bottom. For sources and targets within the mixed layer, sonar conditions will be particularly good; below the layer, however, a large shadow zone is created by the negative gradient, which prohibits detection of any target beyond about 5 km.

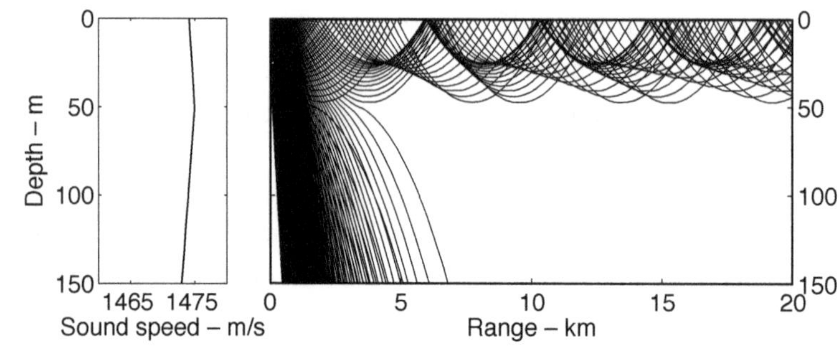


Figure 6.15. Ray trace in a shallow-water ocean having a mixed layer down to 50 m and a negative gradient below.

## 6.7 Caustics and turning points

In this section we briefly describe some of the limitations of classical ray theory, particularly in the vicinity of caustics and turning points. More thorough treatments of these limitations, as well as methods for overcoming them, are found in the books by Officer (1958), Brekhovskikh and Godin (2002), and Jensen et al. (1994).

A turning point is a location where a ray turns from going down to going up, or vice versa. A signal traveling along a ray path that goes through a turning point can be considered to experience a gradual reflection. Shortly after the turning, the ray usually grazes a caustic line that separates an "insonified" region, where two rays arrive, from a shadow zone, where no rays arrive. After grazing the caustic the ray phase is decreased by 90°. This phase change has no effect on the travel time of a signal following the ray, but the shape of the signal will change.

Caustics are defined, as mentioned previously, as the locus of points in space where the derivative of range $r$ with respect to the ray's initial angle $\theta_0$ vanishes, that is, where

$$\frac{dr}{d\theta_0} = 0 \tag{6.75}$$

In principle, equation (6.75) can be used to find the locations of caustics. In general this is a complicated issue, and we therefore choose to examine the simple case illustrated in figure 6.16, where the sound speed increases monotonically with depth according to

$$c(z) = c_0 + gz \tag{6.76}$$

We position the source at zero depth and follow a ray path with initial angle $\theta_0$, which has a radius of curvature of

$$R = \frac{c_0}{g\cos\theta_0} \tag{6.77}$$

After the ray has gone through $n$ turning points and $(n - 1)$ surface reflections, the ray reaches a point at depth $z$, horizontal distance $r$, and ray angle $\theta$. Figure 6.16 shows the geometry for $n = 2$.

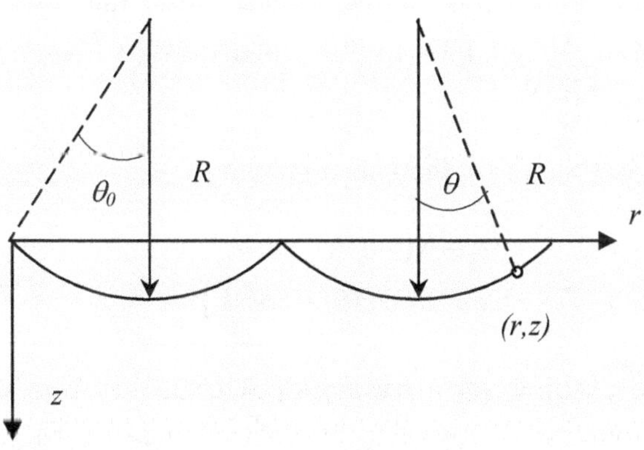


Figure 6.16. Calculating the range and depth to a caustic in a case of a sound speed profile with constant positive gradient. The initial ray angle is $\theta_0$, the source depth is $z = 0$, and the radius of curvature is $R$.

From figure 6.17 we see that the horizontal distance $r$ is given by

$$r = R\left[(2n-1)\sin\theta_0 + \sin\theta\right] \tag{6.78}$$

In order to calculate the intensity at the running point using equation (6.38) we need to determine the derivative of $r$ with respect to the initial angle $\theta_0$, taking into account that both the ray curvature $R$ and the angle $\theta$ are functions of $\theta_0$. It is left as an exercise for the reader to demonstrate that

$$\frac{dr}{d\theta_0} = \frac{R}{\cos\theta_0}\left[(2n-1) + \frac{\sin\theta_0}{\sin\theta}\right] \tag{6.79}$$

Setting the derivative $dr/d\theta_0 = 0$ gives

$$\sin\theta_c = -\frac{\sin\theta_0}{2n-1} \tag{6.80}$$

where $\theta_c$ is the angle of the ray at the caustic. Note that this relationship is a consequence of the circular ray paths, but is independent of the actual values of the sound speed and the gradient.

The coordinates of the caustic are then calculated by

$$\begin{aligned} r &= R(\sin\theta_0 - \sin\theta_c) \\ z &= R(\cos\theta_c - \cos\theta_0) \end{aligned} \tag{6.81}$$

Figure 6.17 gives an example for a waveguide with $c_0 = 1475$ m/s and $g = 0.03$ s$^{-1}$. The figure shows a number of rays and the loci of three caustics, computed using equations (6.80) and (6.81) with $n = 1$, 2, and 3. Figure 6.18 presents a closer look at the ray paths in the vicinity of the caustic. The figure shows two eigenrays connecting a source at zero depth with a receiver at range 12.25 km and depth 50 m. The two eigenrays have been reflected from the surface twice. One eigenray has grazed two caustics, the other eigenray has not yet reached the second caustic.

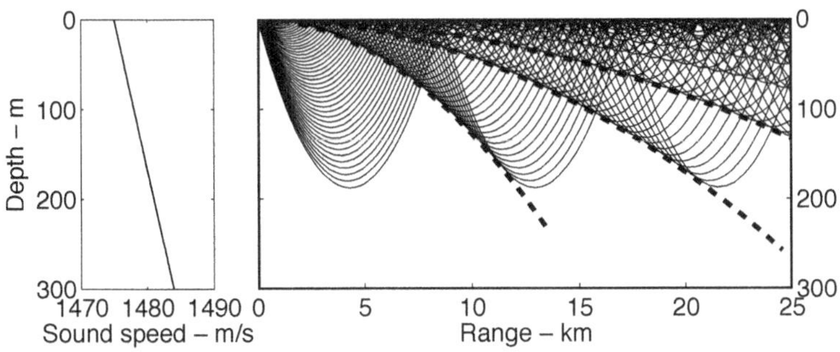


Figure 6.17. Example of turning points and caustics showing ray paths for rays with initial angle from 0.1° to 5° at 0.1° intervals. The caustics are shown with dashed lines.

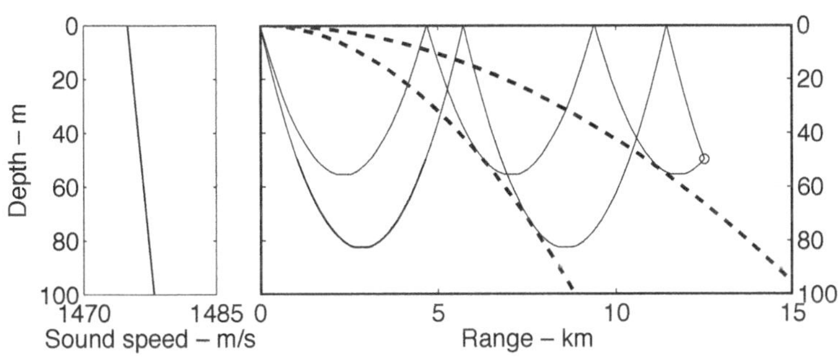


Figure 6.18. Detailed view of some of the ray path near a caustic showing three eigenrays connecting a source at zero depth with a receiver at distance of 12.25 km and a depth of 50 m. The caustics are shown with dashed lines.

## 6.8 Synthesizing the complete sound field

As indicated at the beginning of this chapter, ray theory can be quite efficient for modeling the propagation of broadband time signals. In this section we discuss this in more detail and present some results of broadband pulse propagation in an oceanic waveguide. A more complete discussion and treatment can be found in Westwood and Vidmar (1987), which contains many useful references to other works on this problem.

Consider a source transmitting a short sound pulse. The received signal in any location will, in general, have contributions from several ray paths, each contribution with different signal amplitude and time of arrival. To model the complete sound field we need to find the eigenrays that connect the source and receiver, and then sum the contributions coherently with the correct amplitudes and times of arrivals. Figure 6.19 shows the eigenrays that connect a source and a receiver at a range of 7.5 km in an ocean environment that has the sound speed profile drawn in the left plot and the range-dependent bathymetry drawn in the ray plot. Two of the eigenrays propagate in the sound channel created by the sound speed minimum at approximately 75 m depth. The other eigenrays are reflected from the surface and the bottom one or more times.

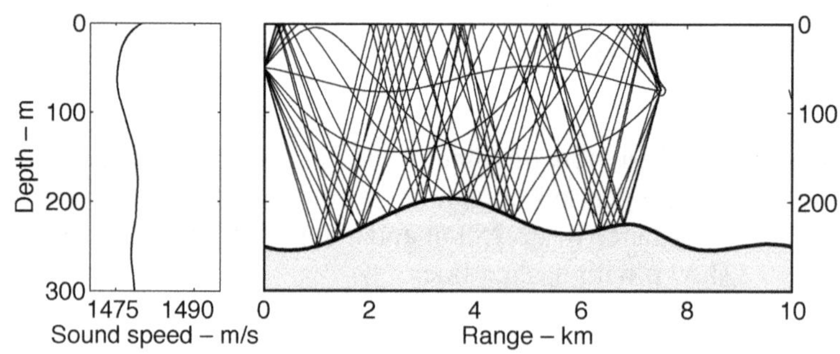


Figure 6.19. Eigenrays for a situation when the sound speed varies with depth and the water depth varies with range. Left: The sound speed profile. Right: The first 17 eigenrays connecting a source at 50 m depth with a receiver at distance of 7.5 km and 75 m depth.

A standard method for determining eigenrays is to start by computing the ray paths for a large number of rays, keeping track of where the rays hit the bottom and the surface or go through turning points. This information, the ray history, is stored in tables for later look-up. Thereafter, a search algorithm must be applied to identify the various ray types and to select the rays that come close to the desired receiver position. This first result may be improved by either applying a recursive search algorithm to find the correct distance within a prescribed tolerance, as done by Dosso and Riedel (2001) and Westwood and Vidmar (1987), or by sorting and interpolation as described in Hovem (2008). In any case, it is important that the interpolation or the search algorithm work on rays having the same type of ray history to ensure that there be a relatively smooth and monotonic relation between initial angle and distance to points where the ray reaches the desired depth.

Having found the eigenrays and their trajectories, the next step is to add all ray contributions coherently, taking into account the effects of bottom and surface reflections and turning points.

Amplitude and phase changes due to reflections from the sea surface and the sea bottom can, at least approximately, be accounted for by use of plane wave reflection coefficients. For a flat sea surface the pressure reflection coefficient is equal to minus one. For the sea bottom, the discussion and examples presented in chapters 4 and 5 are relevant for situations where the bottom is homogenous and can be modeled as a fluid. Chapter 15 covers the case of a solid halfspace. We found that the reflection coefficient is strongly dependent on the incident angle of the ray. In the case where the sound speed in the bottom is higher than the sound speed in the water there is a critical angle given by

$$\theta_{crit} = \cos^{-1}\left(\frac{c_{water}}{c_{bottom}}\right) \qquad (6.82)$$

For angles less than the critical angle there is total reflection and the reflection loss defined by equation 4.31 is zero, or at least quite small if sound absorption in the bottom is included. However, for these angles the reflection coefficient has a non-zero phase angle, which has the effect of displacing the ray along the bottom interface before it is reflected back into the water. This effect is called beam displacement and was mentioned in chapter 3 with general reference to

Brekhovskikh and Lysanov (2003), Jensen et al. (1994), and articles by Tindle and Bold (1981), Tindle (1983), and Westwood and Tindle (1987). These papers demonstrate that the accuracy of ray modeling in shallow water can be improved when beam displacement is included, particularly for very low frequencies. The improvement comes, however, at the price of more complicated calculations and longer computation times since the displacement and travel times become functions of frequency.

For rays at angles steeper than the critical angle, the reflection losses are significant as illustrated in the examples presented in sections 4.3 and 4.4. As a consequence the major contributions to the sound field at longer ranges come from rays with angles smaller than the critical angle.

The synthesis of the received signals is performed in the frequency domain by multiplying the frequency spectrum of the source signal by the transfer function of each of the eigenrays and summing all the significant contributions. Symbolically this operation can be expressed as

$$\Phi(\omega) = Q(\omega) \cdot \sum_{n=1}^{N} A_n \cdot B_n(\omega) \cdot S_n \cdot T_n \exp(i\omega\tau_n) \tag{6.83}$$

In equation (6.83) $\Phi(\omega)$ is the spectral component at angular frequency $\omega$ of the signal received at a given location expressed as a sum over the $N$ eigenrays that are included in the synthesis.

The factors in equation (6.83) are:

| | |
|---|---|
| $Q(\omega)$ | the spectral component at angular frequency $\omega$ of the received signal |
| $A_n$ | the spreading factor, defined in equation (6.40) |
| $B_n$ | the combined effect of all bottom reflections |
| $S_n$ | the combined effect of all surface reflections |
| $T_n$ | the effect of the −90° phase shift associated with each turning point |
| $\tau_n$ | the travel time |

Figure 6.20 shows computer-generated time series at a depth of 100 m and an interval of ranges from 0.5 km to 10 km for the sound speed profile and bottom topography of figure 6.19. The bottom has sound speed of 1700 m/s, a density of 1500 kg/m$^3$, and an absorption of 1.0 dB/wavelength.

The time scale is a reduced time given by

$$t_{red} = t - r / c_{red} \tag{6.84}$$

with the value $c_{red}$ = 1475 m/s. The dashed line indicates the effect of the critical angle $\theta_{crit}$ = 29.6°, which is given by the line

$$t(r) = \frac{r}{c_{red}}\left(\frac{1}{\cos\theta_{crit}} - 1\right) \tag{6.85}$$

The first and the strongest arrivals are the direct rays for ranges less than about 2 km. These arrivals are followed by a series of surface- and bottom-reflected contributions. Arrivals to the right of the dashed line strike the bottom with angles steeper than critical and are therefore greatly reduced in magnitude due to bottom reflection loss. Arrivals to the left of the line are at angles less than the critical angle and experience low bottom reflection loss.

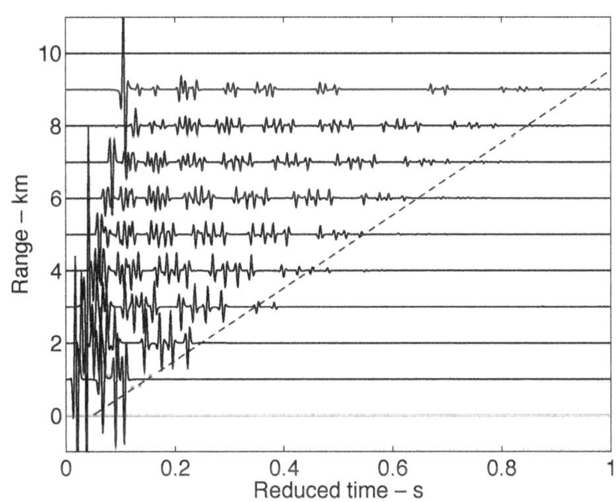


Figure 6.20. Computer-generated broadband time series over an interval of ranges for the environment shown in figure 6.19. The dashed line indicates the effect of the critical angle.

# *Chapter 7*

# *Normal modes*

## *7.1 Introduction*

An efficient, commonly used method for solving the wave equation, and consequently for calculation of sound propagation, is the method of ***normal modes***. The method is particularly applicable to situations involving relatively low frequencies and shallow waters, where the depth can be measured in terms of a limited number of wavelengths. The method considers sound propagation as propagation in a waveguide created by the water surface and seabed. Such a waveguide is not ideal, as there are losses not only in the water and within the seabed, but also especially at the interfaces of the seabed and sea surface. In chapter 17 we see that the normal mode approach is a special case of a more comprehensive and exact theory for solving the wave equation in stratified media.

We first present the general theory behind normal mode propagation, and then illuminate the properties of the solution with the help of a series of examples. Most of these cases are so simple that they can be solved without use of elaborate numerical techniques, but we also demonstrate some examples where we have used state-of-the-art normal mode propagation codes. Finally, we derive the solution in the time domain for a particular example and then use this case to discuss the concepts of phase and group speed and frequency dispersion.

Most textbooks on underwater acoustics have a chapter on normal modes. Alternative reading is found in Medwin and Clay (1998), Kinsler et al. (2000), Williams (1970), Jensen et al. (1994), Tolstoy and Clay (1964), and Frisk (1994). More specialized readings can be found in Brekhovskikh and Godin (1999, 2002) and Boyle (1984).

## 7.2 The wave equation in cylindrical coordinates

Figure 7.1 shows the first problem we consider. The water depth is constant and equal to $D$; there is a source at $z_s$ and a receiver located at a horizontal distance $r$ and depth $z$. We assume that the sound speed $c(z)$ is a function of the depth and also assume that it does not vary with the horizontal distance.

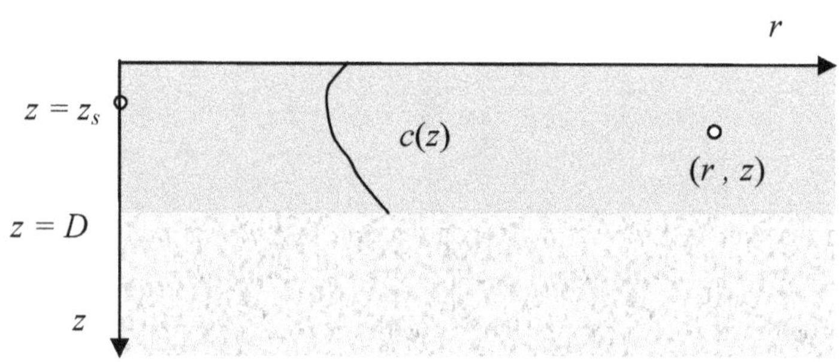


Figure 7.1. Geometry of sound propagation in a sound channel with water depth $D$ and source and receiver at depths $z_s$ and $z$, respectively, separated by the horizontal distance $r$.

The standard normal mode theory requires that no parameter for the water or the bottom can vary with range. In this ***range-independent*** ocean problem, the ocean bottom can be layered with an arbitrary number of layers with different sound speeds and densities. Although there exist more advanced theories that can treat range-dependent problems, they will not be discussed here, but can be found in the books by Jensen et al. (1994) and in Frisk (1994). Furthermore, in our derivation of the theory we will ignore the effect of absorption in the bottom materials. However, we will briefly discus the effect of bottom absorption by using the same perturbation approach as we used in sections 3.3 and 4.3.

Figure 7.2 shows the coordinate system. With no azimuth dependence there is cylindrical symmetry around the $z$-axis at $r = 0$, and therefore we use cylindrical coordinates.

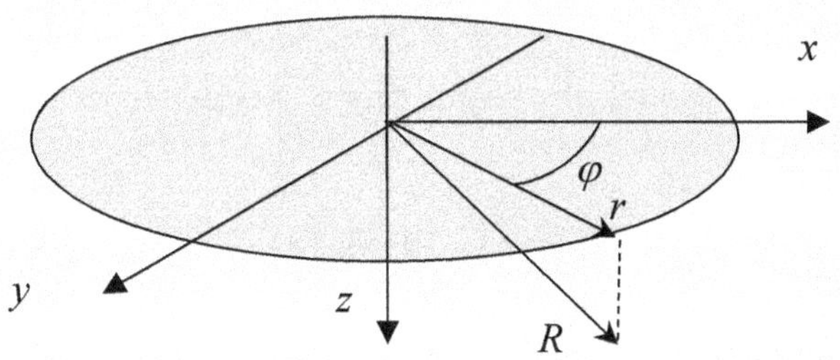


Figure 7.2. Geometry for cylindrical coordinates. The coordinates of a point in space are given by the horizontal range $r$, the azimuth angle $\varphi$, and the depth $z$. The distance from the origin in the figure to the point is $R$.

With a point source of strength $Q$ at depth $z_s$, and at $r = 0$, and assuming that the density is constant in the water and in each layer of the ocean bottom, the inhomogeneous wave equation for the velocity potential $\Phi(r, z)$ is

$$\frac{1}{r}\frac{\partial}{\partial r}\left(r\frac{\partial\Phi}{\partial r}\right)+\frac{\partial^2\Phi}{\partial z^2}+\kappa^2\Phi=-Q\frac{\delta(r)\delta(z-z_s)}{2\pi r} \tag{7.1}$$

In equation (7.1) the wave number is $\kappa = \omega / c$, and we have used the expression of equation 2.43 for the three-dimensional spatial delta function.

The problem is to solve this wave equation with the appropriate boundary conditions at the sea surface and the bottom. For example, we may, for instance, assume that the sound pressure at sea surface is zero and that the bottom is rigid. Consequently the boundary conditions in this case are

$$\begin{aligned} \Phi(r,z)\big|_{z=0} &= 0 \\ \left.\frac{\partial\Phi(r,z)}{\partial z}\right|_{z=D} &= 0 \end{aligned} \tag{7.2}$$

We consider first the homogenous equation, which is equation (7.1) without the source term. Our approach in obtaining the solution in terms of normal modes follows the method of separation of variables described in chapter 2. Therefore the solution is written as a product of two functions, a depth function $\xi(z)$ and a range function $\psi(r)$

$$\Phi(r,z)=\xi(z)\psi(r) \tag{7.3}$$

where $\xi(z)$ is dependent only on the depth $z$ and $\psi(r)$ is dependent only on the horizontal range $r$. We insert equation (7.3) into equation (7.1), and obtain

$$\frac{1}{\psi(r)}\left[\frac{d^2\psi(r)}{dr^2}+\frac{1}{r}\frac{d\psi(r)^2}{dr}\right]=-\frac{1}{\xi(z)}\left[\frac{d^2\xi(z)}{dz^2}+\kappa^2\xi(z)\right] \tag{7.4}$$

Since the left hand side of equation (7.4) depends only on $r$ and the right hand side depends only on $z$, both sides must be equal to the same constant, which we call the separation constant and which we denote $k^2$.

This gives two separate differential equations:

$$\frac{d^2\psi(r)}{dr^2}+\frac{1}{r}\frac{d\psi(r)}{dr}+k^2\psi(r)=0 \tag{7.5}$$

$$\frac{d^2\xi(z)}{dz^2}+\left(\kappa^2-k^2\right)\xi(z)=0 \tag{7.6}$$

The constant $k$ is interpreted as the horizontal component of the wave number and the vertical component is then given by

$$\gamma^2=\kappa^2-k^2 \tag{7.7}$$

The depth equation or the modal equation, equation (7.6), has an infinite number of solutions that are like modes of a vibrating string. The modes are characterized by mode shape functions, $\xi_m(z)$, and horizontal wave number components, $k_m$. The mode functions are the eigenfunctions and the wave numbers are the eigenvalues that are obtained from solving the depth equation with boundary conditions. For a given mode number $m$ there is one particular horizontal wave number $k_m$, and therefore a particular solution, $\psi_m(r)$, of the range equation. The horizontal wave number is constant but the vertical wave number, $\gamma_m(z)$, varies with the sound speed and therefore with the depth. The vertical wave number $\gamma_m(z)$ and horizontal wave number $k_m$ are related by

$$k_m^2 = \left(\frac{\omega}{c(z)}\right)^2 - \gamma_m^2(z) \tag{7.8}$$

Furthermore, the mode functions are orthogonal and can also be normalized to be orthonormal, that is,

$$\int_0^D \xi_m(z)\xi_n(z)dz = \delta_{nm} \qquad \delta_{nm} = \begin{cases} 1 & n = m \\ 0 & n \neq m \end{cases} \tag{7.9}$$

The next step is to assume that the complete solution can be written as a sum of normal modes $\Phi_m(r,z)$,

$$\Phi(r,z) = \sum_m \Phi_m(r,z) = \sum_m \psi_m(r)\,\xi_m(z) \tag{7.10}$$

which inserted in equation (7.1) yields

$$\sum_{m=1}^{\infty}\left\{\frac{1}{r}\frac{d}{dr}\left[r\frac{d\psi_m(r)}{dr}\right]\xi_m(z) + \psi_m(r)\left[\frac{d^2\xi_m(z)}{dz^2} + \kappa^2\xi_m(z)\right]\right\} = -Q\frac{d(r)\delta(z-z_s)}{2\pi r} \tag{7.11}$$

The term in the second squared bracket can be reduced by using equation (7.6) to become

$$\sum_{m=1}^{\infty}\left\{\frac{1}{r}\frac{d}{dr}\left[r\frac{d\psi_m(r)}{dr}\right]\zeta_m(z) + k_m^2\psi_m(r)\xi_m(z)\right\} = -Q\frac{\delta(r)\delta(z-z_s)}{2\pi r} \tag{7.12}$$

Next, we multiply equation (7.12) by $\xi_n(z)$ and integrate over the depth $z$. By using the orthogonal property of the eigenfunctions, equation (7.9), only the $n = m$ term in the sum remains, yielding

$$\frac{1}{r}\frac{d}{dr}\left(r\frac{d\psi_m(r)}{dr}\right) + k_m^2\psi_m(r) = -Q\xi_m(z_s)\frac{\delta(r)}{2\pi r} \tag{7.13}$$

This is an inhomogeneous wave equation with a source at $r = 0$ with source strength $Q\xi_m(z_s)$, and it is a standard differential equation.

The general solution of equation (7.5) was discussed in sections 2.5 and 2.7. Now we impose the radiation condition and retain only the part describing an outgoing wave whereby

$$\psi_m(r) = A_m H_0^{(1)}(k_m r) \tag{7.14}$$

The boundary condition at the source, where $r = 0$, is that the field $\psi_m(r)$ must be finite. We find the value of $A_m$ by integrating equation (7.13) over a small circle enclosing the source with radius $\varepsilon$, which first results in

$$A_m \left. \frac{dH_0^{(1)}(k_m r)}{dr} \right|_{r=\varepsilon} = -\frac{Q\xi_m(z_s)}{2\pi r} \tag{7.15}$$

When $\varepsilon$ goes to zero, the Hankel function can be replaced with its small argument approximation

$$H_0^{(1)}(k_m r) \approx \frac{2i}{\pi} \ln\left(\frac{k_m r}{2}\right) \tag{7.16}$$

and the coefficient becomes

$$A_m = (i/4)Q\xi_m(z_s) \tag{7.17}$$

Inserting equation (7.17) into equation (7.14) yields the solution of equation (7.13) as

$$\psi_m(r) = i\frac{\xi_m(z_s)Q}{4} H_0^{(1)}(k_m r) \tag{7.18}$$

The complete solution of the wave equation, equation (7.1), therefore becomes

$$\Phi(r,z) = \frac{iQ}{4} \sum_{m=1}^{\infty} \xi_m(z_s)\xi_m(z) H_0^{(1)}(k_m r) \tag{7.19}$$

For ranges where $kr >> 1$, we can use the approximation

$$H_0^{(1)}(kr) \approx \sqrt{\frac{2}{\pi kr}} \exp\left[i(kr - \pi/4)\right] \tag{7.20}$$

producing

$$\Phi(r,z) = \frac{iQ}{\sqrt{8\pi r}} e^{-i\pi/4} \sum_{m=1}^{\infty} \xi_m(z_s)\xi_m(z) \frac{\exp(ik_m r)}{\sqrt{k_m}} \tag{7.21}$$

Equations (7.19) and (7.21) show that the acoustic field is given as a sum of modal eigenfunctions $\xi_m(z)$ with eigenvalues $k_m$. For a given sound speed profile and sea bottom, these mode functions and wave numbers need to be determined only once, and thereafter the sound field can be calculated for all positions in a space of interest by using equation (7.21). Each mode is a cylindrical wave, decaying with range as 1 / sqrt($r$).

The transmission loss *TL* is defined as the sound pressure field at the receiver, relative to the sound pressure at a reference distance of $r = 1$ m. A point source in an infinite medium produces a field with amplitude $\Phi_0(r)$, which is defined as

$$\Phi_0(r) = \frac{Q}{4\pi r} \tag{7.22}$$

and the transmission loss, expressed in dB, is therefore

$$\begin{aligned} TL &= -20\log\left|\frac{\Phi(r,z)}{\Phi_0(r=1)}\right| \\ &= -20\log\left|\sqrt{\frac{2\pi}{r}} \sum_{m=0}^{\infty} \xi_m(z_s)\xi_m(z) \frac{\exp(ik_m r)}{\sqrt{k_m}}\right| \end{aligned} \tag{7.23}$$

In equations (7.23) the contributions of the different modes are summed coherently, taking the modal phase into account. Since the modes are propagating at different speeds, as we will see in section 7.3, the sound field may therefore be quite complicated, with the modes interfering constructively at some ranges and destructively at other ranges. If knowledge of the detailed structure of the sound field is not required, one may choose to add the modal contributions incoherently, thereby obtaining the incoherent transmission loss as a measure of the intensity of the average field. The incoherent transmission loss is defined by

$$TL = -20\log\left\{\sqrt{\frac{2\pi}{r}}\sqrt{\sum_{m=0}^{\infty}\left|\xi_m(z_s)\xi_m(z)\frac{\exp(ik_m r)}{\sqrt{k_m}}\right|^2}\right\} \qquad (7.24)$$

The incoherent transmission loss can be interpreted as a smoothed or averaged transmission loss that is useful when the details of the wave field are not required.

### *7.2.1 Dispersion, phase speed, and group speed*

From the previous section we have seen that the frequency domain solution of the wave equation with boundary conditions has the general form of

$$\Phi(\omega,r) = A(\omega)\exp\left[ik(\omega)r\right] \qquad (7.25)$$

The distance between source and receiver is $r$, $A(\omega)$ is a frequency weighting function that may include the frequency spectrum of source, and $k(\omega)$ is the frequency-dependent horizontal component of the wave number in the $r$ direction. The corresponding time function $\phi(t)$ is obtained by the inverse Fourier transform, equation (2.40) in section 2.3, yielding

$$\phi(t) = \frac{1}{2\pi}\int_{-\infty}^{\infty} A(\omega)\exp\left\{i\left[\left(k(\omega)r - \omega t\right)\right]\right\}d\omega \qquad (7.26)$$

The phase speed $v$ is defined by the speed of the displacement of constant phase points, and is given by the combinations of distances $r$ and times $t$ that make the exponent in equation (7.26) equal to zero. Hence the ***phase speed*** is defined by

$$v = \frac{\omega}{k(\omega)} \qquad (7.27)$$

In the general case the wave number $k(\omega)$, and hence the phase speeds, are dependent on frequency. This implies that the different frequency components of the wave propagate with different speeds, causing the shape of the time signal to change with distance. This effect is called dispersion, and is a general feature of waveguide propagation. We have already seen an example of ***intrinsic dispersion*** in section 2.4 where we introduced absorption into the wave equation, which resulted in a frequency dependent wave number (equation (2.45)). In this chapter we have introduced the concept of normal modes, where the eigenvalues and thus the Green's function are highly dispersive; namely, the group speed of individual modes varies with mode number and frequency. This leads to ***geometrical dispersion*** and will be discussed later in this chapter.

Therefore, when evaluating the integral of equation (7.26) we frequently find that $A(\omega)$ varies slowly with frequency as compared with the exponential part that oscillates rapidly with frequency. For most of the times $t$, the integral therefore yields very small values and the time signal is essentially zero for most of the time. The exceptions occur at the points, called ***stationary points***, where the exponent is constant; that is when

$$k(\omega)r - \omega t = \text{constant} \tag{7.28}$$

By differentiation of this equation we find that the stationary points occur for all combinations of $t$ and $r$ that are related by a speed called the ***group speed*** $w$, which is defined by the derivative of the wave number with respect to the angular frequency

$$w = \frac{d\omega}{dk(\omega)} \tag{7.29}$$

The two wave speeds are related and the relationship is found by taking the derivative of equation (7.27) with respect to $\omega$, resulting in

$$w = \frac{d\omega}{dk} = \frac{v}{\left(1 - \dfrac{\omega}{v}\dfrac{dv}{d\omega}\right)} \tag{7.30}$$

When $k$ is constant, the group speed is equal to the phase speed.

## *7.3 Waveguide with constant sound speed*

We now return to the solution of the wave equation. In general, $c(z)$ and $\gamma(z)$ are complicated functions of depth, and the depth equation can only be solved numerically. For insight, however, it is useful to study some simple cases that can be solved analytically.

In the first case we consider the sound speed, and thus the vertical wave number, to be independent of depth. The vertical wave number is constant, $\gamma(z) = \gamma$, and the solution of the depth equation is simply

$$\xi(z) = A_+ \exp(i\gamma z) + A_- \exp(-i\gamma z) \tag{7.31}$$

Equation (7.31) represents two plane waves traveling in the positive and negative $z$ directions. The amplitudes $A_+$ and $A_-$ are determined by the boundary conditions at the sea surface and bottom. The boundary conditions involve the sound pressure and the vertical particle displacements, which have to be continuous at the interfaces. At $z = 0$, which is the sea surface, $A_-$ is the incident wave and $A_+$ is the reflected wave. The waves are related through the plane wave reflection coefficient $R_s$ in such a way that

$$A_+ = R_s A_- \tag{7.32}$$

It now remains to find the vertical wave numbers $\gamma$, which are determined by the boundary conditions at the sea bottom. At the bottom, where $z = D$, the incident wave is $A_+ \exp(i\gamma D)$ and the reflected wave is $A_- \exp(-i\gamma D)$; the two are related by the sea bottom reflection coefficient $R_b$

$$R_b = \frac{A_- \exp(-i\gamma D)}{A_+ \exp(i\gamma D)} = \frac{A_-}{A_+} \exp(-2i\gamma D) \tag{7.33}$$

Equations (7.32) and (7.33) combined produce

$$R_s R_b \exp(2i\gamma D) = 1 \tag{7.34}$$

Equation (7.34) is an example of a ***characteristic equation.*** This particular equation is valid for waveguides that have constant sound speed and are bounded by media that yield the reflection coefficients $R_s$ and $R_b$.

In order to understand equation (7.34), consider figure 7.3, where propagation of a plane wave departing from a certain depth is indicated. The wave is reflected at the bottom and at the surface, and is subjected to the reflection coefficients $R_s$ and $R_b$ before it returns to the source depth. The "transfer function" for this path is $H(\omega)$, which results from the wave crossing the depth twice and therefore being affected by the surface and bottom reflection coefficients

$$H(\omega) = R_b R_s \exp(2i\gamma D) \tag{7.35}$$

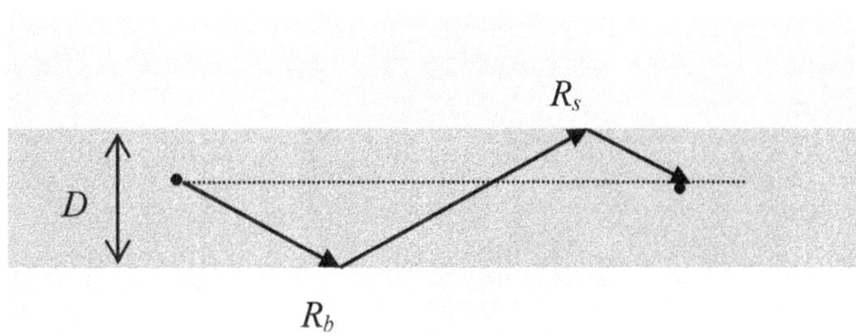


Figure 7.3. The trajectory of a plane wave that crosses the depth twice and is reflected from the bottom and the surface.

The condition for having a propagating wave is that this transfer function must equal unity, which is expressed by equation (7.34). This means that the phase of the wave fronts at the two points at the beginning and end of the trajectory shown in figure 7.3 must be constant.

### *7.3.1 The ideal waveguide*

Next we consider the special case of a totally reflecting surface where $R_s = -1$, and a totally reflecting and infinitely hard seabed where $R_b = 1$. The boundary conditions at the bottom and the sea surface are then given by equation (7.2) and from equation (7.31) we find that the solution of the wave equation is

$$\xi(z) = 2iA_+ \sin(\gamma z) \tag{7.36}$$

$A_+$ is a constant given by the strength of the sound source and is, in the context of this discussion, of no interest. The vertical wave number $\gamma$ is determined by the characteristic equation, which now becomes

$$\exp(2i\gamma D) = -1 \tag{7.37}$$

For this equation to be satisfied, the vertical wave number must assume the discrete values indicated in equation (7.38)

$$\gamma_m = (2m-1)\frac{\pi}{2D} \qquad m = 1, 2, \ldots \tag{7.38}$$

Consequently, the horizontal wave number *k,* which determines propagation in the horizontal direction, must have discrete values or eigenvalues

$$k_m = \sqrt{\left(\frac{\omega}{c_0}\right)^2 - \gamma_m^2} \tag{7.39}$$

Equations (7.38) and (7.39) show that there are modes for which $\gamma_m > \omega / c$ and so the eigenvalues are imaginary. These modes are evanescent and their contribution is in most cases of minor importance, except for very short ranges. In the following we therefore only consider the propagating modes with real eigenvalues.

The total number of propagating modes $M$ is determined by the greatest value of *m,* resulting in a real horizontal wave number $k_m$ in equation (7.39), that is, for

$$\gamma_m \leq \frac{\omega}{c_0} \tag{7.40}$$

With equation (7.38), this expression gives the total number of modes for a given frequency $f$

$$M \leq \frac{2Df}{c_0} + 0.5 \tag{7.41}$$

Having now found the permissible values for the vertical wave numbers, the eigenfunctions are determined from equation (7.36). After the normalization of this equation to satisfy equation (7.9), we find the eigenfunctions for the propagating modes to be

$$\xi_m(z) = \sqrt{\frac{2}{D}} \sin(\gamma_m z) \tag{7.42}$$

The phase and group speeds for this ideal waveguide are, respectively,

$$v_m = \frac{\omega}{k_m} = c_0 \left[ 1 - \left( \frac{c_0}{\omega} \gamma_m \right)^2 \right]^{\left(-\frac{1}{2}\right)} = \frac{c_0}{\cos\theta_m} \tag{7.43}$$

$$w_m = \frac{d\omega}{dk_m} = c_0 \left[ 1 - \left( \frac{c_0}{\omega} \gamma_m \right)^2 \right]^{\left(\frac{1}{2}\right)} = c_0 \cos\theta_m \tag{7.44}$$

In equations (7.43) and (7.44) we have introduced the propagation angle $\theta_m$ for every mode, given by

$$\sin\theta_m = \gamma_m \frac{c_0}{\omega} \tag{7.45}$$

Figure 7.4 shows an example of phase and group speeds as a function of frequency calculated with equations (7.43) and (7.44) using the values $D = 100$ m and $c_0 = 1500$ m/s. The phase speed is always greater than both the sound speed in water and the group speed. As the frequency decreases, there is a frequency for each mode at which the phase speed approaches infinity and the group speed goes to zero. This frequency is known as the cutoff frequency $f_{0,m}$ for mode number $m$. It is found by using equation (7.30) and inserting $k_m = 0$, which gives $\omega_{0,m} = c_0\gamma_m$.

The cutoff frequency corresponds to the conditions that the mode propagation angle is equal to $\pi/2$ and the horizontal wave number $k = 0$. From equation (7.38), we then obtain the cutoff frequency for mode $m$

$$f_{0,m} = \frac{c_0}{4D}(2m-1) \tag{7.46}$$

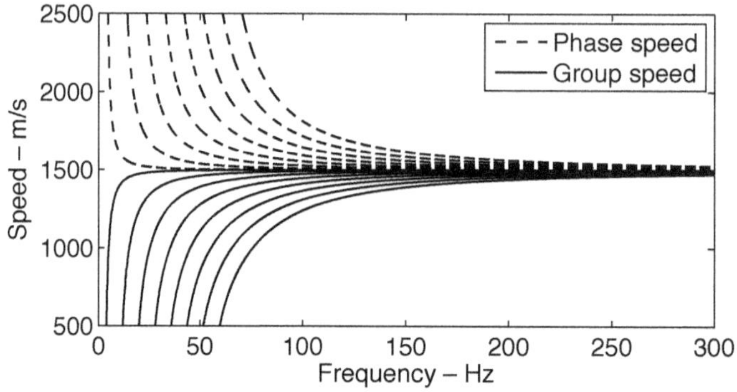


Figure 7.4. Phase and group speeds as a function of frequency for the first 8 modes of an ideal waveguide.

In the example shown in figure 7.4, the first mode has a cutoff frequency of 3.75 Hz and mode 7 has a cutoff frequency of 48.75 Hz, as given by equation (7.46). Frequencies lower than a given mode's cutoff frequency will not be propagated by that mode. At 50 Hz and 100 Hz, equation (7.41) shows that there are 7 and 13 modes, respectively, propagating in this waveguide with a 100 m water depth.

Figure 7.5 shows the amplitudes of the first four mode functions, or eigenfunctions, for the case in which $c_0$ = 1500 m/s, and the water depth is $D$ = 100 m.

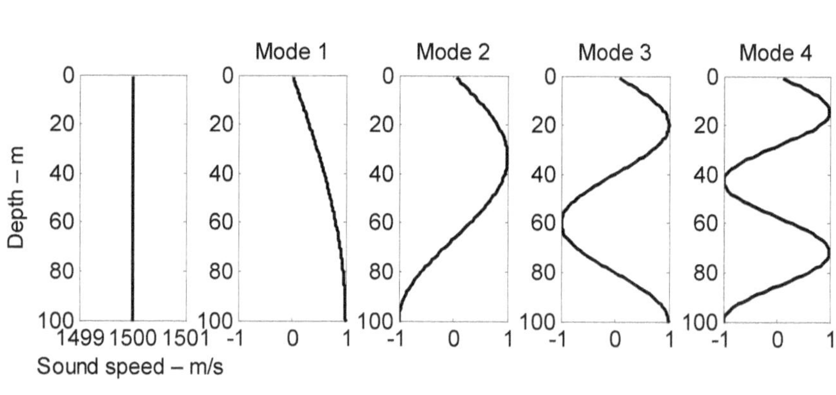


Figure 7.5. The first four mode functions for an ideal waveguide that has a completely reflecting hard bottom and a pressure-release surface.

From this figure and equation (7.36) we note that the mode functions are sinusoidal functions starting with a value of zero at the surface and having zero gradient at the bottom with $m-1$ zero crossings over the depth of the waveguide. Thus, as expected, the boundary conditions are satisfied with $\xi(z)=0$ and $d\xi(z)/dz=0$ at $z=0$ and $z=D$, respectively.

### *7.3.2 Pekeris waveguide*

We now treat a more realistic case with the seabed having finite sound speed and density. We assume constant sound speeds and densities in the water and seabed, with the values $c_0$ and $\rho_0$ for water, and $c_1$ and $\rho_1$ for the seabed. This is a classical case first solved by Pekeris (1948). The case is simple enough to be solved analytically, but is sufficiently realistic to give an understanding of what may happen in more complicated cases. A basic assumption is that the sound speed $c_1$ in the bottom is higher than the speed in water $c_0$. This requirement ensures the existence of a critical angle and total reflection for angles less than critical, which is a condition for propagating modes.

In principle, we can apply the same procedure as used for the ideal waveguide and write the solution as a sum of modal functions. However there is an important difference when the boundaries are not completely reflecting as they were in the previous case. The fact that some of the acoustic energy is transmitted (and lost) through a boundary means that the expansion of equation (7.10) is no longer a closed set. Nevertheless, the results are still useful for most cases of practical applications, remembering that the expansion in normal modes is not an exact solution.

Proceeding therefore as we did for the ideal fluid waveguide we find that the characteristic equation, equation (7.34), is still valid, but now with the bottom reflection coefficient

$$R_b = \frac{\rho_1\gamma_0 - \rho_0\gamma_1}{\rho_1\gamma_0 + \rho_0\gamma_1} \tag{7.47}$$

For the sea surface, we continue assuming a totally reflecting surface where $R_s = -1$.

As explained earlier, $|R_b|^2$ represents that part of the incident wave intensity reflected back into the water, while $(1 - |R_b|^2)$ represents the part being transmitted into the seabed. For the sound signal in water to be transmitted with relatively small losses, it follows that the reflection coefficient must be close to unity or, ideally, $|R_b|^2 = 1$. This implies that the dominant contributions to the sound field at long ranges are the modes with propagation angles less than the critical angle, since for these angles there is total reflection. This presupposes that there is a critical angle and therefore requires the sound speed in the bottom to be higher than the sound speed in the water, as we have noted before.

The vertical wave number in the seabed is

$$\gamma_1 = \sqrt{\left(\frac{\omega}{c_1}\right)^2 - k^2} \tag{7.48}$$

and using $k = (\omega / c_0)\cos\theta_0$, this becomes

$$\gamma_1 = \frac{\omega}{c_0}\sqrt{\left(\frac{c_0}{c_1}\right)^2 - \cos^2\theta_0} \tag{7.49}$$

The critical angle $\theta_{crit}$ occurs when $\gamma_1 = 0$, and thus can be given by $\cos\theta_{crit} = c_0 / c_1$, as long as $c_1 > c_0$. For grazing angles $\theta_0$ below the critical angle, $\gamma_1$ is imaginary and is given by

$$\gamma_1 = +i\frac{\omega}{c_0}\sqrt{\cos^2\theta_0 - \left(\frac{c_0}{c_1}\right)^2} \tag{7.50}$$

where the + sign is required to represent a decreasing amplitude since we use the time variation of $\exp(-i\omega t)$. See section 2.3 and equations (2.34) and (2.35).

For angles less than the critical angle, $|R_b| = 1$, and reflection only produces a phase shift since

$$R_b = \exp\left(-2i\arctan\left|\frac{\rho_0\gamma_1}{\rho_1\gamma_0}\right|\right) \tag{7.51}$$

The characteristic equation becomes

$$\exp(2i\gamma_0 D)\exp\left(-2i\arctan\left|\frac{\rho_0\gamma_1}{\rho_1\gamma_0}\right|\right) = -1 \tag{7.52}$$

which has a number of discrete solutions $m$ that are given by

$$\arctan\left|\frac{\rho_0\gamma_1}{\rho_1\gamma_0}\right| = \left(m - \frac{1}{2}\right)\pi + \gamma_0 D \tag{7.53}$$

We substitute for the wave numbers using

$$\begin{aligned} \gamma_1 &= i\frac{\omega}{v_m}\sqrt{1 - \left(\frac{v_m}{c_1}\right)^2} \\ \gamma_0 &= \frac{\omega}{v_m}\sqrt{\left(\frac{v_m}{c_0}\right)^2 - 1} \end{aligned} \tag{7.54}$$

and then equation (7.53) can be put in a form that is easy to solve numerically

$$\tan\left[\left(m - \frac{1}{2}\right)\pi + \frac{\omega D}{v_m}\sqrt{\left(\frac{v_m}{c_0}\right)^2 - 1}\right] = \frac{\rho_0\sqrt{1 - \left(\frac{v_m}{c_1}\right)^2}}{\rho_1\sqrt{\left(\frac{v_m}{c_0}\right)^2 - 1}} \tag{7.55}$$

Equation (7.55) is the characteristic equation for the eigenvalues and defines a relationship between the frequency and the phase speed for mode $m$ that is known as a dispersion relationship. We see that a necessary condition for a real solution is that

$$c_0 \le v_m \le c_1 \tag{7.56}$$

This implies that the phase speed will have a value located between the sound speed in the seabed and that in water. Equation (7.56) also requires that the sound speeds in the water and the seabed must satisfy $c_1 > c_0$, as established earlier.

The number of modes is determined by the fact that the angles

$$\theta_m = \cos^{-1}\left(\frac{k_m}{\kappa}\right) \tag{7.57}$$

must be less than the critical angle, that is, $\theta_m < \theta_{crit}$. The approximate values of the angles $\theta_m$ are given by equation (7.45), and by using equation (7.38), we find that the number of modes is approximately

$$M \leq \frac{2Df}{c_0}\sqrt{1-\left(\frac{c_0}{c_1}\right)^2} + 0.5 \tag{7.58}$$

which reduces to equation (7.41) when $c_1 \to \infty$. The lowest mode has the cutoff frequency

$$f_1 = \frac{c_0}{4D\sqrt{1-\left(\frac{c_0}{c_1}\right)^2}} \tag{7.59}$$

Equation (7.53) can easily be solved numerically. For a case in which the water depth is 100 m and the parameters are those given in table 7.1, we first find the phase speeds by solving equation (7.55) numerically for $\omega$ as a function of $v_m$. Subsequently, the group speeds can be found by numerical differentiation, using equation (7.44). For this case, figure 7.6 shows plots of the phase and group speeds as functions of the frequency. For a given frequency, there are only a limited number of modes that can propagate. For the relevant parameters, equation (7.58) gives $M$ as six modes at 100 Hz and three modes at 50 Hz.

Table 7.1
Parameters used to illustrated the properties of the Pekeris waveguide.

| | Sound speed | Density |
|---|---|---|
| Water | $c_0$ = 1500 m/s | $\rho_0$ = 1000 kg/m$^3$ |
| Bottom | $c_1$ = 1700 m/s | $\rho_1$ = 1500 kg/m$^3$ |

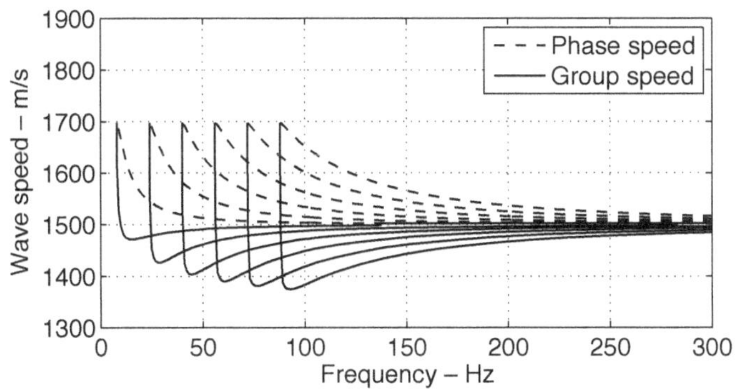


Figure 7.6. Phase and group speeds plotted as functions of frequency for a Pekeris waveguide defined by the parameters of table 7.1.

Comparing figure 7.6 and figure 7.4, we note that there are significant differences between the Pekeris waveguide and the ideal waveguide. In the Pekeris waveguide, we see that the phase speed has an upper limit equal to the sound speed in the seabed, and that the phase speed is higher than the sound speed in the water.

We use the results of figure 7.6, which shows the solution of equation (7.55), to extract the values for the phase speeds to be used in the subsequent discussions as an example. At 50 Hz we find that the three modes have phase speeds of $v_1 = 1513.0$ m/s, $v_2 = 1554.5$ m/s, and $v_3 = 1632.5$ m/s. The modal propagation angles, calculated with the equation (7.43), are found to be 7.5°, 15.2°, and 23.2°. The corresponding vertical wave numbers are (in m$^{-1}$) $\gamma_1 = 0.0274$, $\gamma_2 = 0.055$, $\gamma_3 = 0.0827$. The mode functions are given by

$$\zeta_m(z) = \sin(\gamma_m z) \tag{7.60}$$

These are the only modes at the frequency of 50 Hz. The mode functions are shown in figure 7.7 and should be compared with the mode functions for the ideal waveguide shown in figure 7.5. Also for a Pekeris waveguide the mode functions are sinusoidal functions starting with a value of zero at the surface, but now the gradient at the bottom is not equal to zero as was the case of the ideal waveguide.

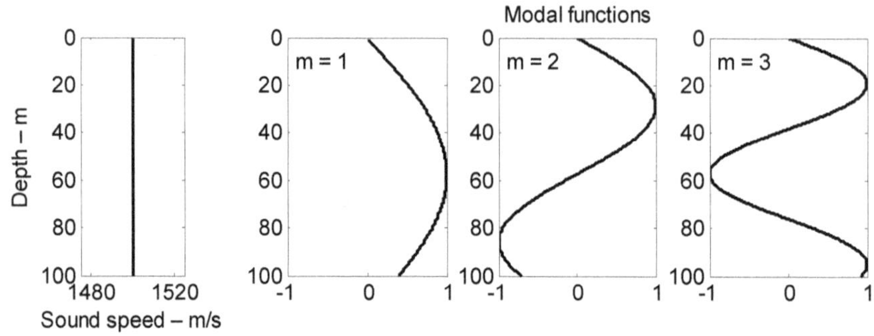


Figure 7.7. The first three modes of a Pekeris waveguide with parameters listed in table 7.1.

For the same example, figures 7.8 and 7.9 show transmission loss *TL* for the frequency of 50 Hz as function of range. Both the coherent and the incoherent *TL* curves are depicted having been derived from equations (7.23) and (7.24), respectively.

In figure 7.8 the source and receivers are at a depth of 50 m, and in figure 7.9 they are at 25 m. In both cases there are significant variations in the transmission loss with range and the variations are almost periodic with different periodicity in the two cases.

The variations and periodicities are examples of mode interference caused by two or more modes that are in phase at some ranges and are out of phase at other ranges. This can be understood by going back to equations (7.21) or (7.23) where the individual terms in the sum are weighted by phases given by the exponents $\exp(ik_m r)$. The coherent sum of these terms gives rise to modal interference.

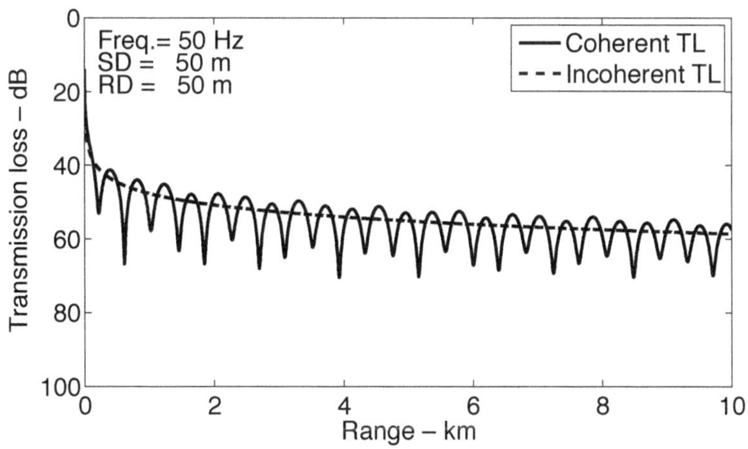


Figure 7.8. Transmission loss plotted as a function of distance for a frequency of 50 Hz. Source and receiver are both at a depth of 50 m. The waveguide parameters are listed in table 7.1.

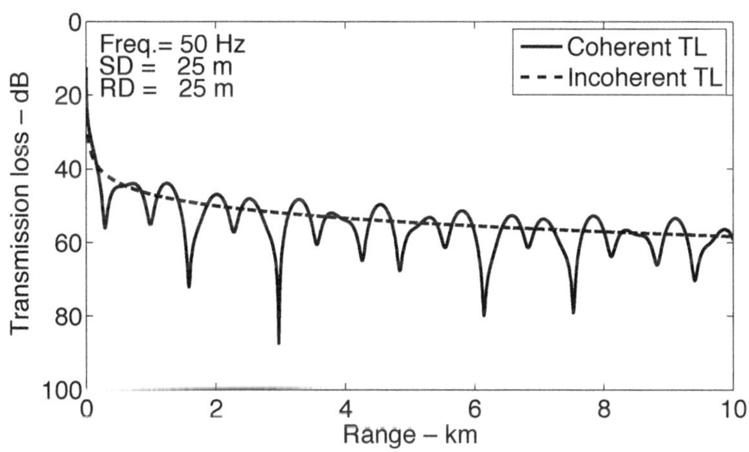


Figure 7.9. Transmission loss plotted as a function of distance for a frequency of 50 Hz. Source and receiver are both at a depth of 25 m. The waveguide parameters are listed in table 7.1.

The horizontal wave numbers are given by

$$k_m = \frac{\omega}{c_0}\sqrt{1-\left(\frac{c_0\gamma_m}{\omega}\right)^2} \tag{7.61}$$

When $\gamma_m << \omega / c_0$, the horizontal wave numbers are approximately

$$k_m \approx \frac{\omega}{c_0}\left[1-\frac{1}{2}\left(\frac{c_0\gamma_m}{\omega}\right)^2\right] \tag{7.62}$$

The difference in horizontal wave number between two modes, $m$ and $n$, is therefore

$$k_n - k_m = \frac{c_0}{2\omega}\left(\gamma_m^2 - \gamma_n^2\right) \tag{7.63}$$

The modes are in phase and add constructively at ranges where the phase difference is an even multiple of $2\pi$. The period of the interference $\Lambda_{nm}$ between the modes $n$ and $m$ is therefore

$$\Lambda_{nm} = \frac{2\pi}{\frac{c_0}{2\omega}\left(\gamma_m^2 - \gamma_n^2\right)} \tag{7.64}$$

With the values for $\gamma_1$, $\gamma_2$, and $\gamma_3$, which we have used before, we obtain $\Lambda_{13} = 432$ m, $\Lambda_{12} = 1157$ m, and $\Lambda_{23} = 690$ m.

With source and receiver at 50 m depth, figure 7.7 shows that mode 2 is less excited because its mode function has a low value at this depth. As a consequence, modes 1 and 3 dominate the interference pattern with a periodicity of $\Lambda_{13} \approx 432$ m as is evident in figure 7.8. With source and receiver at 25 m, all three modes are exited, resulting in a more complicated interference patterns with all three periods represented, as can be seen in figure 7.9.

Figures 7.8 and 7.9 also show that the incoherent transmission loss can be interpreted as a smoothed or averaged transmission loss, which may be useful when the details of the mode interference predicted by the coherent summation are not required or useful.

### *7.3.3 Propagation with constant sound speed gradient*

For a few particular sound speed profile cases, the solution of the depth equation can be expressed by known functions. One of these cases is the constant sound speed profile. Another case occurs when the sound speed as a function of depth is linear in $c^{-2}(z)$; this is expressed as

$$c^{-2}(z) = c_0^{-2} + az \tag{7.65}$$

where $c_0$ is the sound speed at the depth $z = 0$. In a layer between the two depths $z_1$ and $z_2$, with speeds $c_1$ and $c_2$, respectively, the gradient $a$ from equation (7.65) is related to the sound speeds and the layer thickness $d = z_2 - z_1$ by

$$a = \frac{1}{d}\left(\frac{1}{c_2^2} - \frac{1}{c_1^2}\right) \tag{7.66}$$

If realistic values are used for $a$ or if $d$ is small enough, the difference between using equation (7.65) and a constant gradient in sound speed is insignificant.

Given the sound speed from equation (7.65), the solution of the depth equation can be expressed by functions known as Airy functions, Ai(ζ) and Bi(ζ) (Abramowitz and Stegun 1965), with the argument

$$\zeta = \omega^2 a\left(k^2 - \frac{\omega^2}{c^2(z)}\right) \tag{7.67}$$

In figure 7.10, we see that both functions oscillate for negative values of $\zeta$ and exponentially decrease or increase for positive values of $\zeta$.

Ai($\zeta$) and Bi($\zeta$) are two independent solutions of the depth equation; therefore, the complete solution to the depth equation can be expressed as a linear combination of these two functions

$$\xi(z) = C_1\mathrm{Ai}(\zeta) + C_2\mathrm{Bi}(\zeta) \tag{7.68}$$

The coefficients $C_1$ and $C_2$ are determined by the boundary conditions.

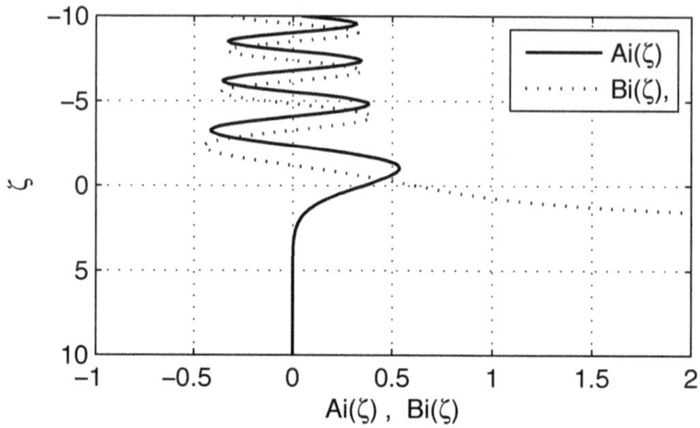


Figure 7.10. Airy functions Ai($\zeta$) and Bi($\zeta$), plotted as functions of $\zeta$.

Next we consider an example with water depth of 500 m, and sound speed defined by equation (7.65) with $c_1$ = 1475 m/s at the surface and $c_2$ = 1525 m/s at the 500-m depth. The reflection coefficients are assumed to be $R_s = -1$ at the surface and $R_b = 1$ at the bottom.

Figure 7.11 shows the mode functions for the first six modes occurring at a frequency of 100 Hz. The ray paths for the same sound speed profile are shown in figure 7.12. The ray picture, figure 7.12, shows shallow angle rays that are trapped near the sea surface.

The mode picture of figure 7.11 shows that these rays correspond to low-order modes, since the mode functions for these modes are concentrated near the sea surface. As with the ray picture, the higher order modes that correspond to larger propagation angles have significant amplitude deeper in the water. We see that the modes are oscillatory in the channel nearest to the surface, with the number of cycles increasing with the mode number, and that the amplitudes of the mode functions decrease monotonically at greater depths. The reader is referred to the papers by Tindle et al. (1978) for a more thorough discussion of this issue.

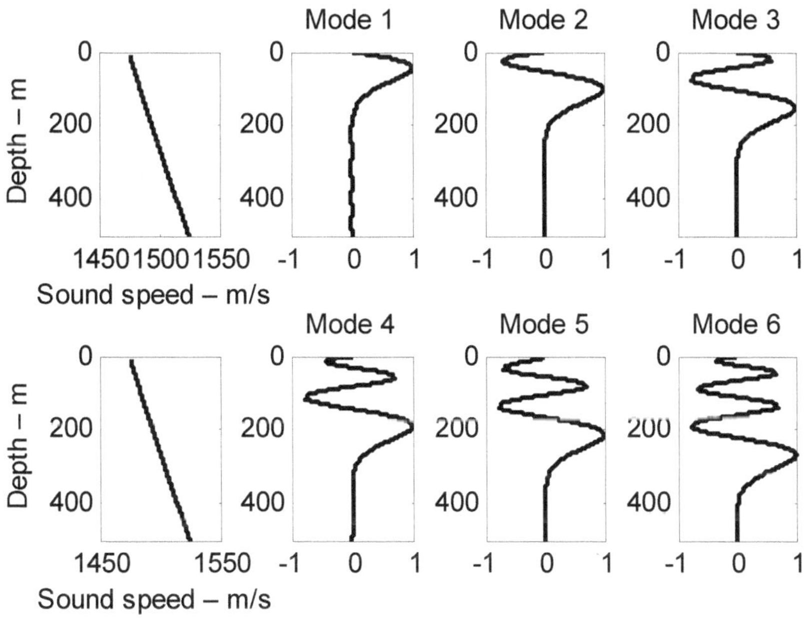


Figure 7.11. The first six mode functions for propagation in an oceanic waveguide where the sound speed is increasing with depth according to equation (7.67).

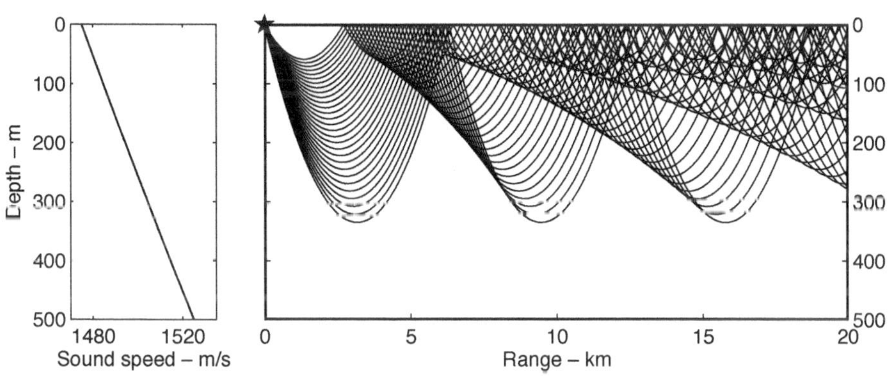


Figure 7.12. Ray paths for propagation with constant sound speed gradient. Starting angles range from 5°–12°.

In principle, there are two different solutions, depending on whether $\gamma^2(z)$ is positive or negative. If $\gamma^2(z)$ is positive, the solution becomes oscillatory and, especially when $\gamma^2(z)$ is a positive constant $\gamma^2_0$, the solution will become a combination of $\cos(\gamma_0 z)$ and $\sin(\gamma_0 z)$, as we saw earlier. If, on the other hand, $\gamma^2(z)$ is negative, $\gamma(z)$ becomes imaginary, with solutions decreasing monotonically with depth. When $\gamma^2(z)$ is a negative constant, the solution is a combination of exponential functions. The conditions for the two different solutions can be found from the expression for the vertical wave number

$$\gamma^2(z) = \frac{\omega^2}{c^2(z)} - k^2 \tag{7.69}$$

For a ray starting with a given angle $\theta_0$ at a given depth $z_0$, it follows from Snell's law that

$$k = \frac{\omega}{c(z)}\cos\theta(z) = \frac{\omega}{c_0}\cos\theta_0 \tag{7.70}$$

Inserting equation (7.70) into equation (7.69), we get

$$\gamma^2(z) = \left[\frac{\omega}{c(z)}\right]^2 \left[1 - \left(\frac{c(z)}{c_0}\cos\theta_0\right)^2\right] \tag{7.71}$$

Thus, whether $\gamma^2(z)$ is positive or negative and, consequently, whether a solution for a given depth is oscillatory or monotonically increasing or decreasing depends upon the sound speed at the depth. If

$$c(z) < \frac{c_0}{\cos\theta_0} \tag{7.72}$$

the solution is oscillatory. However, if the condition of equation (7.72) is not true, the solution is monotonically decreasing with depth. At the depth where $c(z) = c_0 / \cos(\theta_0)$, there is a turning point at which the ray is horizontal; therefore that depth indicates a transition from the oscillatory to the monotonically decaying mode functions. For a given mode characterized by the angle $\theta_0$ and the sound speed $c_0$, the conditions in equation (7.72) along with the sound speed profile determine which depth interval that gives either an oscillatory or a monotonic decreasing solution to the depth equation. The lowest modes have the smallest angles and are closest to the horizontal, and will therefore be oscillatory for the smallest depth interval, as shown in figure 7.11.

## *7.4 The effect of absorption in the seabed*

Energy is always lost in the water and bottom because of absorption, by conversion from the compressional wave in the water to other wave types, such as shear waves, in the bottom, or by scattering from rough surfaces, particularly the sea surface. These losses have so far been ignored in our discussion. For low frequency applications bottom reflection losses are often most important, and in the following we present a simple and approximate calculation for these losses. More thorough treatments of modal attenuation in general can be found in the papers by Koch et al. (1983) and the book by Jensen et al. (1994), page 311.

The standard method to include attenuation is to apply perturbation theory, assuming that the sound speed has a small imaginary component. This approach was used in section 4.3 for calculating the reflection loss of an absorbing bottom, and we apply the same approach here to obtain an estimate of the wave attenuation caused by bottom absorption.

With bottom absorption included, the reflection coefficients become complex and can be written as

$$R_b = |R_b| \exp(-i\phi_b) \tag{7.73}$$

Equation (7.34) with $R_s = -1$ becomes

$$R_b \exp(2i\gamma_m D) = -1 \tag{7.74}$$

When there is no bottom absorption, the reflection coefficient $R_b$ is equal to 1 for incident grazing angles lower than the critical angle. With absorption loss, $|R_b| < 1$ and the horizontal and vertical wave numbers have imaginary components, that is,

$$\begin{aligned} k_m &= k_{r,m} + ik_{i,m} \\ \gamma_m &= \gamma_{r,m} + i\gamma_{i,m} \end{aligned} \tag{7.75}$$

By inserting the second of equations (7.75) into equation (7.74), and splitting the resulting equation into real and imaginary parts we obtain

$$2D\gamma_{r,m} - \phi_b = (2m-1)\pi \tag{7.76}$$

$$\ln|R_b| - 2\gamma_{i,m} D = 0 \tag{7.77}$$

While equation (7.76) is the same result that we found earlier with equation (7.52), equation (7.77) shows that when $|R_b| < 1$, the vertical number acquires an imaginary component

$$\gamma_{i,m} = \ln|R_b|\frac{1}{2D} \tag{7.78}$$

This component results in an imaginary component $k_{i,m}$ of the horizontal wave number

$$k_{i,m} = -\gamma_{i,m} \tan\theta_m = -\ln|R_b|\frac{\tan\theta_m}{2D} \tag{7.79}$$

where $\theta_m$ is the propagation angle of the wave striking the bottom and defined by equation (7.57). We call $k_{i,m}$ the mode attenuation coefficient since it causes a reduction of the amplitude of mode $m$ with range $r$ proportional to $\exp(-k_{i,m}\, r)$. The contribution to the transmission loss in dB is

$$TL_m = 20\log|R_b|\frac{r}{2D}\tan\theta_m \tag{7.80}$$

Equation (7.80) has a simple geometric explanation, which is shown in figure 7.13. If we consider a mode corresponding to the grazing angle $\theta_m$, the distance between each seabed reflection equals $2D / \tan\theta_m$. For each reflection, the wave loses amplitude with a loss in dB equal to $20\log|R_b|$.

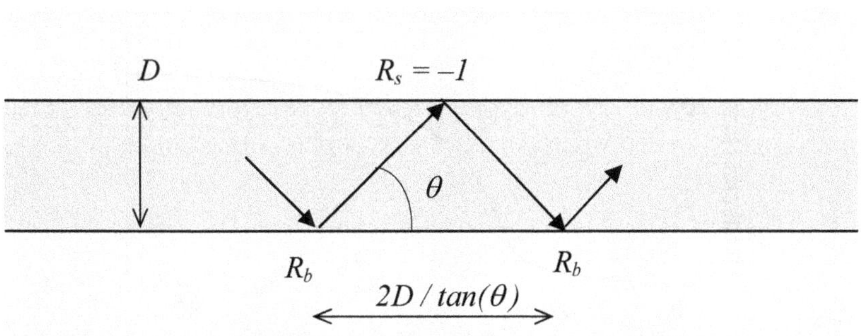


Figure 7.13. Geometric interpretation of the reflection loss at the seabed.

The derivation and the results represented by equations (7.79) and (7.80) are applicable to any type of bottom reflection loss. However, we now concentrate on the effect bottom reflection loss caused by sound absorption in the bottom.

We consider the same Pekeris waveguide as discussed in section 7.3.2 with parameter values given in table 7.1, but now with a bottom absorption coefficient of 1.0 dB per wavelength. By applying the results of section 4.3 we obtain the reflection loss as function of angle shown in figure 7.14.

In section 7.3.2 we found that in this case there are three modes with propagation angles at the bottom of 7.2°, 15.2°, and 23.2°. At these angles figure 7.14 gives the reflection losses of 0.44 dB, 0.80 dB, and 1.32 dB, respectively, for the three modes. The contribution to the transmission loss $TL_m$, defined by equation (7.80), results in an increased loss at the distance of 10 km of 2.8 dB, 10.9 dB, and 28.3 dB, respectively for the three modes.

Figure 7.15 shows the transmission loss as function of range both with and without bottom absorption included. We see that the effect of the seabed reflection loss is quite pronounced—strongly reduced oscillations as function of range, and increased transmission loss of about 10 dB at 10 km.

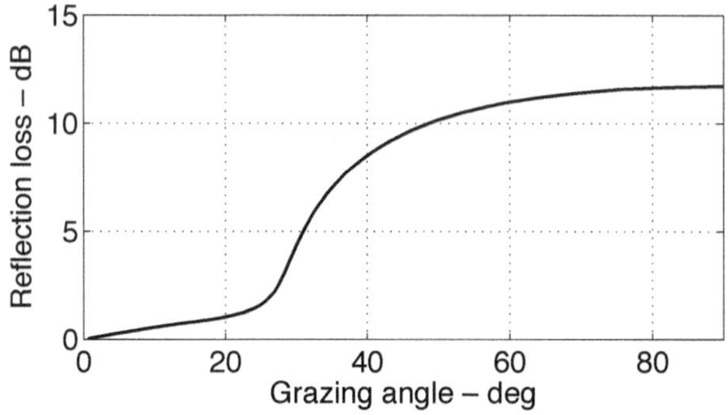


Figure 7.14. Bottom reflection loss in dB as a function of the grazing angle. The parameters are given in table 7.1 and the bottom absorption coefficient is 1.0 dB per wavelength.

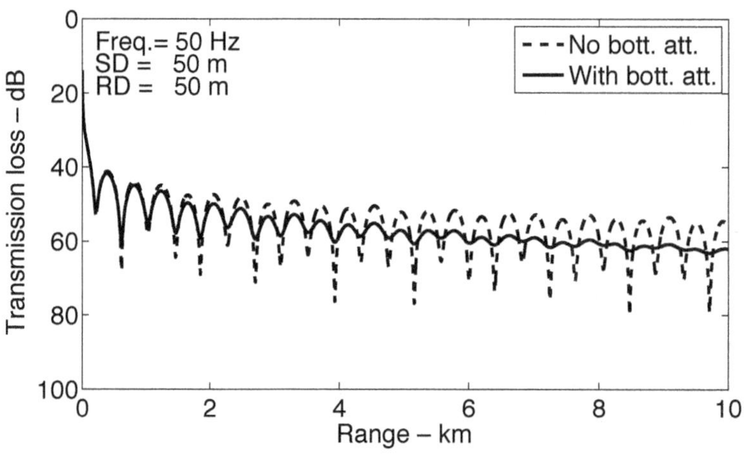


Figure 7.15. Transmission loss as a function of distance at 50 Hz. The solid line was calculated with 1.0 dB / $\lambda$ absorption loss at the seabed; the dashed line was calculated without seabed reflection loss.

From the preceding discussion it is evident that at longer distances modes 2 and 3 are strongly reduced compared with mode 1. The effect of the bottom absorption is therefore to reduce the contributions of the higher modes more than those for the lower modes.

### *7.5 Propagation in deep water*

We consider a deepwater case where the sound speed profile is a Munk profile, given by the formula in equation (6.75) and where $c_0 = 1500$ m/s, $z_0 = 1000$ m, and $\varepsilon = 0.00737$. The sound speed profile is the same as was used in figure 6.12. Figure 7.16 shows the sound speed profile and some of the mode functions. These have been computed with the Kraken normal mode program; see Porter and Reiss (1984) and Porter (1991). In the sound channel, we see that the modes are oscillatory and that the number of nodes increases with mode number. Outside the sound channel the amplitudes of the modes decrease exponentially. Note that the mode function for mode number $n$ has $n - 1$ zero crossings.

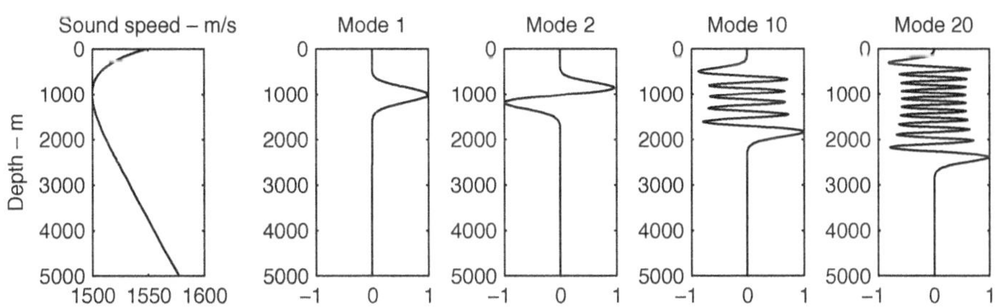


Figure 7.16. Sound speed profile and modes 1, 2, 10, 20, and the frequency of 50 Hz for the deepwater sound field with the Munk sound speed profile.

In section 6.4.2 we considered a similar case with propagation in a symmetric sound channel using ray theory. We found that the rays having the shallowest initial angles are the slowest and that the last arrival is the ray that propagates along the sound channel axis. Since rays with the shallow initial angles may be associated with the lower order modes, it follows that, in this case, the higher modes arrive earlier than the lower order modes.

Figure 7.17 shows the sound field, represented by the transmission loss as function of range for source depth of 1000 m and a frequency of 50 Hz. The figure also shows the ray paths calculated for the same Munk sound speed profile. The similarities between the mode sound field and the ray paths are striking, but there are also important differences. The most noticeable is that the mode solution shows that the sound field diminishes gradually near the turning points, and the field extends deeper in depth.

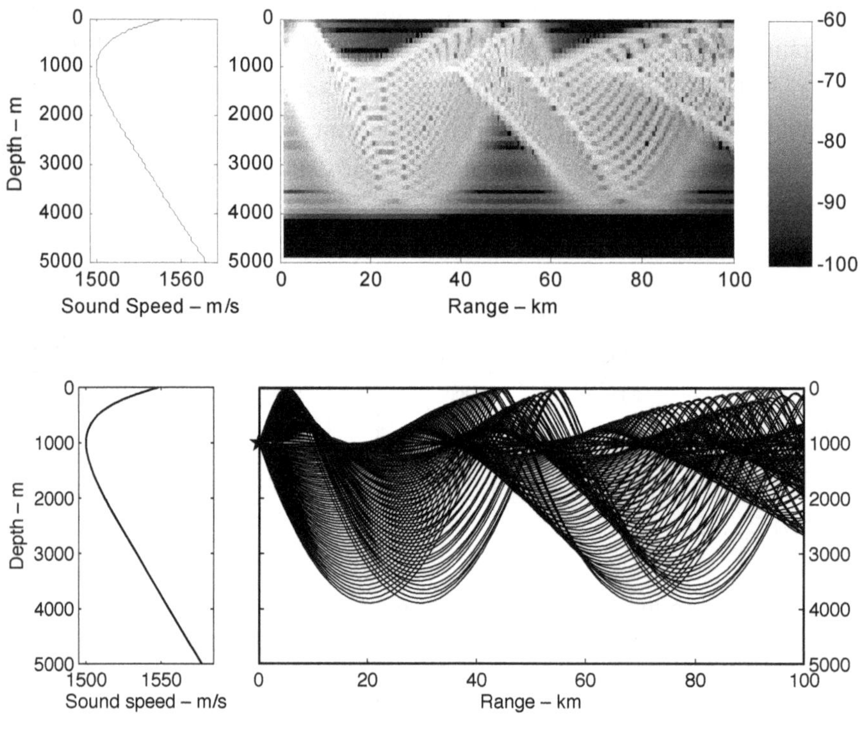


Figure 7.17. Sound propagation in the Munk sound speed profile. Source depth is 1000 m. Top: Contoured transmission loss (dB) as a function of range and depth for a frequency of 50 Hz, calculated by the normal mode method, using 50 modes. Bottom: Ray paths for rays starting between –15° to 15° at 0.25° increments.

## *7.6 Time responses*

Finally, we show some examples of time response in order to demonstrate the effects of frequency dependent wave speed and dispersion. Recall the discussion in section 7.3 where we found that the solution consists of a sum of modes, each propagating with a different wave speed and that the wave speeds are functions of frequency.

We consider the Pekeris waveguide with parameters $c_0 = 1500$ m/s, $c_1 = 1700$ m/s, $\rho_1 / \rho_0 = 1.5$, and $D = 100$ m. The phase and group speeds for these values were calculated earlier and appear in figure 7.6.

We assume that the source signal has a bandwidth of approximately 75 Hz to 125 Hz, and we observe the time response at a distance $r = 20{,}000$ m. Figure 7.18 shows the transmitted signal and the time functions of the four lowest modes as a function of reduced time $t_{red}$, expressed as

$$t_{red} = t - \frac{r}{c_{red}} \tag{7.81}$$

To create figure 7.18, the reduced sound speed $c_{red} = 1500$ m/s is used.

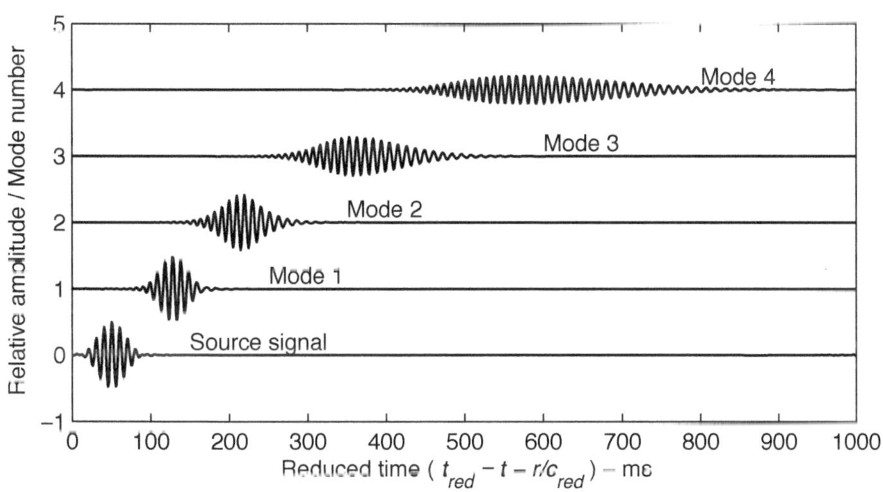


Figure 7.18. The time responses of the first four modes propagating in a Pekeris waveguide.

The various modes arrive at different times, with the lowest mode arriving first and the higher modes later. Note that this arrival structure is different from the deepwater case with the Munk profile (section 7.5) where the higher order modes arrive first and the lower mode last. The dispersion effect is evident in the stretching of the mode signals in time, with the greatest stretching for the highest mode. Notice that the higher frequency components in each mode arrive earlier than the lower frequencies. The differing arrival times and the stretching can be understood from the frequency dependence of group speed, shown in figure 7.6. Although in figure 7.17 the modes are shown separately, in reality we would receive a weighted sum of modes that often overlap in time.

# *Chapter 8*

# *Acoustic transducers*

## 8.1 Introduction

A number of different sound source types are useful in underwater acoustics; the choice of sound source to use depends on the application. For scientific studies of transmission conditions, echo structure, and similar things, explosive charges containing only some hundred grams of TNT are often used (Weston 1960, Chapman 1985). The resulting explosion causes an acoustic shock pulse with a short rise time and wide bandwidth. However, in recent years the use of even small explosive charges has become environmentally unacceptable; for scientific purposes, shattering light bulbs may be an alternative to the use of explosives as acoustic broadband sources. With this method, an ordinary domestic electric light bulb is broken mechanically; the implosion results in a short, acoustic pulse with surprisingly high peak pressure and broad bandwidth (Chapman et al. 1997, Grant et al. 2003).

For seismic investigations and the mapping of seabed stratification, air guns or electric "sparkers" are more commonly used; see Dobrin (1976). These generate pulses that resemble those resulting from explosions, having a sharp initial pulse followed by oscillating bubble pulses.

An air gun contains air compressed at high pressure in a chamber. This air, released by opening a valve, creates a rapidly expanding bubble that results in a very powerful pulse, for which the strength, time signature, and frequency spectrum are determined by the shape and size of the air chamber. The pulse of an air gun usually covers a frequency range of 1–100 Hz, with pulse lengths of 10–100 ms. Air guns are the sound sources in marine seismic exploration but will not work very well in deep water where the ambient pressure becomes comparable to the initial pressure in the air chamber. An alternative combustive source is described by Wilson, Ellzey, and Muir (1994).

The sparker is used in shallow marine seismic surveying, for high resolution within the seabed's upper layers. The sound pulse is generated as a consequence of an electric discharge into the water. The energy being discharged comes from a large electric capacitor. The discharge causes an electric arc, which creates an expanding gas bubble that subsequently generates a strong sound pulse. The sparker releases pulses with durations of 5–20 ms that cover a frequency range of about 50–1000 Hz, depending on the size of the capacitor.

The sources mentioned above are used when there is demand for high sound pressure with relatively low frequencies. They can be controlled or modulated only to a small extent, or not at all. Also, the pulse repetition frequency is relatively low, since it is determined by the time needed to charge the capacitor or to compress the air in the air gun.

For reception of such sources, linear electroacoustic transducers based on piezoelectricity are almost always used. For underwater acoustic applications of sonar and echo sounding, and for navigation and data transmission, piezoelectric transducers are used for both reception and transmission. As transmitters, they are advantageous because they can be modulated and can be constructed for use at high and low frequencies. They are also reciprocal, and can consequently be used for both transmission and reception.

Magnetostrictive transducers are used to a lesser extent, but several new magnetostrictive materials have appeared recently; see the book by Engdahl and Mayergoyz (2000). This has led to a renewed interest in magnetostrictive transducers.

We will, however, limit the following discussion to piezoelectric transducers with general references to Wilson (1988), Stansfield (1991), and Kinsler et al. (2000). Our discussion is similar to the one found in the latter reference.

Generally, an electroacoustic transducer, whether piezoelectric, magnetostrictive, or electrodynamic, can be represented by an electromechanical two-port network with an electrical and a mechanical port, as shown in figure 8.1.

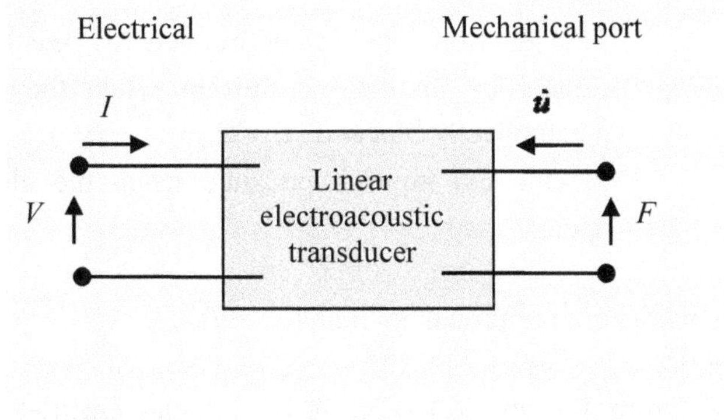


Figure 8.1. Two-port representation of an electroacoustic transducer having both an electrical and a mechanical port.

On the electrical side, there is an electric voltage $V$ and a current $I$, while on the mechanical or acoustic side, there is a force $F$ and a particle velocity $\dot{u}$. Note that in this chapter u is the particle displacement and $\dot{u}$ is the particle velocity. The reason for this is that we are dealing with stress-strain relations in solids and therefore we need a variable for the strain; see also the note on this issue in chapter 2.

If the transducer is linear, the relations between the electrical and mechanical quantities can be expressed by a set of linear equations in the form

$$\begin{aligned} V &= Z_{eh} I + T_{em} \dot{u} \\ F &= T_{me} I + Z_{mo} \dot{u} \end{aligned} \tag{8.1}$$

This is a general representation; in order to derive these equations, we do not need to know anything about the transducer's mode of operation, apart from the fact that it is linear—which means that there is proportionality between current and voltage, as well as between mechanical force and deformation. The coefficients are given their names according to their functions, and are defined as follows:

$Z_{eb} = (V/I)_{\dot{u}=0}$ Electric input impedance when the transducer is mechanically blocked and cannot move ($\dot{u} = 0$)

$T_{em} = (V/\dot{u})_{I=0}$ Transduction coefficient related to electric voltage caused by mechanical displacement when the electric port is open ($I = 0$)

$T_{me} = (F/I)_{\dot{u}=0}$ Transduction coefficient related to mechanical force caused by an electric current when the transducer is mechanically blocked ($\dot{u} = 0$)

$Z_{mo} = (F/\dot{u})_{I=0}$ Mechanical input impedance when the electric port is open ($I = 0$)

Piezoelectric transducers are reciprocal, so that $T_{em} = T_{me}$.

With the help of figure 8.1 and equation (8.1), we can calculate the electrical impedance, seen from the electric side. When the transducer is mechanically unloaded—that is, when $F = 0$—the electrical impedance $Z_i$ derived from equations (8.1) becomes

$$Z_i = \frac{V}{I} = Z_{eb}\left(1 - \frac{T_{me}T_{em}}{Z_{mo}Z_{eb}}\right) = Z_{eb}\left(1 - k^2\right) \tag{8.2}$$

Here, $k^2$ is the electromechanical coupling coefficient, defined as

$$k^2 = \frac{T_{me}T_{em}}{Z_{mo}Z_{eb}} \tag{8.3}$$

The mechanical load impedance $Z_l$ is the ratio between the mechanical force and velocity of the transducer's vibrating surface. When the transducer is mechanically loaded with the mechanical load impedance, which equals $F / \dot{u}$, the electrical input impedance $Z_{in}$ is defined as

$$Z_{in} = \frac{V}{I} = Z_{eb}\left(1 - k^2 \frac{Z_{mo}}{Z_{mo} - Z_l}\right) \tag{8.4}$$

Generally, it is difficult to determine the load impedance, but in the special case of a vibrating surface that is large compared to the wavelength, the load impedance $Z_l$

equals the product of the specific acoustic impedance of the medium outside, here the water, and the transducer's effective radiation area *A*, so that

$$Z_l = \rho_l c_l A \tag{8.5}$$

This is a low frequency approximation; see Kinsler et al. (2000) and Beranek (1954) for more information on how to handle the higher frequency case.

## 8.2 Piezoelectric transducers

In piezoelectric materials, such as piezoelectric crystals or polycrystalline ceramic materials, there is a connection between the mechanical and electric quantities because of the asymmetric lattice structure of the piezoelectric material, in which the centers of the positive and negative domains are displaced. When the material is subjected to an electric field, it will become mechanically deformed—either expanded or contracted, depending on the polarity of the applied voltage. Conversely, a mechanical deformation will create an electric voltage. Note that in an elastic solid, like a piezoelectric material, strain in one direction also results in strain in other directions.

Piezoelectric materials exist naturally, but in practice the materials used most are synthetic ceramic materials, such as barium titanate ($BaTiO_3$) and lead zirconate titanate ($PbZrTiO_3$), also known by the term *PZT*. See Wilson (1988) and Stansfield (1991) for more information about piezoelectric materials and their use in transducers. Piezoelectric elements can be obtained in many different geometrical shapes, including spherical shells, cylindrical tubes, rods, and rectangular or circular plates.

Figure 8.2 shows two frequently used types of ceramic elements. The left side of the figure shows a solid piezoelectric disk with a conductive metal layer attached to both end surfaces. An element of this type is usually used in the ***thickness mode***, which means exploiting the movement in an axial direction. When electric voltage is applied to the element, the disk will be deformed by becoming thicker or thinner, depending on the polarity of the electric field. Correspondingly, a mechanical force will generate an electric voltage between the two electrodes. The direction of the electric field is commonly referred to as the 3-axis, or axis number 3, which in this case is the axis normal to the end surfaces. When exploiting the thickness deformation, the element is said to be working in a ***33-mode***.

One very common piezoelectric transducer is a thin-walled ceramic ring or tube with electrodes, that is, a conductive metal coating, on the inside and outside (see figure 8.2, right). A voltage on the electrodes will cause an expansion or contraction of the ring's radius, thereby generating an acoustic wave in the surrounding medium. The reverse is also true: an external mechanical force will change the radius and produce an electric voltage between the electrodes, which can be used for reception of an acoustic wave. In this case, the mechanical deformation is in the tangential direction and normal to the direction of the electric field; the element is working in a ***transversal mode***, or a ***31-mode***.

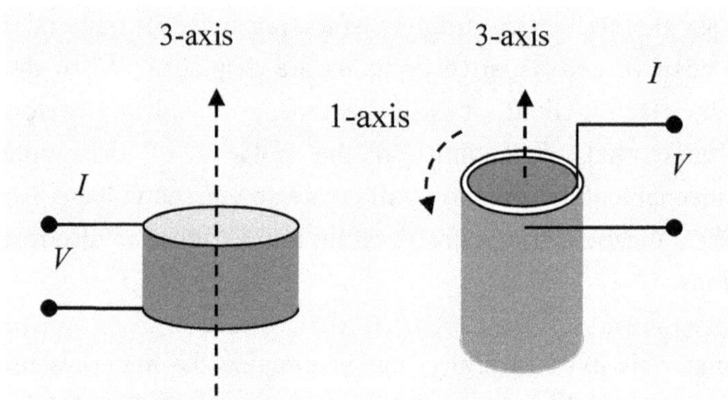


Figure 8.2. Examples of transducer elements: left, a disk, and right, a cylindrical tube.

The two types of transducers shown in figure 8.2 are among those used most in underwater acoustics, and therefore we will analyze and discuss their electrical and mechanical characteristics in some detail. The discussion is mainly aimed at users of transducers, since the primary objective is to describe these transducers' principles of operation and main characteristics.

## 8.3 The piezoelectric ring transducer

The ring can be used both as a transmitter and as a receiver, but is most often used as a hydrophone, which is a receiving element. The useful frequency range of a ring hydrophone is usually limited to frequencies where the wavelength of the acoustic wave is larger than the diameter of the ring. For such frequencies, the whole ring is exposed to the same sound pressure. Usually, the inner part of the ring is insulated so that the sound pressure here is zero.

The intention behind this analysis is to develop a set of equations similar to equations (8.1), but now also including the dimensions of the transducers in the calculations. The mechanical and electrical properties of the piezoelectric material are provided by the manufacturers. For a transducer operating in the transversal 31-mode, a common set of coefficients for the piezoelectric material is defined by the following equations:

$$\begin{aligned} D_3 &= d_{31}T_1 + \varepsilon_{33}^{T}E_3 \\ S_1 &= \frac{1}{Y_{11}^{E}}T_1 + d_{31}E_3 \end{aligned} \tag{8.6}$$

where $Y_{11}^{E}$ is a coefficient of elasticity that relates the strain $S_1$, in the tangential 1-direction, to the mechanical stress $T_1$ when the element is short-circuited and $E_3 = 0$. The electric displacement is $D_3$ and $E_3$ is the electrical field strength in the 3-direction. When the element is not mechanically loaded and $T_1 = 0$, the dielectric constant of the element is $\varepsilon_{33}^{T}$. The coefficient $d_{31}$ connects the electric and mechanical quantities.

Equations (8.6) give the relations between the electric and mechanical variables of the piezoelectric material itself. The notation may seem complicated, but is in accordance with common practice. The indexes 3 and 1 relate to the polarization being in the 3-direction and the movement being in the 1-direction. In equations (8.6), we find that the coefficients have the SI units:

$$\left[d_{31}\right] = \frac{m}{V}$$
$$\left[Y_{11}^{E}\right] = \frac{N}{m^2} \qquad (8.7)$$
$$\left[\varepsilon_{33}^{T}\right] = \frac{F}{m}$$

The electromechanical coupling coefficient $k_{31}$ is determined by $d_{31}$ and normalized by the electric and mechanical coefficients. For an element in the 31-mode, the electromechanical coupling coefficient is expressed as

$$k_{31} = d_{31}\sqrt{\frac{Y_{11}^{E}}{\varepsilon_{33}^{T}}} \qquad (8.8)$$

We will now determine the electrical and acoustical characteristics of a ring transducer, which appears in figure 8.3; it is a thin-walled piezoelectric ring with a radius *a,* height *h,* and wall thickness *t.*

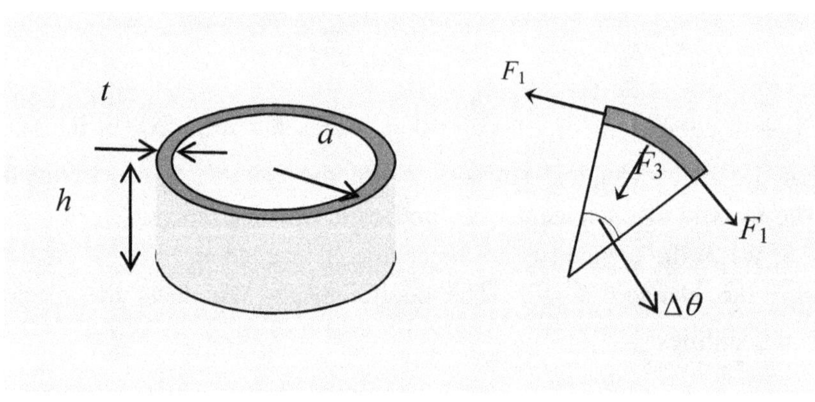


Figure 8.3. A piezoelectric ring transducer with radial polarization. Left, the transducer dimensions: height *h*, radius *a*, and wall thickness *t.* Right, a small segment of the ring, with the mechanical forces indicated.

To begin, first we must find the connection between the mechanical stress $T_1$ and the strain $S_1$ in the circumferential direction. With a small radial expansion $u$ of the radius $a$, the relative expansion of the circumference becomes

$$S_1 = \frac{2\pi(a+u) - 2\pi a}{2\pi a} = \frac{u}{a} \tag{8.9}$$

We now consider the small ring segment spanning the angle $\Delta\theta$ in figure 8.3. A tangential stress $T_1$ in the wall gives a tangential force $F_1 = htT_1$ and a radial force

$$F_3 = -\Delta\theta htT_1 \tag{8.10}$$

Newton's second law then gives us

$$a \cdot \Delta\theta\rho ht\ddot{u} = -\Delta\theta htT_1 \tag{8.11}$$

where $\rho$ is the density of the piezoelectric material, and $\ddot{u}$ is the particle acceleration, given as the second derivative of the particle displacement with respect to time, $\ddot{u} = d^2u/dt^2$. The particle velocity is $\dot{u} = du/dt$. Note that in this chapter we will use the sign convention of electrical circuit theory, which has a time dependence of $\exp(i\omega t)$ and is the opposite of the sign convention used in the other chapters of this book.

Initially, we assumed the segment and therefore $\Delta\theta$ to be very small, but after examining equation (8.11), we see that there is nothing to prevent letting the angle $\Delta\theta$ span the complete circumference of the ring. Therefore we might as well use $\Delta\theta = 2\pi$ in equation (8.11). Furthermore, we will also account for the loss of energy caused by friction associated with mechanical vibrations. To do this, we assume that the energy loss caused by the vibration is proportional to velocity, and that the equation describing the dynamics of the ring will therefore be

$$M\ddot{u} + R_m\dot{u} + 2\pi htT_1 = F \tag{8.12}$$

Here, $R_m$ represents the energy loss mentioned before, $F$ is the external force, and $M$ is the mass of the ring

$$M = 2\pi\rho aht \tag{8.13}$$

So far we have only considered the mechanical effects of the vibrations within the ring; we now include effects of the electrical field in our discussion. From equation (8.6), we have

$$T_1 = Y_{11}^E \left(S_1 - d_{31}E_3\right) \tag{8.14}$$

which, inserted into equation (8.12), gives us

$$M\ddot{u} + R_m\dot{u} + \frac{2\pi ht}{a}Y_{11}^E u = 2\pi htd_{31}Y_{11}^E E_3 + F \tag{8.15}$$

Now equation (8.15) can be written

$$\left(i\omega M + R_m + \frac{K}{i\omega}\right)\dot{u} = 2\pi hd_{31}Y_{11}^E V + F \tag{8.16}$$

where $K$ is the mechanical stiffness of the ring. The voltage across the electrodes is defined by $V = E_3 t$. By comparing equations (8.15) and (8.16), the mechanical stiffness $K$ is found to be

$$K = 2\pi \frac{ht}{a} Y_{11}^E \tag{8.17}$$

$$\begin{aligned} D_3 &= d_{31}Y_{11}^E S_1 + \left(\varepsilon_{33}^T - d_{31}^2 Y_{11}^E\right)E_3 \\ &= d_{31}Y_{11}^E \frac{u}{a} + \varepsilon_{33}^T\left(1 - k_{31}^2\right)E_3 \end{aligned} \tag{8.18}$$

Here we have replaced $S_1$ in accordance with equation (8.9) and introduced the electromechanical coupling coefficient $k_{31}$ from equation (8.8).

The total electric charge $Q$ on the electrodes is $2\pi ahD_3$, and the current $I$ is the time derivative of the electric charge, $I = \partial Q / \partial t$. When we also include the voltage over the electrodes, which is $V = E_3 t$, we find that the equation that describes the current is now

$$I = i\omega \frac{2\pi ah}{t}\varepsilon_{33}^T\left(1 - k_{31}^2\right)V + 2\pi hd_{31}Y_{11}^E \dot{u} \tag{8.19}$$

Together, equations (8.16) and (8.19) describe the piezoelectric ring completely—that is, both electrically and mechanically. These equations can be represented by, or visualized as, the electromechanical equivalent circuit shown in figure 8.4.

On the electric side, we have the current $I$ and the voltage $V$, while on the mechanical side, we have force $F$ and the velocity $\dot{u}$. In equation (8.19) we can identify a capacitor with capacitance $C_p$ given as

$$C_p = \frac{2\pi ah}{t}\varepsilon_{33}^T\left(1 - k_{31}^2\right) \tag{8.20}$$

The capacitance $C_p$ is measured at the electrical terminal when the ring is mechanically blocked, that is, when $\dot{u} = 0$.

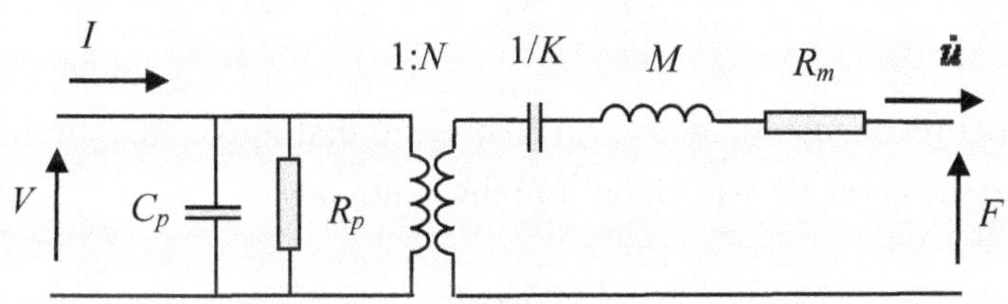


Figure 8.4. The electromechanical equivalent circuit of a piezoelectric transducer.

The "transformation factor" $N$ of the electromechanical transformer in figure 8.4 is derived from equations (8.16) and (8.19), and is found to be

$$N = 2\pi h d_{31} Y_{11}^F \tag{8.21}$$

In figure 8.4 we have inserted a leakage resistance $R_p$ in parallel to $C_p$ that is not included in equation (8.19). It represents the dielectric loss in the piezoelectric ceramic. The mechanical loss is modeled by $R_m$, as observed earlier in the discussion of equation (8.11).

On the mechanical side, there is a series coupling of a mass $M$ and a stiffness $K$ that together give a resonance frequency $f_0$

$$f_0 = \frac{1}{2\pi}\sqrt{\frac{K}{M}} = \frac{1}{2\pi a}\sqrt{\frac{Y_{11}^E}{\rho}} \tag{8.22}$$

Since the sound speed $c_0$ of the piezoelectric ring is

$$c_0 = \sqrt{\frac{Y_{11}^E}{\rho}} \tag{8.23}$$

the resonance frequency can be expressed as

$$f_0 = \frac{c_0}{2\pi a} \tag{8.24}$$

Thus, mechanical resonance occurs for a frequency that makes the circumference of the ring equal to the acoustic wavelength in the ring, $\lambda_0$

$$\lambda_0 = 2\pi a \tag{8.25}$$

Equations (8.16) and (8.19) can be converted into the form of equation (8.1)

$$\begin{aligned} V &= Z_e I - N Z_e \dot{u} \\ F &= -N Z_e I + \left(Z_m + N^2 Z_e\right)\dot{u} \end{aligned} \tag{8.26}$$

where the electrical impedance is

$$Z_e = \frac{1}{i\omega C_p + \dfrac{1}{R_p}} \tag{8.27}$$

and the mechanical impedance is

$$Z_m = i\omega M + \frac{K}{i\omega} + R_m \tag{8.28}$$

The equivalent circuit in figure 8.4, which is described by equations (8.26), describes the transducer both acoustically and electrically.

A piezoelectric ring of this type is commonly used as a hydrophone or as a sound receiver, and we need to know how large an electric voltage we get on the electrical terminals for a given sound pressure. The receiver sensitivity is defined by the sound pressure $p$ on the transducer and the voltage $V$ over the electrical terminal when the current $I = 0$. The force $F$ equals the sound pressure $p$ multiplied by the area $A$ of the transducer's surface exposed to the sound pressure. This is valid approximation for low frequency applications of this hydrophone; see Kinsler et al. (2000) on how to handle the higher frequency case. If we assume that the inner part of the ring is screened off from sound pressure, it follows that $A = 2\pi ah$. From equations (8.26) we now find the receiver sensitivity to be

$$M_V = \left(-\frac{V}{p}\right)_{I=0} = \frac{A}{N}\frac{Z_e}{Z_e + \frac{Z_m}{N^2}} \tag{8.29}$$

The ring transducer is mainly used as a hydrophone for frequencies much lower than the resonance frequency. From equations (8.27) and (8.28) we find that, at low frequencies, $Z_e \approx 1/(i\omega C_p)$ and that $Z_m \approx K/(i\omega)$. Thus, the receiver sensitivity at low frequencies becomes approximately

$$M_V = \frac{A}{N}\left(\frac{C}{C_p + C}\right) \tag{8.30}$$

The capacitance $C_p$ is given in equation (8.20), while the capacitance $C$ can be found by using equations (8.17) and (8.21), so that

$$C = \frac{N^2}{K} = \frac{2\pi ah}{t}\varepsilon_{33}^T k_{31}^2 \tag{8.31}$$

Note that for low frequencies, where the approximation of equation (8.31) is valid, this hydrophone is omnidirectional with the same receiving sensitivity to sound from all directions.

Figure 8.5 shows the simplified equivalent circuit valid for a piezoelectric transducer at frequencies much lower than the resonance frequency. As explained above, this simplified circuit is derived from the equivalent circuit shown in figure 8.4 by ignoring the mass *M* and the resistances $R_p$ and $R_m$.

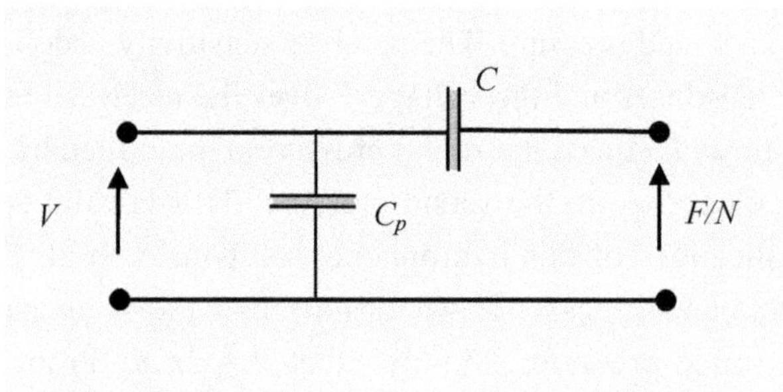


Figure 8.5. Circuit diagram equivalent to the ring hydrophone, for frequencies far below the resonance frequency.

The mechanical quantities have been transformed to electrical quantities by the use of the transformation factor *N* of equation (8.21). Using equations (8.20) and (8.31), we find that the capacitor values are related to the coupling coefficient by

$$\frac{C}{C_p + C} = k_{31}^2 \tag{8.32}$$

When the parameters of the piezoelectric material are inserted, the receiver sensitivity for a ring hydrophone at low frequencies becomes

$$M_V = a\frac{d_{31}}{\varepsilon_{33}^T} \tag{8.33}$$

As an example, we consider a hydrophone with the nominal parameters given in table 8.1. These parameters are typical for PZT ceramic used in hydrophones.

*Table 8.1. Parameters for a ring hydrophone.*

| Material parameter | Dimensions |
|---|---|
| $d_{31} = 125 \cdot 10^{-12}$ m/V | $a = 1.25$ cm |
| $Y_{11}^{E} = 0.08 \cdot 10^{12}$ N/m$^2$ | $h = 1.25$ cm |
| $\varepsilon_{33}^{T} = 1300 \cdot 8.8\ 10^{-12}$ F/m$^2$ | $t = 0.3$ cm |
| $\rho = 7.6 \cdot 10^3$ kg/m$^3$ | $R_p = 10^{10}\ \Omega$<br>$R_m = 10^3\ \Omega$ |

According to these values, the sound speed of the piezoelectric material is $c_0 = 3250$ m/s and the coupling coefficient $k_{31} = 0.33$. The resonance frequency is $f_0 = 41$ kHz and the capacitance values become $C = 0.41$ nF and $C_p = 3.3$ nF.

Figure 8.6 shows a plot of the calculated receiver sensitivity as a function of the frequency for the numerical values given in table 8.1. In the calculations for figure 8.6, we have used the complete expression in equation (8.29).

We see that the receiver sensitivity remains constant until the frequency approaches the resonance frequency, at which point the sensitivity first rises until the resonance frequency is reached; then the sensitivity drops rapidly as the frequency becomes very high. The maximum resonance sensitivity is mainly given by the loss represented by $R_m$ and $R_p$.

When the numerical values from table 8.1 are used in equation (8.33), we find that for this hydrophone the low frequency sensitivity $M_V$ is

$$M_V = 1.37 \cdot 10^{-4}\ \text{V/Pa} \tag{8.34}$$

or in dB, as more usually expressed

$$M_V = -197\ \text{dB re 1 V}/\mu\text{Pa} \tag{8.35}$$

For instance, using a sound pressure of 0.1 Pa with equation (8.34) gives a voltage over the ring of $1.37 \cdot 10^{-5}$ V. By comparing the value of equation (8.33) with the graph in figure 8.6, we see that the low frequency expression of sensitivity in equation (8.33) is a good approximation for a large frequency band.

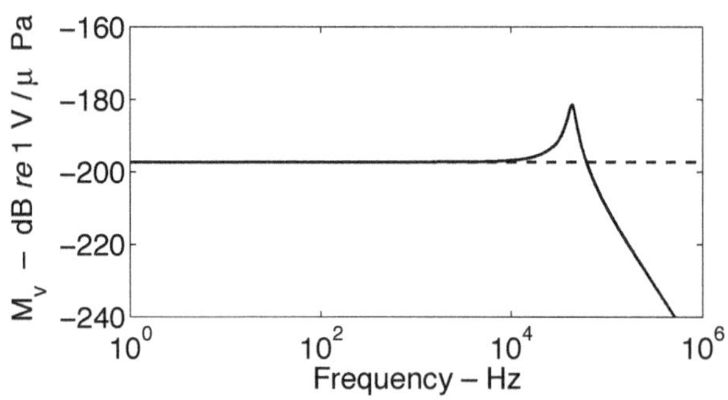


Figure 8.6. Receiver sensitivity calculated as a function of the frequency. The parameters are given in table 8.1. The dashed line shows the approximated low-frequency result, given by equation (8.35).

In view of the preceding discussion, especially when considering equation (8.33), one may get the impression that receiver sensitivity depends only on the material properties of the piezoelectric material and the ring radius, and not on the height and thickness of the ring. This is only partly true, as we also have to take into account the capacity of the electrical cable and the input impedance of the receiver to which the hydrophones are connected. If the transducer is connected to a receiver with a cable with a capacitance of $C_c$ (measured in farads), this capacitance will act in parallel to $C_p$ and reduce the receiver sensitivity to

$$M_V = \frac{V}{P} = \frac{A}{N}\left(\frac{C}{C_p + C + C_c}\right) \tag{8.36}$$

For the best sensitivity, it is essential that the cable's capacitance be small compared with the hydrophone capacitance. Thus, if an application requires a long cable, the size of the hydrophone must be increased to achieve the necessary hydrophone capacitance; see equation (8.31). Alternatively, a special electronic device, known as a charge amplifier can be used to terminate the receiving end of the cable. The use of a charge amplifier eliminates the effect of cable capacitance.

As observed earlier, a piezoelectric transducer can also be used as a transmitter. In the following case, we want to determine the sound pressure the transducer generates at a given distance for a given electrical input. We start with the equivalent circuit of Figure 8.4, into which we introduce the mechanical load impedance $Z_l$ or radiation impedance, representing the acoustical loading of the external medium (water). We shall discuss the radiation impedance more in detail in the next section.

After introducing $Z_l$ we get the circuit equivalent shown in figure 8.7. From this figure, we see that the surface of the transducer is vibrating with a radial velocity of

$$\dot{u} = \frac{NV}{Z_m + Z_l} \tag{8.37}$$

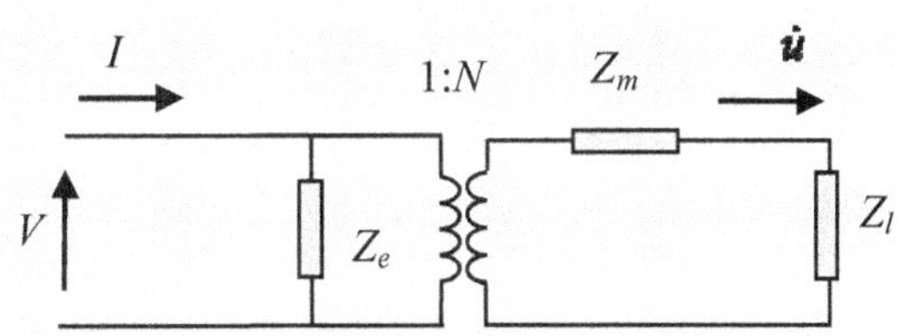


Figure 8.7. Electromechanical equivalent circuit of the transducer as transmitter.

The source strength was defined in chapter 2, and in this case is $Q = A\dot{u}$. According to equation (2.103), the sound pressure at the distance $r$ is then

$$p(r) = V \frac{i\omega\rho_l}{4\pi r} \frac{AN}{Z_m + Z_l} \tag{8.38}$$

The transmitter sensitivity $S_v$ is defined as the ratio of the driving electric voltage to the sound pressure $p_0$ at a distance $r$ in the farfield of the transducer, but with the sound pressure reduced to a reference distance of $r_0 = 1$ m by assuming spherical spreading

$$p_0 = \frac{r}{r_0} p(r) \tag{8.39}$$

Thus, equation (8.37) gives the transmitter voltage sensitivity as

$$S_V = \left( \frac{p_0}{V} \right)_{I=0} = \frac{i\omega\rho_l}{4\pi} \frac{AN}{Z_m + Z_l} \tag{8.40}$$

At low frequencies, since $Z_m \approx K / i\omega >> Z_l$, this expression can be simplified to

$$S_V = \left( \frac{p}{V} \right)_{I=0} = -\frac{1}{2} \omega^2 \rho_l d_{31} \frac{ha^2}{t} \tag{8.41}$$

In figure 8.9, the transmitter sensitivity is plotted as a function of the frequency, calculated according to equation (8.40) with the numerical values given in table 8.1.

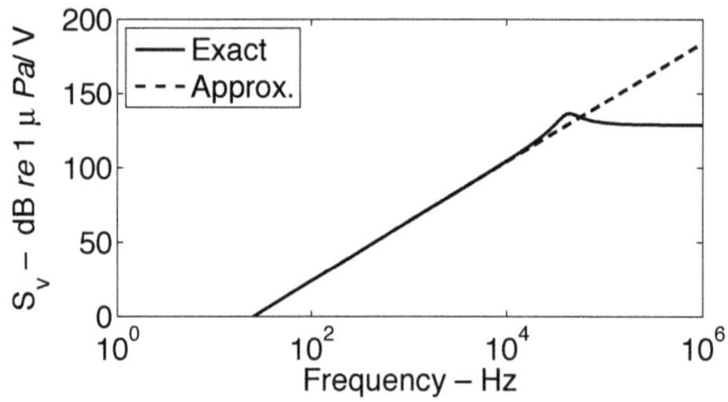


Figure 8.8. The solid line represents transmitter sensitivity, calculated as a function of the frequency of a ring transducer that has the resonance frequency of 40 kHz, using equation (8.40). The dashed line shows the approximated low-frequency result, which was calculated with equation (8.41).

The approximate results, which are indicated by the broken line, are calculated with equation (8.41). The plot shows that the expression is accurate for frequencies below the resonance frequency. Note that the transmitter for frequencies below resonance is omnidirectional with a sensitivity that is proportional to the frequency squared.

## 8.4 Piezoelectric disk in 33-mode

A piezoelectric disk used in thickness mode or 33-mode is shown in figure 8.9 in a cross-sectional view. Metal electrodes are attached on both sides of the disk; the voltage across the electrodes is $V$ and the current is $I$.

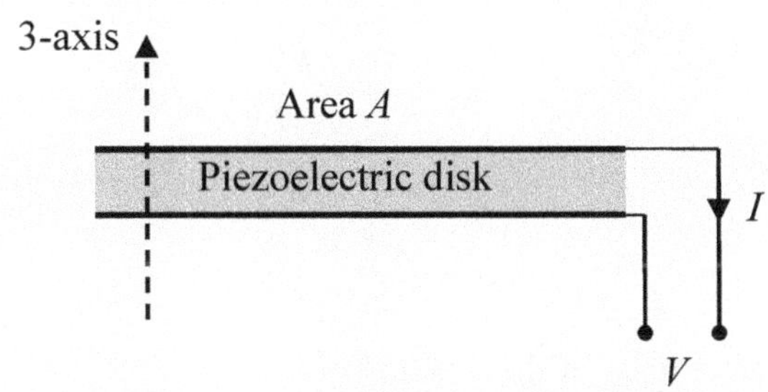


Figure 8.9. Piezoelectric disk in 33-mode, with thickness (element length) $l$ and area $A$. The voltage across the electrodes is $V$ and the current is $I$.

For an element in the 33-mode, the equations are written in either of the following forms:

$$\begin{aligned} S_3 &= \frac{1}{Y_{33}^E} T_3 + d_{33} E_3 \\ D_3 &= d_{33} T_3 + \varepsilon_{33}^T E_3 \end{aligned} \tag{8.42}$$

or

$$\begin{aligned} T_3 &= \alpha_{33} S_3 - h_{33} D_3 \\ E_3 &= -h_{33} S_3 + \beta_{33} D_3 \end{aligned} \tag{8.43}$$

Here, $S_3$ is the relative change in thickness, $T_3$ is the mechanical stress, $E_3$ is the electrical field strength, and $D_3$ is the electric displacement. As mentioned earlier, the index "33" means that both the polarization and oscillation are in the 3-direction. When the element is short-circuited ($E_3 = 0$), the elasticity modulus is $Y_{33}^E$, and $\varepsilon_{33}^T$ is the dielectric constant when the element freely expands and contracts (that is, when $T_3 = 0$).

The relations between the coefficients in equations (8.42) and (8.43) are

$$\begin{aligned} \alpha_{33} &= Y_{33}^E \frac{1}{1-k_{33}^2} \\ h_{33} &= \frac{1}{d_{33}} \frac{k_{33}^2}{1-k_{33}^2} \\ \beta_{33} &= \frac{1}{\varepsilon_{33}^T} \frac{1}{1-k_{33}^2} \end{aligned} \tag{8.44}$$

The coupling coefficient $k_{33}$ of the 33-mode is

$$k_{33} = d_{33} \sqrt{\frac{Y_{33}^E}{\varepsilon_{33}^T}} \tag{8.45}$$

The sound speed $c$ in the piezoelectric disk is given as

$$c = \sqrt{\alpha_{33}/\rho} \tag{8.46}$$

where $\rho$ is the density of the piezoelectric material and $\alpha_{33}$ is the bulk modulus of elasticity.

The desired vibrating mode is in the form of plane waves in the axial direction. This is exactly the situation we treated in section 5.2.1; therefore, we can consider the disk to be an acoustic delay line that can be represented by the equivalent circuit shown in figure 8.10.

The equations governing the mechanical forces and the particle velocities on the two sides of the delay lines in figure 8.10 are

$$\begin{aligned} F_1 &= (Z_a + Z_b)\dot{u}_1 - Z_b \dot{u}_2 \\ F_2 &= Z_b \dot{u}_1 - (Z_a + Z_b)\dot{u}_2 \end{aligned} \tag{8.47}$$

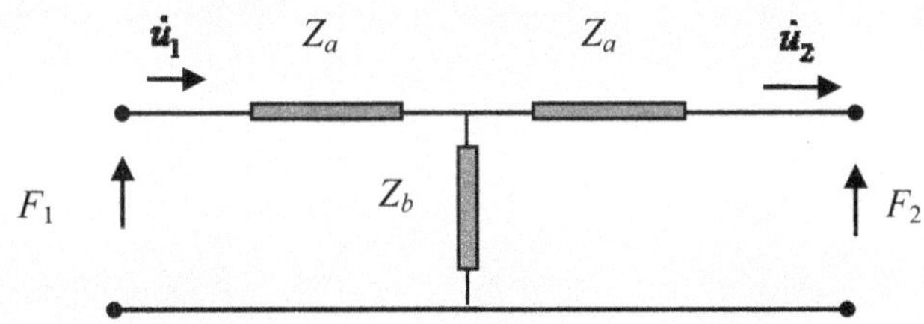


Figure 8.10. Equivalent circuit for an acoustic delay line.

The mechanical impedances $Z_a$ and $Z_b$ are the same as those given by equations 5.31 and 5.33, but with opposite signs, and are now written as

$$Z_a = Z_0 i \tan(\kappa l / 2)$$
$$Z_b = \frac{Z_0}{i\sin(\kappa l)} \tag{8.48}$$

where $\kappa$ is the wave number, $l$ is the length of the delay line and $\kappa l$ represents the phase delay, and $Z_0$ is given as

$$Z_0 = \rho c A \tag{8.49}$$

Note that the signs in equations (8.48) are the opposite of the signs in equation 5.31 and 5.32, by reason of the change in sign convention in this chapter.

Equations (8.47) and (8.48), together with figure 8.10, describe the mechanical or acoustical behavior of the transducer element, given the assumption that only plane waves in the axial direction are propagating in the element.

In order to include the electrical part of the description, we return to the first equation of equations (8.43). We see that the term $-h_{33}D_3$ requires the addition of another term to the mechanical part of equation (8.47). With this term included, equation (8.47) becomes

$$
\begin{aligned}
F_1 &= Z_0\left[\frac{\dot{u}_1}{i\tan(\kappa l)} - \frac{\dot{u}_2}{i\sin(\kappa l)}\right] + \frac{h_{33}}{i\omega}I \\
F_2 &= Z_0\left[\frac{\dot{u}_1}{i\sin(\kappa l)} - \frac{\dot{u}_2}{i\tan(\kappa l)}\right] + \frac{h_{33}}{i\omega}I
\end{aligned}
\tag{8.50}
$$

We next consider the second equation of equations (8.43). The voltage over the element electrodes is the integral of the electric field strength $E_3$, and therefore is expressed as

$$
\begin{aligned}
V &= \int_0^l E_3 dx = \int_0^l \left(-h_{33}\frac{\partial u}{\partial x} + \frac{\beta_{33}}{i\omega A}I\right)dx \\
&= h_{33}\left(u_{x=0} - u_{x=l}\right) + \frac{\beta_{33} l}{i\omega A}I
\end{aligned}
$$

Finally, using the particle velocities on both sides of the element gives the relation between the voltage, the current, and the particle velocities in the form

$$
V = h_{33}\frac{(\dot{u}_1 - \dot{u}_2)}{i\omega} + \frac{\beta_{33} l}{i\omega A}I \tag{8.51}
$$

Equations (8.50) and (8.51) describe the piezoelectric element fully, that is, both mechanically and electrically. The circuit equivalent diagrammed in figure 8.11 represents these electromechanical equations.

The mechanical impedances $Z_a$ and $Z_b$ are given by equation (8.48); they represent distributed circuit elements, not lumped circuit elements, and therefore are described by periodic trigonometric functions, as discussed later.

The electromechanical transformer has the transformation factor $N$

$$
N = h_{33}C_p = \frac{A}{l}d_{33}Y_{33}^E \tag{8.52}
$$

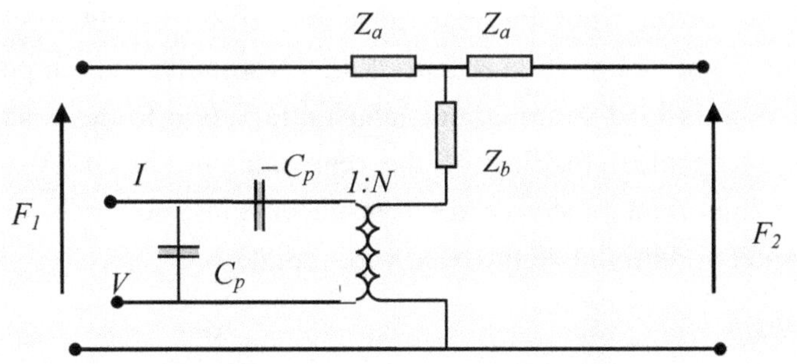


Figure 8.11. The electromechanical circuit equivalent of a piezoelectric element in 33-mode.

The parallel capacitor $C_p$ is

$$C_p = \frac{A}{\beta_{33} l} \tag{8.53}$$

and substituting equation (8.44) for $\beta_{33}$, we find that

$$C_p = C_0 \left(1 - k_{33}^2\right) \tag{8.54}$$

with

$$C_0 = \frac{A}{l} \varepsilon_{33}^T \tag{8.55}$$

We note that $C_p$ is the capacitance value when the element is mechanically blocked in place, that is, when $S_3 = 0$. The capacitance $C_0$ is the value that will be measured at the electrical terminal when the end surfaces are free to move, that is, when $T_3 = 0$. The "negative" capacitor of capacitance equal to $C_p$ in the equivalent circuit diagram of figure 8.11 is mathematically necessary to satisfy equations (8.50) and (8.51), but otherwise has no physical meaning.

The circuit in figure 8.11 can be further transformed by introducing the mechanical load impedances for the front and rear sides, $Z_2$ and $Z_1$, respectively. This produces the equivalent circuit shown in figure 8.12. Normally, the impedance $Z_2$ will represent the water's load—that is, the acoustic radiation to the water—and $Z_1$ will represent the mechanical loading of the rear side, produced by the mechanical mounting. An important objective for transducer construction is to maximize the radiation to water and minimize the backward radiation.

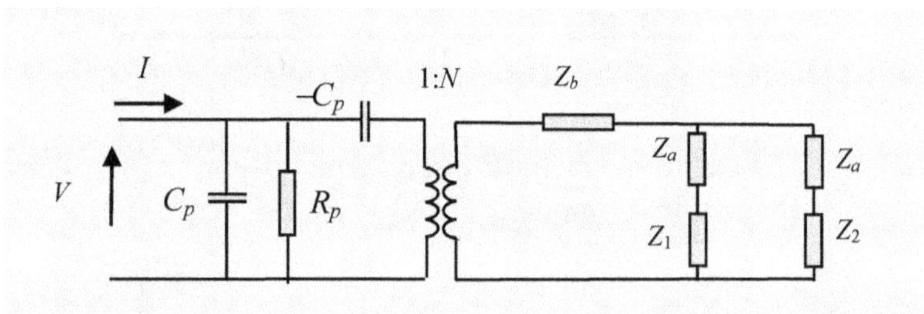


Figure 8.12. Electric circuit equivalent for a piezoelectric element in 33-mode with the mechanical load impedances $Z_2$ and $Z_1$ on the front and rear sides, respectively.

To indicate how this can be accomplished, we neglect for the moment the loss resistances and calculate the electric input admittance $Y_i$ seen from the electric terminal, which is

$$Y_i = i\omega C_p + \frac{1}{\frac{Z_{mec}}{N^2} + \frac{1}{-i\omega C_p}} = \frac{1}{\frac{1}{i\omega C_p} + \frac{N^2}{Z_{mec}\left(\omega C_p\right)^2}} \tag{8.56}$$

The electrical input impedance is therefore

$$Z_i = \frac{1}{i\omega C_p} + \frac{N^2}{\left(\omega C_p\right)^2 Z_{mec}} \tag{8.57}$$

where $Z_{mec}$ is the mechanical impedance seen from the secondary side of the transformer, which consists of a series connection of $Z_b$ and a parallel/serial combination of $Z_1$, $Z_2$, and $Z_a$. From electrical circuit theory, we find that $Z_{mec}$ is expressed as

$$Z_{mec} = Z_b + \frac{(Z_a + Z_1)(Z_a + Z_2)}{Z_1 + Z_2 + 2Z_a} \tag{8.58}$$

By using equations (8.57) and (8.58), and inserting for $Z_a$ and $Z_b$ by using equations (8.48), we can write $Z_i$ as

$$Z_i = \frac{1}{i\omega C_p} + \left(\frac{N}{\omega C_p}\right)^2 \frac{-2Z_0\left[1-\cos(\kappa l)\right] + i[Z_1 + Z_2]\sin(\kappa l)}{Z_0[Z_1 + Z_2]\cos(\kappa l) + i\left[Z_0^2 + Z_1 Z_2\right]\sin(\kappa l)} \tag{8.59}$$

When the element is free from loss and the load impedances are real radiation impedances, the real part of $Z_i$ must represent the acoustic radiation. The real part of equation (8.59) is

$$R_i = \mathrm{Re}\{Z_i\} = \left(\frac{N}{\omega C_p}\right)^2 \frac{(Z_1 + Z_2)}{Z_0^2} \frac{\left(1-\cos(\kappa l)\right)^2 + \dfrac{Z_1 Z_2 \sin^2(\kappa l)}{Z_0^2}}{\left(\dfrac{Z_1 + Z_2}{Z_0}\right)^2 \cos^2(\kappa l) + \left(1 + \dfrac{Z_1 Z_2}{Z_0^2}\right)^2 \sin^2(\kappa l)} \tag{8.60}$$

With a suitable choice of the disk thickness $l$ and the mechanical load impedances $Z_1$ for the rear side, one may obtain maximum radiation into the water. We now consider two ways to obtain this maximum, using a half-wavelength and a quarter-wavelength resonator, respectively.

### *8.4.1 The half-wavelength resonator*

When the load impedance at the rear side of the transducer is very small, so that $Z_1 \approx 0$, for instance, because there is air behind the active element, according to equation (8.60), the real part of the electrical impedance $R_i$ becomes

$$R_i = \left(\frac{N}{\omega C_p}\right)^2 \frac{Z_2\left[1-\cos(\kappa l)\right]^2}{Z_2^2\cos^2(\kappa l)+Z_0^2\sin^2(\kappa l)} \tag{8.61}$$

The resistance $R_i$ has a maximum value when the thickness $l$ is an uneven number of half-wavelengths, that is when $\kappa l = \pi\ (2m-1)$ for $m = 1, 2, 3 \cdots$. In practice, the first mode, where $m = 1$ and $\kappa l = \pi$, is almost always used. The resonance frequency is then given by

$$f_0 = \frac{\omega_0}{2\pi} = \frac{c}{2l} \tag{8.62}$$

where the sound speed in the piezoelectric material, $c$, is given by equation (8.46).

The real part of the electrical impedance then becomes

$$R_i = \left(\frac{N}{\omega_0 C_p}\right)^2 \frac{4}{Z_2} = \frac{4h_{33}^2}{\omega_0^2 Z_2} \tag{8.63}$$

This last expression is obtained by using the values of the transformation factor $N$ and the capacitor $C_p$ from equations (8.52) and (8.53).

It is more convenient for the analysis to modify the equivalent circuit as shown in figure 8.13, where the load impedance $Z_2$ is in a separate branch. The transformation can be verified by considering two cases: (i) with $Z_2 = 0$, and (ii) with $Z_2 = \infty$ and by demanding that the electrical input impedances be the same for the two circuits.

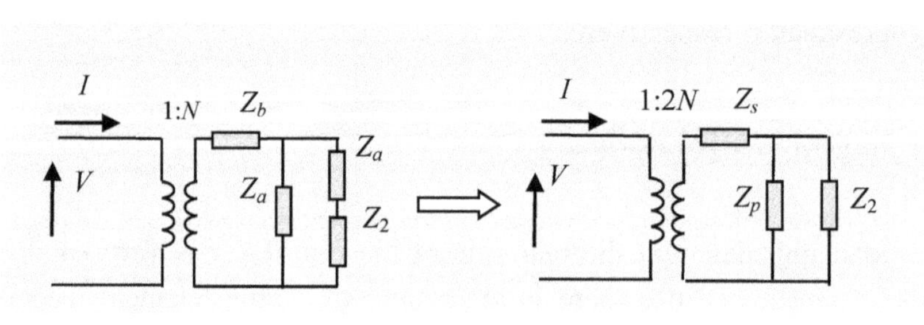


Figure 8.13. Left, the original circuit with impedances $Z_a$ and $Z_b$ and transformation factor 1:$N$. Right, the equivalent circuit with the new impedances $Z_s$ and $Z_p$ and with transformation factor 1:2$N$.

This gives the following two equations:

$$\begin{aligned} &\text{(i)} \qquad Z_{in} = \left(\frac{1}{N}\right)^2 \left(Z_b + \frac{Z_a}{2}\right) = \left(\frac{1}{2N}\right)^2 Z_s \\ &\text{(ii)} \qquad Z_{in} = \left(\frac{1}{N}\right)^2 \left(Z_b + Z_a\right) = \left(\frac{1}{2N}\right)^2 \left(Z_s + Z_p\right) \end{aligned} \tag{8.64}$$

Solving for the new impedances $Z_p$ and $Z_s$ on the right-hand side of figure 8.13 gives

$$\begin{aligned} Z_p &= 2Z_a = 2iZ_0 \tan\left(\frac{\kappa l}{2}\right) \\ Z_s &= 2\left(Z_a + 2Z_b\right) = -\frac{2iZ_0}{\tan\left(\dfrac{\kappa l}{2}\right)} \end{aligned} \tag{8.65}$$

Since we are most interested in frequencies near the resonance frequency, we take equations (8.65) and insert

$$\kappa l/2 \approx \frac{\pi}{2}\left(\frac{\omega - \omega_0}{\omega_0}\right) + \frac{\pi}{2} \tag{8.66}$$

Because of the trigonometric rule stating that $\tan(\kappa l + \pi/2) = -1/\tan(\kappa l)$, equations (8.65) can be transformed into

$$\begin{aligned} Z_p &= -2iZ_0 \frac{1}{\tan\left(\dfrac{\pi}{2}\dfrac{\omega - \omega_0}{\omega_0}\right)} \\ Z_s &= 2iZ_0 \tan\left(\frac{\pi}{2}\frac{\omega - \omega_0}{\omega_0}\right) \end{aligned} \tag{8.67}$$

For frequencies close to the resonance frequency, the parallel impedance $Z_p$ is very high and consequently has little importance and can be removed. Furthermore, in the expression for the series impedance $Z_s$ we replace the tangent function with the argument, and get approximately

$$Z_s \approx iZ_0\pi\frac{\omega-\omega_0}{\omega_0} \tag{8.68}$$

We find the equivalent circuit representation for this impedance by relating it to a series coupling of a mass $M$ and a spring with stiffness $K$, which gives the impedance $Z'$

$$\begin{aligned} Z' &= i\omega M + \frac{K}{i\omega} = i\omega_0 M\left(\frac{\omega}{\omega_0}-\frac{\omega_0}{\omega}\right) \\ &\approx 2i\omega_0 M\left(\frac{\omega-\omega_0}{\omega_0}\right) \end{aligned} \tag{8.69}$$

In making the two impedances in equations (8.68) and (8.69) equivalent, we find that

$$M = \frac{\pi}{2}\frac{Z_0}{\omega_0} = \frac{Al\rho}{2} \tag{8.70}$$

and

$$\frac{1}{K} = \frac{2}{\pi\omega_0 Z_0} = \frac{2l}{\pi^2 A\alpha_{33}} \tag{8.71}$$

In order to get the last values in equations (8.70) and (8.71), we have applied the expressions for $Z_0$ from equation (8.49), $c$ from equation (8.46), and $f_0$ from equation (8.62).

The acoustic impedances $Z_1$ and $Z_2$ represent the acoustic impedances of the back and the front ends, respectively, of the piezoelectric element. In the case of a half wavelength transducer the backing impedances is assumed to be zero, and $Z_2$ represents the acoustic load impedance of the medium. This impedance is in general complex with a resistive real part and an imaginary reactive part with the general form as

$$Z_2 = R_a + iM_a \tag{8.72}$$

where $R_a$ is the radiation resistance and $M_a$ is the radiation reactance.

$R_a$ represents the radiation of acoustic energy to the medium and $M_a$ represents a mass loading having the effect of changing the resonance frequency of the transducer. The radiation impedance can be calculated for simple geometrical forms such as a sphere and rectangular or circular disks, we discuss this more in detail in section 8.6.2 where we demonstrate how the values of $R_a$ and $R_m$ can be determined by measurements.

As a result of these substitutions, the transducer element can be represented by the equivalent circuit shown in figure 8.14. The concentrated, or lumped, circuit elements represent the effective dynamic mass $M$ and stiffness $K$ of the transducer.

We notice that the effective dynamic mass is half of the total mass; this is true because the particle velocity and acceleration are not constant across the length of the element. For the same reason, the effective stiffness is greater than the static stiffness. In addition to the components determined above, we have also included in figure 8.14 a leakage resistance $R_p$, representing the dielectric loss in the element, and a series resistance $R_m$, representing the mechanical loss. The acoustic radiation resistance is represented by $R_a$. We emphasize that this equivalent with concentrated circuit elements is only valid for a single mode of vibration at frequencies near the resonance frequency given by equation (8.62), and setting $Z_2 \approx R_a$ requires that the radius $a$ of the transducer is much larger that the wavelength. A simple half-wavelength transducer is shown in figure 8.15.

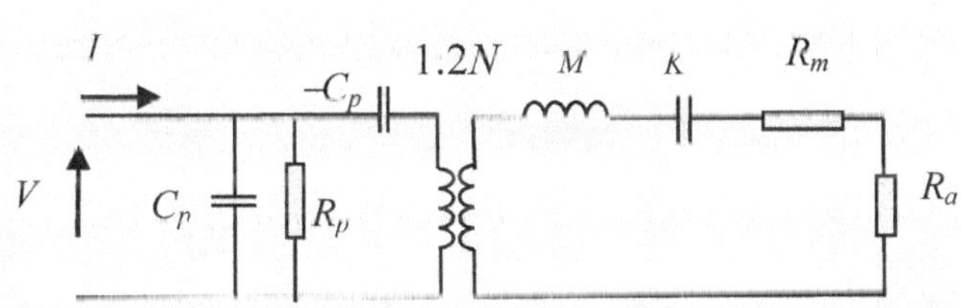


Figure 8.14. Electromechanical equivalent of a half-wavelength transducer at frequencies near the resonance frequency. A leakage resistance $R_p$ and a series resistance $R_m$ have been added to represent physical losses.

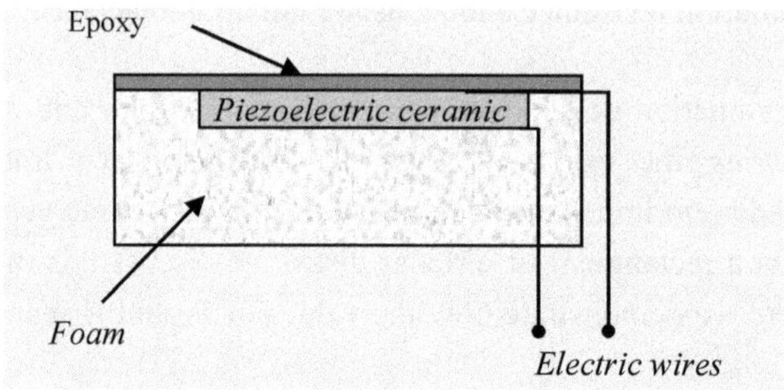


Figure 8.15. Half-wavelength transducer in 33-mode.

The piezoelectric disk is cast in a foamlike plastic material with very low specific acoustic impedance, and therefore the material is acoustically equivalent to air; this, with a good approximation, makes the rear side impedance $Z_1 \approx 0$. The transducer's front side is protected by a thin layer of, for instance, epoxy. The epoxy layer will have an effect on the resonance frequency of the transducer, but the effect will be moderate for a thin layer.

### *8.4.2 The quarter-wavelength resonator*

Another way to limit the backward radiation is to use a very high load impedance on the rear side, so that the load impedance $Z_1 = \infty$. The real part of the electrical impedance, which is given by equation (8.60), then becomes

$$R_i = \left(\frac{N}{\omega C_p}\right)^2 \frac{Z_2 \sin^2(\kappa l)}{Z_0^2 \cos^2(\kappa l) + Z_2^2 \sin^2(\kappa l)} \tag{8.73}$$

The first maximum value occurs when $\kappa l = \pi/2$, that is, when the length is a quarter of a wavelength. This produces the resonance frequency

$$f_0 = \frac{\omega_0}{2\pi} = \frac{c}{4l} \tag{8.74}$$

At the resonance frequency, the real part of the electrical input impedance is

$$R_i = \left(\frac{N}{\omega_0 C_p}\right)^2 \frac{1}{Z_2} = \frac{h_{33}^2}{\omega_0^2 Z_2} \tag{8.75}$$

The electromechanical equivalent circuit for the quarter-wavelength resonator can be reduced in the same way as in equations (8.61) through (8.67). When $Z_1 = \infty$, the mechanical impedances $Z_a$ and $Z_b$ are coupled in series. The sum of $Z_a$ and $Z_b$ is equal to a new series impedance $Z_s$, with

$$Z_s = Z_a + Z_b = -i\frac{Z_0}{\tan(\kappa l)} \tag{8.76}$$

An expansion of this expression around $\kappa l = \pi / 2$ gives

$$Z_s = iZ_0 \tan\left(\frac{\pi}{2}\frac{\omega - \omega_0}{\omega_0}\right) \approx iZ_0 \frac{\pi}{2}\left(\frac{\omega - \omega_0}{\omega_0}\right) \tag{8.77}$$

This impedance is that of a series coupling of a mass $M$ and a stiffness $K$, as earlier, but this time with the values

$$\begin{aligned} M &= \frac{\pi}{4}\frac{Z_0}{\omega_0} = \frac{1}{2}Al\rho \\ 1/K &= \frac{4}{\pi Z_0 \omega_0} = \frac{8}{\pi^2}\frac{l}{\alpha_{33} A} \end{aligned} \tag{8.78}$$

The very high rear load impedance required for the quarter-wavelength transducer can be quite difficult to achieve directly. However, remember that in section 5.2.1 we showed that high impedance can be created by using a quarter-wavelength transformer terminated with very low impedance. Thus, we can construct a quarter-wavelength transducer by attaching a quarter-wavelength piezoelectric disk to a metallic plate (for instance, steel) that is a quarter-wavelength thick. The other side of the metallic plate must face air or some other material with very low specific acoustic impedance. Together, the combined element with the piezoelectric element and the plate results in a transducer element that has the half-wavelength thickness required for resonance. A simple quarter-wavelength transducer constructed in this way is shown in figure 8.16.

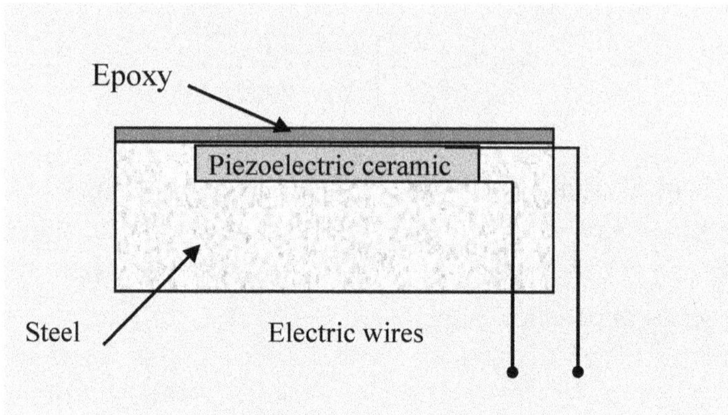


Figure 8.16. A quarter-wavelength transducer in 33-mode.

## 8.5 The piston transducer

Figure 8.17 shows a common construction of an acoustic vibrator element. The driving element is a stack of piezoelectric ceramic disks with electrodes operating in the 33-mode. The radiating head of the element is designed for maximizing the acoustic radiation to water and often has a conical shape or is a cylinder with larger diameter. It acts like an impedance transformer between the ceramic and water load and is often made of lightweight material such as aluminum. The end mass is designed to limit the backward acoustic radiation. An axial bolt, which also ensures a mechanical prestress, keeps the elements together. At its resonance frequency the piezoelectric element acts as a lumped-element spring with vibrator head and back elements acting like lumped-element masses. This design is often referred to as a *Tonpilz* design, which is a German term that means "singing mushroom." For detailed design information on Tonpilz transducers, see Wilson (1958) or Stansfield (1991).

Elements like the one shown on the left in figure 8.17 are normally used to build up the larger sound projectors used in sonar systems. An example that could be an echo sounder transducer is shown on the right. For other applications, shapes other than a rectangular assembly are used—for instance, mounting the elements on the surface of a sphere or a cylinder.

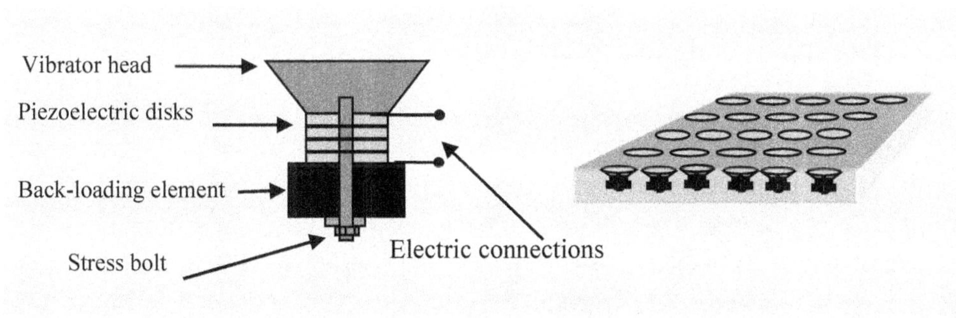


Figure 8.17. A common design for a larger underwater transmitter/receiver. Left, the device is composed of many piston elements; right, the elements are mounted together in a plate. Together they constitute the complete transducer.

While the analysis in section 8.4 is valid, note that the piezoelectric element need not equal one-half or one-quarter wavelength, as long as the head and backing sections have the lengths required to produce the desired resonance frequency for the whole element. We shall not go into the detailed design and dimensioning of such vibrators, but in section 8.6 we shall discuss some measurement techniques that transducer users may apply in order to verify that a transducer performs as expected.

## 8.6 Measurement and characterization of transducers

We have so far dealt with the fundamental principles for use of piezoelectric transducers in underwater acoustics and outlined how they can be analyzed and designed. In this section, we will present some common measurement methods with which to determine the characteristics of a given transducer. For more information see the ANSI standard S1.20-1988, "Procedures for calibration of underwater electroacoustic transducers."

In order to explain the principles, we consider an idealized case with a transducer vibrating in only one mode. The general electromechanical circuit representation of a transducer, valid for frequencies close to the resonance frequency, is shown in figure 8.18. There are two circuit elements coupled by the electrical-mechanical transformer; the electrical element at the left hand side of the transformer and the mechanical element on the right hand side.

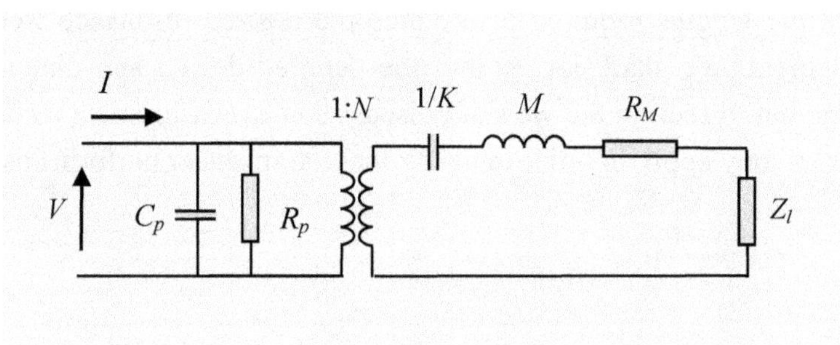


Figure 8.18. The electromechanical equivalent circuit of a piezoelectric transducer.

The mechanical branch consists of the compliance $1 / K$, the mass $M$, the mechanical loss resistance $R_M$ and the radiation impedance $Z_l$. The electrical side is the parallel coupling of the capacitance $C_p$ and the electrical leakage resistance $R_p$ that represents dielectrical losses in the capacitor.

### *8.6.1 The radiation impedance*

The radiation load impedance of the medium is in figure 8.18 denoted $Z_l$. This impedance is, in general, complex with a resistive real part and an imaginary reactive part with the general form as

$$Z_l = R_l + iM_l \tag{8.79}$$

where $R_l$ is the radiation resistance and $M_l$ is the radiation reactance. $R_l$ represents the radiation of acoustic energy to the medium and $M_l$ represents a mass loading having the effect of changing the resonance frequency of the transducer. In the following we assume the reactive components $M_l$ are included in the mass $M$ and we thereby replace $Z_l$ with its resistive part $R_l$ representing the acoustic radiation of the external medium.

The radiation impedance can be calculated for simple geometrical forms such as a sphere and rectangular or circular disk, see for instance Kinsler et al. (2000), Stansfield (1991), and Wilson (1988). In section 2.6 we have seen an example when discussing radiation from a spherical source. For a spherical source with radius $a$, the radiation impedance is the product of the area of the radiation surface and the specific acoustic impedance given by equation 2.103 or 2.104, evaluated at $r = a$. The radiation impedance of a radiating sphere with radius $a$ is therefore

$$Z_l = 4\pi a^2 \rho_w c_w \left[ \frac{(\kappa_w a)^2}{(\kappa_w a)^2 + 1} + i\frac{(\kappa_w a)}{(\kappa_w a)^2 + 1} \right] \tag{8.80}$$

For low frequencies where $(\kappa_w a) << 1$

$$Z_l = 4\pi a^2 \rho_w c_w \left[ (\kappa_w a)^2 + i(\kappa_w a) \right] \tag{8.81}$$

In these equations the values for the sound speed $c_w$, density $\rho_w$, and the wave number $\kappa_w$ are for those for the external medium (water).

The radiation reactance corresponds to a mass $M_r = 3\rho V$, where $V = (4/3)\pi a^3$ is the volume of the sphere. For low frequencies the reactance part is considerably larger than the radiation resistance, but the radiation resistance increases proportionally with frequency squared for low frequencies. In the limit of very high frequencies the reactive part vanishes and the radiation resistance becomes

$$R_l = 4\pi a^2 \rho_w c_w \tag{8.82}$$

Similar results can be obtained for radiation from a plane circular piston with radius $a$; see Kinsler et al. (chapter 7.5, 2000). For low frequencies the result is

$$Z_l = \pi a^2 \rho_w c_w \left[ \frac{1}{2}(\kappa_w a)^2 + i\frac{8}{3}(\kappa_w a) \right] \tag{8.83}$$

The acoustic radiating resistance is therefore equal to

$$R_l = \frac{1}{2\pi} A_p^2 \rho_w c_w \kappa_w^2 \tag{8.84}$$

which is proportional to the square of the surface area and to the square of the frequency. The reactive term corresponds to a mass of a cylinder with radius $a$ and height $8a / 3\pi$.

For high frequencies, the reactive term vanishes and the acoustic radiating resistance is simply the product of the radiation surface area and the acoustic impedance of the radiating medium, that is

$$R_l = \pi a^2 \rho_w c_w \tag{8.85}$$

For many cases of practical interest the radiation surface will be in the intermediate region between high and low frequency regimes. For instance, in the transducer elements shown in figure 8.17, the diameter of the vibrator head would be normally approximately equal to one wavelength, corresponding to $k_w a = \pi$. Consequently, the radiation impedance of a single element will have both a real and an imaginary part. However, a complete transducer will normally consist of many such elements vibrating in phase, and therefore the relevant radiation impedance is the impedance of the total surface area, which may well fall in the regime valid for high frequencies where the radiation is essentially resistive and given by the product of $(\rho_w c_w)$ and the surface area of the transducer.

### 8.6.2 Admittance measurements

In practical cases, we may have a transducer available, but do not know its specifications and properties. We now discuss how to determine the main characteristics of the transducer with simple electrical measurements. It is therefore convenient to transfer the electromechanical equivalent of figure 8.18 to the electrical equivalent circuit of figure 8.19 by using the mechanical-electrical transformer ratio 1 / *N*.

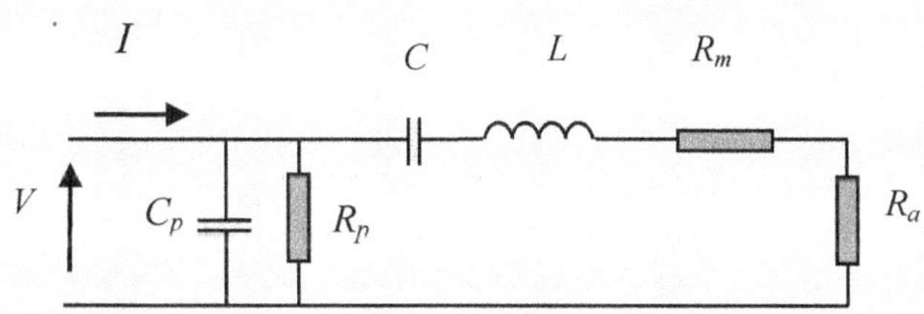


Figure 8.19. Electrical equivalent of a piezoelectric transducer.

In using this representation, we assume that the transducer mainly consists of a series coupling of a mass and a stiffness, which when transformed over to the electrical side are represented by an inductance $L$ and a capacitance $C$. In addition, there is the piezoelectric element's parallel capacitor $C_p$. The mechanical and dielectric losses are represented by the series resistance $R_m$ and the parallel resistance $R_p$. The acoustic radiation is represented by the resistance $R_a$. All of these parameters are initially unknown, and the task is to determine their values by measurements. As we will see, these values can be determined from two sets of electric measurements of the transducer, taken while the transducer is in air or water.

The electrical admittance seen from the electrical side is expressed as

$$Y = i\omega C_p + \frac{1}{R_p} + \frac{1}{i\omega L + \dfrac{1}{i\omega C} + R_m + R_a} \tag{8.86}$$

The left panel in figure 8.20 contains a plot of $Y$ as a function of the frequency. This calculation is made with the parameters given in table 8.2. These values are normalized so that the resonance frequency is equal to unity.

Table 8.2. Parameter values.

| | |
|---|---|
| $R_p = 5$ | $C_p = 3$ |
| $L = 1$ | $C = 1$ |
| $R_m = 0.05$ | $R_a = 0.0$ (air)<br>$R_a = 0.05$ (water) |

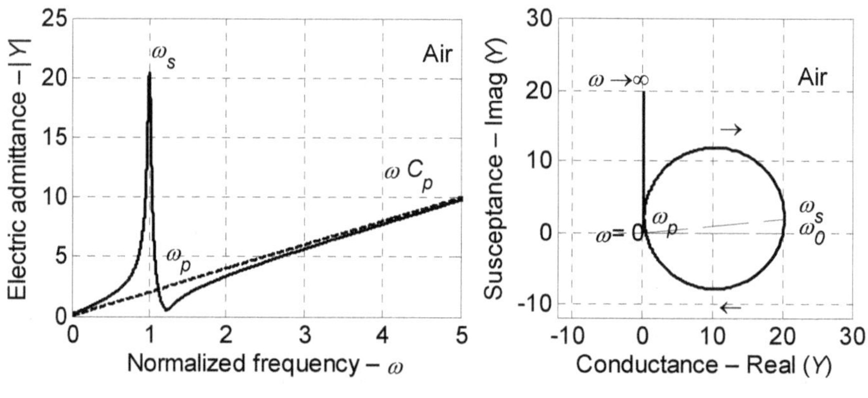


Figure 8.20. Measurement of the admittance of a transducer in air.
Left, absolute admittance value as a function of frequency.
Right, real and imaginary parts, with frequency as the parameter.

The dashed line on the left in figure 8.20 represents the parallel capacitor's contribution to the admittance, $i\omega C_p$. The dashed line on the right represents the maximum amplitude of the electrical admittance at frequency $\omega_s$.

From the left side of figure 8.20, we see that the admittance has pronounced minimum and maximum values. The maximum admittance value occurs at a frequency $\omega_t$ that is very close to the frequency $\omega_s$ where the mechanical series branch has its resonance frequency

$$\omega_s = \sqrt{\frac{1}{LC}} \tag{8.87}$$

In order to find the minimum admittance value, we temporarily exclude loss from our calculations, which is acceptable for an approximate estimate of the resonance frequency. Therefore we insert $R_p = \infty$ and $R_m = R_a = 0$ into equation (8.86), and determine the frequency at which $Y$ becomes equal to zero. This resonance, which is the parallel resonance $\omega_p$, is given by

$$\omega_p = \sqrt{\frac{1}{L}\left(\frac{1}{C} + \frac{1}{C_p}\right)} = \omega_s\sqrt{1 + \frac{C}{C_p}} \tag{8.88}$$

With the values given for air in table 8.2, it follows that $\omega_s = 1$ and $\omega_p = 1.15$ (remember that we are using normalized values).

If the transducer characteristics are unknown, we can measure the electric admittance as a function of the frequency, and subsequently determine $C_p$ from the asymptotic behavior of the admittance, as indicated on the left in figure 8.20. By using equations (8.87) and (8.88), $L$ and $C$ can be calculated from measurements of the two resonance frequencies, $\omega_s$ and $\omega_p$.

Another representation of $Y$ as a function of frequency is shown on the right in figure 8.20. Here, we have plotted the imaginary part of $Y$ as a function of its real part—that is, the admittance is in the complex plane, with the frequency as its parameter. We have marked the maximum and minimum values. We see that the trajectory of $Y$ makes a loop in the complex plane. Ideally, this loop is a circle, but measurements of real transducers will often show deviation from the ideal circle, and other modes and resonances will appear to be small circles or loops on the admittance curve. Still, for well-constructed transducers, it should not be difficult to determine the circle and its parameters. At very high or very low frequencies, as we see from equation (8.86), the real part of the admittance, $\mathrm{Re}\{Y(\omega = 0)\}$, is

$$\mathrm{Re}\{Y(\omega = 0)\} = \mathrm{Re}\{Y(\omega = \infty)\} = \frac{1}{R_p} \tag{8.89}$$

At the series resonance frequency, the real part is

$$\operatorname{Re}\{Y(\omega=\omega_s)\}=\frac{1}{R_p}+\frac{1}{R_m+R_a} \tag{8.90}$$

therefore, the diameter of the circle is

$$D=\frac{1}{R_m+R_a} \tag{8.91}$$

In figure 8.20, we see the result when the transducer is measured in air, where the acoustic load resistance is $R_a \approx 0$. Given the numerical values in table 8.2, the diameter is $D \approx 20$, since $R_m = 0.05$. When the transducer is measured in water, the results become as shown in figure 8.21. Note the difference in scales of figures 8.20 and 8.21.

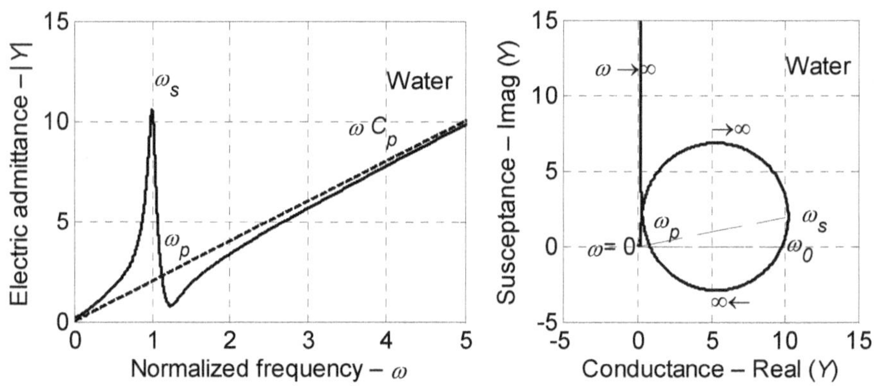


Figure 8.21. Measurement of the admittance of a transducer in water. Left, absolute value of the admittance as function of the frequency. Right, real and imaginary parts, with the frequency as the parameter.

We see there is a significant difference from the curves of figure 8.20, which is caused by the effect of acoustic radiation, represented by the equivalent resistor $R_a$. Since $R_a$ is much larger when the transducer is in water than in air, the circle diameter and the maximum admittance at the resonance frequency $\omega_s$ are much smaller in water. For instance, we see from figure 8.21 that the circle diameter is now $D \approx 10$, since we have used the value $R_a + R_m = 0.1$. From these examples we can see that, by taking electric measurements in air and water, we can determine all the main parameters of a specific transducer. We emphasize that the measurements must be taken in open water, or in a tank that is so large that reflections from boundaries have no influence on the acoustic radiation conditions.

The transducer efficiency can be determined in a similar manner. The total electric power injected into the transducer is distributed between mechanical power $P_m$ and electrical power $P_e$. At the resonance frequency, the power dissipated mechanically is divided into one part that goes to acoustic radiation, represented by the resistance $R_a$, and another part lost to friction, represented by $R_m$. The electrical power is dissipated in the parallel resistance $R_p$. Therefore, the ratio of the mechanical power and the electrical power dissipation at the resonance frequency is given by

$$\frac{P_m}{P_e} = \frac{R_p}{R_m + R_a} \tag{8.92}$$

The electromechanical efficiency factor $\eta_{em}$ thus becomes

$$\eta_{em} = \frac{P_m}{P_e + P_m} = \frac{R_p}{R_p + (R_m + R_a)} \tag{8.93}$$

and the electroacoustic efficiency $\eta_{ea}$ is

$$\eta_{ea} = \frac{P_a}{P_e + P_m} = \frac{R_a}{(R_a + R_m)}\eta_{em} \tag{8.94}$$

We can determine the coupling coefficient $k$ from measurement of the parallel and series resonance frequencies. From equations (8.87) and (8.88), we find that

$$k = \sqrt{\frac{C}{C + C_p}} = \sqrt{1 - \left(\frac{\omega_s}{\omega_p}\right)^2} \tag{8.95}$$

With the values used in this example, the coupling coefficient $k = 0.5$, and the electroacoustic efficiency $\eta_{ea} = 0.49$.

When the transducer's admittance (or impedance) is measured in air, the resonances become sharp and the coupling coefficient can be measured quite accurately. The coupling coefficient of the whole transducer is normally only about 50–60% of the coupling coefficient estimated from the material parameters.

The electrical power in the equivalent circuit of figure 8.19 is dissipated in the resistances $R_p$, $R_m$, and $R_a$ and is therefore given by the real part of $Y^\circ(\omega)$ given by equation (8.86). We can therefore define a power transfer function $|H^\circ(\omega)|$ as

$$|H(\omega)| = \mathrm{Re}\left[Y(\omega)\right] \tag{8.96}$$

Figure 8.22 shows a plot of this function using the values of table 8.2 for water.

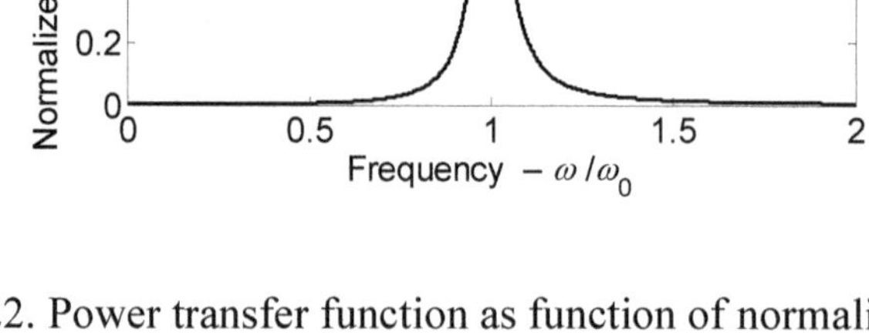


Figure 8.22. Power transfer function as function of normalized frequency.

Generally, a resonator can be characterized by two parameters, the resonance frequency $\omega_0$ and the bandwidth $\Delta\omega$. Another measurement of bandwidth is the quality factor or $Q$-value. $Q$, as a quality factor, is a carryover from electric circuit theory and has little to do with the quality of the transducer, but is another measure of the relative frequency bandwidth. The quality factor $Q$ of a resonant system is, in general, defined as

$$Q = \frac{\omega_0}{\omega_u - \omega_l} \tag{8.97}$$

where $\omega_u$ and $\omega_l$ are the two angular frequencies, above and below resonance, respectively, at which the power has dropped to one-half its resonance frequency value. The maximum value occurs at the resonance frequency $\omega_0$ and is

$$Y_{max} == \frac{1}{R_p} + \frac{1}{(R_m + R_a)} \tag{8.98}$$

If the $Q$-value is high, a series expansion of $|H(\omega)|$ will yield that the bandwidth $\Delta\omega = \omega_u - \omega_l$ is approximately

$$\Delta\omega = \omega_u - \omega_l = \frac{\omega_0}{Q} \tag{8.99}$$

The $Q$-value of the system in figure 8.18 is

$$Q = \frac{\omega_0 M}{R_l + R_m} \tag{8.100}$$

The value in this case is $Q = 20$ and the relative bandwidth is $\Delta\omega / \omega_0 = 0.05$.

Many underwater applications, particularity those involving pulses of short duration or FM transmission require transducers with as wide a relative frequency band as possible. Thus an important design goal is to obtain a sufficient low $Q$-value to ensure the required bandwidth for the actual application. Often this is achieved by connecting electric components to the electrical side of the transducer. Normally, one will choose to neutralize the condenser $C_p$ by coupling an electrical inductor or coil with the inductance $L = 1/(\omega_0^2 C_p)$ in series with the transducer.

### *8.6.3 Reciprocity measurements*

Although commercially available hydrophones specify receiver sensitivity, it may be necessary to verify this. Sometimes, one may wish to use the hydrophone as a transmitter; in such a case, it is also important to determine the transmitter sensitivity. By using the reciprocity principle, which we introduced in section 2.8, we can solve both these tasks.

Figure 8.23 illustrates the principle and the technique. We have a sound source with the source strength $Q$ that, at a distance $r$, produces the sound pressure $p$. The sound pressure is detected by a transducer working as a hydrophone (a receiver) having a receiving sensitivity $M_v$. The measured electric voltage is $V$, as shown in the top drawing in figure 8.23.

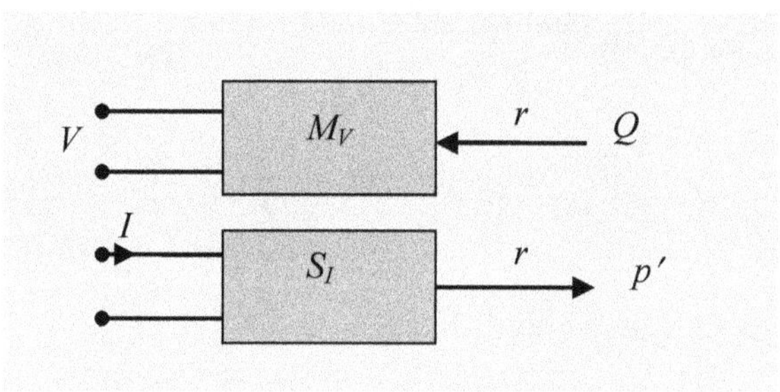


Figure 8.23. The reciprocity principle. Top: the source $Q$ produces sound pressure $p$ at distance $r$. A hydrophone with receiver sensitivity $M_V$ then produces voltage $V$. Bottom: with current input $I$, a transducer with the transmitter sensitivity $S_I$ produces sound pressure $p'$ at distance $r$.

We now assume that the same transducer also is used as a transmitter having a transmitting sensitivity $S_I$. When the transducer is excited with the electrical current $I$, it generates the sound pressure $p'$ at the same distance $r$.

As explained in section 2.8, the reciprocity principle states that

$$\frac{V}{Q} = \frac{p'}{I} \tag{8.101}$$

The reciprocity principle has general validity and applies to all piezoelectric transducers that do not have built-in electric amplifiers.

When we know the sensitivity of voltage $M_V$ of the receiver, we can find the transmitter sensitivity for the current $S_I$. Thus we next assume that, at a distance $r$ in the farfield, the source strength $Q$ produces a sound pressure with the amplitude $p$

$$p = \omega\rho\frac{Q}{4\pi r} \tag{8.102}$$

and the electric voltage is

$$V = M_v \cdot p = M_v \frac{\rho c}{2\lambda r} Q \tag{8.103}$$

With the input current $I$, the transducer produces a sound pressure $p'$

$$p' = S_I I \tag{8.104}$$

By using the reciprocity principle described in equation (8.101), we get the transmitter sensitivity

$$S_I = \frac{p'}{I} = \frac{V}{Q} = M_V \frac{\rho c}{2\lambda r} \tag{8.105}$$

where $\rho c$ is the product of the density and the sound speed of the surrounding medium (water) and $\lambda$ represents the signal wavelength.

The reciprocity factor of spherical waves is defined as

$$J = \frac{2\lambda r}{\rho c} = \frac{2r}{\rho f} \tag{8.106}$$

From equation (8.105), we see that the receiving and transmitting sensitivities are related by

$$S_I = \frac{M_V}{J} \tag{8.107}$$

The condition requiring the spherical propagation of waves and, consequently, of measurements in open space in the farfield was introduced by the assumption in equation (8.102).

In order to determine the sensitivity of voltage of the transmitter, $S_V$, we must take the transducer's input impedance $Z_i$ into account. The input voltage for transmission is $V = Z_i I$; consequently

$$S_V = \frac{M_V}{J|Z_i|} \tag{8.108}$$

As an example, we can take the same hydrophone we used in section 8.3. For frequencies far below resonance, the equivalent circuit in figure 8.5 applies, and since $C << C_p$, it follows that $Z_i \approx (i\omega C_p)^{-1}$. When using the values found in section 8.3 and in equation (8.34), $M_V = 1.37 \cdot 10^{-4}$ V/Pa and $C_p = 3.3$ nF. The transmitter's voltage sensitivity at the reference distance, 1 m, is therefore in this case given as

$$S_V = 1.42 \cdot 10^{-3} f^2 \quad \mu\text{Pa/V} \tag{8.109}$$

or

$$20\log(S_V) = -57 + 40\log f \quad \text{dB re } 1\mu\text{Pa/V} \tag{8.110}$$

where $f$ is the resonance frequency. This is the same as the result we obtained earlier and illustrated in figure 8.8. The transmitter sensitivity, and subsequently the sound pressure, therefore increases with frequency squared. This is a general result for transducers working below the resonance frequency. For higher frequencies, the full expression for input impedance must be used.

Transducer calibration for either transmission or reception is most easily done by using another transducer with a known sensitivity. If this information is not available, we can use the reciprocity principle, as we now demonstrate.

The method presupposes that we have three transducers available: one transmitting transducer ($S$) and one hydrophone ($H$), neither of which needs to be reciprocal, as well as a third transducer ($T$) that must be reciprocal, as it is used both for transmission and reception. In principle, the transducers are positioned in an isosceles triangle, as shown in figure 8.23, but note that other configurations may be required with directional transducers.

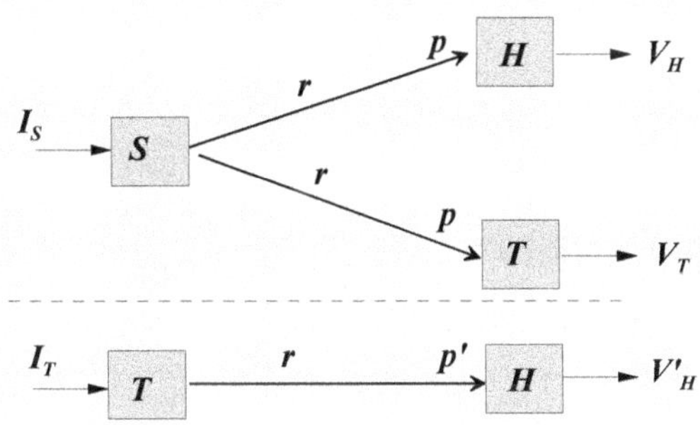


Figure 8.24. Reciprocity calibration. Top: we transmit at $S$ and receive at $H$ and $T$, which are located at the same distance from $S$. Bottom: we transmit from the reciprocal transducer $T$ and receive at the hydrophone $H$.

The task is now to determine the transmitter sensitivity $S_I$ of the transmitting transducer $S$, and the receiving sensitivity $M_V$ of the hydrophone $H$. The transducer $T$ has the receiving sensitivity $M'_V$ and the transmitting sensitivity $S'_V$. We do not have to know these sensitivities. Since $T$ must be reciprocal, it follows that

$$M'_V = JS'_I \tag{8.111}$$

We inject the current $I_s$ to the transmitter $S$, and at the distance $r$ in the farfield we receive the sound pressure $p$, which is expressed as

$$p = p_0 \frac{r_0}{r} = S_I I_S \frac{r_0}{r} \tag{8.112}$$

where $p_0$ is the sound pressure at the reference distance $r_0$. The voltages measured at the hydrophone $H$ and transducer $T$, respectively, become

$$V_H = M_V p = M_V S_I I_s \frac{r_0}{r} \tag{8.113}$$

$$V_T = M'_V p = M'_V S_I I_s \frac{r_0}{r} \tag{8.114}$$

The next step is to transmit from the transducer $T$, using the current $I_T$. Then the measured voltage at hydrophone $H$ is

$$V'_H = M_V S'_I I_T \frac{r_0}{r} \tag{8.115}$$

which together with equation (8.107) gives us

$$V'_H = M_V M'_V \frac{1}{J} I_T \frac{r_0}{r} \tag{8.116}$$

By multiplying equations (8.113) and (8.116), and then dividing their product by equation (8.114), we obtain

$$M_V = \sqrt{\frac{V_H V'_H}{V_T I_T} J \frac{r}{r_0}} \tag{8.117}$$

Therefore, by measuring the voltages $V_H$, $V'_H$, and $V_T$, as well as the current $I_T$, we can calculate the receiver sensitivity $M_V$ with equation (8.117) and, subsequently, the transmitter sensitivity $S_I$ with equation (8.113)

$$S_I = \frac{1}{M_v} \frac{V_H}{I_s} \frac{r}{r_0} \tag{8.118}$$

As mentioned in section 8.6.1, it is assumed that the measurements are taken in open space in the farfield. In reality, taking measurements at low frequencies may require quite long distances for directive transducers. Therefore, calibration techniques based on measurements in the nearfield have been developed (Bobber 1970), but nearfield measurements will not be pursued here.

# *Chapter 9*

# *Acoustic projectors and receivers*

## 9.1 Introduction

In chapter 8 we considered the design and operations of single transducer elements. Normally, many such elements are used together in order to satisfy given requirements for power output and directivity. An example was shown in figure 8.17, where several elements are mounted to form a larger plane projector. Other frequently used configurations mount the transducer elements on a sphere or a cylinder. In some systems, a number of hydrophones are arranged in the form of a line array, used vertically or horizontally. For instance, towed hydrophone arrays with several hundred elements are used in marine seismic applications and in some surveillance systems.

In all such systems and configurations, the signals received or transmitted from the individual element are coupled together after being subjected to changes in amplitude and in time delays, or in phase shifts. This is called ***beamforming*** and ***beamsteering***, having the purposes of achieving certain spatial characteristics in the steered direction with respect to maximum transmission or receiving sensitivity. In this chapter, we shall present some basic beamforming theory and discuss the need to satisfy specified requirements as to beamwidth and sidelobe levels. The same theory applies to both transmission and reception; we will consider each in turn. We shall, however, only examine what can be considered conventional beamforming. For more elaborate techniques of array processing, such as high-resolution techniques or adaptive beamforming, the book by Van Trees (2002) is recommended.

## 9.2 Radiation and directivity of lines and plates

The calculation of the acoustic field formed from a number of sources is based on the principle of superposition, which holds that the total field may be found by summing up the partial contributions of all the separate elements. According to this principle, the transducer is considered to be divided into many elements, each element being much smaller than the wavelength, so that each can be considered to be a point source.

In section 2.7, equation (2.101), we found that the field from a point source at frequency $\omega$ at a range $r$ is given by

$$\Phi = \frac{Q}{4\pi r}\exp(i\kappa r) \tag{9.1}$$

where $\Phi$ is the velocity potential, $Q$ is the source strength, and $\kappa$ is the wave number. The sound pressure $p$ and the particle velocity $v$, both of which have a radial component, are expressed as

$$p = -\rho\frac{\partial\Phi}{\partial t} \tag{9.2}$$

$$v = \frac{\partial\Phi}{\partial r} \tag{9.3}$$

For a collection of discrete point sources, each with source strength $Q_i$ in positions given by the vectors $\mathbf{r}_i$, the sound field at the field point $\mathbf{r} = (x_0, y_0, z_0)$ is the sum of all the contributions (see figure 9.1). Therefore the total sound field at the point $\mathbf{r}$ is the sum $\Phi$, which is expressed as

$$\Phi = \sum_i \frac{Q_i}{4\pi|\mathbf{r}-\mathbf{r}_i|}\exp(i\kappa|\mathbf{r}-\mathbf{r}_i|) \tag{9.4}$$

Note that the time-dependence $\exp(-i\omega t)$ is implicit in the ensuing discussion.

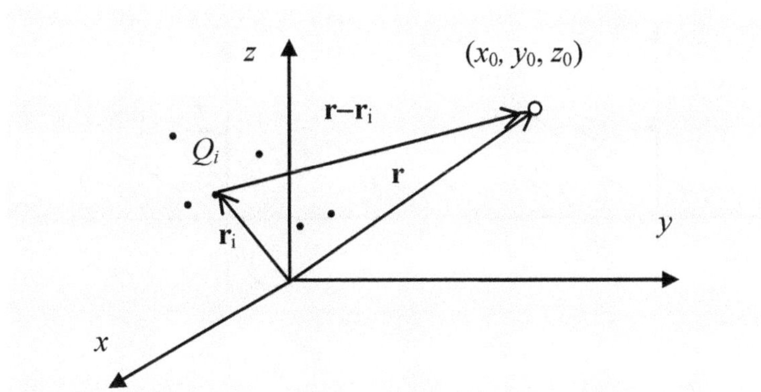


Figure 9.1. All point sources $Q_i$ contribute to the field at $\mathbf{r} = (x_0, y_0, z_0)$.

When the sources are continuously distributed over a surface $S$, the sum is replaced with an integral over all sources, so that

$$\Phi = \int_S \frac{dq}{4\pi|\mathbf{r}-\mathbf{r}_i|}\exp\left(i\kappa|\mathbf{r}-\mathbf{r}_i|\right) \tag{9.5}$$

where $dq$ is the spatial distribution of the source strength.

The simplest example of an acoustic array is the dipole (figure 9.2). The dipole has two sources, which are separated by the distance $d$. The two sources have equal source strengths but opposite polarities, so that $Q_1 = -Q$ and $Q_2 = Q$.

The distance from a source point at position $x$ to the receiver point is

$$|\mathbf{r}-\mathbf{r}_i| \approx r \mp x\sin\theta + \frac{1}{2}\left(\frac{x^2}{r}\right) + \ldots \tag{9.6}$$

The three first terms of the power series expansion of equation (9.6) are

$$|\mathbf{r}-\mathbf{r}_i| = \sqrt{(x_0 \mp x)^2 + z_0^2} = r\sqrt{1 \mp \frac{2x\sin\theta}{r} + \left(\frac{x}{r}\right)^2} \tag{9.7}$$

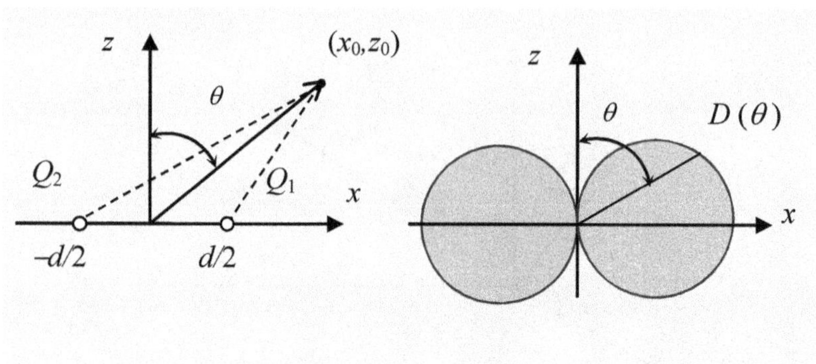


Figure 9.2. Left, the acoustic dipole has two equal-strength sources in opposite phase, with distance separation *d*. Right, the angular distribution of the combined radiation.

When the distance between the two sources at $x = \pm d/2$, is much smaller than the distance to the field point **r**, that is, when $d << |\mathbf{r}|$, we may apply the approximation

$$\left|\mathbf{r} - \mathbf{r}_{1,2}\right| = r \mp \frac{1}{2} d \sin\theta \tag{9.8}$$

This expression is used in the exponent of equations (9.4) and (9.5) in order to account for the phase difference between the signals from the two point sources. In the denominator, it is sufficiently accurate to use

$$\left|\mathbf{r} - \mathbf{r}_{1,2}\right| = r \tag{9.9}$$

We can visualize the approximation as shown in figure 9.3, which shows a point source at height $r$ above a plane surface. The field from the source spreads spherically and the distance from the source to a given point $x$ on the surface is given by equation (9.6). The effect of the wave front curvature is

$$\Delta r \approx \frac{1}{2}\left(\frac{x^2}{r}\right) \tag{9.10}$$

This term is neglected in the farfield approximation expressed by equation (9.8), and for $r >> x^2$, $\Delta r \approx 0$.

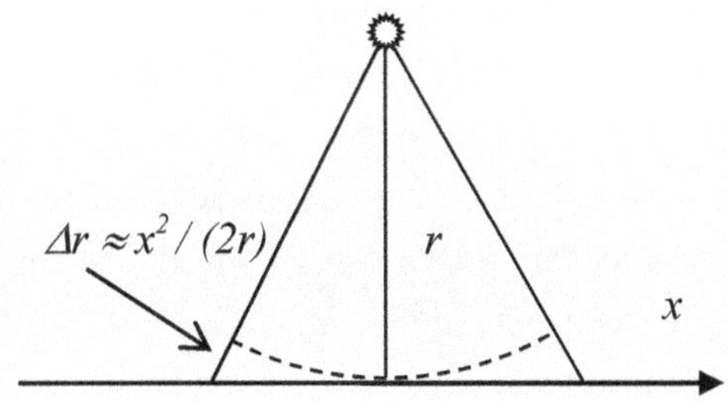


Figure 9.3. A point source at height $r$ over a plane surface. With the farfield approximation, the effect of the field curvature, $\Delta r \approx = x^2 / 2r$, is neglected.

In the farfield the sound field is thereby given as

$$\begin{aligned} \Phi &= \frac{Q}{4\pi r}\left[\exp\left(i\kappa\frac{d}{2}\sin\theta\right) - \exp\left(-i\kappa\frac{d}{2}\sin\theta\right)\right]\exp(i\kappa r) \\ &= i\frac{Q}{2\pi r}\sin\left(\kappa\frac{d}{2}\sin\theta\right)\exp(i\kappa r) \end{aligned} \tag{9.11}$$

This expression can be further approximated if the distance $d$ is considerably smaller than the wavelength, so that $\kappa d << 1$, which gives the acoustic field $\Phi$ as

$$\Phi = i\frac{Q\kappa d}{4\pi r}\sin\theta\exp(i\kappa r) \tag{9.12}$$

and the sound pressure $p$ as

$$p = -\omega\rho\frac{Q\kappa d}{4\pi r}\sin\theta\exp(i\kappa r) \tag{9.13}$$

Equation 9.13 shows that the sound pressure from the dipole decreases proportionally with distance, and, as shown in the right side of figure 9.2, has a directional dependence given by

$$B(\theta) = \sin\theta \tag{9.14}$$

The approximations given by equations (9.8) and (9.9), and used respectively in the exponents and the denominators of equations (9.4) and equation (9.5), are called the ***farfield approximation***. This approximation implies that in the exponent, which represents the phase or time delays, we take into account the two sources' different distances to the field point by using the first two terms of the series expansion. For the denominator, which controls the amplitude, the relatively small distance differences have a lesser impact, so we may use the same distance for all source points. In most of this chapter we assume that the farfield approximation is valid; the exception is in section 9.3, where we shall discuss the farfield approximation and nearfield effects in more detail. Now we apply equations (9.4) and (9.5) to calculate the directivity functions of some common arrays.

### *9.2.1 Line arrays*

Consider a line array in the form of a thin cylinder with length $L$ and a small radius compared to the wavelength, as shown in figure 9.4. The element is characterized by a source strength $q(x)$ per length unit.

When the farfield conditions are satisfied, the acoustic field $\Phi$ becomes

$$\Phi = \frac{\exp(i\kappa r)}{4\pi r} \int_{-L/2}^{L/2} q(x)\exp(-i\kappa x \sin\theta)\,dx \tag{9.15}$$

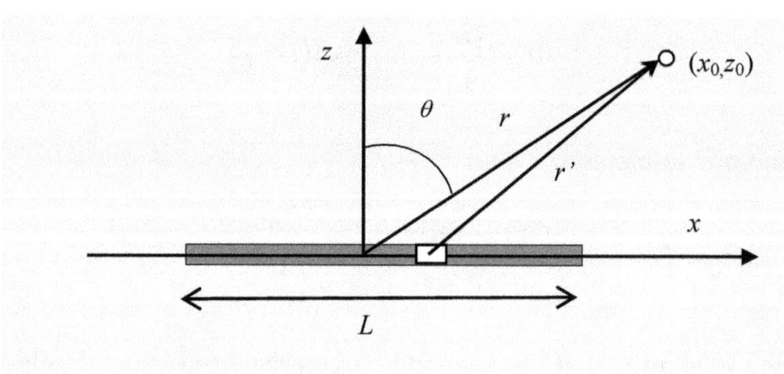


Figure 9.4. Line array with length $L$. The acoustic field is found by integrating the contributions from the whole line.

The directional distribution of the sound field, $B(\theta)$, which may also be referred to as either the ***directivity function*** or the ***beam pattern***, is defined by the integral in equation (9.15). For a line array, the directivity function becomes a body of revolution expressed as

$$B(\theta)=\int_{-L/2}^{L/2} q(x)\exp(-i\kappa x\sin\theta)\,dx \tag{9.16}$$

Note that the weighting or aperture function $q(x)$ describes an array's excitation or source strength where the array is a transmitter or, alternatively, the array's sensitivity distribution where it is a receiver. We assume that $q(x)$ is normalized in such a way that integration over the whole length of the array is equal to one

$$\int_{-L/2}^{L/2} q(x)\,dx=1 \tag{9.17}$$

From equation (9.16), we see that the directivity function of a line array is the Fourier transform of the aperture function. The spatial angular frequency $\Omega$, is defined as

$$\Omega=\kappa\sin\theta=(2\pi/\lambda)\sin\theta \tag{9.18}$$

We now calculate the directivity function for different types of the aperture functions $q(x)$, starting with the simplest function that is constant, or uniform, over the length of the array, which is the aperture function

$$q(x)=\frac{1}{L},\quad |x|\le\frac{L}{2} \tag{9.19}$$

When this expression is inserted into equation (9.16), after some calculation, we obtain the directivity function

$$B(\theta)=\frac{\sin\left(\pi\frac{L}{\lambda}\sin\theta\right)}{\left(\pi\frac{L}{\lambda}\sin\theta\right)} \tag{9.20}$$

so that the uniform aperture function yields a directivity function in the form of a $\sin(x)/x$ function. Note that all beams for a line array or projector are surfaces of revolution. In figure 9.5, we can see that this directivity function has a lobe structure with a maximum, or mainlobe, at $\theta = 0°$, and with a number of secondary maxima (sidelobes).

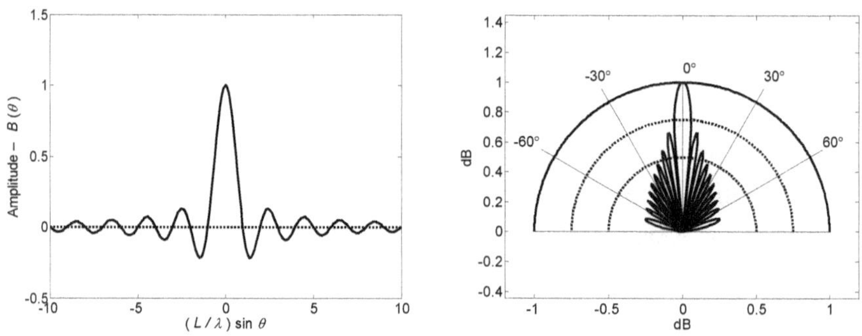


Figure 9.5. The normalized directivity function of a uniformly excited line array with the length $L = 10\lambda$. Left panel: $B(\theta)$ plotted in rectangular coordinates as a function of $(L/\lambda)\sin\theta$. Right panel: $B(\theta)$ plotted in polar coordinates, in dB with a 40 dB range.

Now we discuss the structure of the directivity function in more detail. First, we notice that, because of the normalization of equation (9.17), the mainlobe has amplitude equal to unity. The directivity function has zeros at angles given by

$$\sin\theta = n\frac{\lambda}{L} \qquad n = \pm 1, \pm 2, \ldots \tag{9.21}$$

The first zero of the directivity function occurs at the angle where $(L/\lambda)\sin\theta = \pm 1$. At the angles given by $\sin\theta = \lambda/(2L)$, the amplitude is $2/\pi$. The width of the mainlobe, $\Delta\theta$, can be defined in several ways. It is quite common to define the width of the mainlobe as $\sin(\Delta\theta/2) = \lambda/(2L)$. When $L >> \lambda$, this definition gives the mainlobe width in radians as approximately

$$\Delta\theta \approx \frac{\lambda}{L} \tag{9.22}$$

The sidelobes occur for angles given by

$$\sin\theta_n = (2n+1)\frac{\lambda}{2L} \qquad n = \pm 1,\ 2,\ 3,.... \tag{9.23}$$

At these angles, the amplitudes of the sidelobes are determined from equation (9.20) to be

$$\left|B(\theta_n)\right| = \frac{2}{\pi(2n+1)} \tag{9.24}$$

The first sidelobe ($n$ = 1) consequently has a maximum amplitude of 2 / (3π), and therefore is only 20log[2 / (3π)] = 13.5 dB under the maximum amplitude of the mainlobe. The amplitudes of the other sidelobes are also quite high, as they only decrease linearly with the angle.

The absolute sound pressure disappeared with the normalization of equation (9.16). The sound pressure $p$, as a function of distance $r$ and direction angle $\theta$, can be written

$$p = p_0\frac{r_0}{r}B(\theta) \tag{9.25}$$

where $r_0$ is a reference distance where the sound pressure on the axis $\theta = 0$ is $p_0$. The value of $p_0$ can be determined from the total radiated power $W$, which is found by integrating over the whole sphere. Taking the rotation symmetry into account, as indicated in figure 9.6, for a line array $W$ is given by

$$W - \int_{-\pi/2}^{\pi/2} I(r)2\pi r^2 \cos O dO \tag{9.26}$$

For an omnidirectional source with the same power the intensity $I_{ref}$ at distance $r$ is

$$I_{ref} = \frac{W}{4\pi r^2} \tag{9.27}$$

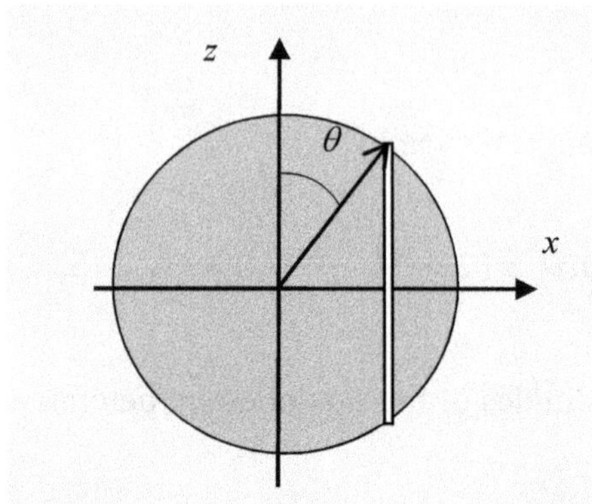


Figure 9.6. Integration to determine a total radiated power.

The directivity factor $D$ is generally defined as the ratio

$$D = \frac{I(\theta = 0)}{I_{ref}} \tag{9.28}$$

This ratio is illustrated in figure 9.7, where we compare the maximum radiation (or sensitivity) of a directional array $I(\theta = 0)$ with the uniform intensity of the point source $I_{ref}$ (or the sensitivity of a point receiver). The directivity index $DI$ is defined by the logarithmic ratio

$$DI = 10 \log D \tag{9.29}$$

Generally, the directivity factor $D$ must be calculated by numerical integration of equation (9.26). For very directive arrays with small beamwidths, the integral will give important contributions only within the beamwidth $\Delta\theta$, so that $D$ becomes approximately

$$D \approx \frac{2}{\Delta\theta} \tag{9.30}$$

The beamwidth of a uniform aperture array is approximately as given by equation (9.22). The directivity factor of a line array is therefore approximately

$$D = \frac{2L}{\lambda} \tag{9.31}$$

With a directive source, the sound pressure $p$ on the axis at a distance $r$ is expressed as

$$p^2 = W \frac{\rho c}{4\pi r^2} D \tag{9.32}$$

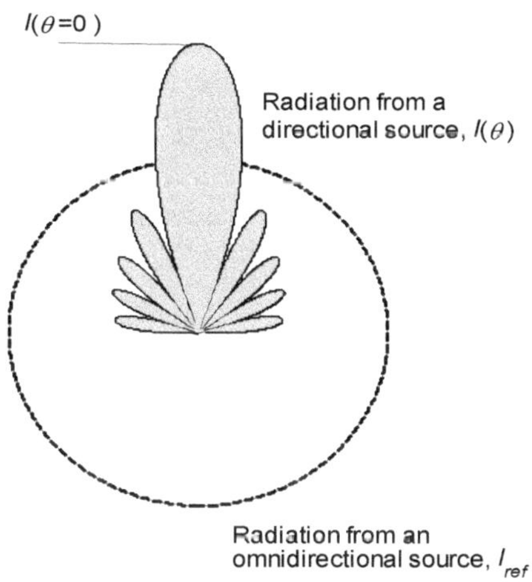


Figure 9.7. The directivity factor is defined by comparing the maximum intensity $I(\theta = 0)$ of a directional array with the intensity of a point source with intensity $I_{ref}$.

In section 2.6, we discussed radiation from a point source and found that sound intensity reduces with range as $10\log I(r) = 10\log I(r_0) - 20\log(r / r_0)$, as given by equation (2.121). In equation (9.32), we have included the directivity of the sound source, which causes the intensity on the sound axis to increase with the factor $D$. By inserting $\rho c = 1.5 \cdot 10^6$ kgs/m$^2$ and changing the unit of measure to decibels, we find the sound level $I$ in dB relative to 1 μPa to be

$$20\log p = 170.8 + 10\log W + DI - 20\log(r) \tag{9.33}$$

The source level *SL* of an acoustic source is defined as the sound pressure level in dB relative to 1 μPa at a reference distance of $r_0$ = 1 m. The source level of a sound source with directivity index *DI* and acoustically radiating watts *W* is therefore

$$SL = 170.8 + 10\log W + DI \tag{9.34}$$

The numerical factor in equation (9.34) applies to water.

For example, a sonar with a 20 dB directivity index and radiating 1 kW acoustic power has a source level of 220.8 dB. Although we use a reference distance of 1 m in the source level definition, the sound level is not necessarily measured at that distance. Usually, it is necessary to measure the sound level at a much greater distance in the farfield; see section 9.3 and figure 9.14.

### *9.2.2 Two-dimensional arrays*

We determine the directional radiation from plane transducers of various shapes, using equation (9.5) as the basis.

Consider first the plane rectangular array in figure 9.8. The array is in the *x*-*y* plane where $z = 0$. The coordinate of a source point is $(x_n, y_n)$ and the rectangular coordinates of the field point are $(x, y, z)$. The rectangular coordinates are related to the spherical coordinates by

$$\begin{aligned} x &= r\sin\theta\cos\varphi \\ y &= r\sin\theta\sin\varphi \\ z &= r\cos\theta \end{aligned} \tag{9.35}$$

The distance between a source point with coordinates $(x', y')$ in the *x*-*y* plane and the field point given by the spherical coordinates $r$, $\theta$, and $\varphi$ are, from figure 9.8, given by

$$r' = \sqrt{(r\sin\theta\cos\varphi - x')^2 + (r\sin\theta\sin\varphi - y')^2 + (r\cos\theta)^2} \tag{9.36}$$

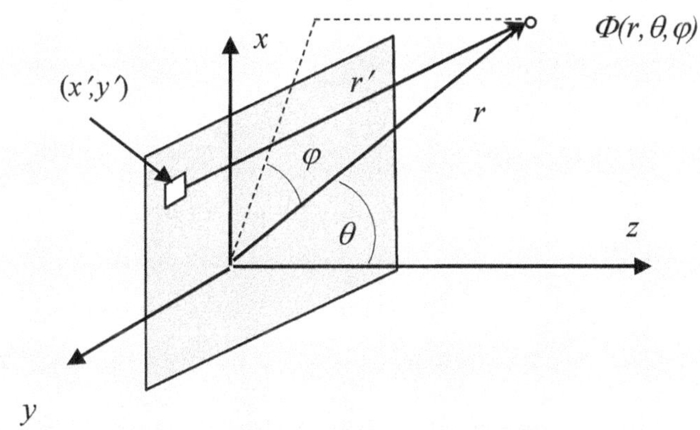


Figure 9.8. A rectangular array. Integration over the whole rectangle gives the acoustic field Φ in the position $(r, \theta, \varphi)$.

The farfield approximation of this expression is

$$\begin{aligned} r' &\approx \sqrt{r^2 - 2rx'\sin\theta\cos\varphi - 2ry'\sin\theta\sin\varphi} \\ &\approx r - x'\sin\theta\cos\varphi - y'\sin\theta\sin\varphi \end{aligned} \tag{9.37}$$

Consequently, the acoustic field at a given point in the farfield can be expressed by

$$\Phi = \frac{\exp(i\kappa r)}{4\pi r}\int_x\int_y q(x,y)\exp\left[-i\kappa\sin\theta(x\cos\phi + y\sin\varphi)\right]dxdy \tag{9.38}$$

The aperture function, or excitation, is $q(x,y)$. The integration is over the whole area of a rectangle with dimensions $L_x$ and $L_y$. Again, the equation shows that the sound field is given as the Fourier transform of the excitation; this time, it is a two-dimensional Fourier transformation.

In a particularly simple case in which the excitation is uniform across a rectangle with dimensions $L_x$ and $L_y$, that is, where

$$q(x,y)=\frac{1}{L_x L_y} \qquad |x|\le\frac{1}{2}L_x \text{ and } |y|\le\frac{1}{2}L_y$$

otherrwise (9.39)

$$q(x,y)=0$$

We find the directivity function to be

$$B(\theta,\varphi)=\frac{\sin\left[\pi\frac{L_x}{\lambda}(\sin\theta\cos\varphi)\right]}{\left[\pi\frac{L_x}{\lambda}(\sin\theta\cos\varphi)\right]}\frac{\sin\left[\pi\frac{L_y}{\lambda}(\sin\theta\sin\varphi)\right]}{\left[\pi\frac{L_y}{\lambda}(\sin\theta\sin\varphi)\right]} \tag{9.40}$$

We see that the directional function is given by the product of two sin($x$)/$x$-functions oriented perpendicular to each other. An example appears in figure 9.9 for the case in which $L_x = 8\lambda$ and $L_y = 4\lambda$.

The directivity factor $D$ is found by integration over all angles; the general expression for the directivity factor for any two-dimensional array is

$$D=\frac{4\pi}{\int_0^{2\pi}\int_{-\pi/2}^{\pi/2}|B(\theta,\phi)|^2\sin\theta\, d\theta\, d\varphi} \tag{9.41}$$

For large arrays with sides $L_x$ and $L_y$ that are much larger than the wavelength, the beamwidths in the two directions are, respectively,

$$\Delta\theta=\frac{\lambda}{L_x}$$
$$\Delta\varphi=\frac{\lambda}{L_y} \tag{9.42}$$

Therefore, the directivity factor becomes approximately

$$D \approx \frac{4\pi}{\Delta\theta\Delta\varphi} = 4\pi\frac{L_x L_y}{\lambda^2} \tag{9.43}$$

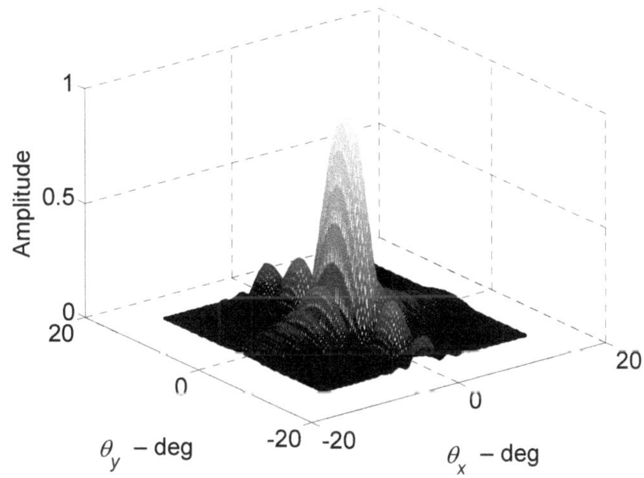


Figure 9.9. The directional function of a rectangular array with dimensions $L_x = 8\lambda$ and $L_y = 4\lambda$.

Since the area $A$ of a rectangle is $A = L_x L_y$, this directivity factor is approximately

$$D \approx 4\pi\frac{A}{\lambda^2} \tag{9.44}$$

This is a general, though approximate, expression of the directivity factor, which with caution can be used for all projectors or receivers with dimensions that are much larger than the signal wavelength.

The next example concerns radiation from a circular disk array with the radius $a$ and with uniform excitation (see figure 9.10). We apply the general expression for the acoustic field in the farfield, equation (9.38). Since there is symmetry about the axis, we will consider only the plane defined by $\varphi = 0$. A narrow strip of the disk, as indicated in the figure, vibrating with an amplitude of $u_0$, has the source strength

$$q(x) = 2au_0 \sin\alpha \tag{9.45}$$

When we use equation (9.38) with $\phi = 0$, we can express the field as

$$\Phi(r,\theta) = \frac{\exp(i\kappa r)}{4\pi r}\int \exp(-i\kappa x \sin\theta) 2au_0 \sin\alpha dx \tag{9.46}$$

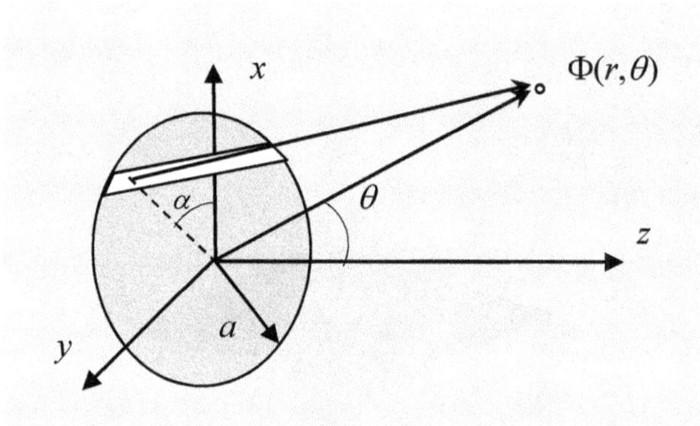


Figure 9.10. Determination of the directivity function of a circular disk array with radius a.

We substitute for $x$ by setting $x = a\cos\alpha$ and $dx = -a\sin\alpha d\alpha$, and find that

$$\Phi(r,\theta) = -\frac{\exp(i\kappa r)}{4\pi r}\int_0^{\pi} \exp(-i\kappa a \sin\theta\cos\alpha) 2a^2 u_0 \sin^2\alpha\, d\alpha \tag{9.47}$$

The integrand has a real part and an imaginary part, but the latter integrates to zero because of the axial symmetry. The real part is a standard integral with the solution

$$\int_0^{\pi} \cos(\kappa a \sin\theta\cos\alpha)\sin^2\alpha d\alpha = -\pi\frac{J_1(\kappa a\sin\theta)}{\kappa a\sin\theta} \tag{9.48}$$

$J_1(\cdot)$ is the Bessel function of the first type and first order. The acoustic farfield of a circular disk is, therefore,

$$\begin{aligned}\Phi(r,\theta) &= \frac{(\pi a^2 u_0)}{4\pi r}\frac{2J_1(\kappa a\sin\theta)}{\kappa a\sin\theta}\exp(i\kappa r) \\ &= \frac{Q}{4\pi r}\frac{2J_1(\kappa a\sin\theta)}{\kappa a\sin\theta}\exp(i\kappa r)\end{aligned} \tag{9.49}$$

The source strength is $Q$ (volume displacement), which in the case of a disk array is given as

$$Q = \pi a^2 u_0 \tag{9.50}$$

The directivity function of a uniformly excited circular disk is, therefore,

$$B(\theta) = \frac{2J_1(\kappa a\sin\theta)}{\kappa a\sin\theta} \tag{9.51}$$

Figure 9.11 shows an example of a two-dimensional array diagram of a circular disk with the radius $a = 5\lambda$, calculated using equation (9.51). The figure shows the array beam pattern expressed in decibels (that is, as $20\log|B(\theta)|$).

By closer study, we see that the distance to the first zero is given by $ka\sin\theta = 3.83$, or approximately by $ka\sin\theta \approx 4$. For small angles, $\sin\theta \approx \theta$, and the width of the mainlobe is approximately

$$\Delta\theta \approx \frac{2\lambda}{\pi a} \tag{9.52}$$

The directivity factor becomes approximately

$$D \approx \frac{4\pi}{\pi(\Delta\theta/2)^2} = 4\pi\frac{\pi a^2}{\lambda^2} \tag{9.53}$$

Since $A = \pi a^2$, this is the same result that we found with equation (9.44) for the rectangular array. This agreement is consistent with the statement that equation (9.44) is approximately valid for the directivity factor of large two-dimensional arrays.

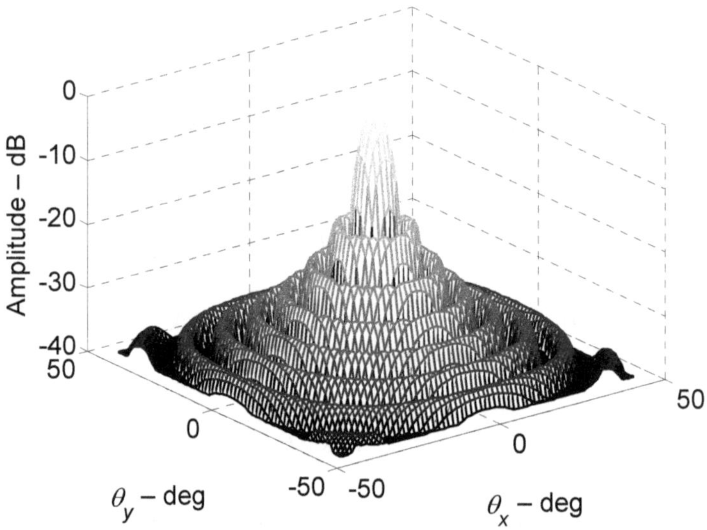


Figure 9.11. The directivity function $20\log|B(\theta)|$ of a plane circular disk array with radius $a = 5\lambda$.

## 9.3 Nearfield, farfield, and focusing

So far, we have assumed that the field point, or receiver, lies so far from the transmitter that we can accurately apply the farfield approximation of equation (9.8) to the exponent and that of equation (9.9) to the denominator of equation (9.5). This is a valid assumption in most practical uses of acoustic systems in the sea. In other cases, we may be interested in the field at distances so short that this assumption is not valid. This rarely happens in marine acoustic applications, but is common in ultrasound applications in medicine and materials research.

In figure 9.12, we consider a circular disk with the radius $a$. An infinitesimally small circular element of the disk has radius $\sigma$ and width $d\sigma$ as indicated in the figure. When the vibration amplitude is constant everywhere on the projector and equals $u_0$, the circular element has the source strength $dQ = 2\pi\sigma u_0 d\sigma$.

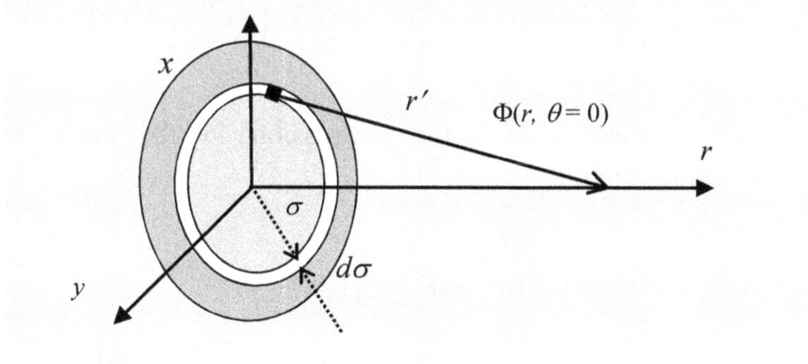


Figure 9.12. Calculation of the field on the axis of a circular piston.

The total field at distance $r$ is obtained by integration over the whole disk. Looking at the general expression of radiation from a source, equation (9.5), we find that it is very difficult to calculate the field in the general case because the distance $|\mathbf{r} - \mathbf{r}_i|$ is different for all points on the source. Therefore to simplify the calculation, we calculate only the field on the axis where $\theta = 0$, as shown in figure 9.12.

The field at distance $r$, which was given in equation (9.5), is now

$$\Phi(r, \theta = 0) = \int_0^a \frac{2\pi u_0}{4\pi\sqrt{r^2 + \sigma^2}} \exp\left(i\kappa\sqrt{r^2 + \sigma^2}\right)\sigma d\sigma \tag{9.54}$$

Compared with the expression of the farfield in equation (9.38), here we are using the exact expression for the distance between the source point and field point, rather than the approximation used earlier. In order to solve the integral, we introduce a new variable $w$ where $w = (r^2 + \sigma^2)^{1/2}$

$$\begin{aligned}\Phi(r, \theta = 0) &= \frac{u_0}{2} \int_r^{\sqrt{r^2+a^2}} \exp(i\kappa w) dw \\ &= \frac{u_0}{2i\kappa}\left\{\exp\left(i\kappa\sqrt{r^2 + a^2}\right) - \exp(i\kappa r)\right\}\end{aligned} \tag{9.55}$$

By introducing the source strength $Q$ as earlier defined in equation (9.50), we can rewrite equation (9.55) as

$$\Phi(r,\theta=0)=\frac{Q}{4\pi r}\exp(i\kappa r)\frac{4r}{2i\kappa a^2}\left\{\exp\left[i\kappa\left(\sqrt{r^2+a^2}-r\right)\right]-1\right\} \quad (9.56)$$

For great distances where $r\to\infty$ the expression in the curly brackets in equation (9.56) is approximately

$$\left\{\exp\left[i\kappa\left(\sqrt{r^2+a^2}-r\right)\right]-1\right\}\approx i\kappa\frac{a^2}{2r} \quad (9.57)$$

Consequently, at great distances the field on the axis becomes asymptotically

$$\Phi(r,\theta=0)=\frac{Q}{4\pi r}\exp(i\kappa r) \quad (9.58)$$

Figure 9.13 shows a plot of $\Phi(r,\theta=0)$ as a function of the distance calculated in equation (9.56). In this case, $\kappa a=2\pi a/\lambda=25$, and $a/\lambda$ is the disk radius, measured in wavelengths. The dashed line in figure 9.13 shows the sound field as given by the farfield approximation, equation (9.58). This means that, when great distances are involved, equation (9.56) will asymptotically approach the expression for the field of a point source. We also see in figure 9.13 that the field will oscillate strongly in the area closest to the projector. Further calculations reveal that the number of oscillations, as well as the range in which they take place, increases proportionally as the product $\kappa a$ increases. The shorter ranges where the sound field oscillates constitute the nearfield, and the farther ranges, where the field decreases linearly with range, are the farfield.

The transition between the farfield and the nearfield can be defined by considering the argument of the second exponential function in equation (9.56). This argument is approximately

$$\kappa\left(\sqrt{r^2+a^2}-r\right)\approx\kappa\frac{a^2}{2r} \quad (9.59)$$

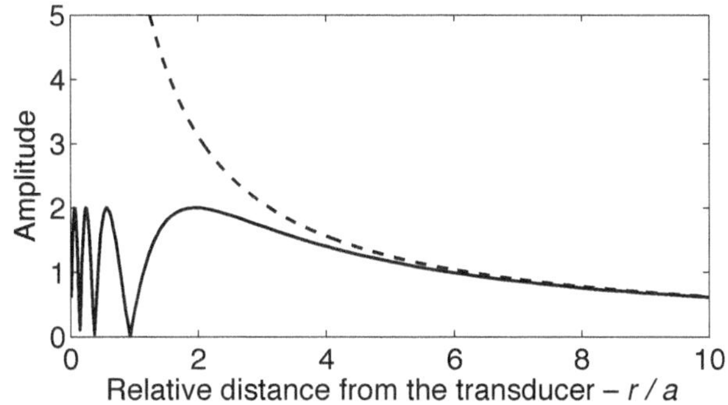


Figure 9.13. The sound field on the axis of a circular disk with radius *a*, and $\kappa a = 25$, plotted as a function of the relative distance *r/a*. The dashed line is the farfield approximation.

When this term is sufficiently small, we are in the farfield; otherwise we are in the nearfield. A common definition of "sufficiently" small is the requirement that

$$\kappa \frac{a^2}{2r} \leq \frac{\pi}{4} \tag{9.60}$$

The transition between the nearfield and the farfield takes place approximately at the Rayleigh distance $R_R$, which is defined as the distance at which the equality in equation 9.57 is valid

$$R_R = \frac{4a^2}{\lambda} \tag{9.61}$$

The criterion in equation (9.60) implies a maximum difference in phase angle of less than 45° for the contributions of the different parts of the array. Within the nearfield, the contributions from the different parts of the array will have different phase angles, thus alternately interfering destructively and constructively, making the field oscillate rapidly with distance, as can be seen in figure 9.13.

It should be clear that this calculation also must be valid for either a line array or a rectangle with side length *L*. By inserting $L = 2a$ into equation (9.61), we get that the Rayleigh distance for a line array with length *L* is

$$R_R = \frac{L^2}{\lambda} \tag{9.62}$$

We can illustrate the transition from nearfield to farfield with a simple sketch, shown in figure 9.14. Here we have drawn the extent of the beam when using the definition of the beamwidth for a line array in equation (9.22) or for a circular piston in equation (9.52).

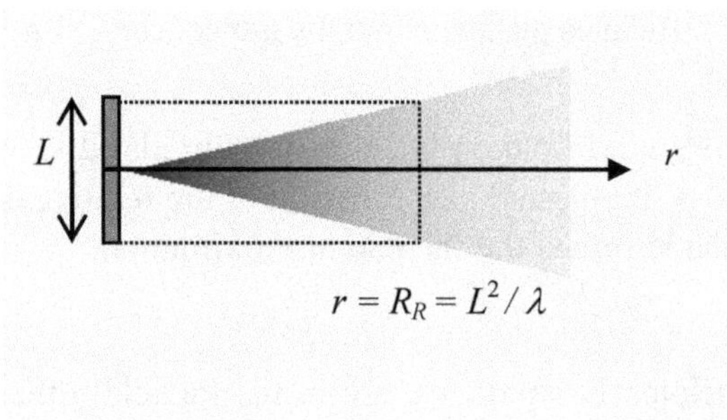


Figure 9.14. Illustration of the Rayleigh distance: at the range $r$, the width of the mainlobe spans the length of the array.

We see that for $r = R_R$, the beam spans a width equal to the array length, or the array diameter, that is

$$\begin{aligned} \Delta\theta \cdot R_R &= L \qquad \text{line or rectangle} \\ \Delta\theta \cdot R_R &= 2a \qquad \text{circular disk} \end{aligned} \tag{9.63}$$

Using the expressions for the beamwidths of equations (9.22) and (9.52), respectively, we get the Rayleigh distances

$$\begin{aligned} R_R &= L^2/\lambda \qquad \text{line or rectangle} \\ R_R &= \pi a^2/\lambda \qquad \text{circular disk} \end{aligned} \tag{9.64}$$

For a line array the Rayleigh distance is given by its length squared; for a two-dimensional array the Rayleigh distance is by the area of transducer surface divided by the wavelength.

### 9.3.1 Focusing

In underwater acoustics, we are primarily interested in the farfield since the Rayleigh distance and the extension of the nearfield are normally quite small. However, in other uses such as ultrasound applications, the uses of focused transducers are relevant.

Focusing can be created by introducing a time delay for the arrival of that part of the sound originating from the center of the projector, relative to the contributions coming from the outer areas of the projector. The time delay will then cause the separate contributions to reach the focal point simultaneously. The principle is illustrated in figure 9.15, where an acoustic lens is attached to the projector surface. This lens, in the shape of a concave disk, must be made from a material with sound speed higher than that of the surrounding medium.

We now consider a circular transducer with radius $a$ and vibration amplitude $u_0$ (see figure 9.15) and we follow a ray path that emerges from the transducer at a position $x$ from the axis. It will reach a point on the axis at a distance $r$. The length of this path is the distance $L(x)$

$$L(x) = \sqrt{r^2 + x^2} \approx r + \frac{x^2}{2r} \tag{9.65}$$

Given that in this approximation $x \ll r$, equation (9.65) shows that the distance $r$ from one point on the lens to a field point on the axis increases proportionally with the square of the distance from the axis, $x^2$.

The spherical lens has the radius of curvature $R_L$ (see figure 9.15), so as a function of radial distance from the axis, the length of the ray path within the lens is

$$L_l(x) = R_L - \sqrt{R_L^2 - x^2} \approx \frac{x^2}{2R_L} \tag{9.66}$$

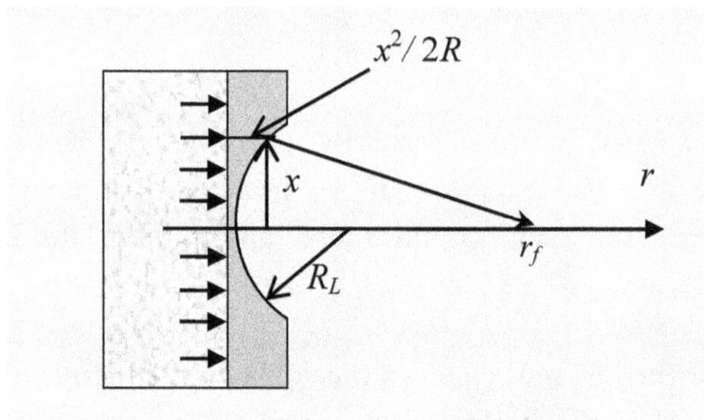


Figure 9.15. A plane piston transducer is focused to the focal length $r_f$ by attaching a spherical cap with radius of curvature $R_L$ to the transducer's front surface.

When the sound speed of the lens material is $c_l$ and the sound speed of the external medium is $c_m$, the total travel time of the ray path is approximately

$$T = \frac{L(x) - L_l(x)}{c_m} + \frac{L_l(x)}{c_l} = \frac{1}{c_m}\left(r + \frac{x^2}{2r} - \frac{x^2}{2R_L}\right) + \frac{1}{c_l}\left(\frac{x^2}{2R_L}\right) \tag{9.67}$$

The acoustic field on the axis therefore approximates

$$\Phi(r) = \frac{1}{4\pi r}\exp\left(i\frac{\omega}{c_m}r\right)\int_0^a 2\pi u_0 x \exp\left[\frac{1}{2}i\omega x^2\left(\frac{1}{rc_m} - \frac{1}{R_L c_m} + \frac{1}{R_L c_l}\right)\right]dx \tag{9.68}$$

The focal length $r_f$ of the lens is defined as the range that makes the exponent in the integrand of equation (9.68) equal to zero, which produces

$$r_f = \frac{R_L}{1 - \frac{c_m}{c_l}} \tag{9.69}$$

Consequently, the field $\Phi(r)$ on the axis can be written as a function of distance $r$ so that

$$\Phi(r)=\frac{Q}{4\pi r}\exp(i\kappa_m r)\int_0^a \frac{2x}{a^2}\exp\left[i\kappa_m\frac{x^2}{2r}\left(1-\frac{r}{r_f}\right)\right]dx \tag{9.70}$$

where we have defined $Q$ and $\kappa_m$ so that $Q=\pi a^2 u_0$ and $\kappa_m=\omega/c_m$.

The integral in equation (9.70) is elementary, and the solution gives the field on the axis as a function of range so that

$$\Phi(r)=\frac{Q}{4\pi r}\exp(i\kappa_m r)\frac{2r}{i\kappa_m a^2\left(1-\frac{r}{r_f}\right)}\left\{\exp\left[i\kappa_m\frac{a^2}{2r}\left(1-\frac{r}{r_f}\right)\right]-1\right\} \tag{9.71}$$

When $r=r_f$, the expression in the curly brackets becomes unity, so that the field at the focal distance $r=r_f$ is expressed as

$$\Phi(r=r_f)=\frac{Q}{4\pi r_f}\exp(i\kappa_m r_f) \tag{9.72}$$

Without the lens the focal length become infinite and the amplitude of the field at range $r$ becomes

$$\Phi(r)=\frac{Q}{4\pi r}\exp(i\kappa_m r)\frac{2r}{i\kappa_m a^2}\left[\exp\left(i\kappa_m\frac{a^2}{2r}\right)-1\right] \tag{9.73}$$

The exponential term in equation (9.73) oscillates between 0 and −1 with the maximum amplitude equal to

$$\left|\Phi(r,r_f=\infty)\right|=\frac{Q}{4\pi r}\frac{2r}{\kappa_m a^2} \tag{9.74}$$

The ratio of the two fields may be defined as the focusing gain and given by

$$G=\frac{\left|\Phi(r=r_f)\right|}{\left|\Phi(r,r_f=\infty)\right|}=\frac{\kappa_m a^2}{2r_f} \tag{9.75}$$

Figure 9.16 shows an example of the field of a focused array, for which $|\Phi(r)|$ of equation (9.71) is calculated as a function of $r$ for the case where $c_m = 1500$ m/s, $c_l = 2500$ m/s, $a = 10$ mm, $R_L = 20$ mm, and the frequency is 5.0 MHz.

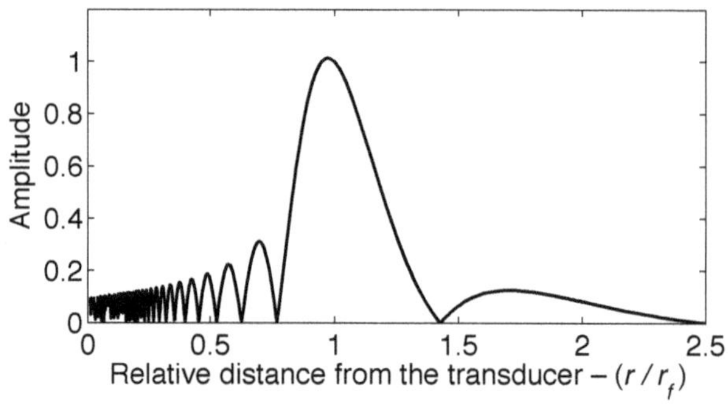


Figure 9.16. Sound pressure plotted as a function of the distance on the axis of a focused transducer. The distance ($r / r_f$) is relative to the focal length $r_f$, which is defined in equation (9.69).

## 9.4 Beamwidth and sidelobes

In section 9.2, we saw that the rectangular aperture with uniform excitation has a fairly high sidelobe level. In some cases, it is not desirable that the array transmits or receives power for angles outside the specified angle of the mainlobe, and therefore it is interesting to find and apply other aperture functions that give smaller sidelobes. Since the directivity function is a Fourier transform of the aperture function, a more rounded shape for the aperture function will produce a directivity function with lower sidelobes. This will, however, lead to a wider mainlobe compared with that produced by a uniform array of the same total size.

The triangular aperture function is given by

$$q(x) = \frac{2}{L}\left(1 - \frac{2|x|}{L}\right) \quad |x| < \frac{L}{2} \tag{9.76}$$

If we imagine achieving this function by convolving two equal rectangles with length $L / 2$, we can easily see that the directivity function becomes the square of the directivity function of a uniform aperture with the length $L/2$. Therefore, the directional function of a line array with the length $L$ and triangular weighting becomes

$$B(\theta)=\left[\frac{\sin\left(\pi\frac{L}{2\lambda}\sin\theta\right)}{\left(\pi\frac{L}{2\lambda}\sin\theta\right)}\right]^2 \tag{9.77}$$

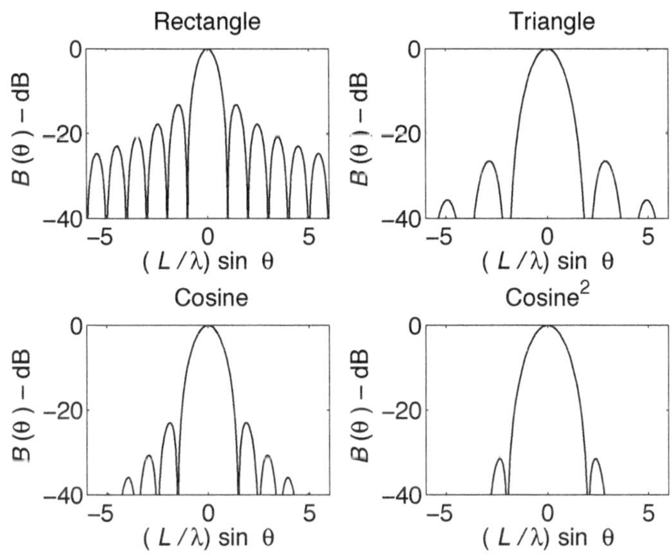


Figure 9.17. Directivity functions as a function of $(L/\lambda)\sin\theta$ for different apertures of length $L$.

The triangle directivity function, shown at the upper right in figure 9.17, has its first zero located at the angle where $\sin\theta = 2\lambda/L$. The width of the mainlobe is therefore approximately

$$\Delta\theta \approx \frac{2\lambda}{L} \tag{9.78}$$

The sidelobes are found at the angles

$$\sin\theta_n = (2n+1)\frac{\lambda}{L} \qquad n = \pm 1, \pm 2, \pm 3 \ldots \tag{9.79}$$

with the values

$$|B(\theta_n)| = \left(\frac{2}{\pi(2n+1)}\right)^2 = \left(\frac{2\lambda}{\pi L \sin\theta_n}\right)^2 \tag{9.80}$$

The first sidelobe has the amplitude $20\log(2/3\pi)^2 = -26.9$ dB, relative to the maximum value of the mainlobe, and the other sidelobes decrease with the square of the angle. The mainlobe of the triangular aperture is about twice as wide as the mainlobe of the rectangular aperture.

A cosine aperture is given by

$$q(x) = \frac{\pi}{2L}\cos\left(\pi\frac{x}{L}\right) \qquad |x| < \frac{L}{2} \tag{9.81}$$

It can also be written as

$$q(x) = \frac{\pi}{4L}\left[\exp\left(i\pi\frac{x}{L}\right) + \exp\left(-i\pi\frac{x}{L}\right)\right] \tag{9.82}$$

Consequently, the directivity function of a cosine aperture is equal to the sum of the directivity functions of two uniform apertures that have been displaced by an angle corresponding to $\sin\theta = \pm\lambda/2L$. The directivity function $B(\theta)$ of an array with cosine weighting is therefore

$$B(\theta) = \frac{\pi}{4}\left\{\frac{\sin\left[\pi\frac{L}{\lambda}\left(\sin\theta - \frac{\lambda}{2L}\right)\right]}{\left[\pi\frac{L}{\lambda}\left(\sin\theta - \frac{\lambda}{2L}\right)\right]} + \frac{\sin\left[\pi\frac{L}{\lambda}\left(\sin\theta + \frac{\lambda}{2L}\right)\right]}{\left[\pi\frac{L}{\lambda}\left(\sin\theta + \frac{\lambda}{2L}\right)\right]}\right\} \tag{9.83}$$

After some calculation, equation (9.83) can be expressed in the form

$$B(\theta)=\frac{\cos\left(\pi\frac{L}{\lambda}\sin\theta\right)}{1-\left(2\frac{L}{\lambda}\sin\theta\right)^{2}} \tag{9.84}$$

The first zero occurs where $\sin\theta = 3\lambda/2L$. The first sidelobe has the maximum value –23.5 dB, as shown on the lower left in figure 9.17.

Similarly, we can also calculate the directivity function of a cosine$^2$ aperture, which is defined as

$$q(x)=\frac{2}{L}\cos^{2}\left(\pi\frac{x}{L}\right)=\frac{1}{L}\left[1+\cos\left(2\pi\frac{x}{L}\right)\right] \tag{9.85}$$

This gives the directivity function

$$B(\theta)=\frac{\sin\left(\pi\frac{L}{\lambda}\sin\theta\right)}{\pi\frac{L}{\lambda}\sin\theta}\frac{1}{1-\left(\frac{L}{\lambda}\sin\theta\right)^{2}} \tag{9.86}$$

The first zero is where $\sin\theta = 2\lambda/L$, and the first sidelobe has the maximum value –32 dB, as seen on the lower right in figure 9.17.

Table 9.1 summarizes the results developed above. The aperture function is denoted as $q(x)$, and the directivity functions $B(\theta)$ are defined by the normalized angle.

$$\zeta=\frac{L}{\lambda}\sin\theta \tag{9.87}$$

The beamwidth $\Delta\theta°$ is the beamwidth in degrees between the –3 dB points and therefore numerically slightly different from the bandwidths defined before. The first sidelobe level is the level of the first secondary peak of the directivity function expressed in dB relative to the level of the main peak. The reason for selecting these apertures as examples is that they are easily calculated. Other apertures can be found numerically as solutions to an optimization problem where special requirements are levied of the directivity function.

Table 9.1 Apertures and directivity properties

| *Apertures** | *Weighting function ** $q(x)$ | *Directivity function* $B(\theta)$ | *Half power beamwidth* $\Delta\theta^\circ$ | *First sidelobe level in dB* |
|---|---|---|---|---|
| Rectangular | $\frac{1}{L}$ | $\frac{\sin \pi\zeta}{\pi\zeta}$ | $50\frac{\lambda}{L}$ | –13.5 |
| Triangular | $\frac{2}{L}\left(1-\frac{2\lvert x\rvert}{L}\right)$ | $\left(\frac{\sin\frac{\pi\zeta}{2}}{\frac{\pi\zeta}{2}}\right)^2$ | $73\frac{\lambda}{L}$ | –26.9 |
| Cosine | $\frac{\pi}{2L}\cos\left(\pi\frac{x}{L}\right)$ | $\frac{\cos(\pi\zeta)}{1-(2\zeta)^2}$ | $69\frac{\lambda}{L}$ | –23.5 |
| Cosine$^2$ | $\frac{2}{L}\cos^2\left(\frac{\pi x}{L}\right)$ | $\frac{\sin \pi\zeta}{\pi\zeta}\frac{1}{1-\zeta^2}$ | $83\frac{\lambda}{L}$ | –32.0 |

*all expressions for $x \le L/2$.

## 9.5 Array with discrete elements

An array with discrete elements is excited only at specific points or areas, and consequently it consists of a number of discrete transmitter or receiver elements. The aperture functions can therefore be expressed by a weighted sum of $\delta$ functions displaced to the points $x_n$ with $n = 0,1,2,3,...N-1$. For an array with $N$ elements with the weights $a_n$, the aperture function $a(x)$ is expressed as

$$a(x) = \sum_{n=0}^{N-1} a_n \delta(x - x_n) \tag{9.88}$$

and the directivity function $B(\theta)$ is

$$\begin{aligned} B(\theta) &= \int \sum_{n=0}^{N-1} a_n \delta(x - x_n) \exp(-i\kappa x \sin\theta)\, dx \\ &= \sum_{n=0}^{N-1} a_n \exp(-i\kappa x_n \sin\theta) \end{aligned} \tag{9.89}$$

In many cases, the distance $d$ between the elements is constant, so that $x_n = nd$. The directivity function then becomes

$$B(\theta) = \sum_{n=0}^{N-1} a_n \exp(-i\kappa nd \sin\theta) \tag{9.90}$$

From equation (9.90) we see that now the directivity function is a discrete Fourier transform (DFT) of the element weights, and therefore the directivity function is a periodic function of the angle $\theta$ as the mainlobe is repeated for all the angles given by

$$\begin{aligned} \sin\theta_p &= p\frac{\lambda}{d} \\ p &= 0, \pm 1, \pm 2, \ldots \end{aligned} \tag{9.91}$$

The number of real solutions for the angular periodic function $\theta_p$ is determined by the ratio between the wavelength $\lambda$ and the distance $d$. The periodic repetition, a result of undersampling the corresponding continuous aperture function, causes repeated mainlobes for real angles. We will soon consider some examples of this.

However, in the simplest case, the aperture weights are constant, $a_n = 1 / N$, and the directivity function consequently becomes

$$\begin{aligned} B(\theta) &= \frac{1}{N} \sum_{n=0}^{N-1} \exp(-i\kappa nd \sin\theta) \\ &= \exp\left[-i\pi\frac{d}{\lambda}(N-1)\sin\theta\right] \frac{\sin\left(\pi\frac{d}{\lambda} N \sin\theta\right)}{N \sin\left(\pi\frac{d}{\lambda}\sin\theta\right)} \end{aligned} \tag{9.92}$$

The first factor is an uninteresting phase factor caused by the choice of phase reference, which can be ignored. Then the directivity function becomes

$$B(\theta)=\frac{\sin\left(\pi\frac{d}{\lambda}N\sin\theta\right)}{N\sin\left(\pi\frac{d}{\lambda}\sin\theta\right)} \tag{9.93}$$

Figure 9.18 shows $|B(\theta)|$ as a function of the normalized angle $(L/\lambda)\sin\theta$. Compare this figure to figure 9.4, which represents the directivity function for the continuous case. The mainlobe is quite similar in the two cases, which means that the discrete line array has the beamwidth of

$$\Delta\theta \approx \frac{\lambda}{Nd}=\frac{\lambda}{L} \tag{9.94}$$

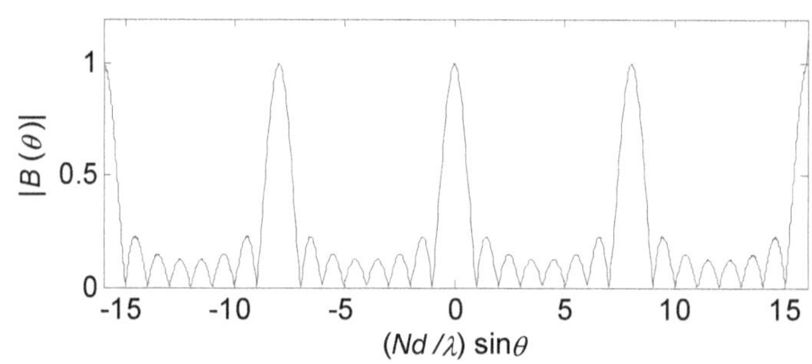


Figure 9.18. The directivity function $|B(\theta)|$ of an array having $N = 8$ elements with spacing $d = 2\lambda$.

The lobe structure produced by the discrete array is different from that of the continuous case because of the repeated mainlobes at angles given in equation (9.91) In figure 9.19, we have used equation (9.93) to calculate the directivity functions for four line arrays with the same total length $L = 16\lambda$, and with uniform weighting, but each having a different number of elements $N$. Therefore, the distance between the elements, measured in wavelengths $d/\lambda$, is different for each array, which makes the lobe structures differ also.

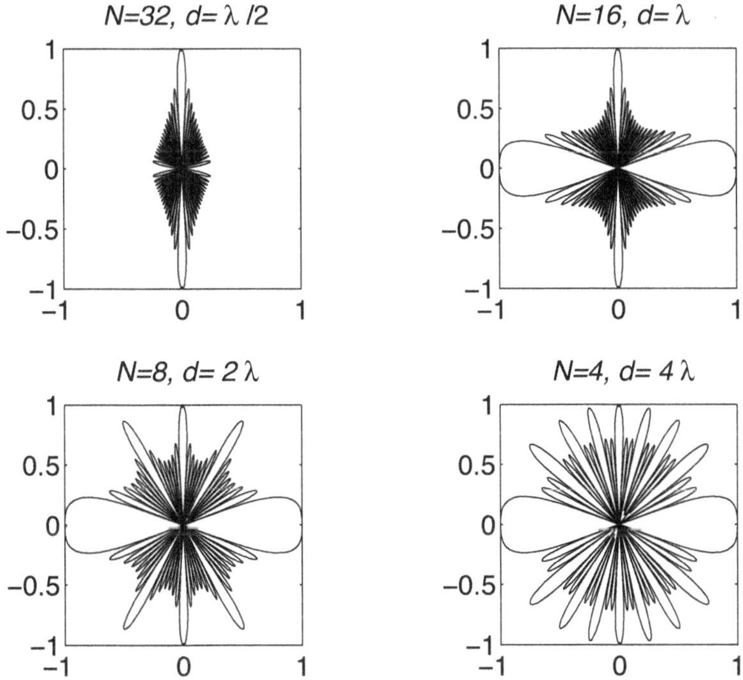


Figure 9.19. Polar representations of the directivity functions for four arrays with length $L = 16\lambda$, but with different numbers of elements ($N$). Upper left: $N = 32$, $d = \lambda/2$. Upper right: $N = 16$, $d = \lambda$. Lower left: $N = 8$, $d = 2\lambda$. Lower right: $N = 4$, $d = 4\lambda$.

To avoid producing repeated mainlobes for real angles, we see from equation (9.91) that the spacing $d$ between the elements must satisfy the requirement

$$d \leq \frac{\lambda}{2} \tag{9.95}$$

The condition defined in equation (9.95) is the Nyquist condition for discretization of spatial functions; this condition is analogous to the requirement in the theory for sampling of time functions that the distance between the samples must be less than the inverse of two times the bandwidth. When this condition is satisfied, as in the upper left example of figure 9.19, the directivity function of a discrete array becomes very similar to that of the corresponding continuous array. In the case of the upper right example, where $d = \lambda$, we get repeated mainlobes for $\theta_p = \pm\pi/2$, as specified by equation (9.91). The polar representations in the lower half of figure 9.19 show the lobe structures resulting from $d = 2\lambda$ and $d = 4\lambda$, respectively.

As in the continuous array case, reduction of the sidelobes is obtained by reducing the weights of the external elements (but at the expense of increasing mainlobe width). It must be emphasized that the weighting has no influence on the repeated mainlobes, as these are caused by the periodicity of equation (9.90).

### *9.5.1 Beamsteering of arrays*

A discrete array may be steered so that it points to any desired direction in space. The steering is done by giving the signal of each element a time delay, or shift in phase angle, in accordance with the element location and the steering angle.

The left-hand side of figure 9.20 shows the operating principle of a line array that uses time-delay beamsteering. The signals from each element are given time delays $T_n$ and the amplitudes are multiplied with the weights $a_n$. Then the sum of all the signals will produce the angular response or beam pattern

$$B(\theta) = \sum_{n=0}^{N-1} a_n \exp(-i\omega T_n) \exp(i\kappa n d \sin\theta) \tag{9.96}$$

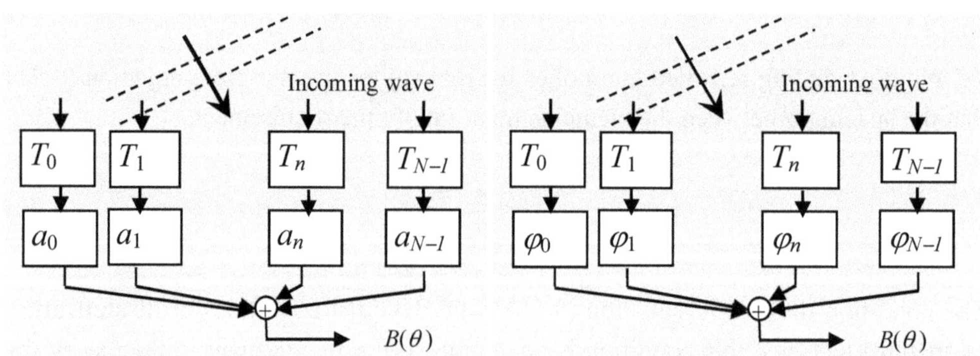


Figure 9.20. Beam steering of a line array. Left, with time delays. Right, with phase shifts.

When the distance between the elements is constant, we choose the time delays as

$$T_n = n\Delta T \tag{9.97}$$

and obtain

$$\begin{aligned} B(\theta) &= \sum_{n=0}^{N-1} a_n \exp(-i\omega n\Delta T)\exp(i\kappa n \sin\theta) \\ &= \sum_{n=0}^{N-1} a_n \exp\left[ i\kappa n d\left( \sin\theta - \Delta T \frac{c}{d} \right)\right] \end{aligned} \tag{9.98}$$

From equation (9.98), it follows that the steered beam pattern can be expressed as

$$B(\sin\theta) = B_0\left( \sin\theta - \Delta T \frac{c}{d} \right) \tag{9.99}$$

where $B_0(\sin\theta)$ is the beam pattern steered to $\theta = 0$ when $\Delta T = 0$.

For instance, the steered beam pattern for the uniformly weighted line array is

$$B(\theta) = \frac{\sin\left[ \pi \frac{Nd}{\lambda}\left( \sin\theta - \Delta T \frac{c}{d} \right)\right]}{N \sin\left[ \pi \frac{d}{\lambda}\left( \sin\theta - \Delta T \frac{c}{d} \right)\right]} \tag{9.100}$$

We see that the steered direction of the array is changed to the angle $\theta_s$

$$\sin\theta_s = \Delta T \frac{c}{d} \tag{9.101}$$

When the bandwidth is relatively small compared to the central frequency, the time delays can be replaced with phase shifts. This replacement results in an implementation that may be easier to realize with electric circuitry. The structure of a phase shift beamformer is shown on the right in figure 9.20.

The incremental phase shift between the elements for steering to a given angle must be

$$\Delta\varphi = \omega\Delta T = 2\pi\frac{d}{\lambda}\sin\theta_s \tag{9.102}$$

Consequently the phase of the signal at element $n$ must be given a phase shift

$$\varphi_n = 2\pi n\frac{d}{\lambda}\sin\theta_s \tag{9.103}$$

The difference between time-delay beamsteering and phase-shift beamsteering can be illustrated by a simulated, idealized example. We consider a case with a line array having $N = 10$ elements with spacing $d = 0.5\lambda$. The signals received at the elements in response to a plane wave arriving at an angle of 30° are shown in figure 9.21. In this example, the carrier or central frequency is $f_0 = 15$ kHz, and the signal consists of five periods of that carrier frequency. The top trace in figure 9.21 is the sum of all the signals received by the elements, divided by the number of elements.

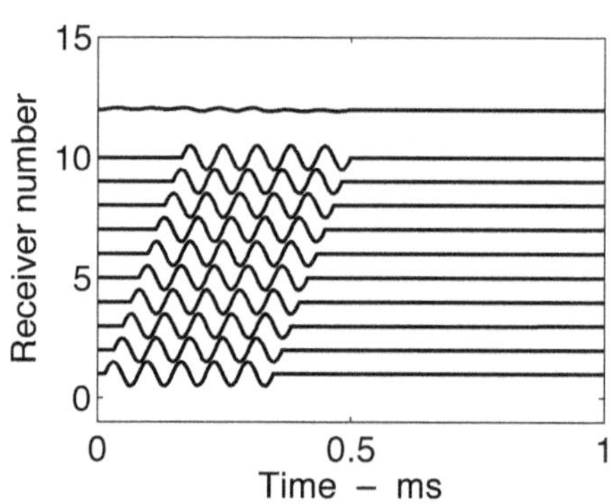


Figure 9.21. Signals received at the elements of a 10-hydrophone array when the incoming plane wave arrives at a 30° angle.

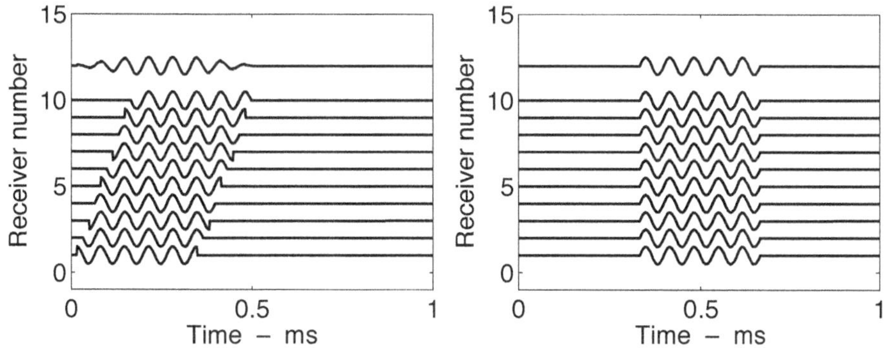


Figure 9.22. The output time signals of the two beamsteering methods: left, time-delay beamsteering and right, phase-shift beamsteering.

Figure 9.22 demonstrates the difference between time-delay and phase-shift beamsteering. On the left, figure 9.22 shows the result of time-delay steering toward the direction of the incoming signal. All signals are added in perfect phase and the sum, which is the top trace, is identical with the incoming signal. On the right, figure 9.22 shows the result of phase-shift steering. Now the phase of the incoming signals is adjusted so that the signals are added in phase, while the time positions of the envelopes are unchanged. In other words, phase steering corrects the phase of the carrier or central frequency so that the signals are added in phase, but the time positions of the signal envelopes are not corrected, which means that the sum is spread out in time relative to the separate signals. However, the maximum value of the summed signal is the same as that for the time-delay steered signal, provided that the time duration of the signal is longer than the time delay across the array. From this example, we can conclude that phase-shift steering can be used for narrowband signals with a pulse duration that is relatively long as compared to a period of the center frequency.

This principle of array steering can be used for both transmission and reception, but as it is easier to create time delays and phase shifts at a low power level, it is preferable to perform beamsteering upon reception rather than transmission.

### 9.5.2 Endfire arrays

For some applications we use endfire arrays, where the main beam is steered in the direction of the array orientation—that is, toward the angle $\theta = 0$. With reference to figure 9.23, we find that between a point in the farfield and an element in position $x'$ on the array, the difference in path lengths, $\Delta r = r - r'$, approximates

$$\Delta r = x' \cos\theta \tag{9.104}$$

In order to steer the array toward $\theta = 0$, the element in the position $x_n = nd$ must be given either a time delay, $T_n = nd / c$, or a corresponding phase shift. Thus, for the directivity function $B(\theta)$, an incoming plane wave with the angle $\theta$ will produce

$$B(\theta) = \sum_{n=0}^{N-1} a_n \exp\left[-iknd(1-\cos\theta)\right] \tag{9.105}$$

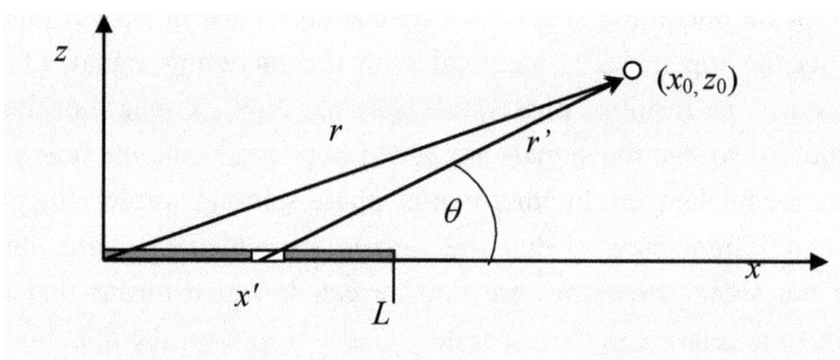


Figure 9.23. An endfire array of length $L$. The angle $\theta$ is defined with reference to the horizontal direction of the array.

With a uniform aperture, this expression can be rewritten as

$$|B(\theta)| = \left| \frac{\sin\left[\pi \frac{Nd}{\lambda}(1-\cos\theta)\right]}{N\sin\left[\pi \frac{d}{\lambda}(1-\cos\theta)\right]} \right| \tag{9.106}$$

An example of this beam pattern is shown in figure 9.24, calculated by using equation (9.106) with the values $N = 16$ and $d = 0.5\lambda$.

Note that all beams for a line array are surfaces of revolution and therefore the beam of the endfire array has a conical shape along the axis of the array. The beam pattern has repeated mainlobes at the angles where the denominator is equal to zero given by

$$\cos\theta_p = 1 \pm p\frac{\lambda}{d} \qquad p = 0, \pm 1, \pm 2, \ldots \tag{9.107}$$

The directivity function in equation (9.106) has a mainlobe in the direction $\theta = 0$; the two nearest zero crossings other than the mainlobe are found at the angles $\theta = \pm\Delta\theta/2$.

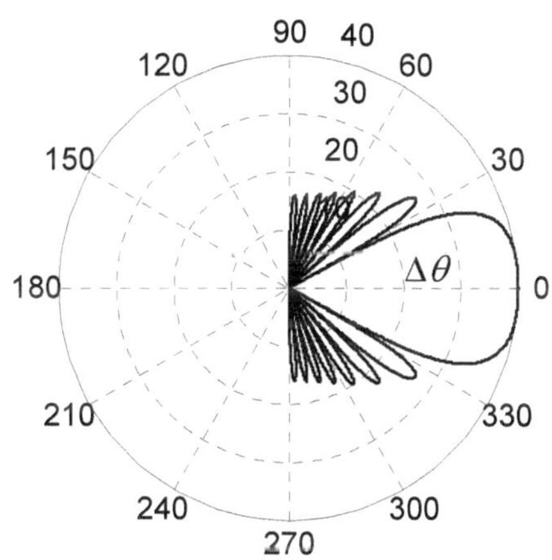


Figure 9.24. Beam pattern for an endfire array with N = 16 elements and spacing $d = 0.5\lambda$. The amplitude scale is in decibels covering the range from 0 dB to 40 dB. The beamwidth is $\Delta\theta$.

Using series expansion, we find that the distance to the first zero crossing, and consequently the beamwidth, is approximately given by

$$\Delta\theta \approx 2\sqrt{\frac{2\lambda}{Nd}} = 2\sqrt{2\frac{\lambda}{L}} \tag{9.108}$$

Thus, the beamwidth of an endfire array decreases with the square root of the array length $L$.

The total spatial angle within the conical beamwidth of the endfire array is

$$\Delta\Omega \approx \frac{\pi}{4}(\Delta\theta)^2 = 2\pi\frac{\lambda}{L} \tag{9.109}$$

and the directivity factor, defined by equation (9.41), is therefore

$$D = \frac{4\pi}{\Delta\Omega} = \frac{2L}{\lambda} \tag{9.110}$$

Equations (9.109) and (9.110) show that the solid-angle beamwidth and directivity function are the same for broadside and endfire arrays with the same length and number of elements.

### *9.5.3 Sector scanning*

Figure 9.25 shows in principle how a steered array can be used in sonar for scanning of a given sector. The transmission projector has a rather broad beam in order to insonify the whole angular sector of interest, but transmits a relatively short pulse with duration $\Delta T$. The angular resolution is achieved by beamforming on reception, when a large number of very narrow beams are formed. The figure shows only one of the receiving beams steered to the direction of $\theta_s$. The signals at these beams' outputs are sequentially scanned at a rate high enough that each beam is scanned at least once within the time of the pulse width, which is a time $\Delta T$.

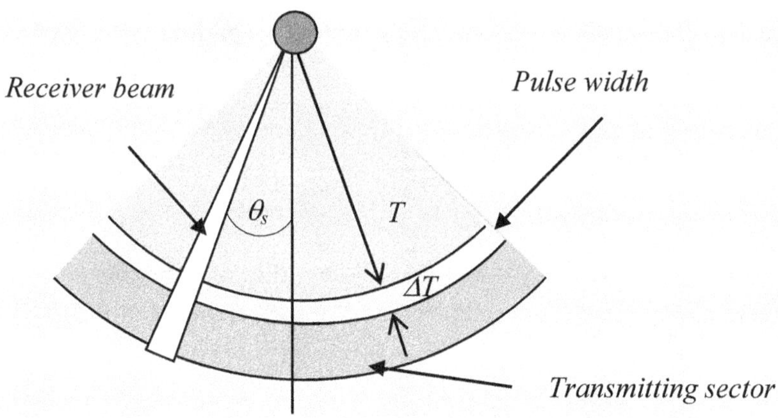


Figure 9.25. A scanning sonar, with its broad transmitted beam and narrow receiver beams. The angular resolution is given by the width of the receiver beam and the radial resolution by the pulse width.

In this way an image of the whole sector's echoes can be built up, with a range resolution given by the transmitted pulse and an angular resolution given by the receiving beamwidths. Thus, a target can be located both by angle (by determining the echo's receiving angle), and by distance (by measuring the echo's arrival time). In an application like this, it is important that repeated mainlobes of the receiver beams not fall within the angular sector of the transmitter.

Thus, the distance *d* between the elements of the receiving array must satisfy the requirement

$$\sin\left(\frac{\theta_s}{2}\right) \le \frac{\lambda}{2d} \tag{9.111}$$

In equation (9.111) $\theta_s$ is the angle of the steering direction. Accordingly, the distance between the elements must be smaller than the smallest distance the wavelength travels in water at the desired angle $\theta_s$ so that

$$d \le \frac{\lambda}{2\sin\frac{\theta_s}{2}} \tag{9.112}$$

The narrowest beamwidth is obtained by a uniform aperture, which produces the beamwidth $\Delta\theta$ given by

$$2\sin\left(\frac{\Delta\theta}{2}\right)=\frac{\lambda}{Nd} \tag{9.113}$$

It follows from equations (9.112) and (9.113) that the total number of independent steered directions is

$$\frac{2\sin(\theta_s/2)}{2\sin(\Delta\theta/2)}=\frac{\lambda/d}{\lambda/(Nd)}=N \tag{9.114}$$

which means that the number of independent steering directions is equal to the number of array elements. Thus, the steered directions can be selected as the angles $\theta_q$, defined by

$$\sin\theta_q=q\frac{\lambda}{Nd} \qquad q=1,2....N \tag{9.115}$$

If we find that the array's sidelobe level is too high, it is suitable to introduce weighting of the elements. However, since this weighting takes place at the expense of the angular resolution, the length of the array and the number of elements must be increased to maintain the same angular resolution.

## 9.6 Angular resolution and steering accuracy

When assessing an array's performance, we must distinguish between its angular resolution and its steering accuracy. Angular resolution is the ability to separate two or more targets found in approximately the same direction from the array. An array's angular resolution is determined by the signal beamwidth, since a sonar cannot separate two targets unless the angular distance between them is at least equal to the beamwidth. Steering accuracy refers to the array's ability to determine the direction of a single target. The steering accuracy is dependent on the noise level, among other things.

Although it is possible to determine steering accuracy by measuring the angle that gives a maximum received signal, in reality another procedure, the "split-beam" technique—or the sum and difference method—is commonly applied. Here we use two receiver arrays, to determine the direction of an incoming wave by measuring the time delay between the signals received on the two arrays. (Note that, for narrowband signals, measurements of time delay can be replaced by phase measurements.)

Consider a receiver array of length $2L$ divided into two equally long parts as shown in figure 9.26. The two array parts are separately beam formed, producing the directional functions $B_1(\theta)$ and $B_2(\theta)$. Since the arrays have their centers at $x = \pm L/2$, the signals from a direction $\theta_0$ will result in a phase contribution to the two signals of $\pm\,\pi(L/\lambda)\sin(\theta - \theta_0)$.

When we use the center point between the array halves as the phase reference point, the array response is expressed as

$$\begin{aligned} B_1(\theta) &= B(\theta)\exp\left[-i\pi\frac{L}{\lambda}\sin(\theta-\theta_0)\right] \\ B_2(\theta) &= B(\theta)\exp\left[i\pi\frac{L}{\lambda}\sin(\theta-\theta_0)\right] \end{aligned} \tag{9.116}$$

$B(\theta)$ is the directivity function of each of the two halves of the array.

The directivity functions of the sum and difference beams are defined as

$$
\begin{aligned}
B_+ &= B_1(\theta) + B_2(\theta) = 2B(\theta)\cos\left[\pi\frac{L}{\lambda}\sin(\theta-\theta_0)\right] \\
B_- &= B_1(\theta) - B_2(\theta) = -2iB(\theta)\sin\left[\pi\frac{L}{\lambda}\sin(\theta-\theta_0)\right]
\end{aligned}
\tag{9.117}
$$

where $B_-(\theta)$ is zero and $B_+(\theta)$ has a maximum value of $\theta = \theta_0$. The incoming wave's strength and direction can be determined by detecting the maximum received signal in the sum beam and the zero crossing of the difference signal when the array sweeps across the relevant angular sector.

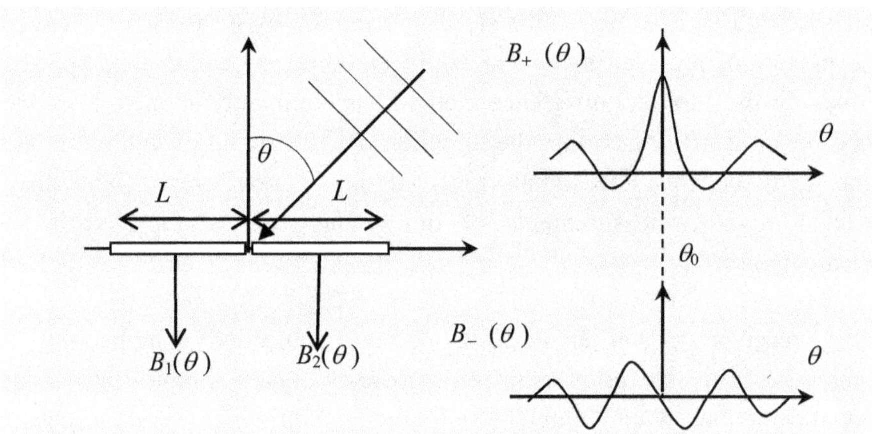


Figure 9.26. Measurement of steering direction with the "split-beam" technique. Left, diagram of receiver array. Top right, directivity function of the sum beam; bottom right, directivity function of the difference beam. The angular direction to the target is $\theta_0$.

# Chapter 10

# Sonar signal processing

## 10.1 Introduction

This chapter describes some of the more common tasks of a sonar signal processor. The term *sonar* covers both active and passive acoustic systems for underwater detection and localization. Other acoustic systems, such as those for navigation and communication, are not discussed explicitly in this chapter, but many of the issues being considered here also apply to them.

Sonars are designed to detect possible targets in a given volume; locate the direction, distance, depth, and velocity of the target; and, if possible, determine certain target properties such as size and shape. An ***active sonar*** transmits acoustic pulses that are reflected by possible targets and returned to the sonar, which tries to determine the above-mentioned target parameters by analyzing the received echo. A ***passive sonar*** system does not transmit a pulse, but seeks to detect the target from the sound it generates. These tasks are made difficult both by variable propagation conditions in the ocean and by noise originating from natural and man-made sources. An active sonar may also receive other, less desirable, echoes from the seabed, water surface, and particles and other objects in the water. When there are many such reflectors, their echoes have a noisy character, referred to as ***reverberation.***

The operating principle of an active sonar is shown in figure 9.25. A transmitter sends a short acoustic pulse to a relatively large angular sector; the sonar's receiver utilizes a number of narrow beams that together cover the transmission sector, and processes the echo in a bandpass filter matched to the frequency band of the target echo signal. When an echo from a target is detected, the target is localized: that is, its bearing is identified by the receiver beam's direction, and its range is obtained from the pulse's travel time. The uncertainty in the target's localized position is determined by the width of the receiver beams and the pulse width. The time duration of the transmitted pulse is the range ***resolution cell,*** also indicated in figure 9.25.

Figure 10.1 presents a schematic drawing of the most important elements in any sonar receiver, whether active or passive. The acoustic signals are received by a number of hydrophones that can either be attached to a hydrophone cable or be installed on the surface of a cylinder or sphere that is attached to the vessel hull.

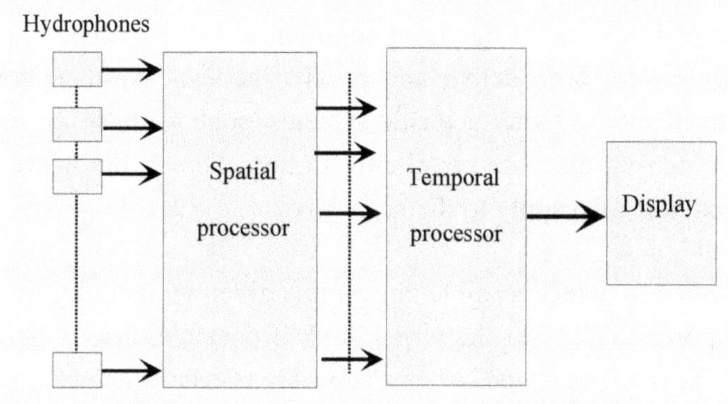


Figure 10.1. Schematic of a sonar receiver. Signals received on the hydrophones are passed to a beamformer, or spatial processor, and then to a temporal processor to produce displayed target information.

The hydrophone signals are passed to a ***beamformer,*** also called a ***spatial processor***, which produces beams directed to the desired angles using the techniques described in section 9.5. If both the spatial and the temporal processors are linear processors, their order will not matter and can be chosen for efficiency considerations; if one or both are nonlinear (such as an adaptive beamformer) the order would matter.

The beamformer's output consists of target echoes, noise, and reverberation from a particular beam direction. These signals are subsequently passed to the temporal processor, which has the primary task of detecting whether there is a target present in the relevant beam.

After a potential target has been detected, the direction can be determined from the direction of the beam in which the target was detected. An active sonar also can determine distance to the target by measuring the round-trip travel time; travel time estimates are not available to passive sonars. More advanced sonars also provide the capability to determine the relative radial velocity of the target by using ***Doppler analysis,*** which we will discuss in section 10.3.

## 10.2 Range resolution

As stated earlier, an active sonar detects the range to the target from measurements of the signal's travel time to and from the detected object. The two-way travel time $T_{max}$ to the sonar's maximum distance $r_{max}$ is expressed as

$$T_{max} = 2\frac{r_{max}}{c} \tag{10.1}$$

where $c$ represents sound speed. For long-range sonar, this two-way travel time can be lengthy, about 20 s for a target at a 15 km range, for example.

***Range resolution*** is the sonar's ability to distinguish between two targets that are in the same direction, but at different distances from the sonar. With an effective pulse length $\Delta T$, the range resolution $\Delta r$ is given by

$$\Delta r = \frac{c}{2}\Delta T \tag{10.2}$$

More advanced sonar systems use modulated signals such as FM (frequency modulated) pulses. After matched filter reception of a modulated signal having a bandwidth $B$, the effective pulse length is approximately $\Delta T \approx 1/B$, and the range resolution of the system is approximately $\Delta r \approx c/2B$.

## 10.3 Doppler shift and Doppler analysis

First, we will define what is meant by a ***Doppler shift*** and then examine the consequences this phenomenon can have for a sonar system. We begin by considering a sonar transmitting a sinusoidal pulse with frequency $f_0$ toward a target approaching the sonar with radial velocity $v_t$. The target will perceive the frequency of the approaching signal $f_t$ to be shifted so that

$$f_t = f_0\left(\frac{v_t + c}{c}\right) \tag{10.3}$$

When the reflected pulse returns to the sonar, the system will perceive its received frequency $f_e$ to be

$$f_e = f_t\left(\frac{c}{c - v_t}\right) \tag{10.4}$$

The Doppler or frequency shift $f_D$ is the difference between the frequency of the received echo and the frequency of the transmitted signal; that is

$$\begin{aligned} f_D = f_e - f_0 &= f_0\left[\frac{c + v_t}{c - v_t} - 1\right] \\ &\approx 2f_0\frac{v_t}{c} \end{aligned} \tag{10.5}$$

The last approximation assumes that the target's velocity $v_t$ is much smaller than the sound speed: $v_t << c$. Note that for a target moving away from the sonar, $f_D$ is negative.

While the Doppler shift occurs in radar systems as well, the ratio between the target speed and the wave speed, $v_t / c$, is much larger for sound in water than for electromagnetic waves in air. Because of sonar's relatively larger Doppler shifts, the maximum possible Doppler shift must be taken into consideration when choosing the sonar receiver's bandwidth. On the other hand, the characteristics of the Doppler shift make possible the determination of the target's radial velocity when a number of bandpass filters, tuned to the frequencies of the expected Doppler-shifted signals, are applied to the received signal. If the frequency resolution, or bandwidth, of each filter is $\Delta f$, the velocity resolution, $\Delta v$, is

$$\Delta v = \frac{c}{2}\frac{\Delta f}{f_0} = \frac{\lambda}{2}\Delta f \tag{10.6}$$

The importance of this consideration is illustrated by the following example, which assumes the sonar is transmitting a sinusoidal pulse with a time duration $\Delta T = 100$ ms and a frequency of $f_0 = 15$ kHz. With the given pulse length, the bandwidth of the transmitted signal will be about $B \approx 1/\Delta T = 10$ Hz. In order to reduce the received noise to a minimum, sonar systems filter the target echoes. Ideally one would use a bandpass filter covering about the same bandwidth as the transmitted pulse. However, according to equation (10.5), a target with the radial velocity $v_t = 5$ m/s will produce a Doppler shift of $f_D = 100$ Hz, which means that an echo from this target will be shifted out of the receiver's passband. In fact, the echoes will be filtered out from all targets except those that are close to stationary or traveling with small radial velocity. Consequently, in order to detect moving targets we are forced to use a receiver bandwidth considerably larger than the transmitted signal's bandwidth. This requirement has a negative effect on sonar performance, since normally the noise level increases with increasing receiver bandwidth.

As an example, let us consider a system with five filters, each having a bandwidth $B$ equal to 100 Hz; the center frequencies of the filters are separated by 100 Hz, as shown in figure 10.2. The sonar uses a center frequency of 15 kHz and performs frequency analysis in five frequency bands, with each 100-Hz band corresponding to a target speed in a 5 m/s interval.

If the vessel carrying the sonar moves with a radial velocity $v_s$ in relative to the target, the total Doppler shift $f_D$ becomes

$$f_D \approx 2 f_0 \frac{v_s + v_t}{c} \tag{10.7}$$

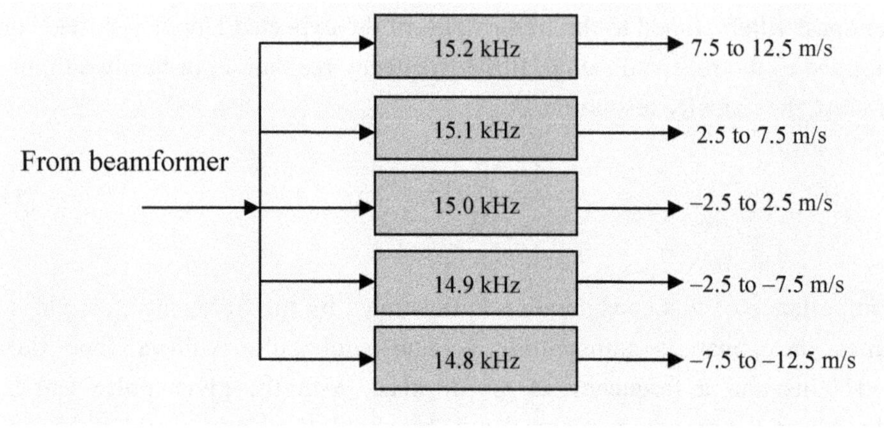


Figure 10.2. A filter bank used for Doppler analysis of the received echo, from which the target's radial speed is determined.

The part of the Doppler shift caused by the speed of the vessel itself can be removed or reduced with a technique called ODN (Own Doppler Nullification). With ODN, the frequency $f_0$ of the transmitted signal is shifted with the transmitting beam's direction in relation to the course direction $\phi$ and the vessel's velocity $v_s$ by transmitting at the frequency

$$f_0(v_s,\phi) = f_0\left(1 - 2\frac{v_s}{c}\cos\phi\right) \tag{10.8}$$

This frequency shift will compensate for the Doppler of a target at the center of the transmitter beam, but not correctly for a target off center. With a beamwidth of $\Delta\theta$ the uncompensated Doppler shift will be

$$\Delta f_D = \pm 2 f_0 \frac{v_s}{c}\left[\cos(\phi \pm \Delta\theta) - \cos\phi\right] \tag{10.9}$$

With this technique the bandwidth required for the receiving system is reduced. Note that ODN can only be used with directional transmission; it is not compatible with omnidirectional transmission.

## 10.4 Normalization

For each beam, and possibly for each Doppler filter, a time signal is produced that acts as a basis for the detection of targets. However, the signal's dynamic range can be significant, with perhaps a 50 – 100 dB variation caused by the difference in transmission loss between a distant and a close target. It is therefore desirable to normalize the signal before further processing. Two normalizing techniques are commonly applied.

With the first technique, Time Variable Gain (TVG), the receiver's amplification is adjusted according to the transmission loss expected as a function of either distance or time after pulse transmission. When spherical propagation is expected, the gain $G(t)$ is varied with time after transmission, $t$, as

$$G(t) = 40\log\left(\frac{c}{2}t\right) + 2\alpha\left(\frac{c}{2}t\right) \tag{10.10}$$

where $\alpha$ is the absorption in dB per meter for the relevant sonar frequency $f_0$. Correspondingly, with expected cylindrical propagation, one should use

$$G(t) = 20\log\left(\frac{c}{2}t\right) + 2\alpha\left(\frac{c}{2}t\right) \tag{10.11}$$

With an extended target—for instance, a shoal of fish—the number of fish that are ensonified by the beam will increase with distance, thereby increasing the echo level. In such cases, when spherical propagation is assumed, it may be more appropriate to use the gain compensation

$$G(t) = 30\log\left(\frac{c}{2}t\right) + 2\alpha\left(\frac{c}{2}t\right) \tag{10.12}$$

TVG is primarily used when we wish to measure the absolute strength of the echo signal, for instance in the case of the acoustic assessment of fish abundance. The technique's accuracy, of course, is dependent on whether the actual transmission loss is equal to the loss included in the gain control.

Alternatively, we can use Automatic Gain Control (AGC), which normalizes by a factor that is inversely proportional to the mean signal level. With this technique, the averaging is done by selecting a time constant, which should be both long enough to achieve a stable estimate without bias and short enough to respond to the average signal reduction with range.

## 10.5 Detection of signals in noise

The most important task of a sonar is detection—that is, determining whether a target is present or not. Figure 10.3 presents a functional diagram illustrating the operating principle of a sonar processor. The signal from the beamformer is passed through a filter with bandwidth $B$ producing the input signal $x(t)$ to the detector. This signal is a sum of a target signal and noise, and the objective of the detector is to enhance the target signal and reduce the noise.

The output signal of the detector, $z(t)$, goes to an operator, or decision device, that decides whether there is a target present or not. The operator will assume that a target is present if the signal magnitude exceeds a certain preset threshold value $\gamma$.

Given the output of the signal processor, the operator must make a decision related to two hypotheses:

$H_0$: No target present

$H_1$: Target present

The operator may be mistaken in two ways: either through giving a false alarm by announcing a target when there is none, or by not indicating a detection when a target is present. This is illustrated by the decision matrix of Table 10.1.

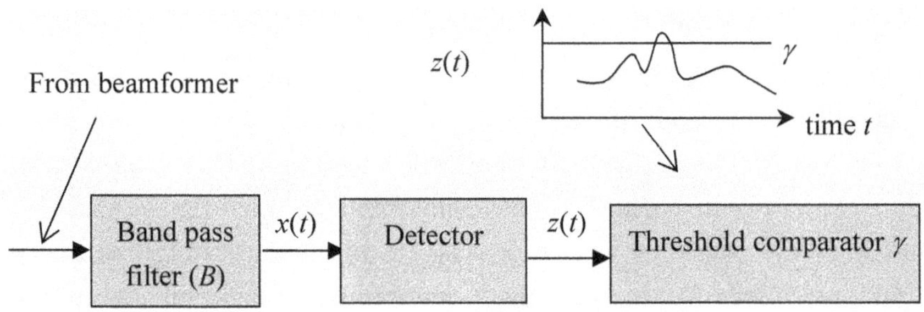


Figure 10.3. Operating principle of a sonar signal processor. The decision of whether a target is present or not is made at the threshold comparator stage.

Table 10.1

Decision matrix showing the possible outcomes of the hypotheses testing.

| | Reality: "No target present" | Reality: "Target present" |
|---|---|---|
| Hypothesis $H_0$: "No target present" | Correct, "No target present" | Missed detection |
| Hypothesis $H_1$: "Target present" | False alarm | Correct, "Target present" |

Because of noise and fluctuations, the signals may reasonably be considered as ***stochastic processes***. Thus, in order to design the signal detector for optimal performance and then to evaluate its operation, we must know the statistical characteristics of the noise, reverberations, and echo signals. The decision on whether there is a target present or not must be taken for each independent resolution cell. The number of resolution cells in angle and range may be considerable, as we shall see in the following example, which is typical of active sonar.

Let us assume that we have an active sonar covering an angular space of $\phi_{cov} = 360°$, with an angular resolution of $\Delta\phi = 15°$. Consequently, the number of angular resolution cells is $N_v = \phi_{cov} / \Delta\phi = 24$. We further assume that the sonar transmits a simple $\Delta T = 100$ ms sine pulse having a range resolution of $\Delta r = c\Delta T / 2 = 75$ m. If the range of the sonar is $r_{max} = 15$ km, the number of range cells becomes $N_r = 15{,}000 / 75 = 200$. Thus, the total number of resolution cells becomes $N = N_v N_r = 4800$.

A common presentation format for such information is the so-called Plan Position Indicator (PPI) display, which uses a polar presentation of the received echo's strength in distance and direction. The maximum pulse repetition interval is the amount of time it takes a pulse to reach $r_{max}$ and return, which in this case is 20 seconds. Consequently, the operator must decide if there is a target present for each of the 4800 resolution cells every 20 seconds.

If the sonar has a Doppler analyzer at one or more of the receiving beams, the number of decisions necessary will become even higher. For instance, a sonar with center frequency $f_0 = 15$ kHz will have a Doppler shift of $f_D = 100$ Hz for a target with a radial velocity $v = 5$ m/s. To cover a velocity interval of $\pm 15$ m/s, $N_D = 5$ Doppler cells are needed, which increases the total number of cells to be resolved to $N = N_v N_r N_D = 24{,}000$.

With such a large number of possible detections and resolution cells, it is natural to apply statistical methods to quantify the performance in terms of the probability of detection, $P_D$, and the probability of a false alarm, $P_{FA}$. In order to do that, we need to know the probability density functions (PDFs) of the signal $z(t)$ under both hypotheses $H_0$ and $H_1$. We refer to $p_0(z)$ as the PDF of the amplitude of the received signal $z(t)$ when no target is present, and to $p_1(z)$ as the PDF of $z(t)$ when a target is present. With threshold value $\gamma$, the probabilities for detection $P_D$ and false alarm $P_{FA}$ are, respectively

$$
\begin{aligned}
P_D &= \int_{\gamma}^{\infty} p_1(z)dz \\
P_{FA} &= \int_{\gamma}^{\infty} p_0(z)dz
\end{aligned}
\tag{10.13}
$$

Thus, the sonar's performance depends on the statistical properties of the signal and noise. In some applications, the received signals have noiselike character under both hypotheses and we may assume, at this time, that the signals are described by Gaussian statistics. Under other circumstances, such as very high reverberation conditions, other types of distributions are used. This may be the case in passive sonar where the task is to detect the noise signal radiated by the target in a background of ambient noise. Consequently, we assume that $p_0(z)$ and $p_1(z)$ are both Gaussian functions with the same variance, $\sigma_0$, but different mean values, $m_0$ and $m_1$

$$
\begin{aligned}
p_0(z) &= \frac{1}{\sqrt{2\pi\sigma_0^2}} \exp\left[-\frac{(z-m_0)^2}{2\sigma_0^2}\right] \\
p_1(z) &= \frac{1}{\sqrt{2\pi\sigma_0^2}} \exp\left[-\frac{(z-m_1)^2}{2\sigma_0^2}\right]
\end{aligned}
\tag{10.14}
$$

The probabilities of detection and false alarm are defined by the areas under the respective curves shown in figure 10.4, in which $m_0 = 4$, $m_1 = 6$, and $\sigma_0^2 = 1$.

The threshold value $\gamma$, which appears as the lower limit of integration in equations (10.14), determines the values of the probabilities. A low value for the detection threshold gives a high probability of detection, but also a high rate of false alarm.

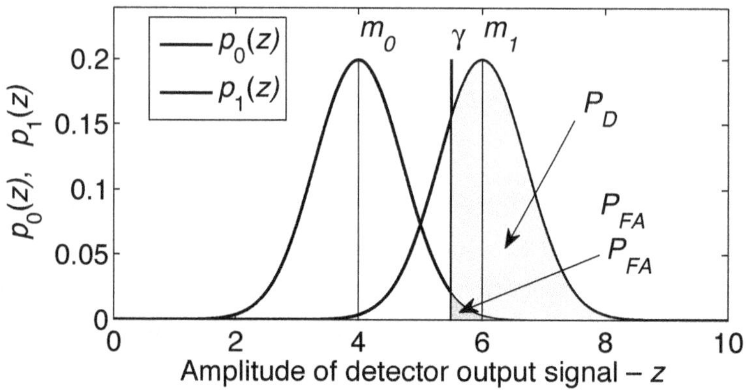


Figure 10.4. Gaussian PDFs for noise-only $p_0(z)$ and signal-plus-noise $p_1(z)$ hypotheses.

Thus, the choice of threshold value is a compromise between low $P_D$ with low $P_{FA}$ versus high $P_D$ with high $P_{FA}$. It is common to select a specific rate of false alarm $P_{FA}$ and then attempt to maximize the probability of detection $P_D$. The choice of $P_{FA}$ is influenced by many conditions, among other things the "cost," or consequences, of making and reacting to a false alarm.

The ***detection index*** $d$ is defined as the square of the difference between the mean values of the signals for the two hypotheses $H_1$ and $H_0$ normalized by the variance of the noise under hypothesis $H_0$

$$d = \frac{(m_1 - m_0)^2}{\sigma_0^2} \tag{10.15}$$

Note that in some texts (for instance Van Trees [1968]) the detection index is defined as the square root of the detection index given in equation (10.15).

Using equations (10.14) and (10.15), the probabilities for detection and false alarm can be calculated as functions of the detection threshold $\gamma$ with the resulting $P_D$ as a function of $P_{FA}$ shown in figure 10.5.

The plots in figure 10.5 are called Receiver Operating Characteristics (ROC) curves because they characterize the sonar system's performance as a function of the signal-to-noise ratio (SNR). The specific curves appearing in figure 10.5 are calculated using the assumption that the signal statistics $z(t)$ are Gaussian, with the same variances under both hypotheses. Other probability densities must be used for other cases. However, the curves in figure 10.5 are also approximately valid for other cases and other detectors, provided a suitable value of $d$ is selected for each individual case.

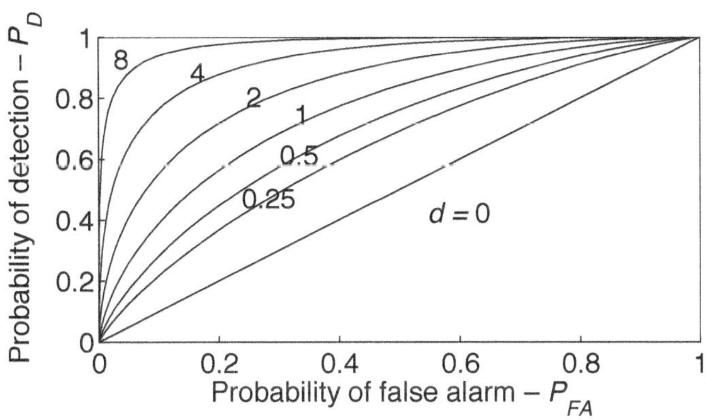


Figure 10.5. Receiver operating characteristic (ROC) curves for a sonar system with Gaussian statistics: $P_D$ as a function of $P_{FA}$ for various values of the detection index $d$.

The performance, in terms of $P_D$ and $P_{FA}$, is determined by the characteristics of the output signal $z(t)$ from the detector. The detection index can be conceptualized as the minimum required SNR present at the detector's output, $(S/N)_{\text{det}}$, to achieve preset values for $P_D$ and $P_{FA}$

$$(S/N)_{\text{det}} - d \tag{10.16}$$

The input signal $x(t)$ (figure 10.3) to the detector is the sum of a target signal and background noise. We assume that the target signal has power $S$ and the background noise has power $N$. Furthermore, we assume that the noise is white within the frequency band $B$, such that $N = BN_0$ where $N_0$ is the noise power in a bandwidth of 1 Hz. The signal-to-noise ratio of the input at the detector is therefore $(S/N)_{in}$.

The ***detection threshold*** $DT$ is defined as the ratio of the mean signal power $S$ in the signal bandwidth $B$ to mean noise power $N_0$ in a 1 Hz frequency band, both measured at the output of the beamformer. In dB, the detection threshold is therefore defined as

$$DT = 10\log\left(S / N_0\right) \tag{10.17}$$

The next step is to relate the detection index to the properties of the input signal x($t$). This relationship is dependent on both the statistics of the input signal and the design and operation of the actual receiver in use. In the following, we consider three different detectors commonly used in sonar systems.

### *10.5.1 The incoherent detector*

We first consider an incoherent detector, which is used when the receiver does not have any *a priori* information about the received signals. This is the kind of detector used in passive sonar, where the task may be described as "detection of target radiated noise in background noise." The sonar accomplishes the task by detecting the small increase in the received signal level that is created by the target signal. A common incoherent detector is the pure energy detector, which consists of a square-law operator followed by an integrator. The operating principle is illustrated in figure 10.6, where $x(t)$ is the output of the bandpass filter, $y(t)$ is the squared signal, and $z(t)$ is the integrated energy output that is passed to the threshold detector.

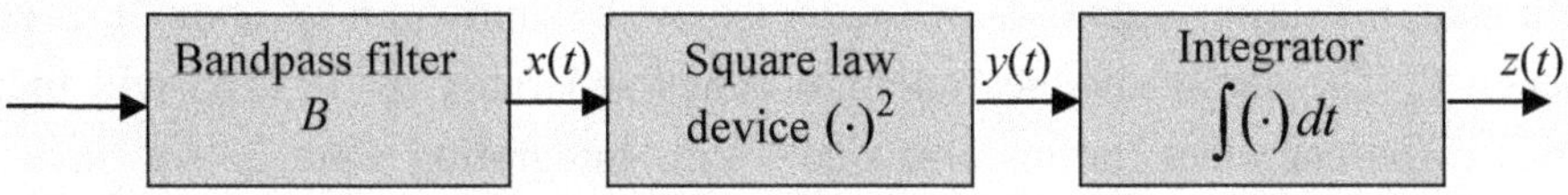


Figure 10.6. Block diagram of an incoherent receiver

With an incoherent detector, the first task is to determine the statistical properties of the signal $z(t)$ after squaring and integration, which requires knowledge of the statistics of the input signal $x(t)$ after bandpass filtering. This is generally a difficult task, except when $x(t)$ has Gaussian statistics. Therefore, we will assume that the signal $x(t)$ belongs to a white Gaussian process with the PDF $p_x(x)$, which is defined as

$$p_x(x) = \frac{1}{\sqrt{2\pi\sigma_x^2}} \exp\left(-\frac{x^2}{2\sigma_x^2}\right) \tag{10.18}$$

where $\sigma_x^2$ is the variance, and where we have assumed that the mean value is zero. After squaring, the signal $y(t)$ is expressed as

$$y(t) = x^2(t) \tag{10.19}$$

and the mean value $m_y$ and the variance $\sigma_y^2$ are given by the expressions

$$\begin{aligned} m_y &= \overline{y} = \sigma_x^2 \\ \sigma_y^2 &= \overline{(y-\overline{y})^2} = 2\sigma_x^4 \end{aligned} \tag{10.20}$$

Note that the first of these equations follows directly from equations (10.18) and (10.19). The result for $\sigma_y^2$ is a property of Gaussian processes (see Papoulis [1984]).

To continue with the determination of the statistical properties of the signal $z(t)$, we assume that the input signal is white noise with bandwidth $B$ about the center frequency $f_0$. The sample values taken at times separated by $1 / (2B)$ will then be statistically independent, or uncorrelated. With integration time $T$, we have now summed $n = 2TB$ independent values. The sum of the squares of Gaussian sample values follows the $\chi^2$ distribution with $n = 2TB$ degrees of freedom.

According to the central limit theorem, the sum of a number of independent variables will approach a Gaussian distribution when the number of variables approaches infinity. For large values of $n$, the integrator's output will approach a Gaussian distribution with the mean value of the data before integration, and with a variance reduced according to the number of independent sample values (that is, $n = 2TB$). Therefore, the integrator output has a Gaussian distribution with mean value and variance

$$\begin{aligned} m_z &= m_y \\ \sigma_z^2 &= \frac{\sigma_y^2}{2TB} = \frac{\sigma_x^4}{TB} \end{aligned} \tag{10.21}$$

We now use these results to calculate the performance of a passive sonar when the background noise has power $N$, the noise signal we wish to detect has power $S$, and both noise and signal have bandwidth $B$.

At the input to the detector, the SNR is

$$(S/N)_{in} = S/N \tag{10.22}$$

After squaring and integration, and given the above-mentioned conditions for Gaussian statistics, the mean values and variances of the detector output $z(t)$ under both hypotheses are

$$\begin{aligned} m_0 &= N \\ m_1 &= S + N \\ \sigma_0^2 &= N^2/TB \\ \sigma_1^2 &= (S+N)^2/(TB) \end{aligned} \tag{10.23}$$

In many practical cases of interest, the SNR at the input is low (that is, $S << N$), and the variances $\sigma_0^2$ and $\sigma_1^2$ are almost equal. We can then apply the general ROC curves illustrated in figure 10.5, computing the detection index $d$ and the SNR at the detector output from equations (10.15) and (10.16) as

$$\left(\frac{S}{N}\right)_{\text{det}} = d = TB\left[\left(\frac{S}{N}\right)_{in}\right]^2 \tag{10.24}$$

The detection threshold [equation (10.17)] is defined as the required ratio of signal power $S$ to noise power $N_0$ measured in a frequency band of 1 Hz. Since we have assumed white noise, $N = BN_0$ and equation (10.24) becomes

$$d = \frac{T}{B}\left[\left(\frac{S}{N_0}\right)_{in}\right]^2 \tag{10.25}$$

The detection threshold in dB for the incoherent receiver is then

$$DT = 10\log\left(\frac{S}{N_0}\right)_{in} = 5\log d - 5\log\left(\frac{T}{B}\right) \tag{10.26}$$

As an example, let us consider a system with bandwidth $B$ = 100 Hz that is trying to detect a weak acoustic signal with $(S / N)_{in}$ = 0.1. This implies that the SNR of the detector input is –10 dB. From ROC curves of the kind shown in figure 10.5, we find that if we require that $P_{FA} = 10^{-4}$ and $P_D > 0.5$, it follows that we must have $d > 15$. When equation (10.25) is solved with respect to $T$, we find that the minimum integration time is $T$ = 15 s. This means that we must integrate the power of the received signal for 15 seconds in order to draw a conclusion about whether a target is present; our conclusion will have a statistical confidence given by the selected values for $P_D$ and $P_{FA}$.

Equation (10.26) shows that we can obtain an arbitrarily high detection index by using a large integration time $T$. In practice, a sonar's performance cannot be increased without limit. As the integration time increases, the coherence of the target signal over time eventually degrades. If the input signal is the output of a beamformer, the target resides in a given beam for a limited time. In addition, large integration times result in delays that may not be acceptable given the sonar's application.

### *10.5.2 The replica correlator or matched filter receiver*

In the preceding section, we considered the detection of an unknown signal in noise. Now we consider the case where the signal we want to detect is completely known. This is an ideal situation, not likely to be encountered in practice, but is interesting because it establishes an upper bound on the performance of a sonar system.

In general, the optimum receiver for detecting a known signal in noise is a ***matched filter*** (Van Trees 1968), whose impulse response is the time inverse of the known signal waveform $s(-t)$. Alternatively, the optimum signal can be realized by use of a ***crosscorrelator*** that multiplies the received signal by a replica of the transmitted signal, and then integrates the results; see figure 10.7. In either case, the receiver calculates the replica correlation

$$w(\tau)=\int_0^T m(t-\tau)s(t)\,dt \tag{10.27}$$

where $s(t)$ is the transmitted signal, $T$ is its duration, and $m(t)$ is the received signal. The time $\tau$ is the time delay between the source waveform and any received echoes.

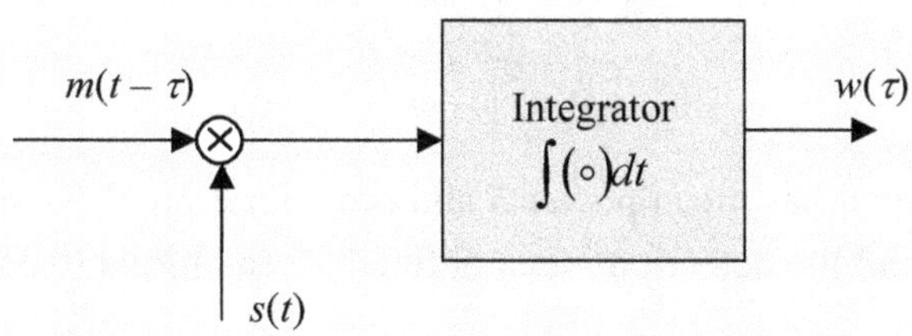


Figure 10.7. Optimal receiver, realized by using a crosscorrelator.

Under ideal conditions, the echo signal is a delayed version of the transmitted signal plus some degree of unwanted noise

$$m(t) = A_m s(t - T_m) + n(t) \tag{10.28}$$

where $A_m$ represents the strength of the echo signal and $T_m$ is the time delay between the signal's transmission from the sonar and the arrival of the received echo and $n(t)$ is noise. We assume that the signal is normalized such that

$$\int_0^T s^2(t)dt = 1 \tag{10.29}$$

After substituting

$$A_m = \sqrt{E} \tag{10.30}$$

where $E$ is the energy in the received echo signal, the replica correlation can be written as

$$w(\tau) = \sqrt{E}\int_0^T s(t-\tau)s(t)dt + \int_0^T n(t)s(t)dt \tag{10.31}$$

Under both hypotheses, it can be shown that the signal at the receiver output is Gaussian, with the mean values and variances given by (see Van Trees [1986])

$$m_0 = 0, \quad m_1 = \sqrt{E}$$
$$\sigma_0^2 = \sigma_1^2 = N_0 / 2 \qquad (10.32)$$

When the echo signal has mean power *S* and time duration *T*, the energy in the echo signal is $E = ST$, and the detection index defined by equation (10.15) becomes

$$d = \frac{2E}{N_0} = \frac{2ST}{N_0} = \frac{S}{N} 2TB \qquad (10.33)$$

The performance of the optimum detector is therefore given by the ROC curves in figure 10.5 for Gaussian statistics using the detection index now given by equation (10.33).

The relationship between the SNR at the input and the output of the optimal detector of a known signal in noise is given by

$$\left(\frac{S}{N}\right)_{det} = 2TB\left(\frac{S}{N}\right)_{in} \qquad (10.34)$$

Conversion of equation (10.34) to decibels results in

$$10\log\left(\frac{S}{N}\right)_{\text{det}} = 10\log\left(\frac{S}{N}\right)_{in} + 10\log(2TB) \qquad (10.35)$$

The value 10log(2*TB*) is called the processing gain and equation (10.35) shows that the processing gain increases with increasing the time-bandwidth product, *TB*. In most cases, however, increasing the bandwidth leads to a corresponding increase in noise level, that is, $N = BN_0$. When this is the case equation (10.35) becomes

$$DT = 10\log\left(\frac{S}{N_0}\right)_{in} = 10\log d - 10\log(2T) \qquad (10.36)$$

which shows that the detection threshold can be reduced by increasing the time duration, $T$, and thereby the energy of the transmitted signal (see equation (10.33)). Increasing the bandwidth $B$ itself will not improve the detection performance.

### *10.5.3 The square-law envelope detector*

In section 10.5.2 we assumed complete knowledge about the received signal, but this is not realistic in most cases encountered in practice. In this section, we consider detection of a sinusoidal pulse whose frequency and length are known, but whose phase is not.

We start by assuming that the transmitted signal $s(t)$ is a sine pulse with frequency $\omega_0 = 2\pi f_0$ and duration $T$, and that $s(t)$ may be written

$$s(t) = \sqrt{2}a(t)\cos\omega_0 t \tag{10.37}$$

where $\omega_0$ is the carrier frequency and $a(t)$ a slowly varying envelope. A fixed point–like target will give an echo that is an attenuated and delayed version of the transmitted signal, which is consequently given as

$$e(t) = A_m\sqrt{2}a(t - T_m)\cos\left[\omega_0(t - T_m) + \theta\right] \tag{10.38}$$

where $A_m$ is an amplitude factor that depends on the sonar's source level, the echo strength, and the transmission loss. The phase of the reflected signal is $\phi = \theta - \omega_0 T_m$, where $T_m$ is the two-way travel time to the target, and $\theta$ represents a possible phase angle in the reflection process.

In most cases, the phase angle of the received signal must be considered unknown and random because it is sensitive to very small changes in the propagation paths to and from the target. For instance, when $f_0 = 15$ kHz, a distance variation of $\Delta r = 5$ cm will lead to a phase shift of 360°. Consequently, it is most realistic to model the received signal's phase angle as being uniformly distributed. Further, without limiting validity, we can shift the time scale so that $T_m = 0$ and normalize the level of the received signal by setting $A_m = 1$.

The energy of the received signal $E$ is given by

$$E = \int_0^T \left[\sqrt{2}a(t)\right]^2 \cos^2(\omega_0 t)\,dt = \int_0^T a^2(t)\,dt \tag{10.39}$$

When the envelope is constant and has amplitude $a_0$, it follows that

$$E = a_0^2 T \tag{10.40}$$

Under the two hypotheses, the signal at the detector's input can be modeled as

$$\begin{aligned} H_0: \quad & x(t) = \sqrt{2}n(t)\cos(\omega_0 t + \phi_n) \\ H_1: \quad & x(t) = \sqrt{2}a(t)\cos(\omega_0 t + \phi_e) + \sqrt{2}n(t)\cos(\omega_0 t + \phi_n) \end{aligned} \tag{10.41}$$

where $\phi_e$ and $\phi_n$ are the uniformly distributed phase angles of the echo signal and the noise signal, respectively.

A common detector used in this case is the square-law envelope detector, shown in figure 10.8. The signal from the beamformer is filtered to the bandwidth $B$ of the signal and thereafter fed to a quadrature demodulator where the signal is multiplied by both $\sin(\omega_0 t)$ and $\cos(\omega_0 t)$, after which the two outputs are integrated to produce the cosine component, $x_c(t)$, and the sine component, $x_s(t)$, as shown in figure 10.8. After squaring, the two components are added to yield

$$z(t) = \left[x_c^2(t) + x_s^2(t)\right] \tag{10.42}$$

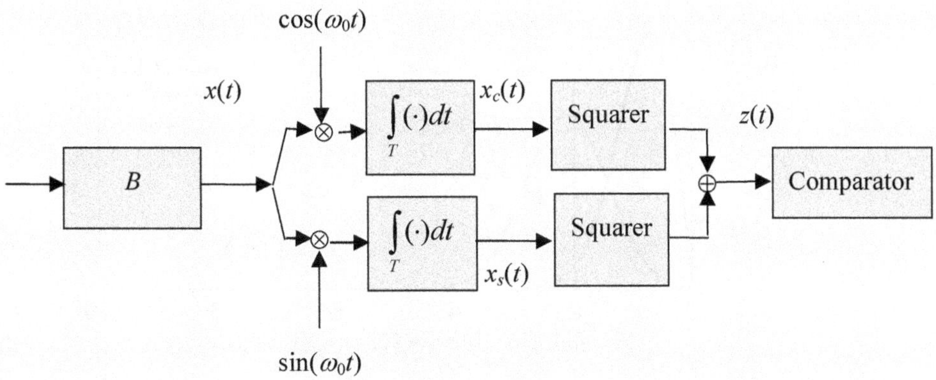


Figure 10.8. Envelope detector for a signal known except for carrier phase.

To determine $P_D$ and $P_{FA}$ we need to know the PDF for the envelope of $z(t)$ under each of the two hypotheses. This is a standard problem in statistical signal analysis, so here we shall only refer to the results. (See Van Trees [1968] for a complete description.). With a rectangular signal envelope, the PDF under hypothesis $H_0$ for the signal $z(t)$ has a Rayleigh distribution, and $p_0(z)$ has the form

$$p_0(z) = \frac{2z}{N_0} \exp\left( -\frac{z^2}{N_0} \right) \tag{10.43}$$

where $N_0$ is the noise power in a 1 Hz band.

Under the hypothesis $H_1$, the signal is the sum of a sine pulse and noise, and the signal envelope has a Rice distribution with $p_1(z)$ given by

$$p_1(z) = \frac{2z}{N_0} \exp\left( -\frac{z^2 + a_0^2}{N_0} \right) I_0\left( \frac{2a_0 z}{N_0} \right) \tag{10.44}$$

where $I_0(\cdot)$ is the modified Bessel function of the first kind and zeroth order. Figure 10.9 shows graphs of the probability densities given by equations (10.43) and (10.44).

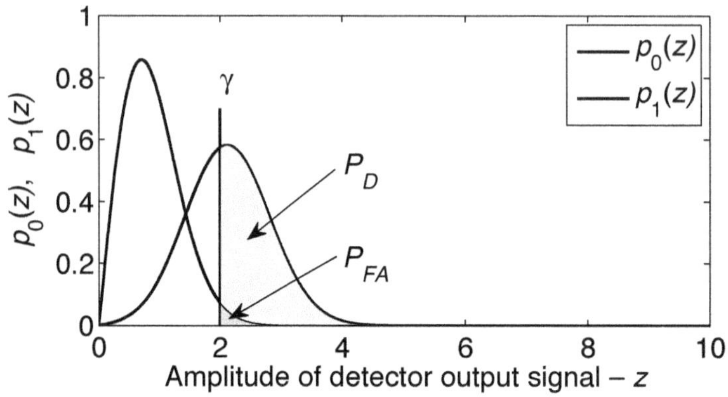


Figure 10.9. The probability densities $p_0(z)$ for only noise being present, equation (10.43), and $p_1(z)$ for both signal and noise being present, equation (10.44).

The detection probability $P_D$ and the false alarm probability $P_{FA}$ are obtained by performing the integrals of equation (10.13) using equations (10.43) and (10.44) for $p_0(z)$ and $p_1(z)$

$$P_{FA} = \int_{\gamma}^{\infty} \frac{z}{N_0} \exp\left(-\frac{2z^2}{N_0}\right) dz = \exp\left(-\frac{\gamma^2}{N_0}\right) \tag{10.45}$$

and

$$P_D = \int_{\gamma}^{\infty} \frac{2z}{N_0} \exp\left(-\frac{z^2 + a_0^2}{N_0}\right) I_0\left(\frac{2a_0 z}{N_0}\right) dz \tag{10.46}$$

If we require a certain value for $P_{FA}$ we can see from equation (10.45) that we must select the threshold

$$\gamma = \sqrt{-\sigma_n^2 \ln\left(P_{FA}\right)} \tag{10.47}$$

Next, the probability of detection is obtained from equation (10.46) using the threshold value of equation (10.45). To solve this problem, the integration must be done numerically, but the result is not very different from the results we obtain using Gaussian statistics, provided that we apply the detection index

$$d = \frac{E}{N_0} = \frac{ST}{N_0} = \frac{S}{N}TB \tag{10.48}$$

which is equal to half the value of the detection index for the known signal case given in equation (10.33). This is because the receiver for the unknown phase signal picks up noise on both the sine and the cosine components of the quadrature demodulator in figure 10.8. Equation (10.48) as well as equation (10.33) demonstrates that the detection index of a coherent detector is equal to the received signal energy divided by the spectral level of the noise.

The SNR at the detector output is

$$\left(\frac{S}{N}\right)_{\text{det}} = TB\left(\frac{S}{N}\right)_{in} = T\left(\frac{S}{N_0}\right)_{in} \tag{10.49}$$

The detection threshold of this detector is therefore

$$DT = 10\log\frac{S}{N_0} = 10\log d - 10\log T \tag{10.50}$$

Note that the difference between the incoherent energy detector (figure 10.6) and the coherent detector (figure 10.8) lies in the processing gain dependence of the *TB* product. For the incoherent detector, the performance increases with $(TB)^{1/2}$ (see equation (10.24)), while for the coherent detector, the performance increases linearly with *TB* (see equation (10.49)).

In principle, we can increase the energy in two ways: by increasing the source level of the sonar or by increasing the pulse length. The former is normally unrealistic, because the source level is usually set as high as possible. The second option is to increase the pulse length while simultaneously reducing the filter bandwidth.

However, lowering the bandwidth results in poorer range resolution, and increasing the pulse length results in higher Doppler sensitivity. The latter consideration implies that a moving target will not be detected unless the receiver has a number of parallel receivers and detectors adjusted to different Doppler shifts, as discussed previously and illustrated in figure 10.2. In the next section, these issues are considered in more detail.

## 10.6 Choice of signal in active sonar

In the last section, we discovered that detection performance is not dependent on signal waveform, but rather on the signal energy. If the transmitted signal's amplitude is constrained by a power limitation in the electrical circuitry or in the transducer, the solution is to increase the transmitted energy by increasing the length of the transmitted signal and reducing the bandwidth correspondingly. In the case of a simple sine pulse, increasing the signal length has the possible negative effect of degrading the range resolution and increasing the sensitivity to Doppler shifts so that even a slowly moving target will produce an echo outside the band of the filter.

This discussion shows that when the active sonar signal is a sine pulse, the pulse length must be selected as a compromise between range resolution and Doppler resolution or Doppler sensitivity. Therefore, in many cases, the sine pulse is not a suitable signal, and a better solution can be obtained by choosing a more complex, modulated pulse, in order to obtain a high $TB$ product. Possible signals include noise from a noise generator or pseudonoise from a shift register, but the most commonly used waveforms are FM pulses, where the frequency sweeps across a frequency band $B$ during a time $T$. The advantages of FM pulses are discussed in the next section.

To illustrate one of the alternative solutions to this problem, we select a more complicated transmitted signal that allows range and Doppler resolution to be specified with greater independence. To demonstrate this solution, we start with the matched filter receiver that correlates the received signal with a copy of the transmitted signal (see equation (10.27)). If the received signal is a time-delayed

($\tau$) and Doppler-displaced ($f_D$) copy of the transmitted signal at range $r$, the absolute value of the replica correlator can approximately be written as

$$\left|\Upsilon(\tau, f_D)\right| = \left| \int_0^T s(t-\tau) s^*(t) \exp(2\pi i f_D t)\, dt \right| \tag{10.51}$$

The time duration of the transmitted signal is $T$. We use the complex conjugated form of $s$, represented by an asterisk ($s^*$), in order to be able to treat bandpass signals. Equation (10.51) represents the correlation of a replica of the transmitted sonar signal $s(t)$ with a reference signal that is identical to the transmitted signal but shifted in frequency by $f_D$. Equation (10.51) is an approximation valid for narrow-band signals since the Doppler effect is replaced with a frequency shift $f_D$ of the carrier frequency and the changes in the signal envelope are neglected.

Figure 10.10 illustrates how the calculation of equation (10.51) can be implemented in practice with parallel processing using a set of reference signals and frequency shifts that span the frequency range of interest (Van Trees 1968, 1971). This figure is an extended version of figure 10.2 showing also the outputs as functions of time delay and frequency.

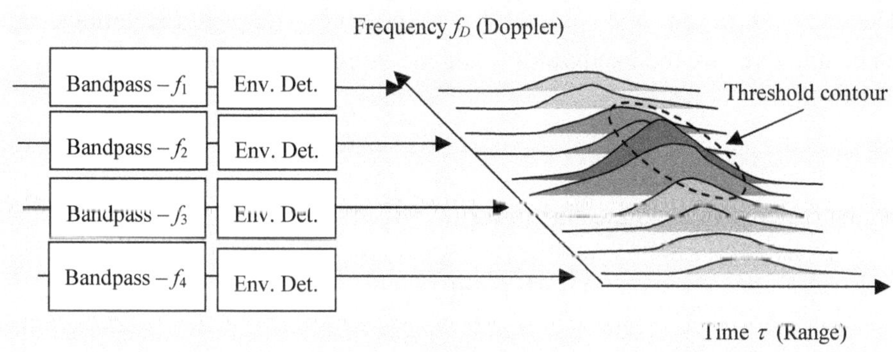


Figure 10.10. Receiver for generation of the time-frequency ambiguity function. The dotted curve indicates the intersection with the detection threshold value.

The combined output of all the channels is a representation of the two-dimensional function $|\Upsilon(\tau, f_D)|$ in equation (10.51), sampled at discrete values of the frequency shifts $f_D$. The output of each channel is compared with a threshold value $\gamma$ and output levels exceeding the threshold value are deemed to be a target detection. The function $|\Upsilon(\tau, f_D)|$ is called the signal ***ambiguity function*** and it provides means for comparing the range and Doppler resolutions achieved from different source waveforms. The function's maximum value is found at $\tau = 0$ and $f_D = 0$, and is equal to the signal energy $E$

$$|\Upsilon(0,0)| = \left|\int_0^T s(t)s^*(t)dt\right| = E \tag{10.52}$$

We assume that $s(t)$ is normalized such that $E = 1$.

Ideally, one would like the ambiguity function to have a single distinct peak in amplitude that decreases smoothly from its peak value. The 3 dB widths of the peak in delay and frequency are denoted $\Delta f$ and $\Delta\tau$, respectively. These values define the range and Doppler resolutions, and it is desirable that $\Delta f$ and $\Delta\tau$ be selected independently. We now demonstrate that this cannot be achieved using a simple sine pulse. However, with an FM pulse, the range and Doppler resolutions are independent and given by the bandwidth $B$ and the integration time $T$, respectively.

Consider first the case of a simple sine pulse $s(t)$ given by

$$s(t) = a(t)\exp(2\pi i f_0 t) \tag{10.53}$$

where the envelope is rectangular and has duration $T$ such that

$$\begin{aligned} a(t) &= \sqrt{\frac{1}{T}} && \text{for} \quad 0 < t < T \\ a(t) &= 0 && \text{for} \quad t < 0 \text{ and } t > T \end{aligned} \tag{10.54}$$

When inserted into equation (10.51), equations (10.54) result in

$$\begin{aligned} \left|\Upsilon(\tau, f_D)\right| &= \frac{1}{T}\left|\int_{\tau}^{T} \exp(2\pi i f_D t)\,dt\right| \\ &= \left|\left(1-\frac{|\tau|}{T}\right)\frac{\sin\left[\pi f_D\left(T-|\tau|\right)\right]}{\pi f_D\left(T-|\tau|\right)}\right| \end{aligned} \tag{10.55}$$

To gain an understanding of the shape of this two-dimensional ambiguity function, we show the two profiles $|\Upsilon(\tau,0)|$ and $|\Upsilon(0,f_D)|$, given by

$$\begin{aligned} \left|\Upsilon(\tau, 0)\right| &= \left|\left(1-\frac{|\tau|}{T}\right)\right| \\ \left|\Upsilon(0, f_D)\right| &= \left|\frac{\sin\left[\pi f_D T\right]}{\pi f_D T}\right| \end{aligned} \tag{10.56}$$

These two functions are depicted in figure 10.11 together with the contour plot of the section in the $\tau - f_D$ plane for $|\Upsilon(\tau, f_D)| = 1/2$ .

Consider the case $f_D = 0$, which is relevant for a stationary target. The value of $\tau$ where $|\Upsilon(\tau, 0)| = 1/2$ occurs for $\tau = \pm 1/(2T)$, and the time resolution is therefore $\Delta t = T$. For two-way transmission, this gives the range resolution of

$$\Delta r = \frac{c}{2}T \tag{10.57}$$

In order to resolve two targets, they would have to be separated in range by at least this difference in range.

For the case where $\tau = 0$, the values where $|\Upsilon(0, f_D)| = 1/2$ are at $f_D \approx \pm T/2$. The frequency (Doppler) resolution is therefore $\Delta f = 1/T$, and using equation (10.6) the velocity resolution becomes

$$\Delta v = \frac{c}{2Tf_0} = \frac{\lambda}{2T} \tag{10.58}$$

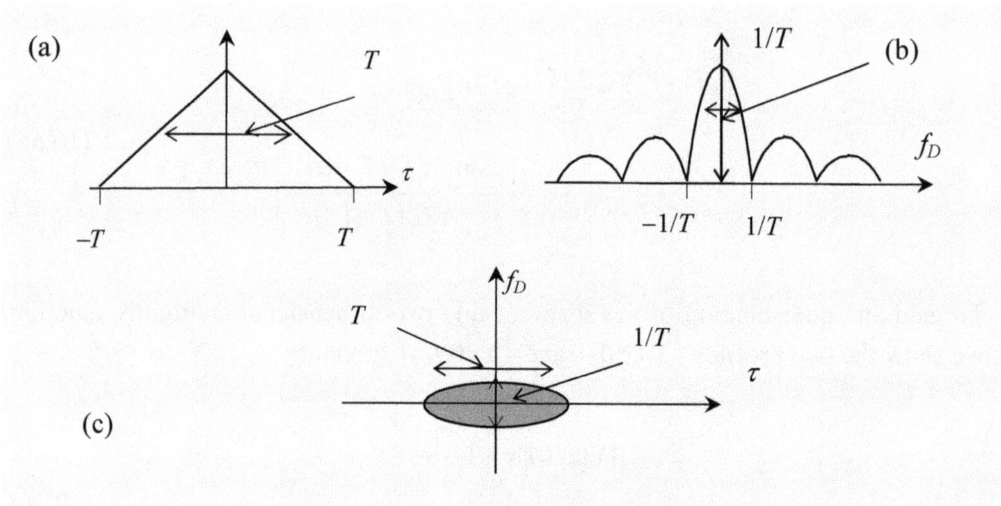


Figure 10.11. Ambiguity function of a sinusoidal pulse:
(a) Delay (range) profile, (b) Doppler profile, and
(c) contour plot in delay and Doppler.

The contour line defined by $|\Upsilon(\tau, f_D)| = 1/2$ defines a region of ambiguity in range and Doppler; targets within this region cannot localized unambiguously since all $(\tau, f_D)$ combinations above the threshold value result in detection. The ambiguity diagram, therefore, indicates the accuracies and resolution in range and Doppler achievable for a given sonar pulse.

Now consider the linearly modulated FM pulse with a transmitted signal having the form

$$s(t) = \sqrt{\frac{1}{T}} \exp\left[2\pi i\left(f_0 t + \frac{1}{2} m t^2\right)\right] \quad \text{for } -T/2 < t < T/2 \qquad (10.59)$$

where $m$ is a parameter that determines the rate of change of frequency of the FM signal. The instantaneous frequency, $f_i$, is

$$f_i = f_0 + mt \tag{10.60}$$

which varies from $f_0 - mT/2$ to $f_0 + mT/2$. Therefore the bandwidth of the time pulse is approximately $B = mT$ and consequently

$$m = B / T \tag{10.61}$$

Equation (10.59) with equation (10.61) defines an up-slope frequency modulation where instantaneous frequency increases linearly with time. A down-slope modulation results with $m = -B/T$ where the instantaneous frequency decreases with time.

Figure 10.12 shows examples of the time signal and frequency spectrum of a linearly modulated FM pulse. Here we have chosen a case where the instantaneous frequency varies from 10 kHz to 20 kHz over the time interval $T = 10$ ms. The time-bandwidth product in this example is therefore $BT = 100$.

The ambiguity function of the up-slope FM pulse is found by inserting equation (10.59) into equation (10.51) and solving the integral. The result is

$$\left|\Upsilon(\tau, f_D)\right| = \left|\left(1 - \frac{|\tau|}{T}\right) \frac{\sin\left[\pi\left(B\frac{\tau}{T} - f_D\right)\left(T - |\tau|\right)\right]}{\pi\left(B\frac{\tau}{T} - f_D\right)\left(T - |\tau|\right)}\right| \tag{10.62}$$

Figure 10.13 provides a three dimensional depiction of $|\Upsilon(\tau, f_D)|^2$ for the case where $T = 0.5$ s and $B = 20$ Hz. The maximum values of this ambiguity function are located on a ridge in the $\tau - f_D$ plane, as indicated in the figure. The ridge has an angle given by the ratio of the bandwidth $B$ to the time duration $T$ as

$$f_D = \frac{B}{T}\tau = m\tau \tag{10.63}$$

This implies that the FM pulse has a time-frequency ambiguity since all targets having combinations of range and Doppler satisfying equation (10.63) result in detections.

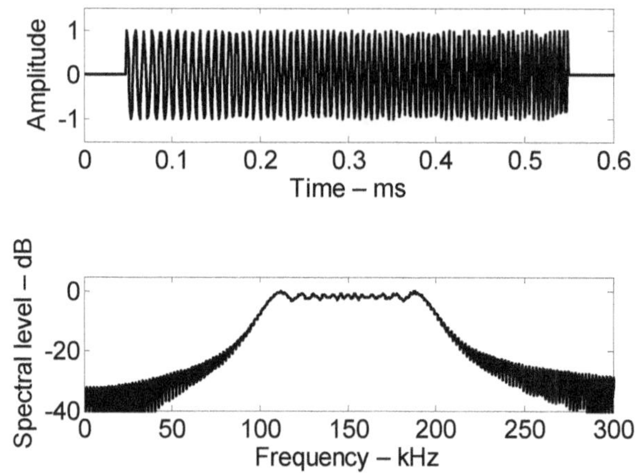


Figure 10.12. Time and frequency functions for an FM pulse. The amplitude of the time signal is normalized to unity, and the maximum spectral level is normalized to zero dB.

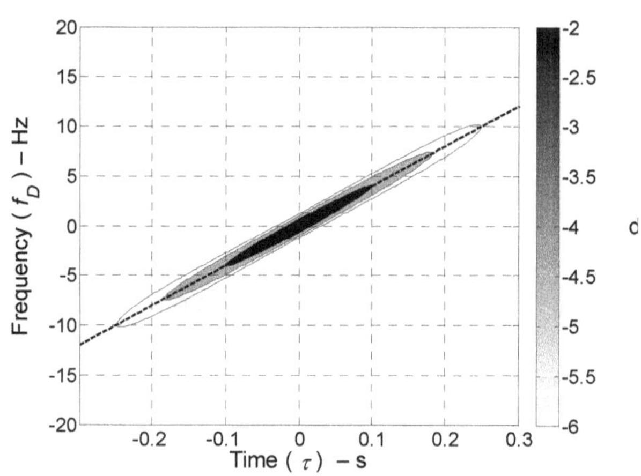


Figure 10.13. Contour plot of the central region of the ambiguity function of an FM pulse with $T = 0.5$ s and $B = 20$ Hz. The dashed line indicates the slope given by equation (10.60).

Figure 10.14 illustrates the same cross sections and contours shown in figure 10.11, but for the up-slope and down-slope modulations. The cross sections are of the type "sin($x$)/$x$" but along the $\tau$ axis there is also triangular weighting, as we can see in equation (10.62). The extent of the function above the threshold value along the $\tau$ axis is about $1/B$ and its extent along the $f_D$ axis is about $1/T$.

From the discussions of the properties of an FM pulse, it follows that Doppler resolution is determined by the pulse length and range resolution is determined by the bandwidth. Since these parameters can be selected independently, we can simultaneously obtain good resolution, in both range and Doppler shift by designing FM pulses with sufficiently high $TB$ products.

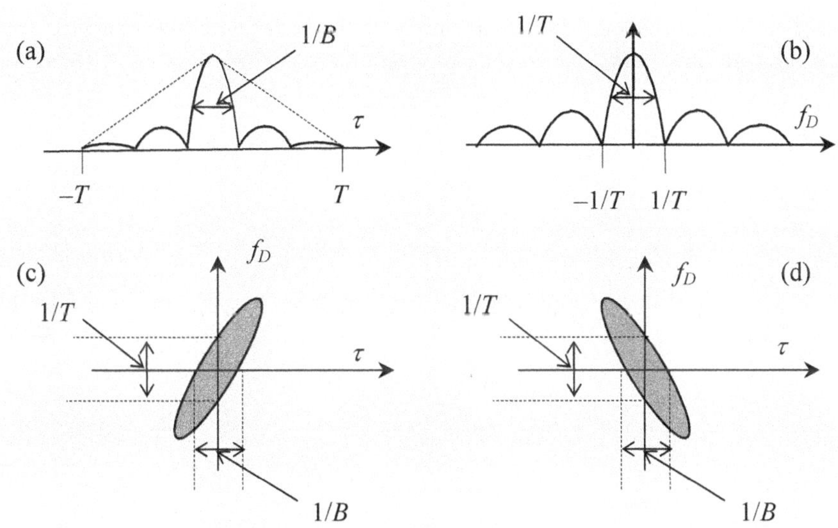


Figure 10.14. Ambiguity function of an FM pulse:
(a) Delay (range) profile, (b) Doppler profile,
(c) up-slope contour plot $|\Upsilon(\tau, f_D)| = 0.5$,
(d) down-slope contour plot $|\Upsilon(\tau, f_D)| = 0.5$.

### *10.6.1 Target detection in the presence of reverberation*

The choice of transmitted signal is a very important consideration relative to the sonar's sensitivity to reverberation. To consider this, observe first that reverberation will always have a distribution in the delay-Doppler space that is generated in the same way as that of the target. For example, ocean surface reverberation has a delay distribution that depends on propagation conditions and a Doppler distribution that depends on the frequency spectrum of the moving waves. For range- and Doppler-spread targets and reverberation, it is convenient to define a two-dimensional scattering function $S_{DR}(\tau, f_D)$ to describe the distribution of scattering in time delay and frequency shift. Under certain conditions it can be shown that the output of the matched filter is given as the convolution of the scattering function and the ambiguity function:

$$\Psi(\tau, f_D) = \int_t \int_f \Upsilon(t-\tau, f-f_D) S_{DR}(t, f)\, dt\, df \tag{10.64}$$

This operation is illustrated by figure 10.15, which shows the ambiguity function matched to a target at time $\tau$ and Doppler $f_D$ in the presence of range- and frequency-spread reverberation.

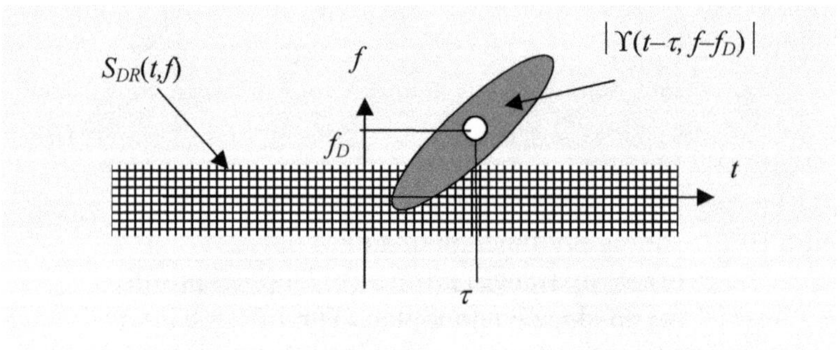


Figure 10.15. The matched filter output for a range- and Doppler-spread target is the convolution of the product of the scattering function and the ambiguity function.

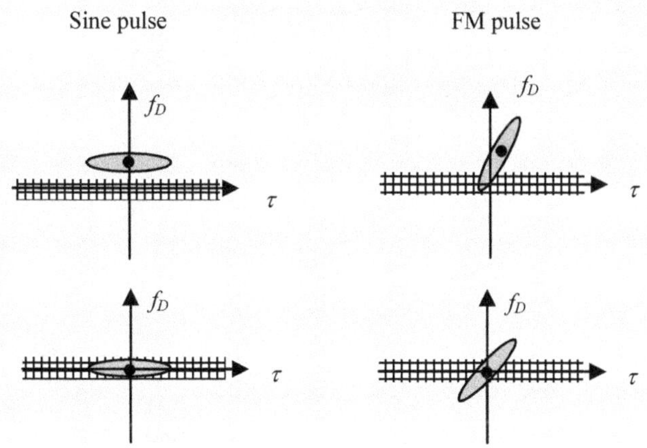


Figure 10.16. Four examples of target detection using two source waveform types, a sine pulse (left) and an FM pulse (right), and in two types of reverberation environments. Target echoes are indicated by black dots, signal waveforms ambiguities by gray shading, and reverberation by hatched areas.

In matched filter detection, only that part of the reverberation $S_{DR}(t,f)$ with an area in common with the signal's ambiguity function $|\Upsilon(t-\tau, f-f_D)|$, will pass through the processor. For a given delay-Doppler distribution of reverberation and targets, we can thus minimize the reverberation and maximize the target by selecting a transmitted signal with an appropriate ambiguity function. In Figure 10.16 the delay-Doppler spaces for two waveforms and two reverberation environments are sketched. The reverberation, which is assumed to be spread in both range and frequency, is indicated by the hatched rectangles. The best reverberation suppression is obtained using a signal with an ambiguity function that shares the smallest common area with the reverberation. Often, advanced sonars are able to transmit combinations of different modulated pulses in order to maximize target detection in reverberant environments.

# *Chapter 11*

# *Sonar systems*

## 11.1 Introduction

In this chapter we consider the principal components of sonar systems. We also discuss some of their limitations and limitation-related problems that may arise. These limiting factors can be environmental, such as sound speed variation, noise, and reverberation; problems also may arise from the parameters of the sonar itself, such as the transmitted power, beamwidth, and so forth. We shall see that all the relevant factors can be described by using what are called the ***sonar equations.*** Discussion will be limited to active sonar; while we will not discuss passive sonar explicitly, much of the discussion also will apply to passive sonar. We shall examine a hypothetical sonar used for detection of objects in the water column as well as a sidescan sonar for seabed mapping and characterization.

The main reference work related to sonar engineering is Urick's *Principles of Underwater Sound* (1983). In addition, *Underwater Acoustic Systems Analysis* by Burdic (1984), *Sonar for Practicing Engineers* by Waite (2002), and *Sonar Signal Processing* by Nielsen (1991) can be recommended. Two books that we often have referred to before, *Acoustical Oceanography: Principles and Applications* by Clay and Medwin (1977) and *Fundamentals of Acoustical Oceanography* by Medwin and Clay (1998), also contain valuable information on several aspects of sonar engineering.

Figure 11.1 provides a sketch of a hull-mounted sonar transmitting a sound beam and receiving an echo from a target. In addition to one or more echoes from distant objects in the water, this sonar will receive noise from distinct and distributed noise sources and reverberation from the sea bottom, sea surface, and the water environment. Figure 11.2 shows a flowchart with the sonar and the environment broken down to their components. The echo can be detected when its sound pressure level can be discerned from noise, reverberation, and other targets. The various factors affecting the detection performance of a sonar can be discussed and evaluated by means of the sonar equations.

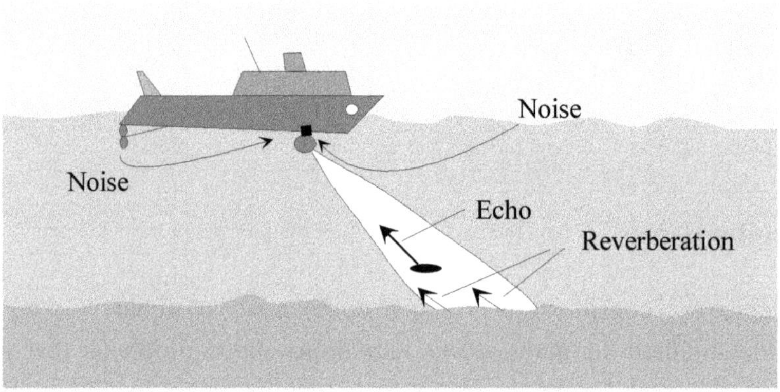


Figure 11.1. Searching for an object with a hull-mounted sonar in an environment having both noise and reverberation.

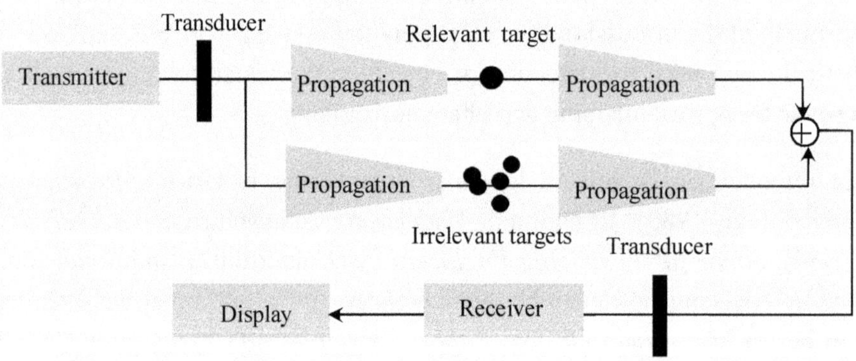


Figure 11.2. Signal flow diagram presentation of an active sonar, broken down into sonar and environmental components.

A comprehensive treatment of the factors depicted in figures 11.1 and 11.2 must include:

**Generation and reception of sound**

—Construction of electromechanical transducers

—Efficiency and sensitivity

—Directivity

**Transmission conditions**

—Sound propagation in water

—Absorption and transmission losses

—Loss by reflection from seabed and surface

**Noise**

—Sources of background noise

—Frequency and environmental dependency

**Reflection capabilities**

—Echo capabilities of targets

—Received noise from the target

—Reverberation from seabed, surface, and volume

**Signal processing**

—Choice of transmission signals

—Processing of received signals

—Presentation

Some of these items have been discussed in earlier chapters; here we will build upon those discussions and, where useful, repeat some of the most relevant subjects.

## 11.2 Sonar equations

The sonar equations, a set of relatively simple equations, allow us to evaluate sonar performance based on knowledge of the acoustic environment and sonar system characteristics. Essentially, we use the sonar equations to determine the level of the signal that is to be detected and to estimate the level of background noise and reverberation.

For an active sonar, the echo level $EL$ is given by

$$EL = SL - 2TL + TS \tag{11.1}$$

where $SL$ is the transmitter's source level, $2TL$ is the two-way transmission loss, and $TS$ is the target strength. All the terms are given in logarithmic units (that is, in dB), so that addition and subtraction replace multiplication and division.

The echo has to compete with noise; the noise level $NL$ is given by

$$NL = NSL + 10\log B - DI_r \tag{11.2}$$

Here $NSL$ is the spectral level of the noise (contained in 1 Hz bandwidth), $B$ is the receiver bandwidth, and $DI_r$ is the directivity index of the receiving system. When the noise is isotropic, $DI_r$ can be obtained from equation (9.30) when it is combined with either equation (9.32) or equation (9.44). In general, the noise has directional properties, which means that the noise level depends on the receiving direction. This means that $DI_r$ must be estimated by integrating the product of the noise's relevant directional distribution and the hydrophone array directivity. We should note that the receiver index $DI_r$ may differ from the transmitter directivity index $DI_s$ if different transducers are used for reception and transmission, or if the noise field is not isotropic.

Often an active sonar will be limited by reverberation because of reflections from the sea surface, seabed, or particles in the water mass. Reverberation is the reradiation of active sonar's transmissions from scatterers such as ocean bottom, surface, and water volume particulates. It is received as a long, slowly decaying tonal blast in the same bandwitch as "legitimate" targets, significantly degrading the signal to noise ratio. The nature of the reverberation level $RL$ depends on whether the reverberation comes from a surface or a volume.

***Surface reverberation***

$$RL = SL - 2TL + S_a + 10\log A \qquad (11.3)$$

Here, $A$ is the size of the surface area that produces the reverberation at a given time and $S_a$ is the ***surface scattering strength,*** a coefficient that characterizes the scattering properties of the surface.

***Volume reverberation***

$$RL = SL - 2TL + S_v + 10\log V \qquad (11.4)$$

The size of the volume that produces the reverberation is represented by $V$, while $S_v$ is the ***volume scattering strength,*** which characterizes the strength of the scattering from the volume. In the two preceding equations, the coefficients $S_a$ and $S_v$ can be thought of as target strength per unit surface and target strength per unit volume, respectively. Clearly, the propagation conditions affect the insonification of the surface and the volume.

For the target to be detected, the echo must exceed the noise or reverberation by a specific detection threshold $DT$. The parameter $DT$ is defined as the difference between the echo level and the noise or reverberation level that is necessary for the probability of detection $P_D$ to be acceptable while at the same time keeping the probability of false alarm $P_{FA}$ sufficiently low (see section 10.5).

If we know the statistical properties of the signal and the noise, we can estimate $P_D$ and $P_F$, functions of the signal-to-noise ratio, and then determine the detection threshold $DT$. Commonly used values are $P_D = 0.5$ and $P_{FA} - 10^{-4}$; such values normally require that $DT$ be in the range of 10 – 15 dB for practical detection of submarine targets.

If the noise dominates over the reverberation, the sonar is ***noise limited*** and the detection range is determined by

$$EL - DT \geq NL \qquad (11.5)$$

If the reverberation is dominant, the sonar is ***reverberation limited*** and the range is determined by

$$EL - DT \geq RL \tag{11.6}$$

Equations (11.1)–(11.6) are sonar equations for an active sonar. They may be used to analyze a given sonar's performance or to determine the specifications for a new sonar. However, we must take into account that several terms in the sonar equations are interdependent, and also that other important requirements and considerations may determine a sonar's specifications—for instance, requirements on the system's resolution and its accuracy in bearing, range, and target velocity discrimination.

The ***figure of merit*** $F_m$ is a parameter often used to characterize the performance of a given sonar. It is defined as the maximum two-way transmission loss $(2TL)_{max}$ that the sonar can sustain for a given performance defined by $P_D$ and $P_{FA}$

$$F_m = (2TL)_{max} = SL + TS - NL + DI - DT \tag{11.7}$$

***Angular resolution*** is defined as the sonar system's capability to distinguish between two equally strong targets at the same distance but with a small difference in bearing. Angular resolution is determined by the beamwidth, as two targets can only be separated from each other if the angular separation between them is at least equal to the beamwidth. We discussed this phenomenon in chapter 9, where we found that a uniformly weighted line array with length $L$ has a beamwidth and angular resolution $\Delta\theta$ of approximately

$$\Delta\theta = \frac{\lambda}{L} \tag{11.8}$$

Resolution must not be confused with accuracy, which is a measure of how correctly it is possible to measure the direction of an individual target. The bearing accuracy is dependent not only on the receiver array, but also on the signal-to-noise ratio, since accuracy arbitrarily improves when the noise is sufficiently low.

***Range resolution*** $\Delta r$ is the ability to distinguish between two equally strong targets lying in the same direction, but with a small difference in distance. For a sonar with pulse length $\Delta T$, the two echoes will not overlap at the time of arrival at the receiver if the radial separation between them is at least

$$\Delta r = \frac{c\Delta T}{2} \tag{11.9}$$

This means we have to apply short pulses for a good range resolution if we are using a simple sinusoidal pulse. However, by using a modulated pulse with bandwidth $B$, for instance, an FM pulse as discussed in section 10.6, the compressed effective pulse length after matched filter reception is $\Delta T \approx 1 / B$, and the range resolution is

$$\Delta r = \frac{c}{2B} \tag{11.10}$$

If a target has a radial velocity relative to the sonar, the echo signal will have a frequency (Doppler) shift that can be measured and then used to determine the target's relative radial velocity. When a radial velocity component $v$ exists between sonar and target, the ***frequency shift*** $\Delta f$ of a transmitted signal at frequency $f_0$ is

$$\Delta f = \frac{2v}{c} f_0 \tag{11.11}$$

In order to distinguish between two targets having a small velocity difference $\Delta v$, the receiver must be able to carry out a frequency analysis of a received signal with a bandwidth (frequency) resolution better than the frequency shift given in equation (11.11). This means that the processing bandwidth $B_{proc}$ must be

$$B_{proc} < \frac{2\Delta v}{c} f_0 \tag{11.12}$$

The time duration $T$ of the transmitted signal then must be

$$T > \frac{c}{2\Delta v f_0} \tag{11.13}$$

We see that a good resolution in both range and velocity can impose contradictory demands. If good resolution in both range and Doppler is required, we will have to use modulated signals with a time-bandwidth product $TB$ that is greater than unity.

## 11.3 The source level

In section 9.2, radiation from directional sources was discussed. There, using equation (9.35), we found that the source level at a reference distance of 1 m, expressed in dB relative to a sound pressure of $1\mu$Pa, is given by

$$SL = 170.8 + 10\log W + DI_s \tag{11.14}$$

Radiated acoustic power is represented by *W*, and the directivity index of the sound projector by $DI_s$. In older literature, the reference pressure is one microbar; in these units the source level is given as

$$SL = 70.8 + 10\log W + DI_s \tag{11.15}$$

To find the directivity index, we can use the general expression we earlier developed with equation (9.41), which produces a general directivity index *DI* that may be used for approximate calculations for a radiating surface with effective area *A,* (equation 9.44) and for a line element with effective length *L* (equation 9.32), respectively:

$$DI = 10\log\left(4\pi\frac{A}{\lambda^2}\right) \tag{11.16}$$

$$DI = 10\log\left(2\frac{L}{\lambda}\right) \tag{11.17}$$

Accordingly, the directivity index increases with projector size and frequency. In principle, we could arbitrarily obtain a high source level by choosing a projector that is very large in proportion to the wavelength. In reality, there will usually be practical limitations on the projector's size and, since acoustic absorption rapidly increases with rising frequency, there will always be a limit on how high the frequencies one can use to attain a given range.

## 11.4 The transmission loss

The transmission loss is determined by the local bathymetry and the sound speed along the propagation track, oceanographic conditions, and attenuation, which is a function of frequency. If sound propagation involves interaction with the sea surface or the seabed, the properties of these boundaries also may have a significant impact on transmission loss.

The dependency of the transmission loss on environmental conditions is a central topic in underwater acoustics; some of the most important elements have been discussed in earlier chapters. Several mathematical propagation models are available for calculating the transmission loss for given environmental conditions (Jensen et al. 1994 and Etter 2001). Two of the most commonly applied methods have been presented in chapters 6 and 7.

In order not to complicate the discussion too much, in this section we will apply very simple transmission models. For transmissions in relatively deep water and at short distances, we can often assume spherical sound propagation, which along with absorption produces the transmission loss *TL*, defined as

$$TL = 20\log\left(r/r_0\right) + \left(r - r_0\right)\alpha_{dB} \tag{11.18}$$

where $\alpha_{dB}$ is the absorption coefficient in dB per meter, $r$ is the distance between the transmitter and receiver, and $r_0$ is a reference distance that is usually selected to be 1 m. When $r_0$ is much smaller than $r$, the transmission loss for spherical propagation becomes

$$TL = 20\log\left(r\right) + r\,\alpha_{dB} \tag{11.19}$$

This expression can be applied with caution to short-distance transmissions in cases where sound refraction does not play a major role and where there are no interfering reflections from the seabed or sea surface. These will usually be the conditions for echo sounders and sonar arrangements where most of the echo is produced from the signal's direct path to the object of detection.

Another simple transmission model that can be applied in cases when there are relatively long transmission distances in shallow water is

$$TL = 10\log(r) + r\,\alpha_{dB} + BL + OL \tag{11.20}$$

Again, we have assumed that the reference distance is 1 m. Here, unlike in the case represented by equation (11.19), we have assumed a cylindrical propagation loss, represented by $10\log(r)$. We also have assumed that additional losses are caused by multiple reflections from the seabed, Bottom Loss ($BL$), and the surface, Ocean Loss ($OL$). The reflection loss for a perfectly calm sea surface is zero because of the high impedance contrast at the water-air interface. However, when ocean waves are present, a coherent reflection loss will be caused by surface roughness, which we treated in section 4.6.

The bottom loss is strongly dependent on the sea floor's geoacoustic properties. Depending on the angles of incidence, the frequency, and the compressional and shear speeds of the bottom, some of the incident acoustic energy may be converted to other wave types in the seabed, resulting in significant contributions to the transmission loss. In addition, the incoherent reflection loss from a rough bottom may affect transmission loss.

In some cases, we may use a combination of spherical and cylindrical spreading: spherical spreading up to a distance $D_w$ equal to the water depth and thereafter, for $r > D_w$, cylindrical spreading

$$\begin{aligned} TL &= 20\log(r) + r\,\alpha_{dB} \qquad r < D_w \\ TL &= 20\log(D_w) + 10\log(r - D_w) + r\,\alpha_{dB} \qquad r > D_w \end{aligned} \tag{11.21}$$

## 11.5 The echo level

The strength of the echo from a target is characterized by its ***target strength*** $TS$ or by the ***scattering cross section*** $\sigma$. These characteristics are defined respectively as

$$TS = 10\log\left(\frac{I_r}{I_i}\right) \tag{11.22}$$

$$\sigma = \frac{P_r}{I_i} \tag{11.23}$$

where $I_r$ is the reflected intensity, $I_i$ is the incident intensity, and $P_r$ is the power of the reflected signal. The target strength $TS$ is related to the scattering cross section $\sigma$ by

$$TS = 10\log\left(\frac{\sigma}{4\pi}\right) \tag{11.24}$$

Reflection and propagation of sound from sonar targets are such complicated matters that here we can only touch upon some of the simpler aspects of the most important characteristics. More general information on these subjects can be found in Freedman (1962a, 1962b), Haslett (1970), Urick (1983), and Medwin and Clay (1998). In practice, echo strength is to be considered an empirical value that must be measured for objects of interest. However, in this section we will consider theoretical models and examine three main categories of objects, for which we will discuss simple cases of echo formation:

- fixed objects, large in proportion to the wavelength;
- fixed objects, small in proportion to the wavelength; and
- resonant objects, large in proportion to the wavelength.

Of course, further categories could be constructed. For example, when working with complex targets, quite possibly a combination of two or more of the categories listed here will be relevant.

For some uncomplicated geometric shapes such as spheres, plates, and cylinders, the calculations are quite simple. For a large, fixed object with a curvature radius that is much larger than the signal wavelength, the reflected echoes can be calculated with the help of methods from geometric optics. To do so, we will determine the target strength of an echo from a large sphere with the radius $a$ (see figure 11.3). The incident power $P_i$ in the thin tube with cross section area $A_i$ is defined as $P_i = A_i I_i$. After total reflection from the sphere's surface, this power is reflected back over the area of reflection, $A_r$, so that the reflected intensity $I_r$ is defined as $I_r = P_i / A_r$.

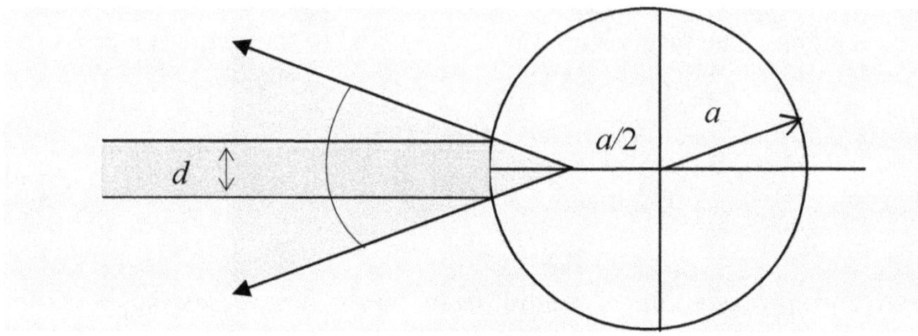


Figure 11.3. Calculation of the target strength of a large sphere.

The calculations of $A_r$ and $A_i$ are geometrical tasks, solved by finding the path of the reflected sound. When the thin tube of incident sound is a cylinder with diameter $d$, $A_i$ is expressed as

$$A_i = \frac{\pi}{4} d^2 \tag{11.25}$$

After reflection, with the output angle equal to the angle of incidence, the reflected sound path will appear to come from a point at a distance $a/2$ from the center of the sphere. Since target strength is defined with respect to a distance of 1 m from the center of the sphere, $A_r$ is defined as

$$A_r = \frac{\pi}{4}\left(\frac{2d}{a}\right)^2 \tag{11.26}$$

Therefore, according to equations (11.22)–(11.24), the target strength of the sphere becomes

$$TS = 10\log\left(\frac{A_i}{A_r}\right) = 10\log\left(\frac{a^2}{4}\right) \tag{11.27}$$

and the scattering cross section becomes

$$\sigma = \pi a^2 \tag{11.28}$$

From equation (11.28), we see that a large sphere has a scattering cross section that is equal to its geometric cross section.

The same result is obtained by observing that the total power intersected by the sphere is $P_i = I_i \pi a^2$. If this power is uniformly reflected to all directions, the scattered intensity at unity distance becomes $I_r = I_i / 4\pi$. By the definitions of equations (11.22) and (11.23) we obtain the target strength and scattering cross section of equations (11.27) and (11.28). This indicates that a large sphere reflects the incident plane wave in the backward direction as though it were isotropic reflector of sound (Urick 1983, 1996).

This result can easily be extended to include plane wave reflection from any curved surfaces. For instance, the target strength of an echo from a general double-curved surface with the main curvature radii $a_1$ and $a_2$ is

$$TS = 10\log\frac{a_1 a_2}{4} \tag{11.29}$$

For an infinitely large plane surface, and with a source at the finite distance $r$, we can no longer expect the wave to be plane across its entire surface. However, this situation is the same as the one in which a plane wave hits a sphere with a radius equal to the distance $r$. The target strength of an infinitely large plane surface consequently is

$$TS = 10\log\frac{r^2}{4} \tag{11.30}$$

where $r$ is the distance to the reflecting surface.

We can also reach this result in another way. For an ideally reflecting and infinitely large surface, the reflected sound wave can be imagined to come from a mirror image of the source, which lies at the same distance $r$ behind the surface, or at the distance $2r$ from the source. Consequently, the geometric transmission loss $TL$ is $20\log(2r)$, while the target strength term presupposes two times the spherical $TL$, which produces $40\log(r)$. The difference between twice the spherical $TL$ and the geometric $TL$ is $40\log(r) - 20\log(2r) = 10\log(r^2/4)$, which is the target strength of the infinitely large surface, equation (11.30).

The next case involves a surface with the infinitely large area $A$. For normal incidence, the incident intensity $I_i$ is expressed as

$$I_i = \frac{P_i}{A} \tag{11.31}$$

The reflected intensity at a 1 m distance from the reflecting surface is

$$I_r = \frac{P_i}{4\pi} D \tag{11.32}$$

where $D$ is the directivity factor defined earlier. Therefore, the target strength as expressed by the directivity factor is

$$TS = 10 \log\left(\frac{DA}{4\pi}\right) \tag{11.33}$$

For a plane surface with an area $A$, we found in equation (9.44) that

$$D = 4\pi \frac{A}{\lambda^2} \tag{11.34}$$

so that its target strength is

$$TS = 10 \log\left(\frac{A^2}{\lambda^2}\right) \tag{11.35}$$

Whether to use equation (11.30) or equation (11.35), that is, whether the surface is to be considered infinite or finite, is determined by whether we are located in the nearfield or the farfield. This question has been discussed already in section 9.3, where in equation (9.62) we found that, for a quadratic surface with side lengths $L$, the transition zone between the nearfield and the farfield is given approximately by the Rayleigh distance $R_R$, so that

$$R_R = \frac{2L^2}{\lambda} \tag{11.36}$$

While equation (11.36) is not entirely in accordance with equation (9.62), the discrepancy can be explained by our present need to take into consideration the length of the diagonal, which is $\sqrt{2}L$. If the relationship between the source-surface distance $r$ and the Rayleigh distance $R_R$ is $r > R_R$, we are in the farfield and the target strength is given by equation (11.35); when $r < R_R$, we are in the nearfield and the target strength is given by equation (11.30). For $r = R_R$, equations (11.31) and (11.35) will give identical results. These same considerations may be used to determine the target strength of other simple objects.

For a straight cylinder with length $L$ and radius $a$, we also can apply the principles represented in figure 11.3 to a small cross section of an acoustic wave with the height $d$. The area of the cylinder receiving sound is now $A = dL$, and the intensity is spread across a solid angle sector $\Omega$, which is approximately given by

$$\Delta\Omega = \left(\frac{\lambda}{L}\right)\left(\frac{2d}{a}\right) \tag{11.37}$$

Since the directivity factor is written as

$$D = \frac{4\pi}{\Delta\Omega} \tag{11.38}$$

and as the cylinder's finite length $L$ and radius $a$ are both considerably larger than the wavelength, its target strength is expressed as

$$TS = 10\log\left(\frac{L^2 a}{2\lambda}\right) \tag{11.39}$$

This expression presupposes that we are in the farfield of the cylinder with regard to the cylinder's length. For an infinitely long cylinder, the effective length is

$$L_{eff} = \sqrt{\lambda r} \tag{11.40}$$

so that the target strength of an infinitely long cylinder is

$$TS = 10\log\left(\frac{ra}{2}\right) \tag{11.41}$$

The results derived so far assume that the target is large compared to the signal wavelength; therefore, the findings are high frequency approximations. When this assumption is not the case, analysis becomes considerably more difficult.

Although in chapter 13 we shall discuss such analyses in more detail, here we will continue by presenting the result of scattering from a small, rigid, fixed sphere. In figure 11.4, the scattering cross section for a rigid sphere with radius a is represented as a function of normalized frequency, so that $ka = 2\pi a / \lambda$. Also, the scattering cross section $\sigma$ is normalized by the geometrical cross section $\pi a^2$.

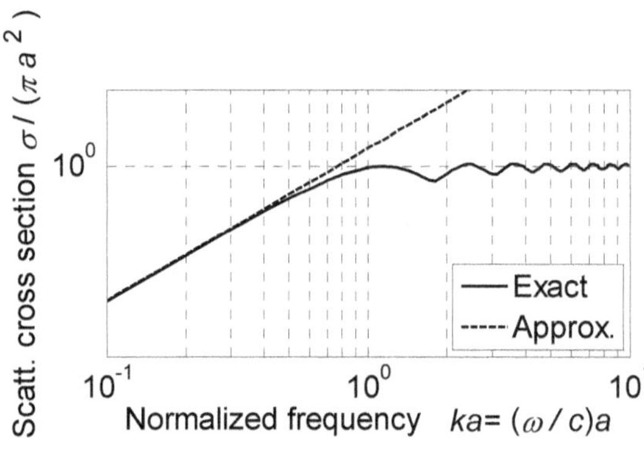


Figure 11.4. The scattering cross section of a rigid sphere, represented as a function of normalized frequency. The solid line represents the exact result (derived in section 13.3; see figure 13.5) and the dashed line is the approximation of equation (11.42).

For low frequencies, the scattering cross section increases with the fourth power of the frequency as given by

$$\sigma = \left(\pi a^2\right)\frac{25}{4}(ka)^4 \tag{11.42}$$

This result is also plotted in figure 11.4.

At very high frequencies, or for a very large sphere, the scattering cross section oscillates and approaches the size of the geometrical cross section. This agrees with our earlier findings from equation (11.28).

Another interesting case is that of the scattering from a gas-filled bubble. Such a bubble will have a resonance frequency given by the density of the surrounding water and the stiffness of the gas encapsulated inside the bubble. The scattering cross section $\sigma$ of a pulsating bubble is given as

$$\sigma = \frac{4\pi a^2}{\left[\left(\frac{f_r}{f}\right)^2 - 1\right]^2 + \delta^2} \tag{11.43}$$

(This expression will be derived in section 13.5.) The resonance frequency $f_r$ is determined from the bubble's radius $a$ and the ambient pressure. At zero water depth, the ambient pressure is one atmosphere and the resonance frequency is

$$f_r = \frac{3.2}{a} \tag{11.44}$$

The numerical factor in equation (11.44) presupposes that $a$ is given in meters and $f_0$ in hertz. In equation (11.43), the attenuation coefficient $\delta$ accounts for losses caused by various mechanisms, including acoustic radiation and conversion to heat. Figure 11.5 shows the scattering cross section of a spherical gas bubble as a function of frequency; the value of $\delta$ is equal to 0.1, which may be considered a typical value for an air bubble in water.

At the resonance frequency, the scattering cross section becomes

$$\sigma = \frac{4\pi a^2}{\delta^2} \tag{11.45}$$

Near the resonance frequency, the scattering cross section is much larger than the geometrical cross section (see both equation (11.43) and figure 11.5). At frequencies higher than the resonance frequency, the scattering cross section is also considerably larger than the geometrical cross section.

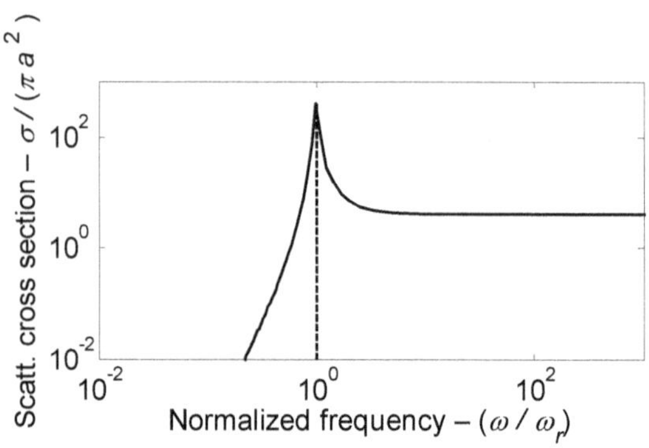


Figure 11.5. The scattering cross section of a gas bubble, plotted as a function of normalized frequency, indicated by the dashed line. The damping coefficient is in this example $\delta = 0.1$.

This phenomenon explains why fish with swim bladders have considerably greater target strength than would be indicated by their size, despite their resonance frequencies being significantly lower than the frequency of echo sounders and sonars used to search for fish.

It is useful to understand the theory and principles of echo formation but, in reality, most targets are too complicated for theoretical treatment. We therefore have to rely on measurements in order to obtain approximate estimates of target strength. Such measurements have been carried out for most targets of interest. For instance, careful measurements have been obtained of many fish species; see Horne (2000) and MacLennan (1990) for an overview of acoustical abundance estimation. For instance, a simple relation between a fish's length and its target strength is an equation of the form

$$TS = 20\log(l) + A \tag{11.46}$$

where $l$ is the length in centimeters and $A$ is a constant (Foote 1987). For a target of a school of cod, when the sonar frequency is 38 kHz, the appropriate value of $A$ is found to be –67 dB. Using equation (11.46), the target strength of the echo from cod, plotted as a function of fish length, appears in figure 11.6.

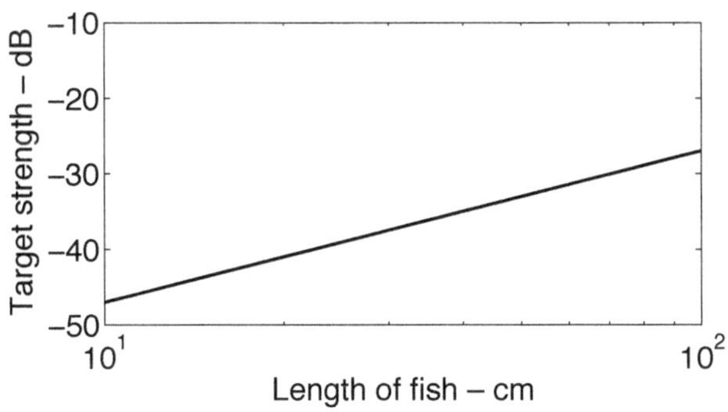


Figure 11.6. Target strength of cod, calculated as a function of fish length.

Target strengths of submarines show great variability with aspect angle and also in measurements of individual echoes (Urick 1983). However the average target strength at relevant sonar frequencies may have a butterfly shape as shown in figure 11.7. Here, the angles 90° and 270° correspond to the echoes from the beam aspect, while 0° and 180° correspond to the echoes from the bow and stern.

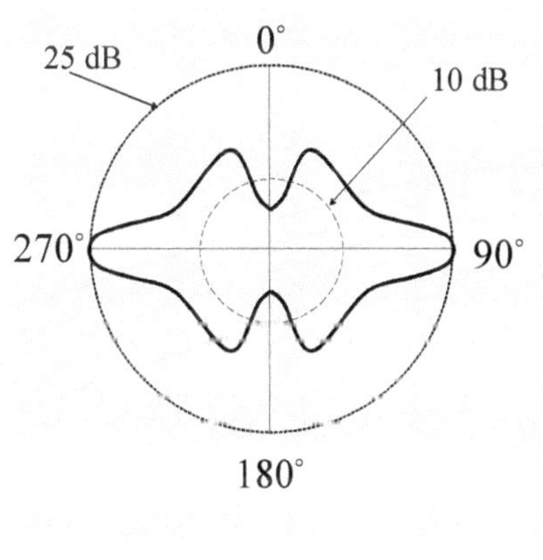


Figure 11.7. Polar diagram showing the target strength of a submarine. Bow and stern aspects are at 0° and 180°. Beam aspects are at 90° and 270°.

## 11.6 Noise

Oceanic noise is generated from a number of different types of acoustic sources with different frequency bands. The noise that originates from ocean waves, rain, but no other definable sources is considered to be natural background noise, called ***ambient noise***. Such noise is usually characterized by the noise spectral level (NSL), that is, the acoustic intensity of the noise in decibels when measured in a 1 Hz frequency band relative to a reference measurement of the intensity of a plane sound wave with a 1 $\mu$Pa sound pressure.

In order to find the total noise level, we must integrate the noise spectrum across the receiver's bandwidth. However, for relatively small bandwidths we can replace integration with multiplication by the bandwidth, which yields the same result as the addition done using the decibel scale in equation (11.2).

The mean ambient noise level often follows a group of curves called Knudsen curves. In figure 11.8, we have simplified versions of Knudsen curves calculated by Wenz (1962) as functions of frequency and wind speed for typical NSLs of noise generated by wind-driven ocean waves.

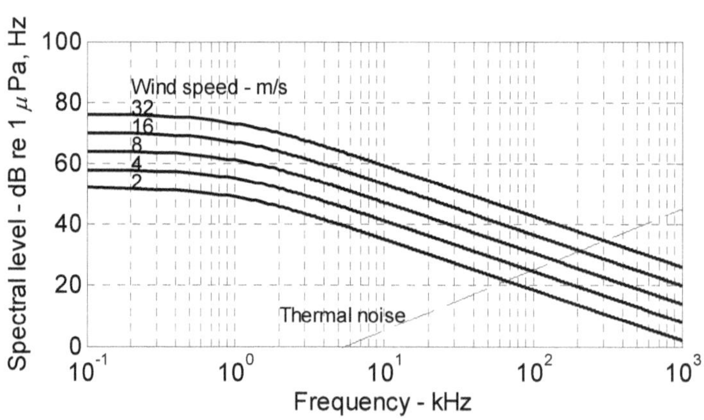


Figure 11.8. Spectral levels of ambient underwater noise, plotted as a function of frequency.

These curves indicate that with increasing frequency the ambient noise declines at about 17 dB per decade of frequency. At high frequencies, the noise is often dominated by thermal noise, for which the spectrum is also shown in figure 11.8. Thermal noise may dominate at higher frequencies, that is, above 100–200 kHz, since as the frequency increases, the thermal noise increases by 20 dB per decade.

In water with ship traffic or other human activity, the noise produced from such activity probably will dominate. In the lower frequency band, noise from passing ships can dominate, and ship noise can have clear spectral lines created by the shaft and blade frequency of the propeller and associated harmonics. At higher frequencies, the noise caused by cavitation may dominate; cavitation noise decreases with the frequency by approximately 20 dB per decade.

In addition to the ambient noise, there will also be self-noise generated by the engine and propellers of one's own vessel and by the movement of the vessel. Noise is also hydrodynamically generated by water flow across the transducers. The noise generated by these sources varies greatly from vessel to vessel so that it is difficult to give specific values. With hull-mounted sonars and echo sounders, self-noise normally will dominate as compared to noise from other sources when vessel velocity is high.

The effect of noise on a sonar is created not only by the noise level, but also by its directional dependence and the directivity of the receiving antenna. Noise generated from concentrated sources such as a passing vessel will have a strong directional dependence, while thermal noise will be omnidirectional. Noise from waves and wind will also exhibit directional properties in the vertical plane.

For nondirectional noise the directivity factor (or directivity index) of a receiver is a measure of noise suppression by the receiver as compared to a nondirectional transducer. Consequently, the directivity index is not the appropriate measure to use when noise has directional characteristics—for instance, for noise that arrives mainly from one particular direction. In such cases, the amount of noise reduction is obtained from the beam pattern of the receiver and the angular distribution of the noise, and each case must be evaluated separately.

## 11.7 Reverberation

In addition, an active sonar will receive reverberation. Commonly, we distinguish between volume reverberation and surface reverberation. Volume reverberation is caused by reflection and scattering from particles and living organisms such as plankton and fish in the water volume. If we are not searching for fish or plankton, their echoes will be regarded as noise that may mask the targets we want to detect. In the context of a sonar system, reverberation is modeled by assuming that it is generated by scattering from a large number of particles with random positions in space or the water volume. Figure 11.9 shows a simple sketch for the calculation of volume reverberation.

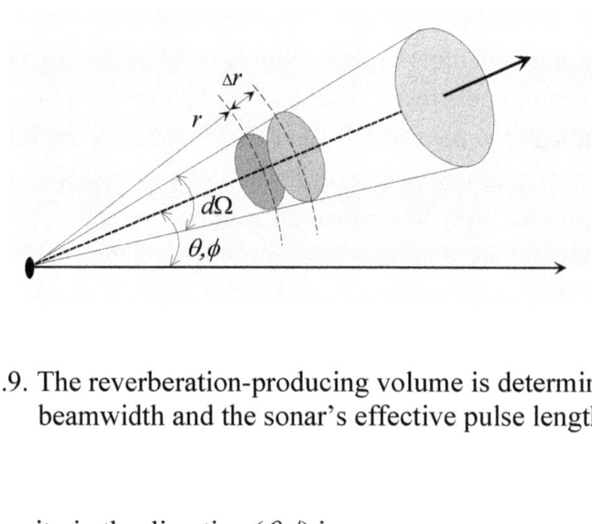


Figure 11.9. The reverberation-producing volume is determined by the beamwidth and the sonar's effective pulse length.

The incident intensity in the direction $(\theta, \phi)$ is

$$I_i = I_0 \frac{r_0^2}{r^2} \tag{11.47}$$

where $r_0$ is a reference distance chosen to equal 1 m.

The volume scattering strength is given by a volume scattering coefficient $m_v$ defined as

$$m_v = \frac{P_r}{I_i} = \frac{\text{Scattered power from a volume of } 1 \text{ m}^3}{\text{Incident intensity}} \tag{11.48}$$

When directivity is the same for transmission and reception, a small water volume $dV$ at the distance $r$ gives a contribution $dI_v$ to the volume reverberation that equals

$$dI_v = I_0 \frac{1}{r^2} \frac{m_v}{4\pi r^2} dV \tag{11.49}$$

This is the reverberation intensity that is received at a post-pulse-transmission time equal to the two-way travel time, $t = 2r / c$.

The total intensity is found by integrating over the volume that produces the reverberation. If we call this effective volume $V$, the total reverberation intensity $I_v$ is expressed as

$$I_v = I_0 \frac{1}{r^4} \frac{m_v}{4\pi} V \tag{11.50}$$

With reference to the intensity of a plane wave with a sound pressure of 1 $\mu$Pa, and with the unit of measure transformed to dB, the volume reverberation level $RL_v$ becomes

$$RL_v = SL - 2TL + S_v + 10\log V \tag{11.51}$$

where $S_v$ is a volume scattering coefficient that can be thought of as target strength per volume unit (in this case, $m^3$). This coefficient is expressed as

$$S_v = 10\log\left(\frac{m_v}{4\pi}\right) \tag{11.52}$$

In the ocean, volume reverberation is caused by fluctuations in water density and temperature in addition to the already mentioned contributions from particulate pollution and from living organisms such as plankton, fry, and fish. Large shoals of fish with swim bladders can sometimes produce very strong reverberation, which creates the "deep scattering layers." This reverberation can be strongly frequency dependent, in particular for frequencies near the swim bladder resonance frequency. Fish change depth according to time of day; when they do so, both the reverberation level and frequency distribution may change significantly.

When the number of scatterers is large, each particular scatterer's characteristic frequency tends to average out. Experience shows that, as the frequency increases, $S_v$ often increases at a uniform rate of approximately 17 dB per decade of frequency, usually lying around –70 dB to –90 dB.

Under simple geometrical conditions, the approximate volume can be calculated as follows for a simple pulse with the pulse length $\Delta T$. With sound speed $c$ and travel time $t$, at a distance between $ct / 2$ and $c(t + \Delta T) / 2$, the volume is limited so that

$$\Delta r = \frac{c\Delta T}{2} \tag{11.53}$$

In angle, the volume is limited by the effective beamwidth $\Omega$ of the sonar transmitter-receiver, so that the total effective volume is

$$V = \frac{c\Delta T}{2} r^2 \Omega \tag{11.54}$$

Therefore, normally the volume reverberation will increase with increasing pulse length. In free space, the volume $V$ and the two-way transmission loss (except for loss from absorption) will increase with $r^4$, causing the volume reverberation to decrease with $r^2$. Under the same conditions, the echo level $EL$ decreases with $r^4$, so that eventually the reverberation level becomes higher than the echo level.

Surface reverberation from the seabed and the sea surface may be treated in the same way. As an example, in figure 11.10 we have a sonar that, at a given point in time after the transmission, insonifies an area $A$ on the seabed.

In the same way in which we calculate the volume reverberation, we can find the reverberation level at $t = 2r / c$, the posttransmission point in time, so that

$$RL = SL - 2TL + S_a + 10 \log A \tag{11.55}$$

where $S_a$ is the surface scattering coefficient, which can be thought of as the target strength per square meter. The insonified area is determined by the sonar's pointing direction and beamwidth, as well as the geometry and the propagation conditions.

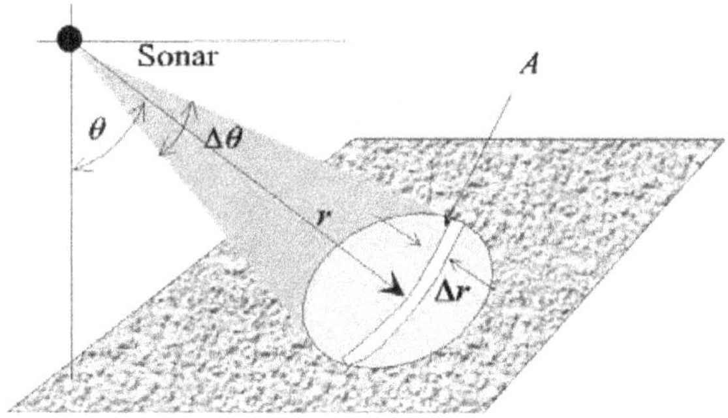


Figure 11.10. The area $A$ of the seabed producing reverberation is given by the pulse length, the sonar system's beamwidth, and the incident angle at the seabed.

In the simplest case, the size of the insonified area $A$ is limited by the beamwidth $\Delta\theta$ and the pulse length, as shown in figure 11.10, and is defined as

$$A \approx \frac{c\Delta T}{2} r\Delta\theta \tag{11.56}$$

We can conclude from equations (11.55) and (11.56) that, with respect to spherical propagation loss, the surface reverberation will decrease with $r$. In reality, we will rarely find this dependence, in part because $S_a$ is dependent on the angle of incidence, and also because the propagation conditions exert a significant influence on the reverberation level.

Reverberation from the ocean surface is dependent on wind (which affects the height and length of the ocean waves), the acoustic frequency, and the angle of incidence. For angles of incidence near the horizontal, surface reverberation may be affected by the near-surface air bubbles generated by breaking waves. There are several theoretical and empirical equations for determining the sea surface scattering strength. A result much quoted in the literature (Chapman and Scott 1964) is

$$S_a = -10\log(8\pi\alpha^2) - 2.17\alpha^{-2}\cot(\theta) \tag{11.57}$$

where $\theta$ is the grazing angle in degrees and $\alpha^2$ is the mean square slope of the surface waves. The latter is given as a function of the wind speed $W$ in m/s as

$$\alpha^2 = 0.003 + 5.12 \cdot 10^{-3} \cdot W \tag{11.58}$$

Figure 11.11 shows the ocean surface's scattering strength as a function of grazing angle and wind speed, according to equations (11.57) and (11.58).

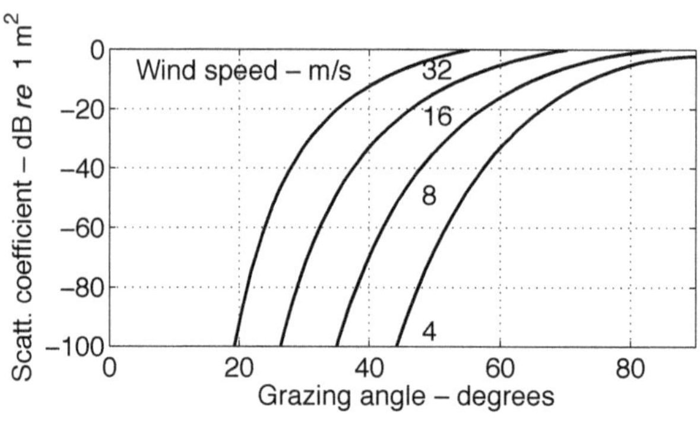


Figure 11.11. Scattering strength of the ocean surface as a function of grazing angle and wind speed (Chapman and Scott 1964).

Another practical expression for sea-surface scattering was obtained by Marsh (1963), who used Eckart's theoretical result (1953) and also assumed a surface roughness spectrum to be that of a so-called fully developed sea. He expressed the surface scattering coefficient as

$$S_a = -36 + 40\log\left[(\tan(\theta)\right] \tag{11.59}$$

This expression for scattering strength is dependent only on the incident angle at the sea surface, not on the signal frequency. We will return to this theory and equation (11.59) in chapter 14, where we shall discuss scattering from rough surfaces in more detail.

Reverberation from the seabed is even more complicated than that from surface scattering because impedance conditions also play a role in seabed reverberation. In addition, the conditions are complicated by the signal's penetration and reflection from the seabed's deeper layers. In practice, we should use measured values, if such exist in the actual area of interest. If measurements are not available, with a certain degree of caution we can apply Lambert's law, which is found to be reasonably accurate when it comes to scattering from very uneven surfaces. Lambert's law states that scattering strength is proportional with the square of the sine of the grazing angle and, when all the energy is radiated back, it follows that the scattering strength $S_a$ will be

$$S_a = 10\log\mu + 20\log(\sin\theta) \tag{11.60}$$

In this expression $\mu$ is a variable whose value changes with geographical location and common values are $10\log(\mu) = -10$ to $-40$ dB. If all of the incident acoustic energy is redistributed into the water, with none lost by transmission into the bottom, the value of $\mu$ can be shown by integration to be $\mu = 1 / \pi$.

Figure 11.12 shows $S_a$ as a function of the grazing angle $\theta$, according to Lambert's law using the value $\mu = 1 / \pi$

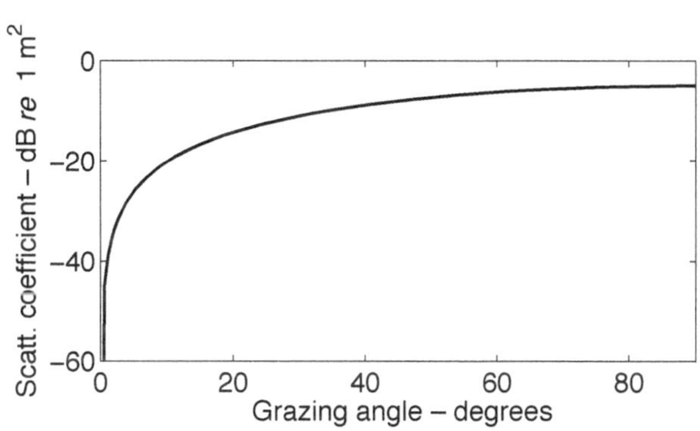


Figure 11.12. A plot describing scattering strength $S_a$ from an uneven surface, in accordance with Lambert's law with the value $\mu = 1 / \pi$.

## 11.8 Example analyses of active sonar

We now apply the sonar equations to analyze some typical sonar systems. The examples are constructs, not representations of any existing sonars, but both the parameters and the performances are realistic, and typical of real sonar systems. Table 11.1 provides the parameters and their values for a sonar intended for fish-finding.

Table 11.1. Parameters for a midrange sonar

| *Parameter* | *Value* | *Parameter* | *Value* |
|---|---|---|---|
| Frequency $f_0$ | 50 kHz | Wavelength $\lambda$ | 3 cm |
| Beam angle $\Delta\theta$ | 10° | Directivity index $DI$ | 25 dB |
| Transducer area $A$ | 230 cm$^2$ | Circular disk with radius $a$ | 8.5 cm |
| Pulse power $W_f$ | 2 kW | Acoustic power $W_a$ | 1 kW |
| Pulse length $\Delta T$ | 1 ms | Range resolution $\Delta r$ | 0.75 m |

With these values, according to equation (11.14) the source level $SL$ is

$$SL \approx 170.8 + 10\log(1000) + DI \approx 225 \text{ dB } re\, 1\,\mu\text{Pa} \tag{11.61}$$

In real environments, the actual noise levels can be greater than the levels predicted by the Knudsen curves in figure 11.8. To account for this, we increase the Knudsen levels by 10 dB.

We also assume that there is a wind speed of 30 m/s. For a 50 kHz frequency, we estimate the noise spectrum level $NSL$ to be

$$NSL = 55 \text{ dB } re \text{ } 1\,\mu\text{Pa, Hz} \tag{11.62}$$

With a 1 ms pulse length, the signal's bandwidth $B$ must equal at least 1 kHz, so that with reference to equation (11.2), the expected noise level $NL$ becomes

$$NL = 55 + 10 \cdot \log(1000) - 25 = 60 \text{ dB } re \, 1\,\mu\text{Pa} \tag{11.63}$$

The 1 kHz bandwidth used above is a minimum bandwidth that leaves no room for Doppler shift caused by target or own-vessel speed. The own-vessel Doppler can partially be compensated for, as discussed in section 10.3, but probably the bandwidth will have to be chosen to be wider than the minimum given by the reciprocal length of the transmitted pulse.

The sonar may receive reverberation from the sea bottom, the sea surface, or from the water volume. As this example sonar is intended to detect targets in the water column, here we shall assume that reverberation is mainly from the water volume.

With an effective circular beamwidth $\Delta\theta = 10^\circ$, the water volume $V$ producing reverberation at the distance $r$ is

$$V = \Delta r \pi \left[ r \tan\left(\frac{\Delta\theta}{2}\right) \right]^2 \approx 0.018 r^2 \tag{11.64}$$

If we further assume that the scattering strength of the volume is $S_v = -90$ dB $re$ 1 m$^3$, we get from equation (11.51) that

$$RL = SL - 2TL - 107.5 + 20 \log r \tag{11.65}$$

We assume that the sonar will be used to detect fish. For an individual cod with a length of 30 – 40 cm, the target strength is about –30 dB, which at the distance $r$ produces an echo level $EL$ of

$$EL = SL - 2TL - 30 \text{ dB } re \text{ } 1\,\mu\text{Pa} \tag{11.66}$$

In order to estimate the transmission loss, we need to know the sound speed profile, but to simplify the example we will assume constant sound speed and no interaction with the seabed or the sea surface, which produces a spherical propagation loss. The absorption coefficient is given in figure 3.5 and for 50 kHz we find that the absorption is $\alpha_{dB} \approx 0.01$ dB/m. With spherical geometrical spreading, this absorption gives a two-way transmission loss as the function of range *r*, so that

$$2TL = 0.02r + 40\log r \tag{11.67}$$

Figure 11.13 illustrates the results of calculating the echo level, noise level, and reverberation level as functions of the distance from the sonar to the target.

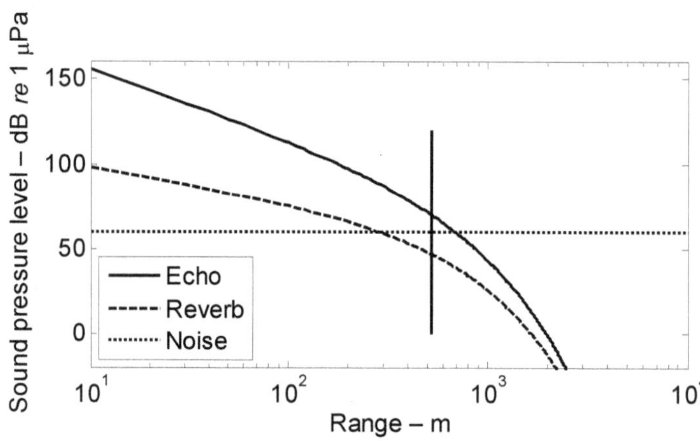


Figure 11.13. Echo level *EL*, reverberation level *RL*, and noise level *NL* as functions of the distance to a target for which *TS* = –30 dB.

The calculations show that the sonar will be reverberation-limited, that is, the reverberation level will be higher than the noise level out to a distance of about 250 m. Thereafter, the level of the noise will be higher than that of the reverberation. As explained in section 9.5, the echo level needs to be higher than the noise and reverberation levels by a certain margin called the detection threshold. Using a detection threshold of 10 dB, we find that the maximum range for detection of a target with target strength *TS* = –30 dB is approximately 500 m (see figure 11.13).

Several changes can be made in order to increase this sonar detection range. Some of the possible changes are discussed below, but one ought to remember that such alterations also may have consequences that increase the sonar's price and size to prohibitive levels.

One possible change might be to increase the source level by increasing the transmitted power. A doubling of the power will increase the source level by 3 dB; for a target for which $TS = -30$ dB, this will increase the range to 600 m. Alternatively, one may increase the size of the transducer, thereby reducing the beamwidth and increasing the source level through the increase in directivity index. For instance, a 3 dB increase in directivity requires a doubling of the transducer area. This change also will lead to a reduction of noise and reverberation, and therefore will have significant impact on the detection range. This solution has, however, several negative consequences. A narrower beamwidth will reduce the insonified water volume, thus leading to a less efficient search operation. The narrower beam may also require more elaborate arrangements for stabilization against roll and pitch. In addition, a larger transducer may be much more expensive and require more space than can be accommodated.

The noise level can be reduced by decreasing the signal bandwidth, but then we must simultaneously increase the pulse length, which means reverberation will also increase.

In figure 11.13, we can see clearly that, at distances larger than approximately 500 m, the absorption loss starts to become significant: at 50 kHz we have an absorption loss of about 20 dB/km for a two-way transmission. Therefore, a very obvious solution for increasing the range is to reduce the frequency so that the absorption loss becomes smaller. In the relevant frequency band, the absorption loss increases approximately with frequency squared. If the frequency is reduced to 20 kHz, the absorption loss decreases to about 1.6 dB/km, which gives a considerable improvement in the sonar range.

To make this technique clearer, we will now analyze a 20 kHz sonar with the parameters given in table 11.2. At this frequency, the ambient noise will become slightly higher, but if we assume that the bandwidth is reduced to about 0.1 kHz, we can again apply the expression

$$NL = 55 \text{ dB } re \text{ } 1 \text{ } \mu\text{Pa} \tag{11.68}$$

Apart from the change in the transmission loss, the other terms of the sonar equation remain unchanged. Figure 11.14 shows echo level, noise level, and volume reverberation as functions of range.

Table 11.2. Parameters for a 20 kHz sonar

| *Parameter* | *Value* | *Parameter* | *Value* |
|---|---|---|---|
| Frequency $f_0$ | 20 kHz | Wavelength $\lambda$ | 7.5 cm |
| Beam angle $\Delta\theta$ | $10^{\circ}$ | Directivity index $DI$ | 25 dB |
| Transducer area $A$ | 1450 cm$^2$ | Circular disk with radius $a$ | 21.2 cm |
| Pulse power $W_f$ | 2 kW | Acoustic power $W_a$ | 1 kW |
| Pulse length $\Delta T$ | 1 ms | Range resolution $\Delta r$ | 0.75 m |

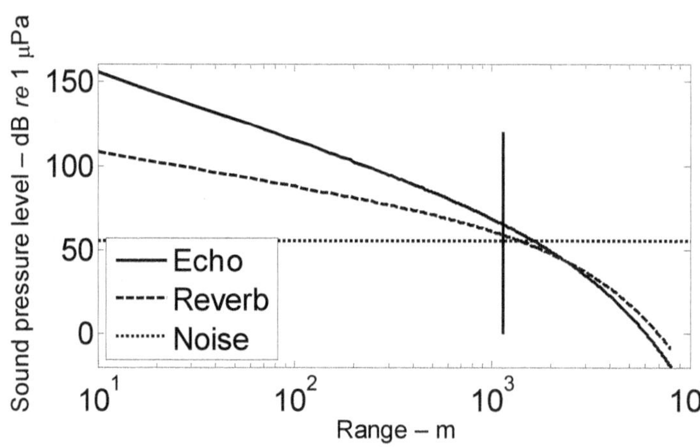


Figure 11.14. Echo level $EL$, reverberation level $RL$, and noise level $NL$ as functions of the distance to a targets having the target strength $TS = -30$ dB.

## 11.9 Sidescan sonar

Sidescan, or side-looking, sonars are used for mapping of the sea bottom and to locate objects like pipelines, shipwrecks, and other objects.

Typically, a sidescan sonar is a tow fish in the shape of a slim cylinder, which is towed relatively close to the sea bottom. This sonar uses separate transducers for transmission and reception. The long, thin transmitting transducer produces an acoustic beam that is very narrow in the along-track direction, but broad in the transversal plane. The small receiving transducer has near-omnidirectional receiving sensitivity. Range resolution is achieved by using a short-duration sine pulse for transmission. Figure 11.15 shows a sketch of the instrument and its operational geometry. Normally there are two sonars in the tow fish so that both sides of the tow track can be surveyed, but only one side, and thus one sonar, is shown in figure 11.15.

The transmitted signal is usually a short sine pulse having pulse length $T$ which is transmitted on the carrier frequency $f_0$. The resolution in the $y$-direction is achieved by virtue of the very narrow beamwidth $\theta_y$ in the along-track direction, and the resolution in the cross-track direction is obtained by using a short pulse length $T$. The slant range $r$ is determined from the measured travel time $\tau$, and the horizontal range $x$ is determined by the sonar's height $H$ above the sea floor

$$\begin{aligned} r &= c\tau / 2 \\ x &= \sqrt{r^2 - H^2} \end{aligned} \tag{11.69}$$

The resolution area $A$ on the sea floor, also called the footprint, is given by

$$A = \left(\frac{cT}{2}\right) r\theta_y \tag{11.70}$$

The intensity of the signal received from the bottom is proportional to the seabed's scattering strength. In some sonar applications, measurement of the bottom scattering strength may be the survey's purpose. In other situations, such as when one is interested in looking for objects, bottom scattering is a limitation.

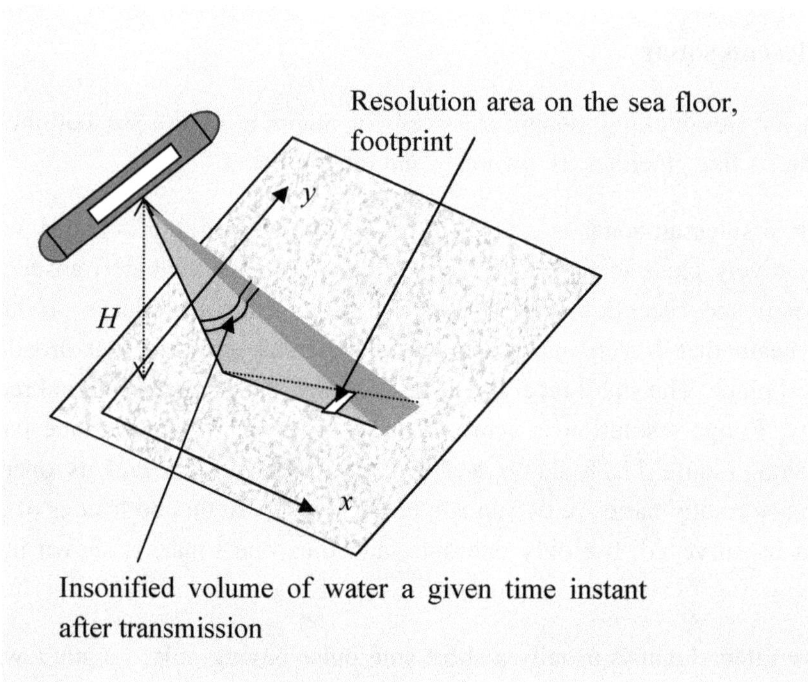


Figure 11.15. The operating principle of a sidescan sonar, which is being towed at a height $H$ above the seafloor, with an acoustic beam that is very narrow in the tow (or $y$) direction and broad in the cross-track (or $x$) direction.

We now use the sonar equations to determine the potential for a sidescan sonar to detect and localize a small object, which in this case is a sphere on the seafloor. The following parameters, typical for many sidescan sonars, are assumed in the subsequent examples.

Vertical beamwidth $\theta_x$: 30°
Along track beamwidth $\theta_y$: 1°
Pulse length $T$: 0.5 ms
Transmitted acoustic power $W$: 1 kW (1000 watts)
Radius of sphere to be detected, $a$: 0.05m
Height above the bottom $H$: 50 m

We also assume that the projector has a 30° tilt with respect to vertical direction. Sidescan sonars for such applications normally work with a carrier frequency in the range of 100 kHz to 1 MHz, so we also will assume this frequency range for our example. However, we shall let frequency be a free parameter in order to study the effect of frequency on detection range and resolution. In our examples only the frequency is changed; the other parameters remain the same as before and as listed above.

The values listed for this sonar give a directivity index of $DI = 31.4$ dB and a source level of $SL = 232.2$ dB *re* 1 $\mu$Pa. The sphere's target strength, in accordance with equation (11.27), is $TS = -32$ dB.

We assume that the bottom scattering strength follows Lambert's law, which was used in equation (11.60) and in figure 11.12. We assume also that the transmission loss is a spherical spreading loss, and that the absorption is that given in figure 3.5.

However, the noise level is very difficult to assess. For the higher frequencies in the actual frequency range, the theoretical minimum level is the thermal noise level as shown in figure 11.8. However, it is quite possible that self-noise generated by the tow-fish and the towing vessel may exceed that level. Since we have no information about this possibility, we shall simply assume that the noise level in the whole frequency region of interest is constant, and equals $NSL = 50$ dB *re* 1 $\mu$Pa, Hz.

With this assumption, we have now all the information required to calculate the expected echo level, bottom reverberation, and noise level. Figure 11.16 shows the results as received sound pressure level as a function of travel time $\tau$.

The calculation has been done for two frequencies: (a) 200 kHz and (b) 1000 kHz. For both frequencies, there is a deep null at about 25 meters that is caused by the directivity null, which occurs at about 60°.

At 200 kHz, we have good coverage and a comfortable echo level over the noise and bottom reverberation from about 50 meters to 350 meters. The 1000 kHz sonar performs more poorly in terms of coverage, reaching only to about 120 meters. With the assumptions we have made, this difference in range is entirely the result of the absorption differences between the two frequencies: 0.062 dB/m at 200 kHz and 0.34 dB/m at 1000 kHz.

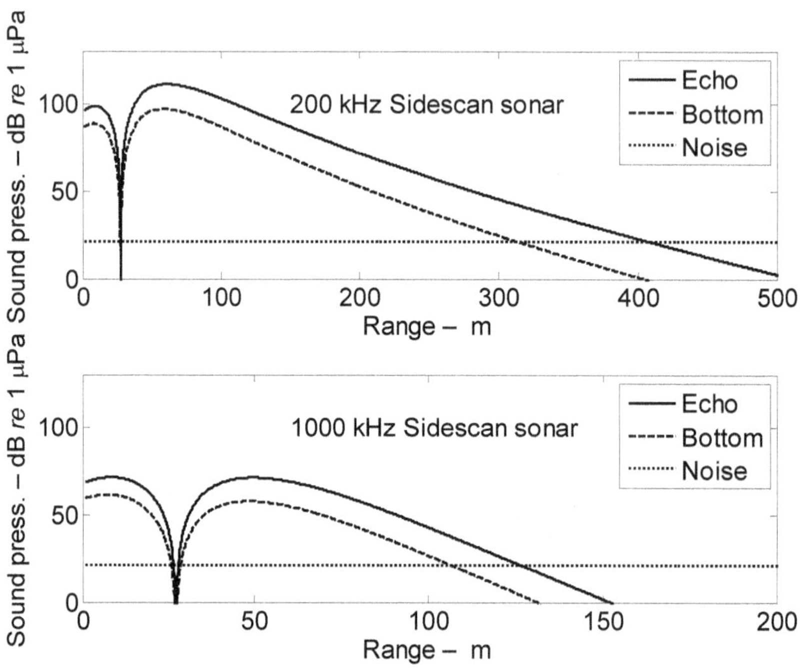


Figure 11.16. Estimated echo, bottom reverberation, and noise levels as functions of target range for two hypothetical sidescan sonar systems working on (top) 200 kHz and (bottom) 1000 kHz.

We have in this section considered the sidescan sonar only as a target-detection sonar. It should be pointed out that many applications of the sidescan sonar are for survey applications where the objective is to visualize bottom features. Even for target detection, the shadow of the target is often more important than the direct target echo. In these cases the bottom backscatter is treated as the "signal" in the sonar equation, and the sonar coverage is determined by the bottom reverberation to noise ratio.

## 11.10 Optimal sonar frequency

In the preceding example, we have seen that, when the parameters are otherwise comparable, a low-frequency sonar has a greater range than a higher-frequency sonar. This difference in range indicates that it may be possible to find an optimal frequency for a sonar system. We will demonstrate this, using some simple considerations in which we assume that the only frequency-dependent components of the sonar equations are the noise level and the absorption part of the transmission loss. We assume that absorption in water increases with the power $n$ and that the noise spectral density decreases with the frequency in the first power. The sonar's bandwidth, source level, and directivity are kept constant, which requires that the sonar's dimensions be regulated with the choice of frequency; however, in practice changes to sonar dimensions may not be possible.

However, with these assumptions, the echo level can be expressed in the form

$$EL = C_1 - r\alpha_0 \left(\frac{f}{f_0}\right)^n \tag{11.71}$$

where $C_1$ is a constant that is not significant in this context and $\alpha_0$ is the absorption coefficient at the reference frequency $f_0$. We saw that absorption is a function of frequency in figure 3.5. Depending on the frequency, $n = 1.5$ and $n = 2$ are realistic absorption values.

In the same way, the noise level can be expressed as

$$NL = C_2 - \log\left(f^m\right) \tag{11.72}$$

where $C_2$ is another insignificant constant.

Typical noise spectra values plotted as a function of frequency are shown in figure 11.8. As remarked earlier, the noise decreases with the frequency at about 17 dB/decade, which corresponds to $m = 1.7$.

The signal-to-noise ratio $S/N$ is expressed as

$$S/N = C_1 - C_2 - r\alpha_0 \left( \frac{f}{f_0} \right)^n + m \cdot \log f \tag{11.73}$$

which has a maximum value that can be found by differentiating, resulting in an expression for the optimal frequency $f_{opt}$

$$\left( \frac{f_{opt}}{f_0} \right)^n = \frac{m}{n\alpha_0 r} \tag{11.74}$$

With the given assumptions, equation (11.74) means that the optimum frequency for detection of an echo signal in noise is proportional to the distance $r$ taken to the power $1 / n$; that is

$$f_{opt} \propto (r)^{-\frac{1}{n}} \tag{11.75}$$

To illustrate equation (11.75), we take a case in which $f_0 = 10$ kHz, where $\alpha_0 \approx 1.0 \cdot 10^{-3}$ dB/m and $n \approx 2$. Inserted in equation (11.74), these values produce

$$f_{opt} \approx 10^4 \sqrt{\frac{1.7}{2 \cdot 1.0 \cdot 10^{-3} \cdot r}} \approx 9 \cdot 10^3 \sqrt{\frac{1}{r_{km}}} \quad \text{Hz} \tag{11.76}$$

Here, $r_{km}$ is the distance given in km. For a distance $r_{km} = 1$ km, equation (11.76) gives the optimal frequency $f_{opt} = 9$ kHz; for a distance $r_{km} = 10$ km, we get $f_{opt} \approx 3$ kHz. Note that these frequencies are still much lower than those that actually would have been chosen, which can easily be explained by several conditions.

Perhaps the most important condition is that we have assumed that the directivity for both transmission and reception are independent of the frequency, which means that the area of the transducer is allowed to increase with lower frequencies. The costs, space, and weight of such an increase can make this assumption quite unrealistic. These deliberations should not be taken too literally, as several other considerations also should be taken into account. But the results illustrate that frequency-dependent absorption in water and the desired range of the sonar decisively influence the choice of frequency.

# Chapter 12

# Nonlinear acoustics and parametric sources

## 12.1 Introduction

By its nature, acoustic wave propagation is a nonlinear process; nonlinearity was first mentioned in chapter 2, in the discussion of the derivation of the wave equation. We pointed out that all the governing equations for the derivation of the acoustic wave equation, that is, equations (2.2), (2.3), and (2.4) are nonlinear equations. However, when the sound field is weak, usually we can ignore the nonlinear parts of these equations, as we have done up to this point in order to make use of the linear wave equation. In this chapter, however, we will discuss circumstances where nonlinear effects can be significant. In such situations, nonlinear acoustics can be useful for practical applications.

There is an extensive literature on nonlinear acoustics, including the books by Beyer (1974) and Hamilton and Blackstock (1997), and there are several articles on the subject in *Encyclopedia of Acoustics*, edited by Crocker (1997). In addition, most books on acoustics include a chapter about nonlinear acoustics (Pierce 1989, Kinsler et al. 2000).

The interest in nonlinearity for underwater acoustic applications began in the late 1950s with Westervelt (1957a, 1957b, 1963), introducing the idea that nonlinear effects can be used to generate low-frequency sound with higher directionality than can be obtained by conventional sources. A nonlinear source of this type, called a parametric source, gained much attention during the 1970s, when the majority of the theories and literature on the topic were developed. The articles by Blackstock (1964), Berktay (1965), Berktay and Leahy (1974), Hobæk (1967), Muir and Blue (1969), and Muir (1974) are some of the most relevant work from that period.

In the present chapter we will emphasize the parametric sonar, which has interesting properties for specific applications in underwater acoustics, particularly for characterization and examination of the upper parts of the seafloor. For such applications, the preference is to use the lowest frequencies possible because of their penetration depth, and to use narrow beams because of their superior resolution. There are also other possible applications in situations where low frequencies and high directivity are required, for instance in underwater communication.

The use of nonlinear acoustics and parametric sources can be of significant practical applications for very long-distance command and control communications of independent, extremely long-range autonomous underwater vehicles (AUVs). Typically, AUVs have limited "real estate" on which to mount transmitter/receiver arrays for detection and acoustic communications. The limited space available relegates the installed arrays to smaller sizes which translates to operating at higher frequencies. At the higher frequencies, the seawater attenuation coefficients will be higher, approximately proportional to frequency squared, and the distances for useful communication and detection will be limited. If, however, the high-frequency arrays are used as parametric source arrays, the transmitted outputs of the parametric arrays can be created at very low frequencies, with very narrow beams, and with no sidelobes. These low-frequency transmissions ensure attenuation coefficients that are very small thus facilitating low transmission loss, meaning significantly great distances for communication. Further, the very narrow spatial beams enhance the Low Probability of Detection (LPI) characteristics of the transmitted signal.

## 12.2 Nonlinear wave propagation

One assumption used in deriving the linear wave equation was that the amplitude is small. Another way to put this is to say that the Mach number is low. The Mach number $M$ is defined as the ratio between the amplitude of the particle velocity in the wave $v$ and the sound speed $c_0$

$$M = \frac{v}{c_0} \tag{12.1}$$

To this point, we have used only the first-order term in the Mach number in deriving

the acoustic wave equation and ignored the higher-order terms. Since the Mach number always will be much smaller than unity this omission seems to be reasonable. But in fact some of the terms that we have so far ignored can result in nonlinear effects that accumulate during propagation and eventually turn out to be quite important.

There are two contributions to nonlinearity in acoustic wave propagation. The first nonlinear effect has to do with elastic properties of the medium and comes from equation (2.5), the equation of state, $p = f(\rho)$, which describes the relationship between the sound pressure $p$ and the density $\rho$. This relationship is always nonlinear for all materials found in nature, and when the nonlinearity is relatively weak, the equation of state can be expanded in a mathematical power series. By retaining the second-order terms, we can write the approximation of the sound pressure as

$$p = p_0 + A\left(\frac{\rho - \rho_0}{\rho_0}\right) + \frac{B}{2}\left(\frac{\rho - \rho_0}{\rho_0}\right)^2 \tag{12.2}$$

The expansion coefficients are

$$\begin{aligned} A &= \rho_0\left(\frac{\partial p}{\partial \rho}\right)_0 \\ B &= \rho_0^2\left(\frac{\partial^2 p}{\partial \rho^2}\right)_0 \end{aligned} \tag{12.3}$$

The subscript 0 indicates that the differential should be calculated at equilibrium density, $\rho = \rho_0$.

The general expression for the sound speed was given in equation (2.9), where we find that

$$c_0^2 = \left(\frac{\partial p}{\partial \rho}\right)_0 \tag{12.4}$$

For waves with infinitely small amplitudes, the speed is given by the linear term of equation (12.2) and can be written

$$c_0 = \sqrt{\frac{A}{\rho_0}} \tag{12.5}$$

which is identical to the result we obtained in section 2.2, since the volume bulk modulus $K$ defined there is the same as the coefficient $A$.

For a wave with finite amplitude, we include the second-order term of equation (12.2) in equation (12.4), whereby we obtain in the first instance

$$\begin{aligned}\tilde{c}^2 &= \left(\frac{\partial p}{\partial \rho}\right)_0 = \frac{A}{\rho_0} + \frac{B}{\rho_0}\frac{(\rho - \rho_0)}{\rho_0} \\ &= c_0^2\left[1 + \frac{B}{A}\frac{(\rho - \rho_0)}{\rho_0}\right]\end{aligned} \tag{12.6}$$

Since the second term in equation (12.6) is much smaller than the first, this speed can be approximated as

$$\tilde{c} = c_0\left[1 + \frac{B}{2A}\frac{(\rho - \rho_0)}{\rho_0}\right] \tag{12.7}$$

In section 2.2, we introduced the condensation, $s = (\rho - \rho_0) / \rho_0$, where $s$ represents the relative change in density. For a plane progressive wave, condensation also equals the relative particle velocity and the Mach number, so that we can rewrite the relationship as

$$\frac{\rho - \rho_0}{\rho_0} = \frac{v}{c_0} = M \tag{12.8}$$

Therefore, the wave speed $\tilde{c}$ is a function of the particle velocity, and is given by

$$\tilde{c} = c_0\left[1 + \frac{B}{2A}\frac{v}{c_0}\right] \tag{12.9}$$

The nonlinear parameter $B / A$, according to equation (12.3), is given by

$$\frac{B}{A} = \frac{\rho_0}{c_0^2}\left(\frac{\partial^2 p}{\partial \rho^2}\right)_0 \tag{12.10}$$

It is also common to define another nonlinear parameter, $\beta$, as

$$\beta = \left(1 + \frac{B}{2A}\right) \tag{12.11}$$

The value of $B/A$ has been found for several fluids and solids (Beyer 1960, Bjørnø 1977, Hovem 1979, Everback 1997). For seawater at normal temperatures, these values are $B/A \approx 5$ and $\beta \approx 3.5$.

The second nonlinear effect derives from the equation of continuity (equations (2.2) and (2.12)) with the interpretation that a stationary observer will "see" the wave speed as the sum of $\tilde{c}$ and the particle velocity $v$. For a finite amplitude progressive plane wave the speed is

$$c = \tilde{c} + v = c_0\left[1 + \left(1 + \frac{B}{2A}\right)\frac{v}{c_0}\right] \tag{12.12}$$

Thus, we can express the propagation speed of the wave as a function of the particle velocity $v$ and the Mach number by writing

$$c = c_0 + \beta v = c_0(1 + \beta M) \tag{12.13}$$

equation (12.13) shows that the propagation speed depends on particle velocity caused by advection and nonlinearity of the state equation. A particular wave part with the particle velocity $v$ propagates with the speed $c_0 + \beta v$. The wave is progressively distorted as it propagates since the parts of the wave with higher amplitudes, in sound pressure or particle velocity, propagate with higher wave speed than that of the low amplitude parts of the wave. This is illustrated in figure 12.1, in which an original sine wave propagates along the $x$-axis with a sound speed as given by equation (12.13). The pulse shape is distorted, gradually approaching the shape of a sawtooth wave, but the distortion generates higher frequency components that are rapidly attenuated since the attenuation increases rapidly with frequency. Therefore the wave at long distances will again become a sine wave, propagating at the original frequency, but with reduced amplitude.

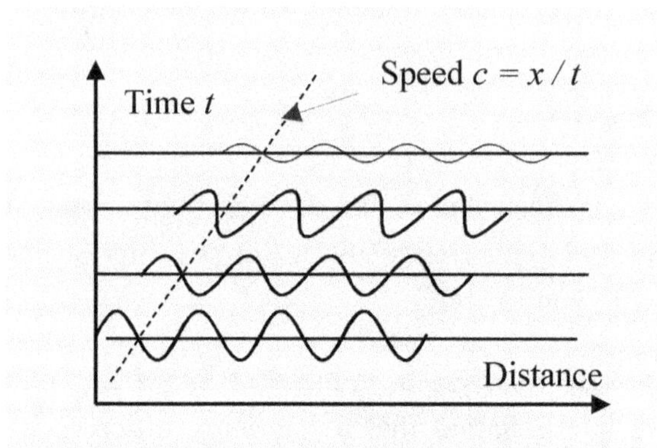


Figure 12.1. Snapshot of an original sine wave propagating in the $x$-direction. The pulse shape is distorted by nonlinear effects, gradually approaching the shape of a sawtooth before attenuation reduces the higher frequency components and the wave again becomes a sine wave with the original frequency, but with reduced amplitude.

We investigate this phenomenon in more detail by considering a plane harmonic wave that starts at $x = 0$ and propagates in the positive $x$-direction. At distance $x$, the particle velocity is given by the implicit equation

$$v(x,t) = V \sin\left[\omega\left(t - \frac{x}{c_0 + \beta v(x,t)}\right)\right] \tag{12.14}$$

The wave angle of inclination of the wave front is given by the derivative of the particle velocity with respect to distance. Therefore, the angle of inclination at the zero-crossings, where $v = 0$, is

$$\left(\frac{\partial v(x,t)}{\partial x}\right)_{v=0} = V\left[-\frac{\omega}{c_0} + \frac{\omega}{c_0^2} x\beta \frac{\partial v(x,t)}{\partial x}\right] \tag{12.15}$$

Solving for the angle of inclination at the zero-crossings produces

$$\left(\frac{\partial v(x,t)}{\partial x}\right)_{v=0} = -\frac{V\dfrac{\omega}{c_0}}{1-\dfrac{\omega}{c_0^2}V\beta x} \tag{12.16}$$

From equation (12.16) we see that the angle of inclination of the wave front becomes infinite at a distance we shall define as the shock formation distance $L_s$, which is given by

$$L_s = \frac{1}{\beta M \kappa_0} \tag{12.17}$$

The wave number $\kappa_0 = \omega / c_0$ and the Mach number is $M = V / c_0$.

Consider the example of a wave having sound pressure amplitude equal to one atmosphere, $10^5$ Pa. The amplitude of the particle velocity is then $V = 0.067$ m/s, and the Mach number is $M = 0.45 \cdot 10^{-4}$ resulting in shock formation distance $L_s$ of about 15.2 m when we use the value of $\beta = 3.5$. Therefore, the wave, although beginning as a pure sine wave, becomes sawtooth-shaped at a distance of about 15.2 m. This example illustrates that even when $M \ll 1$ over distance, the accumulated effect of nonlinearity can create significant nonlinear distortion on a signal.

From the above discussion it may appear that nonlinearity will always lead to significant distortion and ultimately even to shock generation; however, this is not so. Other effects that cause diminishing of the signal as it propagates in space will limit the distortion. There are at least two effects that diminish the amplitude as the wave propagates. The first of these is the normal absorption loss discussed in chapter 2; this causes the amplitude to reduce with $\exp(-\alpha x)$, where $\alpha$ is the attenuation coefficient in nepers/m. The absorption length $L_a$, a characteristic measure to describe absorption, is defined as

$$L_a = \frac{1}{\alpha} \tag{12.18}$$

The ratio of the absorption length $L_a$ to the shock formation distance $L_s$ is a measure of the relative importance of nonlinear effects; this ratio is known as the Goldberg number $\Gamma$, which is given by

$$\Gamma = \frac{L_a}{L_s} = \frac{M\beta\kappa_0}{\alpha} \tag{12.19}$$

When $\Gamma < 1$, absorption will dominate and shocks cannot be generated.

Another effect that can slow down the shock formation is spherical spreading. The field of a plane transducer source can, in linear theory, be divided into a nearfield and a farfield. In the nearfield, the outgoing wave is essentially a ***collimated wave***—a plane wave propagating in the axial direction and extending over a cross section equal to the area of the transducer. In the farfield, the wave propagates spherically with a reduction in amplitude proportional to the distance, *r*. Most of the wave intensity stays within a cone having angles equal to those of the beamwidth, as illustrated in figure 12.2.

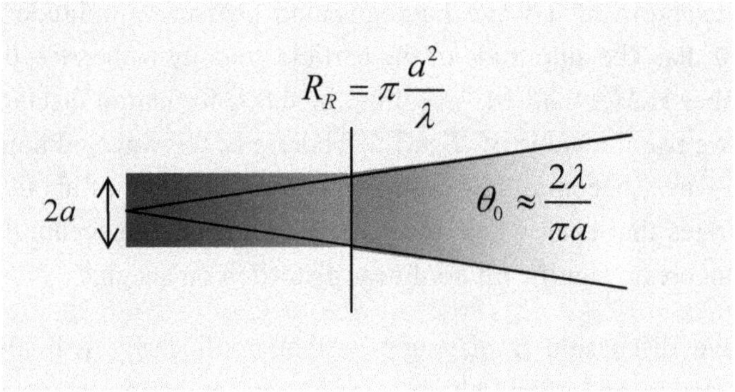


Figure 12.2. The acoustic radiation field of a planar transducer source. In the sketch, the nearfield stretches out toward the Rayleigh distance $R_R$; the collimated wave front has a transverse extent equal to the area of the source. Beyond the nearfield is the farfield, where the wave propagates with spherical radiation; most of the wave intensity is within a conical area having angles equal to those of the beamwidth.

The transition between the farfield and the nearfield of a linear projector occurs approximately at the Rayleigh distance. In general, for a planar transducer with area *S*, the Rayleigh distance is defined as

$$R_R = \frac{S}{\lambda} \tag{12.20}$$

For a circular transducer with radius $a$, the Rayleigh distance becomes

$$R_R = \frac{\pi a^2}{\lambda} = \frac{1}{2}\kappa a^2 \tag{12.21}$$

since the beamwidth is approximately

$$\Delta\theta = \frac{2\lambda}{\pi a} \tag{12.22}$$

In the farfield, the amplitude of the wave decreases rapidly because of spherical spreading, thereby weakening the nonlinear interaction.

Waveform distortion leads to the generation of higher frequencies and therefore to an extra nonlinear attenuation, caused by the transfer of energy from the original transmitted wave to the higher harmonics (Blackstock 1964). The impact of this phenomenon is defined by the saturation parameter $\sigma_0$, defined as the ratio between the Rayleigh distance $R_R$ and the shock formation distance $L_s$:

$$\sigma_0 = \frac{R_R}{L_s} \tag{12.23}$$

We can also express $\sigma_0$ as

$$\sigma_0 = \frac{\beta \kappa_0 p_0 R_R}{\rho_0 c_0^2} \tag{12.24}$$

In the next section we shall see that the Rayleigh distance $R_R$, the shock formation distance $L_s$, and the saturation parameter $\sigma_0$ are essential considerations in the design and performance of parametric sources.

## 12.3 Parametric sound generation

As mentioned earlier, Westervelt introduced the idea of using nonlinearity for generation of sound; this concept is referred to as ***parametric sound generation.*** Parametric sound generation is based on nonlinear interaction. Consider an ordinary transducer transmitting a ***primary wave*** composed of the sum of two sine waves with relatively high frequencies $f_1$ and $f_2$ that are centered on the mean primary frequency $f_p$, where $f_p = 0.5(f_1 + f_2)$. In the water volume in front of the transducer, the nonlinear effect causes some of the primary wave's energy to be converted into a ***secondary wave*** having the new frequency components $2f_1$, $2f_2$, $f_1 \pm f_2$, and so on.

The most interesting frequency for applications in underwater acoustics is the secondary or difference frequency $f_s$, where $f_s = (f_1 - f_2)$. The secondary wave will be a low-frequency wave if the two primary frequencies are almost equal. This low-frequency wave will experience only a small absorption loss, and therefore such a secondary wave has the potential to propagate for long distances. Furthermore, the beamwidth of the secondary wave may be quite narrow, actually approaching the beamwidth of the primary wave, because the secondary wave is generated in a water volume in front of the projector. This volume, which can have a considerable extent, is a ***virtual source*** that acts as an endfire array. Because the directivity of the secondary wave is determined by the extension of the interaction volume, the volume can produce a secondary wave that has a narrow beamwidth at the low frequency of the secondary field.

### *12.3.1 The source function*

In order to arrive at an equation for the secondary wave, we use a perturbation approach by expressing the sound field as the sum of a primary first-order field ($p$) and a second-order field ($s$). The sound pressure and the particle velocity are thereby expressed as

$$\begin{aligned} p &= p_p + p_s \\ v &= v_p + v_s \end{aligned} \tag{12.25}$$

We assume that the secondary field is much weaker than the primary field

$$p_p >> p_s$$
$$v_p >> v_s \tag{12.26}$$

If we require that the hydrodynamic equations be fulfilled independently for each order and separately collect the primary and the secondary fields, we obtain two wave equations (Westervelt 1963)

$$\nabla^2 p_p - \frac{1}{c_0^2}\frac{\partial^2 p_p}{\partial t^2} = 0 \tag{12.27}$$

$$\nabla^2 p_s - \frac{1}{c_0^2}\frac{\partial^2 p_s}{\partial t^2} = -\rho_0 \frac{\partial f(\mathbf{r},t)}{\partial t} \tag{12.28}$$

where the source term at the right-hand side of equation (12.28) is given as

$$\frac{\partial f(\mathbf{r},t)}{\partial t} = \frac{\beta}{\rho_0^2 c_0^4}\frac{\partial p_p^2}{\partial t^2} \tag{12.29}$$

Equation (12.27) is the ordinary linear wave equation for the primary or original sound field, and equation (12.28) is the wave equation for the secondary wave created by the nonlinear propagation of the primary wave. We see that this secondary field is generated by a nonlinear interaction of the primary field, resulting in a volume-distributed source strength $f(\mathbf{r},t)$ as in equation (2.35).

Equation (12.29) establishes that the source term of the secondary field is proportional to the square of the primary field. Notice also that, according to equation (12.27), the wave equation for the primary field is unaffected by the generation of the secondary field, and therefore the equation does not account for the energy transferred to the secondary field. Thus, the solution represented in equations (12.27) and (12.28) can be no more than approximations.

The formal proof of the Westervelt source function is rather complicated, but it is interesting to note that the correct result is obtained by using the expression of equation (12.13) and substituting into the ordinary linear wave equation, which gives

$$\nabla^2 p = \frac{1}{c_0^2\left(1+\beta\frac{v}{c_0}\right)^2}\frac{\partial^2 p}{\partial t^2} \tag{12.30}$$

We now use equations (12.25) and (12.26), and assume that $p_p$ and $v_p$ are related by the specific impedance of a plane progressive wave, $p_p = (\rho_0 c_0)\, v_p$. After separately collecting the primary and the secondary fields we obtain the results of equations (12.27) and (12.28).

To determine the secondary sound field, we first solve the wave equation for the primary field for all positions in space, and then calculate the virtual source function of equation (12.29). We then apply the Green's functions (see section 2.6) and integrate the virtual source functions over the whole volume. The secondary field at the difference frequency can then be expressed as

$$p_s(\mathbf{r},t) = -\frac{\rho_0}{4\pi}\int_V \frac{\partial q(\mathbf{r}')}{\partial t}\frac{\exp\left[(i\kappa_s - \alpha_s)\left|\mathbf{r}-\mathbf{r}'\right|\right]}{\left|\mathbf{r}-\mathbf{r}'\right|}dV' \tag{12.31}$$

In this equation, the wave number and the absorption coefficient for the secondary field are $\kappa_s$ and $\alpha_s$, respectively. The position of an elementary volume element is $\mathbf{r}'$ and the receiver position is $\mathbf{r}$, as shown in figure 12.3.

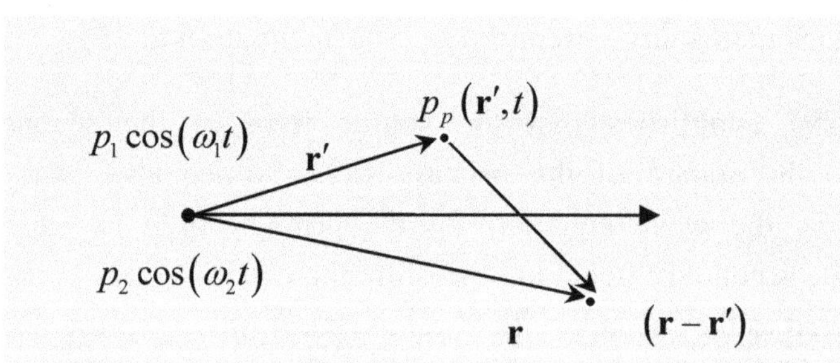


Figure 12.3. A parametric sound source, with the primary wave being the sum of two sinusoidal signals with frequencies $\omega_1$ and $\omega_2$. Nonlinear interaction creates a volume distribution of virtual sources. The secondary wave, at the difference frequency $\omega_1 - \omega_2$, results from integrating the contributions over the whole interaction volume.

In principle, the secondary sound field can be calculated by equations (12.28) and (12.29), using integration as suggested in figure 12.3. This is no easy task, as the primary field from the source is a very complicated function. We can, however, describe some of the characteristics of the solution and thereby some features of the nonlinear interaction, by using the Rayleigh distance $R_R$, the saturation parameter $\sigma_0$, and the primary attenuation coefficient $\alpha_p$.

According to their values, we can distinguish among three cases:

**Case 1**. The interaction is mainly in the nearfield, and is limited by absorption

$$\sigma_0 < 1 \quad \text{and} \quad 2\alpha_p R_R > 1$$

**Case 2**. The interaction is mainly in the nearfield, and is limited by nonlinear absorption

$$\sigma_0 > 1 \quad \text{and} \quad 2\alpha_p R_R < 1$$

**Case 3**. The interaction is mainly in the farfield, and is limited by spherical spreading

$$\sigma_0 < 1 \quad \text{and} \quad 2\alpha_p R_R < 1$$

## 12.4 Modeling parametric arrays

The model that Westervelt originally treated in 1963 is based on the assumption that the two primary waves are collimated. A further assumption is that ordinary small-signal absorption reduces the primary amplitudes to essentially zero before the Rayleigh distance is reached, and therefore the entire nonlinear interaction volume is in the nearfield. These conditions are rarely met in reality and therefore several approaches have been proposed to describe parametric nonlinear sound generation in more general cases.

This section reviews some of these methods and particularly describes the model developed over several years by Mellen and Moffett (1971) and Moffet and Mellen, (1977, 1981). A useful feature of this model is that it gives insight into how the various properties of the primary source affect the level and directivity of the secondary sound field. The model is useful when constructing parametrical sources.

Modeling of parametric arrays can also be done accurately by direct numerical integration of a parabolic equation called the Khokhlov-Zabolotskaya-Kuznetsov (KZK) equation. A treatment of this approach is however outside the scope of this text. The reader can consult the article by Hamilton (1997) for a description and for further references.

### *12.4.1 Westervelt's model for interaction in the nearfield*

The primary wave $p_p$, which is the sum of two collimated plane waves with angular frequencies $\omega_1$ and $\omega_2$, is defined as

$$p_p = p_1 \exp\left[\left(i\kappa_1 - \alpha_1\right)x - i\omega_1 t\right] + p_2 \exp\left[\left(i\kappa_2 - \alpha_2\right)x - i\omega_2 t\right] \qquad (12.32)$$

where the difference frequency $\omega_s$ and the wave number of the difference frequency component $\kappa_s$ are defined as follows

$$\omega_s = \omega_1 - \omega_2 \qquad (12.33)$$

$$\kappa_s = \kappa_1 - \kappa_2 \qquad (12.34)$$

The secondary source term first defined in equation (12.29) is now expressed as a function of the distance $x$ from the source

$$q(x) = -\frac{\beta}{\rho_0^2 c_0^4}\omega_s^2 p_1 p_2 \exp\left[\left(i\kappa_s - \alpha_1 - \alpha_2\right)x\right]\exp\left(-i\omega_s t\right) \qquad (12.35)$$

Next we simplify the integral in equation (12.31) since, at a great distance from the source, we can make the approximation

$$\frac{\exp\left[\left(i\kappa_s - \alpha_s\right)\left|\mathbf{r} - \mathbf{r}'\right|\right]}{\left|\mathbf{r} - \mathbf{r}'\right|} \approx \exp\left[i\kappa_s\left(r - x\cos\theta\right)\right]\frac{\exp[-\alpha_s(r - x\cos\theta)]}{r} \qquad (12.36)$$

Figure 12.4 shows a representation of Westervelt's model in which the interaction volume is a cylinder with a cross section equal to the area $S$ of the transducer and where the secondary source function decays exponentially with the distance $x$. Thus, there remains just a single integration along distance $x$ that produces

$$p_s(r,\theta) = \frac{\beta p_1 p_2 \omega_s^2 S}{4\pi\rho_0 c_0^4}\left\{\int_0^\infty \exp\left[i\kappa_s x(1-\cos\theta)-(\alpha_1+\alpha_2-\alpha_s\cos\theta)x\right]dx\right\} \times \frac{\exp[r(i\kappa_s-\alpha_s)]}{r}\exp(-i\omega_s t) \quad (12.37)$$

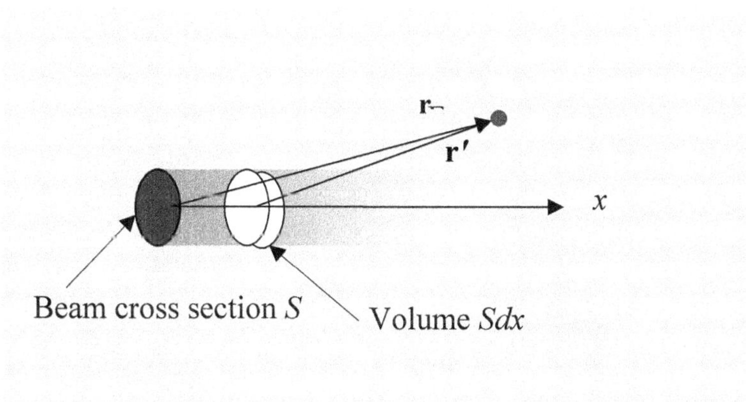


Figure 12.4. Westervelt's model, with absorption limiting the interaction volume in the nearfield.

Equation (12.37) is the expression for an end-fire array with an exponential weight or aperture function, also called a taper function. Furthermore, since the attenuations of the two high-frequency primary waves $\alpha_1$ and $\alpha_2$ are normally much higher than the attenuation of the low-frequency secondary wave, we can approximate the exponential term as:

$$\exp\left[-(\alpha_1+\alpha_2-\alpha_s\cos\theta)x\right] \approx \exp(-2\alpha_p x) \quad (12.38)$$

where the mean primary wave attenuation $\alpha_p$ is

$$\alpha_p = \frac{1}{2}(\alpha_1+\alpha_2) \quad (12.39)$$

After integration, equation (12.37) defines the amplitude $|p_s(r,\theta)|$ of the secondary field in terms of functions of range and angle

$$\left|p_s(r,\theta)\right| = \left[1+\left(\frac{\kappa_s}{\alpha_p}\sin^2\left(\frac{\theta}{2}\right)\right)^2\right]^{-\frac{1}{2}} \frac{\beta p_1 p_2 \omega_s^2 S}{8\pi\rho_0 c_0^4 \alpha_p} \frac{\exp(-\alpha_s r)}{r} \tag{12.40}$$

On the axis where $\theta = 0$, the secondary field sound pressure has the amplitude $p_s(r,0)$

$$\left|p_s(r,0)\right| = \frac{\beta p_1 p_2 \omega_s^2 S}{8\pi\rho_0 c_0^4 \alpha_p} \frac{\exp(-\alpha_s r)}{r} \tag{12.41}$$

By using equations (12.20) and (12.24), the secondary sound pressure on the axis can also be written in the form of

$$\left|\frac{p_s(r,0)r}{p_p R_R}\right| = \left(\frac{\kappa_s}{\kappa_p}\right)^2 \frac{\sigma_0}{4\alpha_p R_R} \exp(-\alpha_s r) \tag{12.42}$$

Here we have assumed that the sound pressures of the two primary waves have the same amplitude by setting $p_1 = p_2 = p_p$.

The angle-dependent part of equation (12.40) is the directivity function $B_s(\theta)$ for the secondary field at the difference frequency

$$B_s(\theta) = \left[1+\left(\frac{\kappa_s}{\alpha_p}\sin^2\left(\frac{\theta}{2}\right)\right)^2\right]^{-\frac{1}{2}} \tag{12.43}$$

and the 3 dB beamwidth $\Delta\theta_s$ is approximately expressed as

$$\Delta\theta_s \approx 4\sqrt{\frac{\alpha_p}{\kappa_s}} \tag{12.44}$$

When the two primary waves have the same acoustic power, that is, when $p_1 = p_2 = p_p$, the total primary power $W_p$ from the transmitter is

$$W_p = \frac{S}{2\rho_0 c_0}\left(p_1^2 + p_2^2\right) = \frac{Sp_p^2}{\rho_0 c_0} \tag{12.45}$$

When equation (12.45) is inserted into equation (12.41), we find that now the amplitude $p_s(r,0)$ can be expressed as

$$p_s(r,0) = \frac{\beta\pi}{2c_0^3}\frac{f_s^2 W_p}{\alpha_p}\frac{\exp(-\alpha_s r)}{r} \tag{12.46}$$

Therefore, at the reference distance $r_0 = 1$ m, we can write the expression defining the source level $SL_s$ of the secondary sound field (neglecting $\exp(-\alpha_s r_0)$ as

$$\begin{aligned} SL_s &= 20\log\left(\frac{p_s}{\sqrt{2}}\right) = 20\log\left(\frac{\beta\pi}{2\sqrt{2}c_0^3}\right) + 20\log\left(\frac{f_s^2 W_p}{\alpha_p}\right) \\ &= -58.8 + 20\log\left(\frac{f_s^2 W_p}{\alpha_p}\right) \quad \text{dB } re \text{ } 1\mu\text{Pa} \end{aligned} \tag{12.47}$$

From these equations, we see that the sound pressure of the secondary wave increases as the square of the primary pressure and the square of the difference frequency increase. This increase is, however, reduced because the attenuation coefficient $\alpha_p$ is also increasing with increasing primary frequency.

### *12.4.2 Nonlinear attenuation*

The Westervelt model does not include the effect of nonlinear attenuation of the primary field, which is caused by the conversion of wave energy to the secondary field. Several models seek to correct that, albeit in an approximate manner.

Bartram (1972) replaced the attenuation factor $\exp(-\alpha_p x)$ with another term that, in addition to regular absorption, takes nonlinear damping into consideration. This is done by substitution, so that

$$\exp(-\alpha_p x) \Rightarrow \frac{\exp(-\alpha_p x)}{1 + \chi\left[1 - \exp(-\alpha_p x)\right]} \tag{12.48}$$

where $\chi$ is a factor that is closely related to the Goldberg number $\Gamma$, equation (12.19), since

$$\chi = \frac{L_a}{\pi L_s} = \frac{\Gamma}{\pi} \tag{12.49}$$

therefore, $\chi$ is another measure of nonlinear attenuation's relative importance within the Rayleigh region. The result of this is that Bartram (1972) proposed to multiply the source level of the Westervelt model with a factor $F$, which is defined as

$$F = 2\chi^{-2}\left[\chi - \ln(1+\chi)\right] \tag{12.50}$$

When the primary sound pressure is very low, nonlinear absorption becomes insignificant and $\chi \to 0$, which means that $F \to 1$, giving the Westervelt result. However, when the primary pressure is high, $\chi$ becomes so large that $F \to \chi^{-1}$, which leads to $F$ becoming inversely proportional to the primary sound pressure. This condition, called saturation, has the consequence that the secondary sound pressure will tend to be proportional to $p_p$ rather than to $p_p^2$, as when the primary sound pressure is very low.

### *12.4.3 Interaction in the farfield*

The underlying assumption of the Westervelt model, that nonlinear interaction takes place in the nearfield, is not realistic for most underwater acoustic applications, since it requires a very large directional transducer. More likely, a significant part of the interaction volume will be in the farfield, where spherical spreading reduces the primary sound pressure proportionally with the distance.

When all interaction takes place in the farfield, it is relatively easy to carry out the integration described by equation (12.31), since the farfield is better behaved and easier to describe mathematically than the nearfield. For instance, for a circular transducer with radius $a$ in the primary field, the field sound pressure $p_p(r,\theta)$ is

$$p_p(r,\theta) = \sum_{i=1}^{2} p_0 \frac{r}{r_0} B(\kappa_i a \sin\theta)\exp(-\alpha_i|\mathbf{r}|)\exp\left[i(\kappa_i|\mathbf{r}| - \omega_i t)\right] \tag{12.51}$$

The source's directivity $B(\kappa_i a\sin\theta)$ was found in section 9.2.2 to be

$$B(\kappa_i a\sin\theta)=2\frac{J_1(\kappa_i a\sin\theta)}{(\kappa_i a\sin\theta)} \tag{12.52}$$

Using equations (12.51) and (12.52), equation (12.31) can be solved numerically, which has been done by Muir and Willette (1972), among others.

More general models for interaction in the farfield have been made, for instance the Berktay-Leahy model (1974). This model predicts that when the interaction is in the farfield the secondary directivity is approximately the same as the product of the two primary directivities, so that $B_s(\theta)$ is

$$B_s(\theta)\approx B_{p1}(\theta)B_{p2}(\theta)\approx B_p^2(\theta) \tag{12.53}$$

According to the approximation of equation (12.53), the secondary beamwidth is somewhat smaller than the primary beamwidth.

### *12.4.4 The Mellen-Moffett model*

A useful and simple model has been developed over several years by Mellen and Moffett (1971) and Moffet and Mellen (1977, 1981). This model takes into consideration not only the effects of nonlinear attenuation, but also that the interaction may be in both the nearfield and the farfield. This is accomplished by approximating the primary field with smoothing or taper functions so that the integral of equation (12.31) can be computed. In the following paragraphs, we present the MM model for the secondary source level, but for the secondary beam pattern and directivity we shall only quote the asymptotic result for the three interaction regimes.

There are two taper functions in the model, $T_1(r')$ and $T_2(r')$, describing the primary sound field. The taper function that accounts for both the linear and nonlinear attenuations, $T_1(r')$, is expressed as

$$T_1(r') = \frac{\exp(-2\alpha_p r')}{1 + \left[\frac{\sigma_0}{2}\sinh^{-1}\left(\frac{r'}{R_R}\right)\right]^2} \tag{12.54}$$

In principle, this function is the same as the one Bartram proposed (see equation (12.48)), but in a different functional form.

The second taper function $T_2(r')$ provides a smooth transition between the nearfield and the farfield

$$T_2(r') = \frac{1}{R_R\left[1 + \left(\frac{\kappa_s r'}{\kappa_0 R_R}\right)^2\right]^{1/2}} \tag{12.55}$$

The secondary field, at the difference frequency and at the axis where $\theta = 0$, is described by the integral

$$\left|\frac{p_s(r,0)r}{p_p R_R}\right| = \frac{\sigma_0}{2}\left(\frac{\kappa_s}{\kappa_p}\right)^2 \int_0^\infty \frac{\exp(-2\alpha_p r')dr'/R_R}{\left\{1 + \left[\frac{\sigma_0}{2}\sinh^{-1}\left(\frac{r'}{R_R}\right)\right]^2\right\}\left\{\left[1 + \left(\frac{\kappa_s r'}{\kappa_p R_R}\right)^2\right]^{1/2}\right\}} \tag{12.56}$$

In equations (12.54)–(12.56), $\sigma_0$ is the saturation parameter previously defined in equations (12.23) and (12.24), $R_R$ is the Rayleigh distance for the transition between the nearfield and the farfield, and $\kappa_p = \omega_p / c$ is the wave number of the mean primary frequency.

The parametric gain $G$ is defined as

$$G = \left|\frac{p_s(r,0)r}{p_p R_R}\right| \tag{12.57}$$

and gives the ratio of the amplitude of the farfield secondary sound pressure $p_s$ to the primary source level $p_p$.

Equation (12.56) easily can be solved numerically to give the parametric gain as a function of the saturation parameter $\sigma_0$ or as a function of the primary source level $SL_p$. The relationship between the saturation parameter and the primary source level is developed from equation (12.24)

$$SL_p = 20\log\left(\frac{p_p}{\sqrt{2}}\right) = 20\log\left(\frac{\sigma_0}{R_R}\right) + 20\log\left(\frac{\rho_0 c_0^3}{\sqrt{2}\omega\beta}\right) \tag{12.58}$$

Using the numerical values $\rho_0 = 1000$ kg/m$^3$, $c_0 = 1500$ m/s and $\beta = 3.5$ in the last term of equation (12.58) produces

$$SL_p = 20\log\left(\frac{\sigma_0}{R_R}\right) + 280.7 - 20\log\left(f_{p,kHz}\right) \quad \text{dB } re \ 1\mu\text{Pa} \tag{12.59}$$

where the primary frequency $f_{p,kHz}$ is measured in kHz.

The Mellen-Moffett model gives the secondary sound pressure as function of a scaled primary source pressure $SL_p{}^*$; defined as

$$SL_p{}^* = SL_p + 20\log f_{p,kHz} \quad \text{dB } re \ 1\mu\text{Pa} \tag{12.60}$$

Figure 12.5 shows examples of parametric gain, expressed in dB, as a function of the scaled primary pressure $SL_p{}^*$ for different values of $\alpha_p R_R$ expressed in dB. These results are obtained by numerical integration of equation (12.57). The top plot is calculated for the downshift ratio $f_p/f_s = 10$ and the bottom, for $f_p/f_s = 20$. Similar plots are also found in Moffett and Konrad (1997).

Figure 12.5 shows that for low primary source levels parametric gain increases proportionally with the pressure of the primary source and, consequently, the secondary sound pressure level increases with the square of the primary sound pressure. For very high-pressure source levels, saturation and nonlinear attenuation become significant, and the parametric gain tends to be constant and independent of the primary source level. This behavior occurs approximately when the scaled primary source level exceeds 280 dB, which corresponds to $\sigma_0 \approx 1$. For such high source levels, the secondary sound pressure level is proportional to the primary sound pressure level. We see also that the parametric gain decreases with increasing downshift ratio $f_p/f_s$.

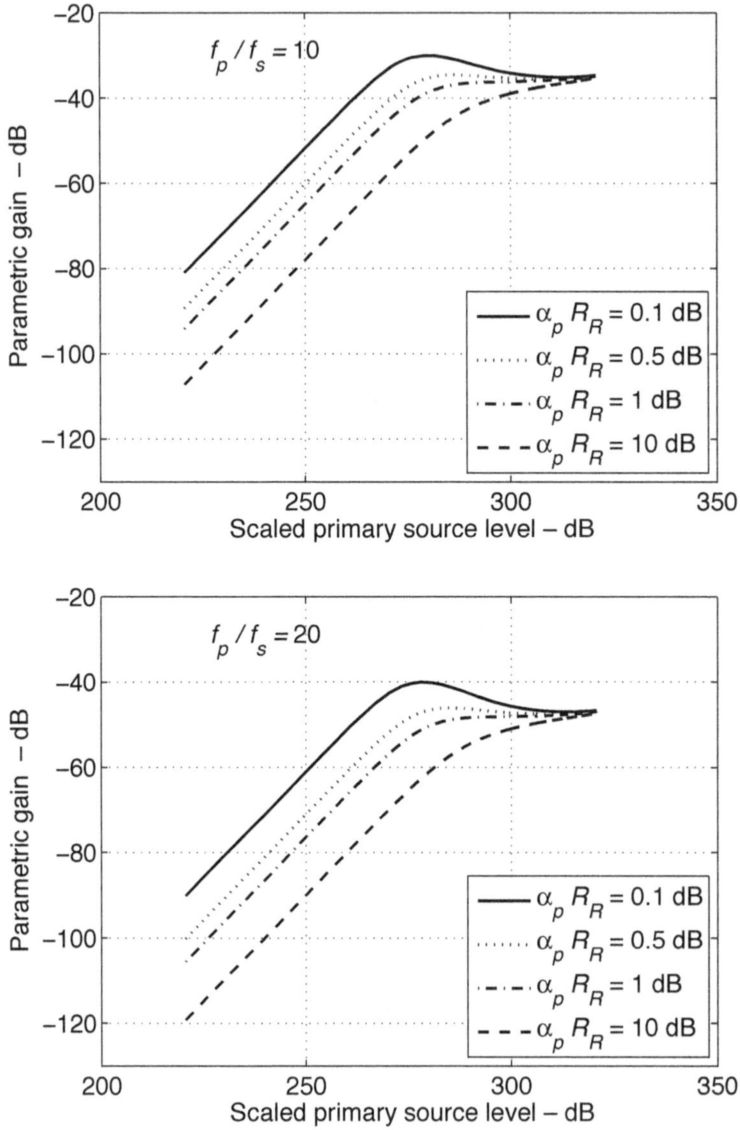


Figure 12.5. Parametric gain, 20log(G), as function of scaled primary pressure $SL_p$* for the values $\alpha_p R_R$(dB) = 0.1, 0.5, 1.0 and 5.0. Top, downshift ratio $f_p / f_s = 10$; bottom, downshift ratio $f_p / f_s = 20$.

The asymptotic result of the Mellen-Moffett model for the three different interaction cases can be summarized as follows.

**Case 1.** $\sigma_0 < 1 \qquad 2\alpha_p R_R > 1$

$$\left|\frac{p_s(r,0)r}{p_p R_R}\right| = \sigma_0 \left(\frac{\kappa_s}{\kappa_p}\right)^2 \frac{1}{4\alpha_p R_R}$$
$$B_s(\theta) = \left[1 + \left(\frac{\kappa_s}{\alpha_p}\sin^2\frac{\theta}{2}\right)^2\right]^{-\frac{1}{2}} \tag{12.61}$$

The interaction in the nearfield is limited by absorption; this model gives us the same result as that of Westervelt. The secondary sound pressure increases proportionally with the square of the primary pressure and with the square of the difference frequency increase. Note that the directivity function has no sidelobes as a consequence of the infinitely long and continuous exponential aperture function.

**Case 2.** $\sigma_0 > 1 \qquad 2\alpha_p R_R < 1$

$$\left|\frac{p_s(r,0)r}{p_p R_R}\right| = \frac{\pi}{2}\left(\frac{\kappa_s}{\kappa_p}\right)^2$$
$$B_s(\theta) \approx \exp\left(-\frac{2\kappa_p R_R \theta^2}{\sigma_0}\right) \tag{12.62}$$

The interaction is in the nearfield and is limited by nonlinear absorption. The secondary sound pressure increases linearly with increasing primary pressure and with the square of the difference frequency. The directivity function has the shape of a Gaussian function, without a sidelobe structure.

**Case 3.** $\sigma_0 < 1 \qquad 2\alpha_p R_R < 1$

$$\left|\frac{p_s(r,0)r}{p_p R_R}\right| = \frac{\sigma_0}{2}\left(\frac{\kappa_s}{\kappa_p}\right) E\left(2\alpha_p R_R\right)$$
$$B_s(\theta) \approx B_p^2(\theta) \tag{12.63}$$

The interaction in the farfield is limited by spherical spreading. The secondary sound pressure is proportional both to the difference frequency and to the square of the primary pressure. In agreement with the results found by Berktay and Leahy (1974), the secondary directivity function is approximately the product of the two directivity functions for the two primary frequencies, which means that the secondary field has a sidelobe structure, but the secondary field's sidelobes will be lower than those of the primary field. Note however that the results summarized above are asymptotic results for large distances from the source.

The function $E(2\alpha_p R_R)$, which accounts for the effect of the attenuation of the primary sound field, is given by the exponential integral

$$E\left(2\alpha_p R_R\right) = \int_{2\alpha_p R_R}^{\infty} \frac{\exp(-z)}{z} dz \tag{12.64}$$

Figure 12.6 shows 20log $[E(2\alpha_p R_R)]$ as a function of the nearfield attenuation $2\alpha_p R_R$.

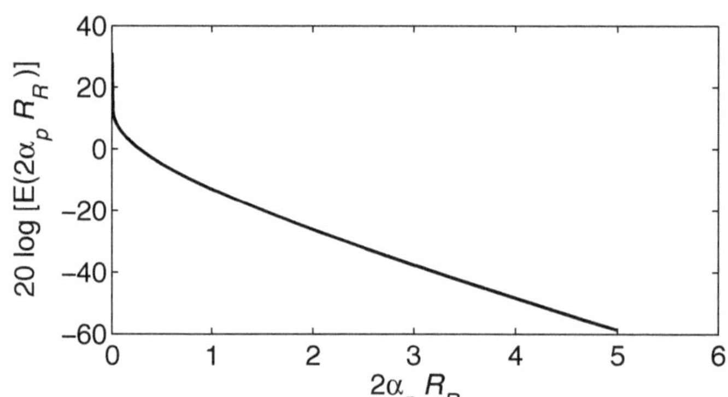


Figure 12.6. Exponential integral plotted as a function of $2\alpha_p R_R$.

## 12.5 Example of a parametric source

Next we shall illustrate some key characteristics of a parametric source by considering an example of a hypothetical parametric sonar. Let the source be a circular transducer with the radius $a = 0.125$ m and a mean primary frequency of 100 kHz. With these values, we have a 15 mm wavelength, and the primary beamwidth and directivity index are 4.3° and 34.4 dB, respectively. The surface area of the transducer is 490 cm$^2$. The transmitted acoustic power at the primary frequency is 1 kW and the primary source level is therefore 235 dB *re* 1 $\mu$Pa. The acoustic intensity at the surface of the transducer is about 2.0 W/cm$^2$, the sound pressure is 1.7510$^5$ Pa, and the particle velocity is $v = 0.12$ m/s, which is approximately the cavitation limit of 100 kHz (Urick 1983).

The wave that exits the transducer is the sum of two signals at the frequencies $f_1 = 105$ kHz and $f_2 = 95$ kHz, so that the difference frequency is $f_s = 10$ kHz. The absorption loss at the mean primary frequency is $\alpha_p \approx 0.04$ dB/m $= 0.0046$ Np/m, while for the secondary frequency $\alpha_s = 0.93 \cdot 10^{-3}$ dB/m $= 0.10 \cdot 10^{-3}$ Np/m (see figure 3.5). We also assume that $\beta = 3.5$.

With these values, the characteristic lengths are

$$L_a = \frac{1}{\alpha_p} = 217 \text{ m}$$
$$R_R = \frac{\pi a^2}{\lambda} = 3.27 \text{ m} \qquad (12.65)$$
$$L_s = \frac{1}{\beta M \kappa_0} = 8.5 \text{ m}$$

The other parameters, which we have defined before, are

$$\sigma_0 = \frac{R_R}{L_s} = 0.38$$
$$2\alpha_p R_R - 0.03 \text{ Np} \qquad (12.66)$$
$$\alpha_p R_R = 0.13 \text{ dB}$$

From the values in equations (12.66) we see that the interaction is mainly in the farfield and is limited by spherical spreading (Case 3). In this situation, nonlinear attenuation plays a bigger role in the interaction than does small-signal absorption, which means that the conditions permitting use of the Westervelt model are not present. Nonetheless, we can use that model to illustrate some important points.

In the left-hand part of figure 12.7, we see the secondary sound pressure level plotted as a function of the acoustical power in the primary signal. This figure contains both the results of the Westervelt model, as represented by equation (12.47), and those for Bartram's model, characterized by equation (12.50).

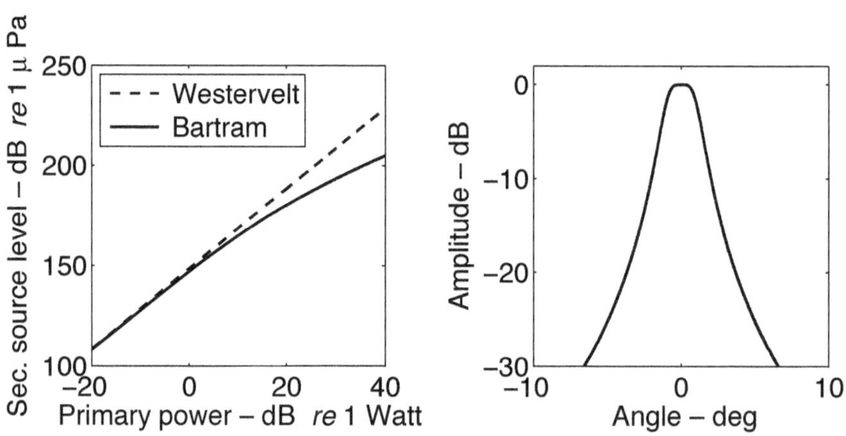


Figure 12.7. Left, the secondary source level function of a primary power source for the Westervelt and Bartram models. Right, the secondary directivity function according to the Westervelt model. Numerical values for the example are given in the text

Results of calculating the Westervelt directivity function using equation (12.43) appear as the right half of figure 12.7, which shows a monotonic decreasing function without sidelobes. The beamwidth is extremely narrow, being 2.4° according to equation (12.44) and figure 12.7. This beamwidth is typical for an infinitely long aperture, which we would expect for the current value of the absorption length $L_a$.

To determine the secondary sound pressure using the Mellen-Moffett model, we return to the graphs of figure 12.5. In this case, the scaled primary sound pressure is

$$SL^* = SL + 20\log 10 = 275 \quad \text{dB } re \ 1\mu\text{Pa} \qquad (12.67)$$

For the value of $\alpha R_R = 0.13$ dB, the top graph of figure 12.5 shows that the parametric gain is approximately –42 dB for $f_p = 10 f_s$. Thus, the secondary source level $SL_s$ at the difference frequency is expected to be

$$SL_s = 235 - 42 = 193 \quad \text{dB } re\ 1\mu\text{Pa} \tag{12.68}$$

This result is approximately 16 dB lower than those obtained using Westervelt probably because the Mellen-Moffett model takes spherical spreading into account, which is significant for the values of this example. In the same way, we find from the bottom graph in figure 12.5 that the parametric gain for a secondary frequency of 5 kHz is about –52 dB, which yields a secondary source level of about 183 dB *re* 1 $\mu$Pa.

Figure 12.8 shows the primary beam pattern calculated by using equation (12.52), as well as the secondary directivity function calculated with the approximation of equation (12.63). We see that the beamwidths of the primary and secondary fields are almost the same, and therefore the secondary directivity index will be about the same as the primary directivity index, that is, about 34 dB. For comparison, the same power output of 1000 W radiated from a transducer of the same size and at the frequency of 10 kHz, would result in a directivity index of 14 dB and a source level of approximately 214 dB. These figures indicate that we have "lost" about 14 dB from the source level and "gained" 20 dB in directivity by using the transducer in a parametric mode.

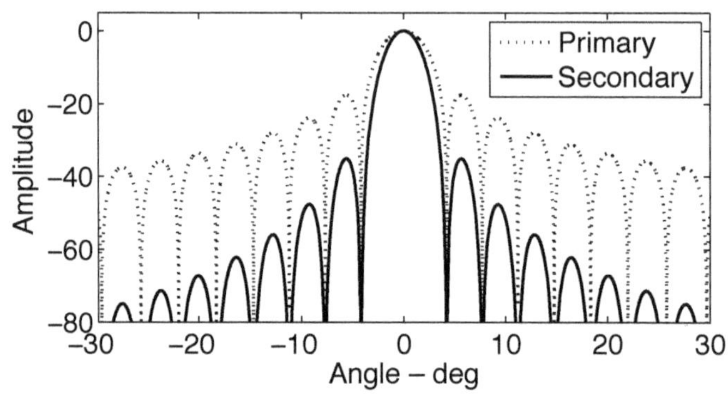


Figure 12.8. Primary and secondary beam patterns for a circular transducer with 12.5 cm radius. The primary frequency is 100 kHz and the secondary frequency is 10 kHz. The primary beam pattern is calculated using equation (12.52); the secondary beam pattern is calculated using the approximation of equation (12.63).

## 12.6 Example of a practical system

We will now present some results from the construction and testing of an experimental system that later was developed into a commercial product. This system was constructed with the aim of mapping the upper seabed strata. For this purpose, use of the lowest possible frequencies is necessary in order to obtain deep penetration into the bottom, and a narrow beamwidth is required for obtaining fine depth resolution. A parametric system can be advantageous for such usage. The system and its construction are described in more detail in Pettersen et al. (1977).

The transmitting transducer is a 35 cm × 70 cm rectangle with an area of 0.25 $m^2$. The mean primary frequency is 38 kHz, and the difference frequency can be selected within the range of 1.6–6.7 kHz. With a 5 kW acoustic output, the primary sound level is about 230 dB *re* 1 $\mu$Pa, and the primary beamwidth is 4°. For a difference frequency of 3.3 kHz, the measured and modeled sound pressures and the asymptotical sound pressures for the primary and secondary fields are as shown in figure 12.9.

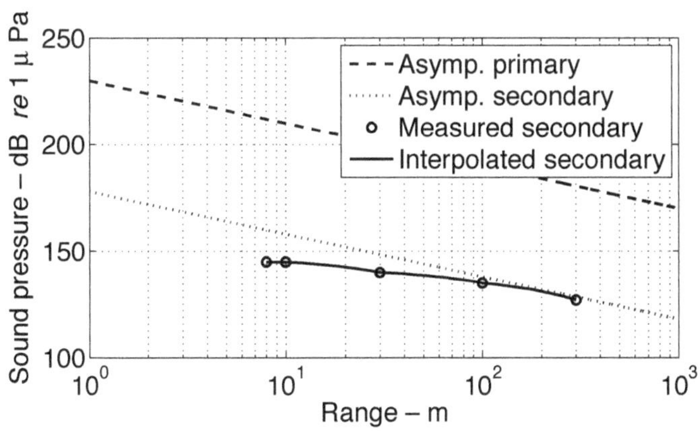


Figure 12.9. Measured and modeled sound pressures, plotted as functions of their distances from the experimental parametric sound source.

Unfortunately, the measured values are quite few and only go up to 300 m, which is hardly sufficient to include the whole nonlinear interaction range. This limitation is caused by experimental difficulties. However, the measurements indicate that the primary sound pressure decreases proportionally with $r$, corresponding to spherical spreading. Out to a distance of 100 m from the source, the secondary sound level only decreases slowly as a consequence of the transfer of the primary wave's energy into the secondary wave. At longer ranges, the decay approaches that of spherical spreading. Extrapolating the measured values at long ranges back to the reference distance of 1 m indicates a secondary sound pressure level of about 188 dB *re* 1 $\mu$Pa, but this value is uncertain since there are few measured points. While the version of the Mellen-Moffett model presented here is not adapted for use with a rectangular-shaped transducer, for a circular transducer of the same area the model predicts a secondary sound pressure level to be about 190 dB *re* 1 $\mu$Pa, which is within reasonable agreement with the measurements.

In figure 12.10, we see how the sound level varies as a function of the angle $\theta$, off the axis, for both the primary and secondary fields; the directivity measurements are taken 28 m from the source.

The primary beamwidth is about 3°, in accordance with the transducer's size, and the wavelength for 38 kHz. The beamwidth at the difference frequency is somewhat broader, about 4°. To create such narrow beams at the difference frequency of 3.3 kHz with a conventional source, the area of the transducer would have to be about 25 $m^2$, a size that exceeds any practical possibility.

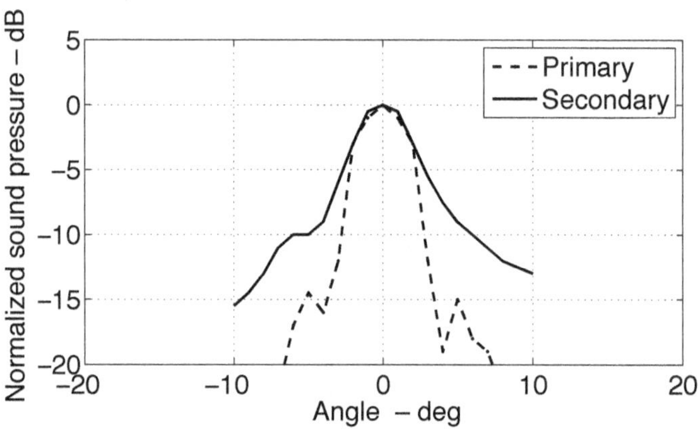


Figure 12.10. Measured directivity for the primary and secondary sound fields generated by the experimental parametric source.

The experimental transducer is mounted in a towed body. The same transducer is also used to receive the primary frequency, but because it is not very sensitive at lower frequencies, the difference frequency is received by a sensor located on a hydrophone cable. During testing, the experimental system was suspended from a slow-moving vessel. Figure 12.11 contains an echogram made during one of the tests, which was conducted in a water depth of about 90 m. The subbottom structure, which was composed of relatively soft sediments, was detectable down to a depth of about 40 meters when the parametric source used a signal with a 3.3 kHz difference frequency.

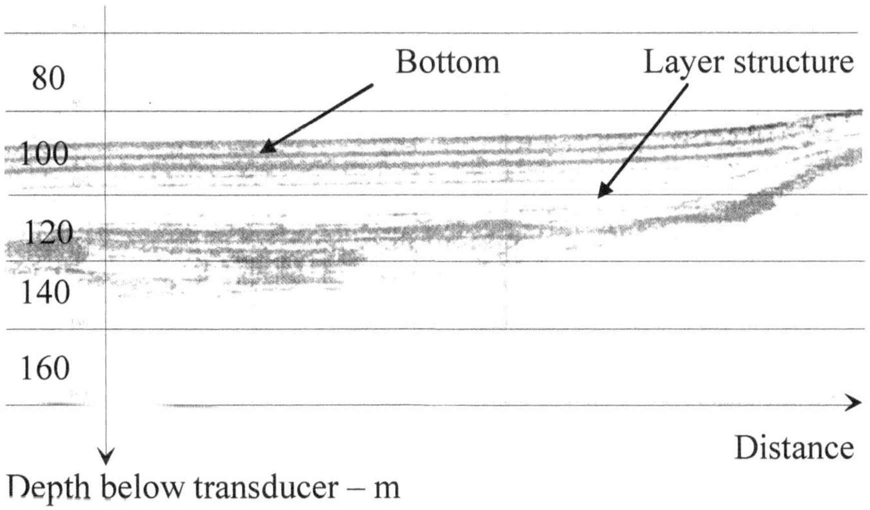


Figure 12.11. Imaging the seabed's structure with the experimental parametric echo sounder, as recorded by an echogram. The bottom lies at approximately 90 m; the signal penetrates about 35–40 m into the seabed.

However, at the 38 kHz primary frequency, no subbottom structure could be seen because of the strong absorption loss that took place at the frequency of the primary wave. The results, obtained with the 3.3 kHz difference frequency, illustrate the advantage of using parametric sonar for certain applications.

# Chapter 13

# Reflection and scattering from particles and bubbles

## 13.1 Introduction

Objects having various shapes and compositions reflect and scatter acoustical waves differently; these differences are important for most applications of underwater acoustics, particularly in applying sonar to detect and classify objects. In this chapter, we present theoretical models for acoustical scattering from particles and bubbles. By applying these models to some simple objects, we obtain insight into the reflection and scattering of acoustical waves from more complicated objects. Note, however, that for most real-life objects, we must determine the reflection characteristics by taking measurements rather than relying solely on theoretical models.

After first defining some basic concepts related to acoustical scattering, we develop the expression for scattering from a rigid sphere. We also study the effects of frequency and of the sphere material's density and sound speed on the scattered acoustical field.

Another discussion topic in this chapter is acoustical scattering from a sphere embedded in a viscous fluid; we use the resulting solution to develop a simple expression for the viscosity's effects on sound absorption in a suspension of solid particles in water. Finally, we consider acoustical scattering by gas bubbles and then determine the speed of sound in bubbly water; this is relevant in situations such as sound propagation in breaking water near the sea's surface and propagation through the bubble clouds generated by a cavitating propeller.

## 13.2 Concepts and definitions

Figure 13.1 shows a situation in which a source with source strength $Q$ is transmitting an acoustic wave that strikes a target object after spherical propagation over a distance $r_S$. The receiver is located at a point that lies a distance $r$ from the object and in the direction given by the angles $\theta$ and $\varphi$ as indicated in the figure.

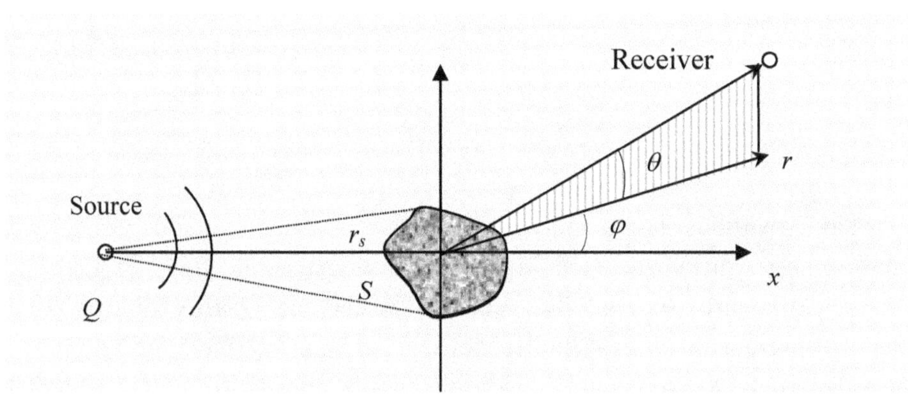


Figure 13.1. Echo measurement of an object presenting an area $S$ to the propagating wave. The incident acoustic wave is scattered toward a receiver at a distance $r$ and in the directions given by the angles $\varphi$ and $\theta$.

The incident field on the object $\Phi_i$ is

$$\Phi_i = \frac{Q}{4\pi r_s}\exp\left[i(\kappa r_s \cos\theta - \omega t)\right] \tag{13.1}$$

where $Q$ is the source strength, $k$ is the wave number given by $\kappa = \omega / c$, and the time dependency is $\exp(-i\omega t)$. Seen from the vantage point of the source, the target exhibits an area $S$, which naturally is dependent on the target's orientation with respect to the direction of the incident field. Where the wave is intersected by the target, the wave intensity is proportional to $|\Phi_i|^2$, and the total power intersected by the target is proportional to $S|\Phi_i|^2$. Some parts of the power received by the bubble or particle may be transformed into heat, while other parts will be scattered again, returning to the medium with a spatial distribution determined by the target's shape and composition.

The scattered intensity per unit of solid angle is termed $I_s(\theta, \varphi)$, where $\theta$ and $\varphi$ are the angles in a polar coordinate system that has its center in the middle of the target, as indicated in figure 13.1. The scattered intensity returning toward the source position is

$$I_{bs} = I_s(\theta = \pi, \varphi = 0) \tag{13.2}$$

As a measurement of the reflected signal's strength, the ***backscattering cross section*** $\sigma_{bs}$ is defined as the ratio between backscattered intensity and incident intensity, multiplied by the square distance $r^2$

$$\sigma_{bs} = r^2 \frac{I_{bs}}{I_i} \tag{13.3}$$

Multiplication by $r^2$ compensates for the spherical spreading that takes place during the signal's return to the receiver. The target's total scattered intensity $I_{tot}$ is found by integration over all spatial angles $\Omega$, so that

$$I_{tot} = \iint I_s(\theta, \varphi) d\Omega \tag{13.4}$$

and the ***total scattering cross section*** $\sigma_s$ is defined as the ratio between the total scattered intensity and the incident intensity, multiplied by the square distance $r^2$

$$\sigma_s = r^2 \frac{I_{tot}}{I_i} \tag{13.5}$$

Both $\sigma_{bs}$ and $\sigma_s$ are measured in square meters, and their sizes are related to the size of the area $S$.

We can postulate an ideal spherical scatterer that has the property of scattering all the intercepted power with equal intensity in all directions. When the ideal sphere has the radius $a$, the area presented by the sphere to the propagating wave is $S = \pi a^2$, and the power intercepted by the target is proportional to $\pi a^2 |\Phi_i|^2$. Then the scattered intensity is

$$I_s(\theta,\varphi) = \frac{\pi a^2}{4\pi r^2}|\Phi_i|^2 \tag{13.6}$$

For an ideal spherical scatterer, it follows that

$$\begin{aligned}\sigma_{bs} &= \frac{a^2}{4} \\ \sigma_s &= \pi a^2\end{aligned} \tag{13.7}$$

Therefore, we can conclude that the ideal sphere's total scattering cross section is equal to the sphere's geometric cross section.

In echo measurement situations, we need to know the value of $I_r$, the scattered acoustic intensity at the receiver position. To obtain $I_r$, we shall assume that the receiver shares the source's location; this means that $r_s = r$. We assume that the source has directivity $D$ and acoustically outputs the power $W$. After spherical propagation, the incident intensity $I_i$ at the target is expressed as

$$I_i = \frac{W}{4\pi r^2}D \tag{13.8}$$

The scattered intensity returning to a receiver collocated with the source $I_r$ is

$$I_r = I_i\frac{\sigma_{bs}}{r^2} = \left(\frac{WD}{4\pi r^2}\right)\frac{\sigma_{bs}}{r^2} \tag{13.9}$$

It is also common practice to use the ***sonar scattering cross section*** $\sigma$, which is defined as

$$\sigma = 4\pi\sigma_{bs} \tag{13.10}$$

When this measure is used, the received scattered intensity is expressed as

$$I_r = \frac{I_{bs}}{r^2} = \left(\frac{WD}{4\pi r^2}\right)\frac{\sigma}{4\pi r^2} \tag{13.11}$$

It is necessary to distinguish between the sonar scattering cross section $\sigma$, defined by equation (13.10), and the total cross scattering section $\sigma_s$, defined by equation (13.5). The sonar scattering cross section $\sigma$ is just another measure of the intensity scattered back to the sonar, whereas the total scattering cross section is a measure of the total scattered intensity integrated over all directions. Thus, the two scattering cross sections are equal only for the ideal spherical scatterer.

In the literature about sonar, normally the logarithmic versions of equations (13.9) and (13.11) are used. This means that sound pressure $p$ replaces intensity in the expressions, so that

$$p^2 = (\rho c) I_r \tag{13.12}$$

After conversion to decibels, we obtain

$$20\log p = 10\log(\frac{\rho c}{4\pi}) + 10\log(W) + 10\log D - 40\log r + 10\log(\frac{\sigma}{4\pi}) \tag{13.13}$$

This is recognizable as one of the sonar equations discussed in chapter 11. The last term of equation (13.13) represents the target strength $TS$

$$TS = 10\log\left(\frac{\sigma}{4\pi}\right) \tag{13.14}$$

There are two other types of cross sections that we need to define. The first of these is the ***absorption cross section*** $\sigma_a$, which is equal to the ratio between the absorbed power—for instance, the loss of that portion of the signal power converted to heat—and the incident intensity. The second new cross section type is the ***extinction cross section*** $\sigma_e$, which represents the total scattered and absorbed intensity

$$\sigma_e = \sigma_s + \sigma_a \tag{13.15}$$

This is the cross section that causes attenuation when an acoustic wave propagates through a cluster of scatterers. For instance, a plane wave in the $x$-direction has the intensity $I(x)$, which is reduced proportionally to the number of scatterers $N$ and the extinction cross section, so that

$$\frac{dI(x)}{dx} = -I(x)N\sigma_e \tag{13.16}$$

Therefore, as a function of the distance $x$ that the wave has propagated through the medium, the intensity $I$ is

$$I(x) = I_0 \exp(-N\sigma_e x) \tag{13.17}$$

where $I_0$ is the incident intensity at $x = 0$ and $N$ is the number of scatterers per unit volume.

## 13.3 Reflection from a rigid sphere in fluid

In this section, we discuss scattering and reflection created when plane acoustic waves hit spherical objects embedded in a fluid medium. The scattering from such objects is essential to a number of acoustical and ultrasound applications. In addition, we apply a mathematical method that is useful also in several other contexts. Consequently, we develop the necessary definitions and expressions in some depth. However, we only consider the fluid case, in which both the ambient medium and the particle are fluids, but each having a different density and sound velocity. While this problem has been thoroughly dealt with in the literature, here we concentrate on the classical articles of Anderson (1950) and Faran (1951). The books by Clay and Medwin (1977) and Medwin and Clay (1989) are also useful references.

We begin with a sphere-shaped particle with the radius $a$, density $\rho_1$, sound speed $c_1$, and the volume stiffness, $K_1 = \rho_1 c_1^2$. The particle is fixed, or rigidly, located in a fluid medium that has the density $\rho_0$, sound speed $c_0$, and volume stiffness

$K_0 = \rho_0 c_0^2$. The incident wave is a plane harmonic wave with angular frequency $\omega$. The wave numbers are $\kappa_0 = \omega / c_0$ and $\kappa_1 = \omega / c_1$, respectively. We assume that the ambient fluid is nonviscous.

We also assume that the incident wave striking the target, $\Phi_i$, can be considered as being locally plane and, therefore, the mathematical expression representing the wave is

$$\Phi_i = \Phi_{inc} \exp\left(i\kappa_0 r \cos\theta\right) \tag{13.18}$$

Since the object is spherical, we need to express the incoming plane wave in spherical coordinates. This requires that the plane wave be expanded in a sum of modes involving spherical Bessel functions and Legendre polynomials (Morse and Ingard 1968) by using the mathematical identity

$$\Phi_i = \Phi_{inc} \sum_{m=0}^{\infty} i^m (2m+1) P_m(\cos\theta) j_m(\kappa_0 r) \tag{13.19}$$

where $P_m(\cos\theta)$ represents the Legendre polynomials of order $m$ and with the argument $\cos\theta$, where $\theta$ is an angle of scattering relative to the incident wave's direction, defined as shown in figure 13.1.

The first three Legendre polynomials are given as

$$\begin{aligned} P_0(\cos\theta) &= 1 \\ P_1(\cos\theta) &= \cos\theta \\ P_2(\cos\theta) &= \frac{1}{2}\left(3\cos^2\theta - 1\right) \end{aligned} \tag{13.20}$$

Subsequent polynomials can be calculated by the recursion

$$P_m(\cos\theta) = \frac{2m-1}{m} \cos\theta P_{m-1}(\cos\theta) + \frac{1-m}{m} P_{m-2}(\cos\theta) \tag{13.21}$$

The Legendre polynomials represent the angular directivity of the radiation of the various modes. The directivity functions for the first four modes are shown in the polar diagrams in figure 13.2.

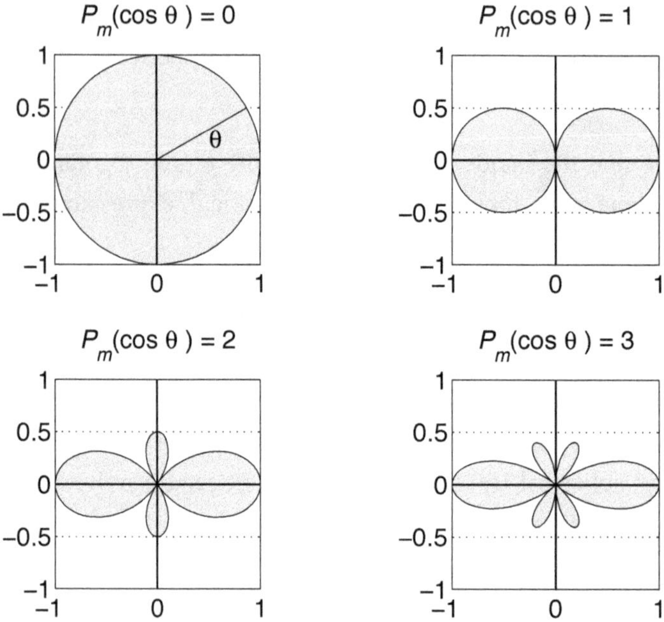


Figure 13.2. Polar diagrams of the directivity functions for the modes $m = 0,1,2,3$, which are represented by the Legendre polynomials $P_m(\cos\theta)$. The angle $\theta$ is defined in the plot of $P_0(\cos\theta)$.

The spherical Bessel and Neumann functions are $j_m(kr)$ and $y_m(kr)$, respectively. These are related to the cylindrical Bessel and Neumann functions $J_m(kr)$ and $Y_m(kr)$ by

$$
\begin{aligned}
j_m(\kappa r) &= \sqrt{\frac{\pi}{2kr}} J_{m+\frac{1}{2}}(\kappa r) \\
y_m(\kappa r) &= \sqrt{\frac{\pi}{2kr}} Y_{m+\frac{1}{2}}(\kappa r)
\end{aligned}
\tag{13.22}
$$

The first spherical Bessel and Neumann functions, $j_m(\kappa r)$ and $y_m(\kappa r)$, are given in equation 2.89 (chapter 2). Numerical routines for calculation of these functions are also readily available.

We also need the derivatives of the spherical Bessel and Neumann functions. These are given by the expressions

$$
\begin{aligned}
j_m'(\kappa r) &= \frac{\partial\left[j_m(\kappa r)\right]}{\partial(\kappa r)} = \frac{m}{2m+1} j_{m-1}(\kappa r) - \frac{m+1}{2m+1} j_{m+1}(\kappa r) \\
y_m'(\kappa r) &= \frac{\partial\left[y_m(\kappa r)\right]}{\partial(\kappa r)} = \frac{m}{2m+1} y_{m-1}(\kappa r) - \frac{m+1}{2m+1} y_{m+1}(\kappa r)
\end{aligned}
\tag{13.23}
$$

As we did for the incident field, we next express the acoustic field inside the particle and the scattered field as infinite sums of spherical modes.

The field inside the spherical particle is expressed as

$$
\Phi_1 = \sum_{m=0}^{\infty} B_m P_m(\cos\theta) j_m(\kappa_1 r) \tag{13.24}
$$

and the scattered field as

$$
\Phi_s = \sum_{m=0}^{\infty} A_m P_m(\cos\theta)\left[j_m(\kappa_0 r) + i y_m(\kappa_0 r)\right] \tag{13.25}
$$

In equations (13.24) and (13.25), we have two temporarily unknown coefficients $A_m$ and $B_m$, both of which are determined by the boundary conditions at the particle surface. These boundary conditions are connected to sound pressure and radial particle velocity, generally given by

$$
\begin{aligned}
p &- i\omega\rho\Phi \\
v_r &= \frac{\partial\Phi}{\partial r}
\end{aligned}
\tag{13.26}
$$

At the interface $r = a$, the sound pressure and radial particle velocity must be continuous. Therefore, two conditions must be satisfied.

First, the internal and external sound pressures must balance

$$
p_i(a) + p_s(a) = p_1(a) \tag{13.27}
$$

Second, the radial particle velocities must balance

$$v_i(a) + v_s(a) = v_1(a) \tag{13.28}$$

Equations (13.27) and (13.28) can be used to determine the unknown coefficients $A_m$ and $B_m$. We concentrate on $A_m$, which describes the scattered field, since that is our main interest here. After some algebra, we find that $A_m$ can be written in the form

$$A_m = -\Phi_{inc} i^m (2m+1) \frac{1}{1+iC_m} \tag{13.29}$$

where the coefficients $C_m$ are given as

$$C_m = \frac{\left[\dfrac{j'_m(\kappa_1 a)}{j'_m(\kappa_0 a)}\right]\left[\dfrac{y_m(\kappa_0 a)}{j_m(\kappa_1 a)}\right] - gh\left[\dfrac{y'_m(\kappa_0 a)}{j'_m(\kappa_0 a)}\right]}{\left[\dfrac{j'_m(\kappa_1 a)}{j'_m(\kappa_0 a)}\right]\left[\dfrac{j_m(\kappa_0 a)}{j_m(\kappa_1 a)}\right] - gh} \tag{13.30}$$

We have introduced the particle's relative density $g$ and sound speed $h$, which are defined as

$$g = \frac{\rho_1}{\rho_0}$$
$$h = \frac{c_1}{c_0} \tag{13.31}$$

In the limiting case of an infinitely hard and heavy sphere, $g$ and $h$ become large and equation (13.30) now approaches

$$C_m = \frac{y'_m(\kappa_0 a)}{j'_m(\kappa_0 a)} \tag{13.32}$$

Normally, we are most interested in that part of the scattered field that is located in the farfield at distances at least some wavelengths from the target. In the farfield, the

Bessel and Neumann functions with the argument ($\kappa r$) may be replaced with approximations valid for large arguments

$$j_m(\kappa r) \xrightarrow[kr\to\infty]{} \left(\frac{1}{kr}\right)\cos\left[\kappa r-(m+1)\pi/2\right]$$
$$y_m(\kappa r) \xrightarrow[kr\to\infty]{} \left(\frac{1}{kr}\right)\sin\left[\kappa r-(m+1)\pi/2\right] \tag{13.33}$$

When we do so, we find that

$$j_m(\kappa r)+iy_m(\kappa r) \xrightarrow[kr\to\infty]{} (-i)^{m+1}\frac{\exp(i\kappa r)}{\kappa r} \tag{13.34}$$

Consequently, the scattered field $\phi_s$, located at a position far from the sphere, is found to be

$$\Phi_s = \frac{i}{\kappa_0 r}\Phi_{inc}\exp(i\kappa_0 r - i\omega t)\sum_{m=0}^{\infty}(2m+1)P_m(\cos\theta)\frac{1}{1+iC_m} \tag{13.35}$$

The angular distribution of the scattered field is given by the sum in equation (13.35) and is given by the form function $F(\theta)$

$$F(\theta)=\left|\sum_{m=0}^{\infty}(2m+1)P_m(\cos\theta)\frac{1}{1+iC_m}\right| \tag{13.36}$$

In figure 13.3, we see the function $F(\theta)$ as calculated by equation (13.36) for scattering from infinitely hard spheres of four different sizes, so that the values of $\kappa_0 a$ = 1, 3, 5, 10. In all these plots the incident sound comes from the left and the center of the sphere is at the origin as indicated by the open circle.

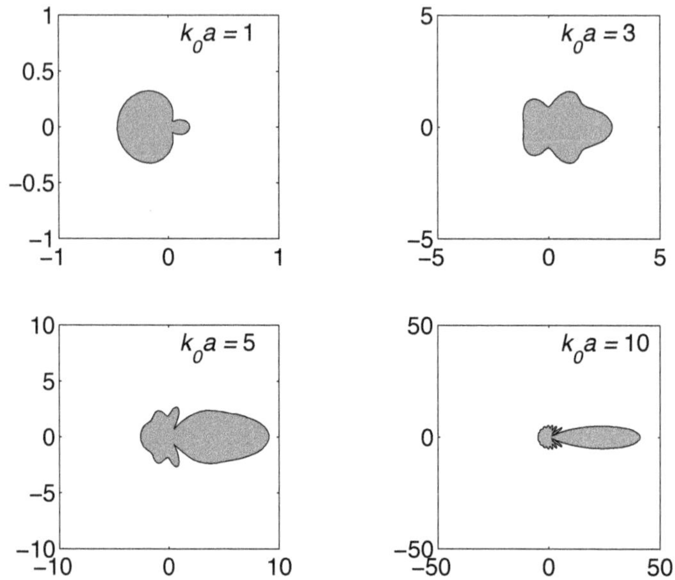


Figure 13.3. Angular distribution of scattering from rigid spheres of different sizes, where $\kappa_0 a = 1, 3, 5$, and 10. The position of the sphere is at the origin. Note the differences in scales used for the plots and that only the scattered fields are shown.

It is important to note that the figure only shows the scattered field, and that the total field is the coherent sum of the scattered field plus the incident field. The effect of the forward-scattered field is to extinguish, completely or partly, the incident field, thereby creating an acoustic shadow zone behind the object. Naturally, the shadow zone extends farther behind a large sphere than behind a small sphere.

The field intensity scattered back toward the source (that is, in the direction $\theta = \pi$), $I_{bs}$, is proportional to

$$I_{bs} = \frac{|\Phi_{inc}|^2}{(\kappa_0 r)^2} \left| \sum_{m=0}^{\infty} (-1)^m (2m+1) \frac{1}{1+iC_m} \right|^2 \tag{13.37}$$

Therefore, according to equation (13.3), the backscattering cross section of a sphere is

$$\sigma_{bs} = \frac{1}{\kappa_0^2}\left|\sum_{m=0}^{\infty}(-1)^m(2m+1)\frac{1}{1+iC_m}\right|^2 \tag{13.38}$$

In equations (13.37) and (13.38), it is taken into account that $P_m(-1) = (-1)^m$ for all mode numbers *m*, which were first seen in the definitions of the Legendre polynomials in equations (13.20) and (13.21).

We can normalize the backscattering cross section in equation (13.38) with the backscattering cross section of the ideal sphere defined in equation (13.7), which produces the normalized backscattering cross sections

$$\tilde{\sigma}_{bs} = \frac{\sigma_{bs}}{a^2/4} = \left(\frac{2}{\kappa_0 a}\right)^2\left|\sum_{m=0}^{\infty}(-1)^m(2m+1)\frac{1}{1+iC_m}\right|^2 \tag{13.39}$$

The total scattered power $P_s$ is found by integration over all angles

$$P_s = \int_0^{\pi}|\Phi_s|^2\, 2\pi r^2 \sin\theta d\theta \tag{13.40}$$

By using equation (13.35) and the orthogonal properties of the Legendre functions, the total scattering cross section can be shown to be

$$\sigma_s = \frac{4\pi}{\kappa_0^2}\sum_{m=0}^{\infty}\frac{2m+1}{1+C_m^2} \tag{13.41}$$

Note that, in general, $\sigma_{bs} \neq \sigma_s$.

Although the scattered field theory we have discussed here generally applies to spheres of all sizes, it is valid only when the sphere material is a fluid and where compressional waves, but not shear waves, may exist inside the sphere. Faran (1951) has developed the scattered field theory for the more general case of an elastic sphere.

As an example, we consider the special instance of scattering from an infinitely hard and heavy sphere. We use equation (13.39) with equation (13.32) to calculate the normalized backscattering cross section as a function of $\kappa_0 a$. The calculation result is shown in figure 13.4, along with an approximate result for a small sphere, which will be derived shortly.

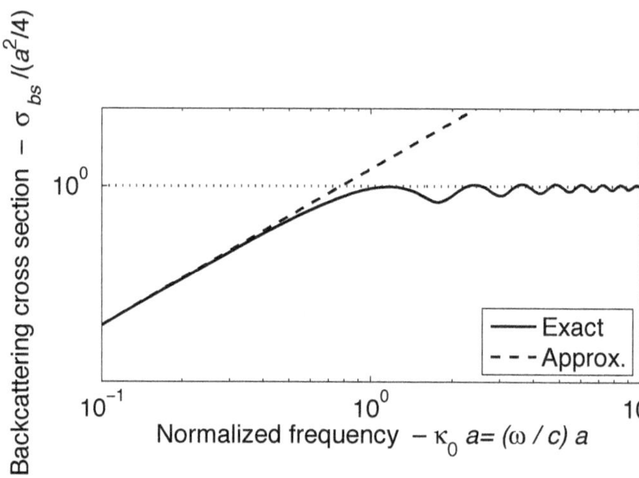


Figure 13.4. Exact and approximate results for the normalized backscattering cross section of an infinitely hard and rigid sphere with radius *a*, from equations (13.39) and (13.50), respectively.

For the calculation to be accurate for high $\kappa_0 a$ values, it is necessary to include many modes. To create figure 13.4 we have included 20 modes. When $\kappa_0 a >> 1$, for a sphere large in comparison with the signal wavelength, we see from the figure that the backscattering cross section approaches the geometrical cross section $\pi a^2$. This means that the target strength of the large sphere is

$$TS = 10 \log\left(\frac{a^2}{4}\right) \tag{13.42}$$

which is the same result we found in section 11.5 by developing a simple argument.

### *13.3.1 Limiting values for low frequencies*

We shall now derive the approximate results for low frequencies; these results are useful later in the discussion. When $k_0a << 1$ for small spheres and low frequency signals, we can develop an asymptotic solution by the following substitution

$$
\begin{aligned}
j_m(\kappa a) &\xrightarrow[\kappa a \to 0]{} \frac{(\kappa a)^m}{1\cdot 3\cdots(2m+1)}\left[1-\frac{(\kappa a)^2}{2(2m+3)}\right] \\
y_m(\kappa a) &\xrightarrow[\kappa a \to 0]{} -\frac{1\cdot 3\cdots(2m-1)}{(\kappa a)^{m+1}}\left[1-\frac{(\kappa a)^2}{2(1-2m)}\right]
\end{aligned}
\tag{13.43}
$$

(Abramowitz and Stegun 1965). Furthermore, we find that the first two modes, $m = 0$ and $m = 1$, dominate; by using equation (13.43) for these modes, we find that

$$
\begin{aligned}
j_0(\kappa a) &= 1 & j_1(\kappa a) &= \frac{\kappa a}{3} \\
j_0'(\kappa a) &= -\frac{\kappa a}{3} & j_1'(\kappa a) &= \frac{1}{3} \\
y_0(\kappa a) &= -\frac{1}{\kappa a} & y_1(\kappa a) &= -\frac{1}{(\kappa a)^2} \\
y_0'(ka) &= \frac{1}{(\kappa a)^2} & y_1'(\kappa a) &= \frac{2}{(\kappa a)^3}
\end{aligned}
\tag{13.44}
$$

When we insert the values of equation (13.44) into equation (13.30), we find that $C_m >> 1$ and, therefore, that for the first two modes the amplitudes of equation (13.29) become

$$
A_m = -\Phi_{inc} i^{(m+1)}(2m+1)\frac{1}{C_m}, \quad m = 0,1 \tag{13.45}
$$

With these approximations, the form function $F(\theta)$ for a small sphere is roughly

$$
F(\theta) = 2(k_0 a)^2\left[\frac{K_0 - K_1}{3K_1} - \left(\frac{\rho_0 - \rho_1}{\rho_0 + 2\rho_1}\right)\cos\theta\right] \tag{13.46}
$$

The first term inside the square brackets on the right side of equation (13.46) is called the ***monopole scattering term,*** $\gamma_m$; it represents the omnidirectional scattering. This term is determined by the contrast, or difference, in elasticity between the sphere and the surrounding medium. From equation (13.46), we see that

$$\gamma_m = \frac{K_0 - K_1}{3K_1} \tag{13.47}$$

The second term inside the square brackets represents the ***dipole scattering contribution,*** $\gamma_d$, which has a cosine-shaped directivity; this term is given by the difference in the densities of the particle and the surrounding medium. In equation (13.46), which is valid for a sphere in a nonviscous fluid, the dipole contribution is given by

$$\gamma_d = \frac{\rho_0 - \rho_1}{\rho_0 + 2\rho_1} \tag{13.48}$$

In the same way with only the first two modes the normalized backscattering cross section of the equation becomes

$$\tilde{\sigma}_{bs} = 4(\kappa_0 a)^4 \left[ \frac{K_0 - K_1}{3K_1} - \left( \frac{\rho_0 - \rho_1}{\rho_0 + 2\rho_1} \right) \right]^2 \tag{13.49}$$

Equation (13.49) demonstrates that the backscattering cross section of a small sphere is proportional to $(k_0 a)^4$, that is, to the frequency in the fourth power. Scattering that has this frequency dependency is referred to as ***Rayleigh scattering.*** For a small hard sphere where $K_0 << K_1$ and $\rho_0 << \rho_1$, the normalized backscattering cross section is

$$\tilde{\sigma}_{bs} \approx \left( \frac{5}{3} \right)^2 (\kappa_0 a)^4 \tag{13.50}$$

In figure 13.4, the approximated result is quite good for $\kappa_0 a << 1$.

For finite values of density and sound speed, the scattering can be higher or lower, depending on the differences in the values and the angles. In figure 13.5, we show plots of backscattering calculations for three different cases, for which the values are given as

Case A. Relative density $g = 0.75$, relative sound speed $h = 0.75$
Case B. Relative density $g = 1000$, relative sound speed $h = 1000$
Case C. Relative density $g = 2.0$, relative sound speed $h = 2.0$

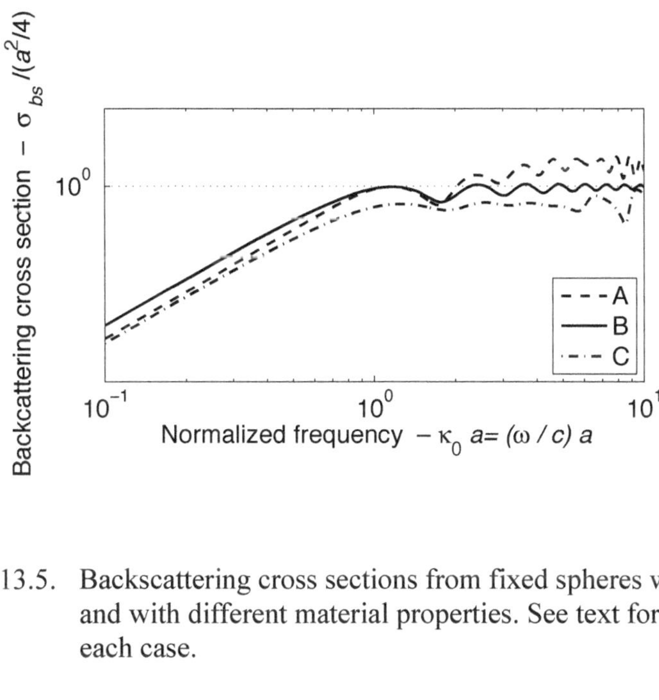


Figure 13.5. Backscattering cross sections from fixed spheres with radius $a$ and with different material properties. See text for values of each case.

These values are selected to illustrate how the scattering cross sections can vary with sound speeds and densities. Case A represents an object with values of sound speed and density 75% lower than for the same parameters in the surrounding medium, Case C with sound speed and density 100% higher than the surrounding medium, and Case B with infinite sound speed and density. The horizontal dotted line represents the limiting value for very high $k_0a$ values. We notice that the scattering cross sections are dependent on the differences in sound speed and density independently, and not on their product (the impedance).

### 13.3.2 Spherical particle in viscous fluid

In the preceding discussion, we considered the scattering from a rigid spherical particle embedded in a nonviscous fluid. Now we consider the situation of particles embedded in a viscous fluid and the consequences that arise in such situations. First, we notice that, even if the particle is rigidly fixed, there is a relative movement between the particle and the surrounding fluid since the fluid oscillates back and forth with the acoustic wave action. This gives the same effect as if the particle is oscillating in the fluid, giving rise to directional scattering, particularly to dipole directional scattering. Furthermore, the relative movement of the fluid with respect to the particle will be impaired and some of the incident acoustic energy will be converted to heat. Consequently, a propagating acoustic wave will experience an additional attenuation. Measurement of this additional acoustic attenuation may be used to determine the concentration of particles in the fluid.

In the preceding section, we found that scattering from a particle that is small compared to the signal's wavelength can be written in the form

$$\Phi_s = \Phi_{inc}(\kappa_0 a)^3 \left(\gamma_m - \gamma_d \cos\theta\right)\frac{1}{\kappa_0 r}\exp\left[i(\kappa_0 r - \omega t)\right] \tag{13.51}$$

Here, we have used the asymptotic result from equation (13.46) and removed the normalization $(a/2)$. We recall that in equation (13.51), $\gamma_m$ is the monopole contribution with omnidirectional scattering, which is determined by the elasticity contrast (or difference in elasticity) between the sphere and the surrounding medium. From equation (13.46), we know that the term is defined as

$$\gamma_m = \frac{K_0 - K_1}{3K_1} \tag{13.52}$$

The second term is the dipole contribution, which has a cosine-shaped directivity; this term is given by the density contrast. For a sphere in a nonviscous fluid, we have found that the dipole contribution is given by

$$\gamma_d = \frac{\rho_0 - \rho_1}{\rho_0 + 2\rho_1} \tag{13.53}$$

Calculation of scattering where viscosity is taken into account is rather complicated and there are different approximate solutions, such as those found in Pierce (1981) and in Harker et al. (1991). However, we use a result from Urick and Ament (1949) that defines the dipole term $\lambda_d$ as

$$\lambda_d = -\frac{1}{3}\frac{qG}{q+G} \tag{13.54}$$

where $q$ is the relative difference in density and is defined as

$$q = \frac{\rho_1 - \rho_0}{\rho_0} \tag{13.55}$$

$G$ is a complex function of the frequency with a real and an imaginary part and can be expanded in a power series. Urick and Ament (1949) gave the first two terms of the expansion as

$$G = \frac{3}{2}\left(1 + \frac{3}{2\beta a}\right) + i\frac{9}{4\beta a}\left(1 + \frac{1}{\beta a}\right) \tag{13.56}$$

In this expression, $\beta$ is given by the frequency $\omega$, the viscosity $\eta$, and the density $\rho_0$ of the fluid

$$\beta = \sqrt{\frac{\omega\rho_0}{2\eta}} \tag{13.57}$$

The inverse value, $\beta^{-1}$, is the viscous skin depth.

We consider two extreme situations

**Case A.** If the viscosity is small or the frequency is high, we see from equation (13.56) that $G \to 3/2$ and that, as found earlier, the dipole contribution will be

$$\gamma_d = \frac{\rho_0 - \rho_1}{\rho_0 + 2\rho_1} \tag{13.58}$$

**Case B.** When the viscosity is very large or the frequency is low, $G \to \infty$, and the dipole term becomes

$$\gamma_d = \frac{\rho_0 - \rho_1}{3\rho_0} \tag{13.59}$$

Equations (13.58) and (13.59) represent the extreme values of the dipole term for very high frequencies and very low frequencies, respectively. For intermediate values of viscosity and frequency, the dipole contribution lies between these boundaries, as expressed by $G$ in equation (13.56).

## 13.4 Effective medium theory

Now we use the results developed to describe the scattering of particles in a viscous fluid to calculate sound speed and absorption in suspensions made up of solid particles in water. Such calculations are relevant to sound propagating in turbid water, and in some industrial applications, as well (Harker et al. 1991).

We apply a theory, referred to as the ***effective medium theory***, that is often used to describe acoustic or seismic waves in composite media. *Effective medium theory* is an omnibus term covering a number of theories, all based on scattering theory, that in different ways attempt to calculate acoustic material parameters for composite materials. Many such theories have been developed for different applications (Kuster and Toksös 1974a and 1974b, Berryman 1980a and 1980b). We describe the main principles of effective medium theory and then apply the theory to the calculation of sound speed and attenuation in suspensions.

To begin, we consider a suspension that is a viscous medium, having in positions $\mathbf{r}_i$ a number of spherical particles with the radius $a_i$. Let us imagine an experiment where we measure the scattering from a large sphere in the suspension, as shown in figure 13.6.

The total field in position $\mathbf{r}$ is the incident field $\Phi_{inc}(\mathbf{r})$ plus the sum of the scattered fields from all particles

$$\Phi(\mathbf{r}) = \Phi_{inc}(\mathbf{r}) + \sum_i \Phi_{s,i}(\mathbf{r}, \mathbf{r}_i) \tag{13.60}$$

In equation (13.60) $\Phi_{s,i}(\mathbf{r}, \mathbf{r}_i)$ is the scattered field in position $\mathbf{r}$ from the particle at position $\mathbf{r}_i$. This field depends on the concentration and the properties of the particles, and most important, the unknown total field $\Phi(\mathbf{r}_i)$. Equation (13.60) is thus an implicit equation where the unknown field appears on both sides of the equality sign.

To calculate the scattered field from a cloud of particles, we must therefore take into consideration that the phase and amplitude of the incident field on each particle are modified as the wave propagates through the suspension. The incident field on a given particle will also include contributions from the scattered fields of the other particles.

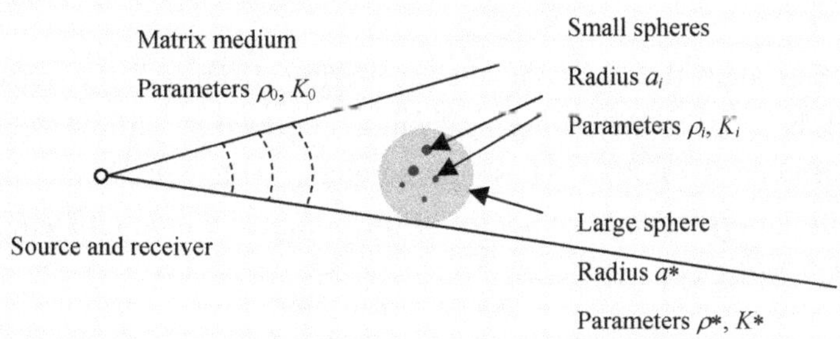


Figure 13.6. The scattered field from the collection of particles ($i$) is postulated to be equal to the scattered field from the large particle with radius $a^*$ and parameters $\rho^*$ and $K^*$.

Consequently, it is almost impossible to calculate accurately the scattering from a cluster of particles if we want to take all the interactions into consideration. Therefore, we must introduce a number of approximations in order to simplify the calculations. We assume that

(1) The incident field is the same for all particles and is equal to the field that would have impinged on one separate particle; and

(2) The distances $r_i$ and angles $\theta_i$ of a particle in relation to the point of observation are the same for all particles.

These assumptions mean that the incident field on every particle is the same and equal to $\Phi_{inc}$. The consequences are that multiple scattering is ignored and that the validity of the result is limited to cases with small particle concentrations and low frequencies.

Under these conditions, the scattered field at the point of observation, $\Phi_s(r)$, is the sum of the scattering from all the particles and according to equation (13.51) includes only the first two terms of

$$\Phi_s(r) = \Phi_{inc} \sum_i (\kappa_0 a_i)^3 \left[ \gamma_{m,i} - \gamma_{d,i} \cos\theta \right] \frac{1}{\kappa_0 r} \exp\left[ i\left( \kappa_0 r - \omega t \right) \right] \qquad (13.61)$$

We now postulate that the sum of the scattering from all the particles, per equation (13.61), is equal to the scattering from a single large particle with radius $a^*$ and material parameters $\rho^*$, $K^*$. These parameters, which at this point are not specified, we call the ***effective parameters*** of the composite medium. The scattering from a single such particle is

$$\Phi_s(r) = \Phi_{inc} (\kappa_0 a^*)^3 \left[ \gamma_m^* - \gamma_d^* \cos\theta \right] \frac{1}{\kappa_0 r} \exp\left[ i(\kappa_0 r - \omega t) \right] \qquad (13.62)$$

In accordance with the effective medium theory, we postulate that the sum of scattered fields from the collection of particles is equal to the scattering from a single particle having the effective material parameters that have been marked with an asterisk (*). Therefore, it follows that the results found with equations (13.61) and (13.62) must also be identical. Since the scattering is assumed to be the same in all directions, the independent contributions from the monopole and dipole must be the same as well. Therefore, we get the two following equations to determine the effective bulk modulus $K^*$ for the monopole and dipole contributions, respectively

$$\gamma_m^* = \sum_i n_i \gamma_{m,i} \qquad (13.63)$$

$$\gamma_d^* = \sum_i n_i \gamma_{d,i} \qquad (13.64)$$

From equation (13.63) for the monopole terms $\gamma_m^*$, we find that effective bulk modulus $K^*$ is

$$\frac{K_0 - K^*}{3K^*} = \sum_i n_i \frac{K_0 - K_i}{3K_i} \tag{13.65}$$

which has the solution

$$\frac{1}{K^*} = \sum_i \frac{n_i}{K_i} + \frac{1}{K_0}\left(1 - \sum_i n_i\right) \tag{13.66}$$

where the effective compressibility $1/K^*$ is the volume average of the compressibility of the constituents, $1/K_i$ .

For the dipole terms, we have several possibilities since the expression used for the dipole term depends on the magnitude of the viscosity. We have chosen to use equation (13.59) for the dipole term of the large sphere and equation (13.54) for the dipole terms for the small spheres. These choices then produce the following equation for the determination of the effective density $\rho^*$

$$\frac{\rho_0 - \rho^*}{3\rho_0} = -\sum_i n_i \frac{q_i G_i}{3(q_i + G_i)} \tag{13.67}$$

where, according to equations (13.55), (13.56), and (13.57)

$$G_i = \frac{3}{2}\left(1 + \frac{3}{2\beta_i a_i}\right) + i\frac{9}{4\beta_i a_i}\left(1 + \frac{1}{\beta_i a_i}\right) \tag{13.68}$$

$$q_i = \frac{\rho_i - \rho_0}{\rho_0} \tag{13.69}$$

$$\beta_i = \sqrt{\frac{\omega\rho_i}{2\eta_i}} \tag{13.70}$$

Now we find that the solution of equation (13.67) is

$$\rho^* = \rho_0 + \rho_0 \sum_i n_i \frac{q_i G_i}{(q_i + G_i)} \tag{13.71}$$

Equation (13.71) provides the frequency-dependent effective density of the composite material. For very low frequencies, $G_i \to \infty$, and the effective density becomes

$$\rho^* = \rho_0 \left(1 - \sum_i n_i\right) + \sum_i n_i \rho_i \tag{13.72}$$

This equation shows that the effective density for low frequencies is equal to the volume average of the densities of the constituents.

The effective bulk modulus of equation (13.66) and the effective density of equation (13.72) produce the effective wave number $\kappa$ of the suspension, which is defined as

$$\kappa^* = \omega \sqrt{\frac{\rho^*}{K^*}} \tag{13.73}$$

The real part of $\kappa^*$ determines the wave speed $c^*$ and the imaginary part of $\kappa^*$ determines the attenuation $\alpha^*$, so that

$$\kappa^* = \frac{\omega}{c^*} + i\alpha^* \tag{13.74}$$

and therefore

$$c^* = \frac{\omega}{\mathrm{Re}\{\kappa^*\}}$$
$$\alpha^* = \mathrm{Im}\{\kappa^*\} \tag{13.75}$$

This result can be considered to be an extension of Wood's equation (see Wood and Weston [1964]), as it gives the sound speed in suspensions with several types of particles; it also takes into account the sound absorption caused by viscosity.

### 13.4.1 Sound speed and attenuation in suspensions

We now apply the effective medium theory to suspensions, where we have particles of type 1 imbedded in a fluid of type 0. Equation (13.66) produces

$$\frac{1}{K^*} = \frac{n_1}{K_1} + \frac{1-n_1}{K_0} \tag{13.76}$$

where $n_1$ is the relative volume of solid particles and $1 - n_1$ is the relative volume of fluid. From equation (13.71), we obtain for the effective density

$$\rho^* = \rho_0 \left( 1 + n_1 \frac{q_1 G_1}{q_1 + G_1} \right) \tag{13.77}$$

At low frequencies where $G_1 \to \infty$, the effective density becomes equal to the density's volume average, that is

$$\rho^* = \rho_0 (1 - n_1) + n_1 \rho_1 \tag{13.78}$$

As an example of this process, we consider a case in which there are mineral particles suspended in water; the volume concentration of particles is $n_1 = 0.1$. The other parameters are given in table 13.1. By applying equations (13.74) and (13.75), we can calculate the sound speed and attenuation for the suspension as functions of the frequency and particle concentration. In figures 13.7 and 13.8, the sound speed and attenuation (in dB/m) are presented as functions of the frequency for the values given in table 13.1.

Table 13.1. Parameters of a suspension of mineral particles in water

| *Parameter* | *Value* | *Parameter* | *Value* |
|---|---|---|---|
| Radius $a$ | $1.0 \cdot 10^{-5}$ m | Density $\rho_0$ | $\rho_0 = 1000$ kg/m$^3$ |
| Viscosity $\eta$ | $1.0 \cdot 10^{-3}$ Pa·s | Sound speed $c_1$ | $c_1 = 4000$ m/s |
| Sound speed $c_0$ | 1500 m/s | Density $\rho_1$ | $\rho_1 = 2650$ kg/m$^3$ |

We see that the sound speed increases as the frequency increases, which means that the composite medium is dispersive. At low frequencies, when $\omega \to 0$, the effective density and volume stiffness are as given by equations (13.78) and (13.76) and consequently the sound speed of the mixture becomes

$$c^* = \sqrt{\frac{K^*}{\rho^*}} = \frac{1}{\sqrt{\left[n_1 K_1^{-1} + (1-n_1)K_0^{-1}\right]\left[n_1\rho_1 + (1-n_1)\rho_0\right]}} \tag{13.79}$$

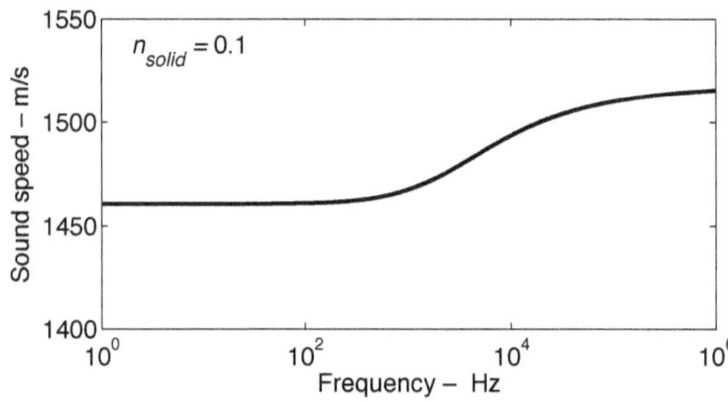


Figure 13.7. Results from calculating sound speed as a function of frequency in a suspension with solid particles with a volume concentration of $n = 0.1$.

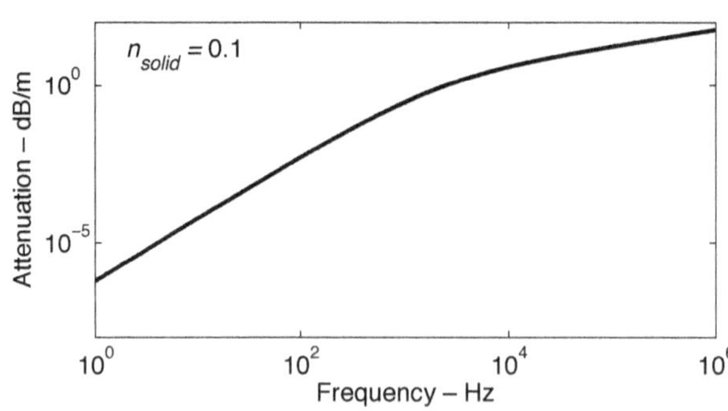


Figure 13.8. Results from calculating attenuation as a function of frequency in a suspension with solid particles with a volume concentration of $n = 0.1$.

Equation (13.79) is Wood's equation for sound speed in a two-component medium, which we have developed already in section 3.4, but using an entirely different approach.

At higher frequencies the sound speed increases. Because the effective high-frequency dynamic mass is smaller than the low-frequency static mass, part of the mass is unable to follow the rapid movements associated with the high-frequency acoustic wave. However, this dispersion is quite small and, in most cases, insignificant.

At low frequencies, attenuation increases proportionally as the square frequency increases, and at high frequencies, as the square root of the frequency increases. The transition takes place at the point where the product is $\beta a \approx 1$. In other words, the transition takes place when the viscous skin depth is about equal to the particle radius.

In figure 13.9, we see attenuation as a function of particle concentration for a 1000 kHz frequency. Note that the plot reflects only the additional attenuation caused by the suspended particles, not the ordinary attenuation of pure salt water, which according to equation (3.24) and figure 3.5 is 0.4 dB/m at 1 MHz.

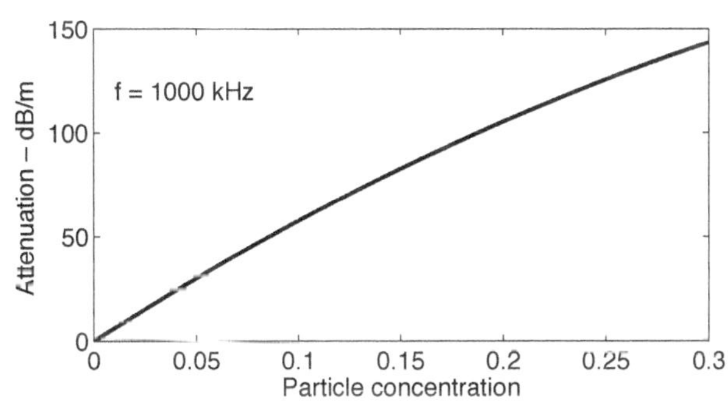


Figure 13.9. Attenuation of sound in a suspension, plotted as a function of particle concentration for a 1000 kHz frequency.

## 13.5 Scattering from gas bubbles

Scattering from gas bubbles in fluids is an important factor in many acoustical applications; a large body of literature exists on the subject. *The Acoustic Bubble* by Leighton (1996) summarizes most of the literature on the topic. In this section, we just present a simplified model for scattering from small gas-filled bubbles in a liquid. The frequency of oscillations of a pulsating spherical bubble was first calculated by Minnaert (1933). A central reference for sound propagation in bubbly liquids is the paper by Commander and Prosperetti (1989) containing both theoretical derivations and comparison of model predictions with experimental data.

We assume that the surrounding fluid medium has the density $\rho_0$ and sound speed $c_0$, and that the bubble has an undisturbed radius $a$, which is considerably smaller than the wavelength of the incident acoustic wave. Now we can set up a differential equation having the form "force equals mass times acceleration"

$$m\ddot{\xi} + R_m\dot{\xi} + s\xi = 4\pi a^2 p_i(t) \tag{13.80}$$

In this equation $\xi$ is the small perturbation of the sphere's radius $a$ caused by the sound pressure of the plane incident wave external pressure $p_i(t)$.

For a small bubble such as the one in this example, the sound pressure is constant over the entire surface. Therefore, the right-hand side of the expression is the external force working to compress the bubble. This external force is balanced by the bubble's elastic and dynamic forces represented by the dynamic mass, $m$, and the effective stiffness, $s$, respectively, which we will now discuss.

### *The stiffness term*

The gas-filled bubble has the volume $V$ and external pressure $p$. When the bubble is subjected to an external pressure increase $dp$, its volume will change according to the gas law. For the adiabatic case $pV^\gamma$ is constant. By differentiation, we find that

$$\frac{dp}{dV} = -\gamma\frac{p}{V} \tag{13.81}$$

Here, $dV$ is the change in volume $V$ and $\gamma$ is the ratio of the specific heats.

For a spherical bubble with radius $a$, $V = 4\pi a^3/3$ and $dV = 4\pi a^2 \xi$, where $\xi$ represents the radial velocity. The pressure increase is equal to the sound pressure, which means that $dp = p_i$ and $p = p_a$, where $p_a$ is the ambient static pressure. Hence the relationship between the sound pressure $p_i$ and the change in the radial extension of the gas bubble is

$$p_i = -\gamma \frac{3p_a}{a} \xi \tag{13.82}$$

The total force on the surface $F_r$ is

$$F_r = -4\pi a^2 p_i = (12\pi\gamma p_a a)\xi \tag{13.83}$$

which means that the equivalent stiffness, or elasticity modulus, $s$, is caused by the internal gas pressure and is defined as

$$s = 12\pi\gamma p_a a \tag{13.84}$$

***The mass term***

A spherical source pulsating with the radial displacement $\xi$ gives a sound pressure $p_s$ at distance $r$ as

$$p_s = -\omega^2 \rho_0 \frac{a^2}{r} \xi \exp[i(\kappa r - \omega t)] \tag{13.85}$$

The total force at the surface where $r = a$ is

$$F_r = -\omega^2 4\pi a^3 \rho_0 \xi \tag{13.86}$$

which means that the dynamic mass $m$ is

$$m = 4\pi a^3 \rho_0 \tag{13.87}$$

Note that $\rho_0$ is the density of the ambient fluid, not the density of the gas inside the bubble. In equation (13.80), the losses, which are assumed to be proportional to the speed $\dot{\xi}$, are given by an unspecified movement resistance $R_m$.

We make a Fourier transform to the frequency domain and get

$$\left(-m\omega^2 - i\omega R_m + s\right)\xi(\omega) = 4\pi a^2 p_i(\omega) \tag{13.88}$$

For an incident harmonic sound pressure with frequency $\omega$, $\dot{\xi}(\omega)$, the radial pulsation amplitude, is given by the frequency function

$$\xi(\omega) = \frac{4\pi a^2 p_i(\omega)}{\left(-m\omega^2 - i\omega R_m + s\right)} \tag{13.89}$$

We insert equation (13.87) for mass $m$ into equation (13.89), and obtain the radial displacement of the bubble walls

$$\xi(\omega) = \frac{p_i(\omega)\dfrac{1}{\rho_0 a\omega^2}}{\left(\dfrac{\omega_r}{\omega}\right)^2 - 1 - i\delta} \tag{13.90}$$

which represents an attenuated resonance. By comparing equations (13.89) and (13.90), we find that the resonance frequency $\omega_r$ is given by the ratio between the elastic modulus $s$ and the mass $m$, so that

$$\omega_r = \sqrt{\frac{s}{m}} \tag{13.91}$$

Using equations (13.84) and (13.87) for $m$ and $s$, we find that

$$\omega_r = \frac{1}{a}\sqrt{\frac{3\gamma p_a}{\rho_0}} \tag{13.92}$$

This resonance or natural frequency of a spherical gas bubble in a liquid was first developed by Minnaert (1993) and is sometimes called the Minnaert frequency.

For instance, under atmospheric pressure an air bubble will have the parameters $\gamma = 1.4$, $\rho_0 = 1000$ kg/m$^3$, and $p_a = 10^5$ Pa. The resonance frequency of an air bubble with the radius $a$ becomes

$$f_r = \frac{3.2}{a} \tag{13.93}$$

At a depth *d*, an air bubble in water with the same radius will have the resonance frequency

$$f_r = \frac{3.2}{a}\sqrt{1 + \frac{d}{10}} \tag{13.94}$$

since the atmospheric pressure equals the pressure of a 10 m water column. Note that in equations (13.93) and (13.94), frequency is measured in Hertz when the radius $a$ and depth $d$ are measured in meters.

In this derivation, we have not taken into account surface tension, a possible shell around the bubble, or deviations from an adiabatic condition, all of which are possible factors for small bubbles in situations where the thermal conductance will lead to an approximately isothermal condition.

The attenuation coefficient $\delta$ in equation (13.90) is expressed as

$$\delta = \frac{R_m}{m\omega} \tag{13.95}$$

given by the movement resistance $R_m$, the mass $m$, and the frequency $\omega$. We must assume that there are at least three reasons for loss: (1) acoustic radiation, (2) viscosity and friction, and (3) net heat conduction that takes place when the temperature of the bubble varies with contraction and expansion, that is, thermal loss.

Accordingly, the attenuation is described by three terms representing these three effects, all of which are dependent on the frequency

$$\delta(\omega) = \delta_r + \delta_v + \delta_t \tag{13.96}$$

where $\delta_r$ represents loss from acoustic radiation, $\delta_v$ represents loss caused by viscosity and friction, and $\delta_t$ represents thermal loss. See the articles by Devin (1959) and by Commander and Prosperetti (1989) for more information about the causes of attenuation.

Now we turn our attention to the acoustic scattering cross section for this case. From equation (13.85), we know that at distance $r$, a pulsating bubble with radius $a$ and radial displacement $\xi(\omega)$ will produce a sound pressure with the amplitude $p_s(\omega)$, which can be defined as

$$p_s(\omega) = -i\omega\rho_0 \frac{a^2}{r}\dot{\xi} = -\omega^2 \rho_0 \frac{a^2}{r}\xi \tag{13.97}$$

Using equation (13.90) in equation (13.97), the scattered sound pressure $p_s(\omega)$ at the bubble surface at $r = a$ becomes

$$p_s(\omega) = -\frac{p_i(\omega)}{\left(\dfrac{\omega_r}{\omega}\right)^2 - 1 - i\delta} \tag{13.98}$$

Thus, the total scattering cross section $\sigma_s$ of a gas bubble is

$$\sigma_s = 4\pi a^2 \left|\frac{p_s(\omega)}{p_i(\omega)}\right|^2 = \frac{4\pi a^2}{\left[\left(\dfrac{\omega_r}{\omega}\right)^2 - 1\right]^2 + \delta^2} \tag{13.99}$$

The acoustic scattering from the bubble is omnidirectional, or equally strong in all directions. The scattering back to the position of the source and transmitter which is represented by the backscattering cross section $\sigma_{bs}$ is

$$\sigma_{bs} = \frac{\sigma_s}{4\pi} = \frac{a^2}{\left[\left(\dfrac{\omega_r}{\omega}\right)^2 - 1\right]^2 + \delta^2} \tag{13.100}$$

Figure 13.10 shows an example of a normalized scattering cross section for which $\sigma_s / \pi a^2$ is a function of $\kappa a$, which is the product of the wave number and the bubble radius, that can be considered a normalized frequency. (This figure also appears in Chapter 11.5). At resonance, the size of the scattering cross section, mainly determined by the attenuation term $\delta$, is considerably larger than the extent of the geometric cross section.

The total scattered and absorbed power $P_e$ is found by integrating the real part of the complex product of the radial speed and the sound pressure over the entire sphere surface, so that $P_e$ is expressed as

$$P_e = 4\pi a^2 \operatorname{Re}\left\{ p_i(\omega)\dot{\xi}^*(\omega) \right\} \tag{13.101}$$

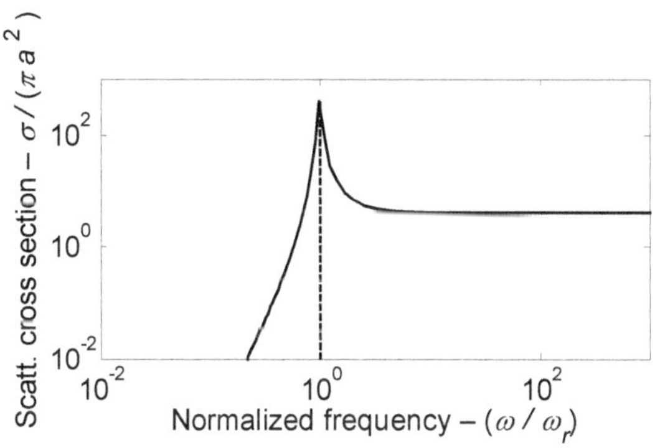


Figure 13.10. Normalized scattering cross section $\sigma_{bs} / \pi a^2$ for an air bubble in water; the damping constant is $\delta = 0.1$.

From equation (13.90), we find that the radial velocity of the bubble oscillations is

$$\dot{\xi}(\omega) = -i\omega\xi(\omega) = -\frac{p_i(\omega)\dfrac{i\omega}{\rho_0 a\omega^2}}{\left(\dfrac{\omega_r}{\omega}\right)^2 - 1 - i\delta} \tag{13.102}$$

Inserting equation (13.102) into equation (13.101) yields the total scattered and absorbed power as

$$P_e = \frac{4\pi a^2 p_i^2(\omega)\dfrac{\delta}{\rho_0 a\omega}}{\left[\left(\dfrac{\omega_r}{\omega}\right)^2 - 1\right]^2 + \delta^2} \tag{13.103}$$

The ***extinction cross section*** $\sigma_e$ is defined as the total scattered and absorbed power, divided by the incident intensity $I_i$ which is

$$I_i = \frac{p_i^2}{\rho_0 c} \tag{13.104}$$

The extinction cross section therefore becomes

$$\sigma_e = \frac{4\pi a^2 \dfrac{\delta}{ka}}{\left[\left(\dfrac{\omega_r}{\omega}\right)^2 - 1\right]^2 + \delta^2} \tag{13.105}$$

### *13.5.1 Sound speed for a fluid with gas bubbles*

Gas bubbles in a fluid alter the effective volume stiffness and reduce the density; consequently, the sound speed changes. To evaluate how free gas bubbles in a fluid affect sound speed and signal attenuation, we consider a water volume in which there is a volume fraction of gas $n_g$ and a volume fraction of pure water $n_0$ where $n_g + n_0 = 1$. The effective density of the mixture is

$$\rho^* = n_g \rho_g + n_0 \rho_0 \tag{13.106}$$

When this mixture is subjected to pressure $p$, a spherical bubble undergoes a change in volume $dV_g$ which can be expressed as

$$dV_g = 4\pi a^2 \xi(\omega) \tag{13.107}$$

here, $\xi(\omega)$ is given by equation (13.90).

Since the volume of the sphere is

$$V_g = \frac{4}{3}\pi a^3 \tag{13.108}$$

the relative volume change of a single bubble is

$$\frac{dV_g}{V_g} = 3\frac{\xi(\omega)}{a} = \frac{3p_i(\omega)}{\rho_0 a^2 \omega^2} \frac{1}{\left(\frac{\omega_r}{\omega}\right)^2 - 1 - i\delta} \tag{13.109}$$

The effective compressibility of the mixture, $(K^*)^{-1}$, becomes

$$(K^*)^{-1} = \frac{3n_g}{\rho_0 a^2 \omega^2} \frac{1}{\left[\left(\frac{\omega_r}{\omega}\right)^2 - 1\right] - i\delta} + (1 - n_g) K_0^{-1} \tag{13.110}$$

In this equation, $n_g$ is the gas volume fraction and $(1 - n_g)$ is the fluid volume fraction of the gas-fluid mixture. The volume stiffness of the fluid is $K_0 = \rho_0 c_0^2$. The mixture's effective wave number $\kappa^*$ is defined as

$$\kappa^* = \omega\sqrt{\frac{\rho^*}{K^*}} \tag{13.111}$$

This wave number is complex; its real part gives the sound speed, while the imaginary part gives the attenuation, which means that

$$\kappa^* = \frac{\omega}{c^*} + i\alpha^* \tag{13.112}$$

Therefore, the sound speed and attenuation are expressed as

$$c^* = \frac{\omega}{\operatorname{Re}\{\kappa^*\}} \qquad \alpha^* = \operatorname{Im}\{\kappa^*\} \tag{13.113}$$

In figure 13.11 we have two examples of sound speed and attenuation for water containing air bubbles; the examples have volume concentrations of 0.10% and 0.01%, respectively. In both cases, the bubbles are equally large, with a radius $a = 10^{-6}$ m, and, the bubble attenuation chosen to be $\delta = 0.01$. The resonance frequency is approximately 3.2 MHz, according to equations (13.92) and (13.93).

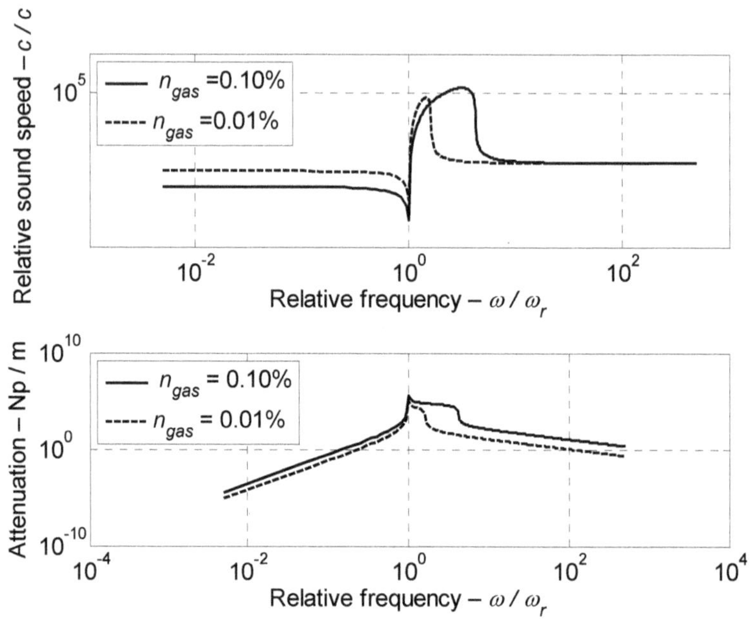


Figure 13.11. Sound speed (top) and attenuation (bottom) in water with free air bubbles, with volume concentrations of 0.10% and 0.01%.

We see from these examples that sound speed varies considerably for frequencies close to resonance. Below resonance, the sound speed in the gas mixture is lower than the sound speed in water, but above resonance, it is higher. Figure 13.11 also shows that the attenuation in bubbly water is very high, especially for frequencies in the vicinity of the resonance frequency.

Next, we find the limiting values for low and high frequencies. For very low frequencies, the effective compressibility approaches a limiting value that can be calculated from equation (13.110) by first inserting $\omega = 0$, and thereafter inserting $\omega_r$ from equation (13.92).

The result for low frequencies is

$$(K^*)^{-1} \approx \frac{n_g}{\gamma p_a} + \left(1 - n_g\right) K_0^{-1} \tag{13.114}$$

The volume stiffness of the gas is $K_g = \gamma p_a$, so that at low frequencies, the effective compressibility equals the volume average of the compressibilities of the water and air

$$(K^*)^{-1} \approx n_g K_g^1 + \left(1 - n_g\right) K_0^{-1} \tag{13.115}$$

The low frequency limit of the sounds speed is therefore

$$c^* = \sqrt{\frac{K^*}{\rho^*}} = \frac{1}{\sqrt{\left[n_g K_g^{-1} + (1-n_g)K_0^{-1}\right]\left[n_g \rho_g + (1-n_g)\rho_0\right]}} \tag{13.116}$$

which is Wood's equation for the sound speed in fluids with gas bubbles (see also equation (13.79). Since $K_g << K_0$ and $\rho_g << \rho_0$, equation (13.116) shows that for low-frequency signals the sound speed in fluids with gas bubbles can be considerably lower than the sound speed in fluid without bubbles.

For very high frequencies, we see from equation (13.110) that

$$(K^*)^{-1} \approx \left(1 - n_g\right) K_0^{-1} \tag{13.117}$$

For realistic gas volume values, this speed is very close to the sound speed in water devoid of gases. Thus, the sound speed in water containing gas bubbles will be strongly dependent on the signal frequency, particularly around the resonance frequency. When bubbles with different sizes are present, the resonance frequency will be distributed across a more or less broad frequency band. Our analysis shows that when we measure sound speed acoustically in water and other fluids, we must use frequencies far above the highest resonance frequency to prevent the measurement results from being affected by the presence of free gas.

# Chapter 14

# Reflection and scattering from surfaces

## 14.1 Introduction

This chapter expands on theory pertinent to reflection and scattering from surfaces, whether of finite or infinite size, and that may be perfectly even and smooth or corrugated and rough. We relate this information to reflection and scattering from the sea's surface and bottom, which we must understand in order to predict the characteristics of reverberation in underwater acoustics.

We start by deriving the Helmholtz-Kirchhoff integral, and then use it to determine the reflected signal from a plane surface of finite size. We show that the reflected signal from such a finite surface generally has two components: a specular component plus a diffracted component generated at the plane's edge. We illustrate this by calculating the reflection and diffraction from a circular disk and from a planar strip.

For the rough-surface scattering problem, there are several approaches. We consider the Helmholtz-Kirchhoff method, which is a physical optics approach, and the Rayleigh-Rice solution, which is a perturbation technique. The Helmholtz-Kirchhoff method is applicable to a large-scale roughness where the rough surface has a radius of curvature much greater than the length of the acoustic wavelength. The Rayleigh-Rice perturbation approach applies to surfaces that are slightly rough, with height deviations that are small compared to the extent of the acoustic wavelength.

For slightly rough surfaces, the two approaches yield similar results, except that the perturbation approach treats the effect of the surface reflection coefficient differently, and more appropriately, than does the optical approach. Since the two methods have different regions of validity in terms of surface roughness, use of a composite model that combines the two approaches may be advantageous; we briefly discuss this model at the end of the chapter.

## 14.2 Derivation of the Helmholtz-Kirchhoff integral

We start with the Gauss theorem, which states that the flux from a vector field can be expressed by either a surface integral or a volume integral so that

$$\int_S \mathbf{F}\cdot d\mathbf{S} = \int_V \nabla\cdot\mathbf{F}dv \tag{14.1}$$

If $\Phi_1$ and $\Phi_2$ are two arbitrary scalar fields, equation (14.1) can be rewritten as

$$\int_S \left(\Phi_1\nabla\Phi_2 - \Phi_2\nabla\Phi_1\right)\cdot d\mathbf{S} = \int_V \left(\Phi_1\nabla^2\Phi_2 - \Phi_2\nabla^2\Phi_1\right)dV \tag{14.2}$$

We now consider a medium with sound speed $c$ and assume that both fields are harmonic waves with the same frequency $\omega$ and satisfy the same wave equation, which is expressed as

$$\nabla^2\Phi_j = -\frac{\omega^2}{c^2}\Phi_j \qquad j=1,2 \tag{14.3}$$

By inserting equation (14.3) into equation (14.2), we find that the right-hand side equals zero when the volume $V$ does not contain singularities or sources. Consequently, it follows that

$$\int_S \left(\Phi_1\nabla\Phi_2 - \Phi_2\nabla\Phi_1\right)\cdot d\mathbf{S} = 0 \tag{14.4}$$

Furthermore, the product $\nabla\Phi\cdot d\mathbf{S}$ can be replaced by the derivative of the field with respect to the normal surface by

$$\nabla\Phi_j\cdot d\mathbf{S} = \frac{\partial\Phi_j}{\partial n}dS \qquad j=1,2 \tag{14.5}$$

so that now equation (14.4) can be rewritten as

$$\int_S \left( \Phi_1 \frac{\partial \Phi_2}{\partial n} - \Phi_2 \frac{\partial \Phi_1}{\partial n} \right) dS = 0 \tag{14.6}$$

In integrating equation (14.6), we use the integration path shown in figure 14.1. Therefore, the contribution to the surface integral has two parts, one part from the semicircle $S$ and the other part from $S'$, which is a small circle with radius $a$, located around the field point at $\mathbf{r}$. Furthermore, we assume that the field on the periphery of the infinitely large semicircle is vanishingly small, and so does not contribute to the integral.

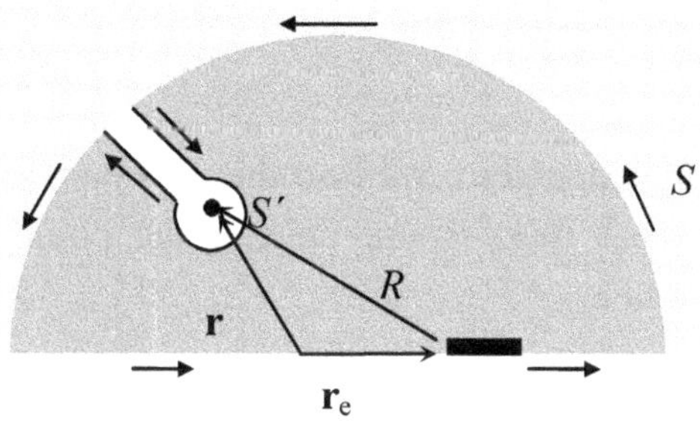


Figure 14.1. Derivation of the Helmholtz-Kirchhoff integral and the integration path for the integral used in equation (14.6).

The integral in equation (14.6) can then be written as

$$\int_S \left( \Phi_1 \frac{\partial \Phi_2}{\partial n} - \Phi_2 \frac{\partial \Phi_1}{\partial n} \right) dS + \int_{S'} \left( \Phi_1 \frac{\partial \Phi_2}{\partial n} - \Phi_2 \frac{\partial \Phi_1}{\partial n} \right) dS' = 0 \tag{14.7}$$

We let $\Phi_1$ be the field to be calculated and $\Phi_2$ be an auxiliary field that must satisfy the wave equation of equation (14.3), but which otherwise we can choose according to our preference. We choose the field $\Phi_2$ to be the direct field from the surface element located at point $\mathbf{r}_e$ in figure 14.1, and we can drop the subscript for $\Phi_1$, which is no longer needed. At the position $\mathbf{r}$, the field emitted from a point source at position $\mathbf{r}_e$ is given by the Green's function as

$$\Phi_2(\mathbf{r}) = \frac{\exp(i\kappa R)}{R} \tag{14.8}$$

where $\kappa = \omega / vc$ and $R = |\mathbf{r} - \mathbf{r}_e|$. Now equation (14.7) can be written as

$$\int_S \left\{ \Phi \frac{\partial}{\partial n}\left[\frac{\exp(i\kappa R)}{R}\right] - \frac{\exp(i\kappa R)}{R}\frac{\partial \Phi}{\partial n} \right\} dS = -\int_{S'} \left\{ \Phi \frac{\partial}{\partial n}\left[\frac{\exp(i\kappa a)}{a}\right] - \frac{\exp(i\kappa a)}{a}\frac{\partial \Phi}{\partial n} \right\} dS' \tag{14.9}$$

On the surface $S'$, $\partial / \partial n = \partial / \partial a$ and $dS' = a^2 d\Omega$, where $a$ is the radius of $S'$ and $\Omega$ is the solid angle, so that the right-hand side of equation (14.9) can be rewritten as

$$-\int_{S'} \left\{ \Phi \left[\frac{i\kappa a - 1}{a^2}\right] \exp(i\kappa a) - \frac{\exp(i\kappa a)}{a}\frac{\partial \Phi}{\partial n} \right\} a^2 d\Omega \tag{14.10}$$

Letting the radius $a$ approach zero, we find that the expression in equation (14.10) approaches $4\pi\Phi(\mathbf{r})$ where $\Phi(\mathbf{r})$ is the field at the point $\mathbf{r}$. Equation (14.9) can subsequently be rewritten as

$$\Phi(\mathbf{r}) = \frac{1}{4\pi}\int_S \left\{ \Phi(\mathbf{r}_e) \frac{\partial}{\partial n}\left[\frac{\exp(i\kappa R)}{R}\right] - \frac{\exp(i\kappa R)}{R}\frac{\partial \Phi(\mathbf{r}_e)}{\partial n} \right\} dS \tag{14.11}$$

This is the Helmholtz-Kirchhoff integral equation for harmonic fields, which shows how the field at a given point can be determined when we know the field at all points on a closed surface $S$. However, the problem with solving this integral equation is that the unknown field is on both sides of the equation as $\Phi(\mathbf{r}_e)$ and $\Phi(\mathbf{r})$.

In theory, we might find this integral equation's exact solution by making successive approximations where we first assume that a field $\Phi^{(0)}(\mathbf{r}_e)$ inside the integral of equation (14.11) has given the result of a field $\Phi^{(1)}(\mathbf{r})$ for all points in the space. After this field is inserted into the right-hand side of the equation, we obtain a new and improved solution. In principle, we can continue this process as long as we wish, ending it only when an accurate solution is found. The first approximation for the scattered amplitude, obtained by substituting the unperturbed wave field in the integral expression, gives us the *Born approximation,* which in acoustics is also called the *Kirchhoff approximation*.

We use this approximation to determine the reflection from a surface that does not necessarily have to be flat. We therefore assume that the scattered field from an element on this surface is equal to the incident field $\Phi_{inc}$ multiplied by a reflection coefficient $\Re$. Consequently, we obtain

$$\Phi = \Re\Phi_{inc} \tag{14.12}$$

$$\frac{\partial \Phi}{\partial n} = -\Re\frac{\partial \Phi_{inc}}{\partial n} \tag{14.13}$$

When equation (14.13) is inserted into equation (14.11), we find the field at point **r** to be

$$\Phi(\mathbf{r}) = \frac{1}{4\pi}\int_S \Re\frac{\partial}{\partial n}\left(\Phi_{inc}(\mathbf{r}_e)\frac{\exp(i\kappa R)}{R}\right)dS \tag{14.14}$$

Equation (14.14) is a mathematical formulation of Huygens' principle, which states that the field is a sum (or integral) of contributions from elementary point sources, each of which gives a contribution proportional to $\exp(i\kappa R) / R$, where $R$ is the distance from the surface to the reception point. In the next section, this equation is our starting point for determining the reflected acoustic field from flat and curved surfaces and, later in this chapter, for developing a mathematical expression of the scattered field from rough surfaces.

## 14.3 Reflection from a plane surface of finite size; edge diffraction

In this section we discuss reflections for a plane surface of finite size. We show that the signal reflected from a finite surface has, in general, two components: the specular reflection and the diffractions from the surface edges.

We start by applying equation (14.14) to determine the reflected wave from a flat surface of finite size but arbitrary shape. We simplify the discussion by limiting it to consider only a situation where the source and the receiver are at the same position $z$ over the $x$-$y$ plane, with an arbitrarily shaped surface in the plane (see figure 14.2). The distance between the source-receiver position and an element on the surface is $R$; the element's position is determined by the angle $\theta$, which is in the $x$-$y$ plane. The minimum and the maximum distances to the surface are $R_1$ and $R_2$, respectively, and these distances depend on the shape of the surface.

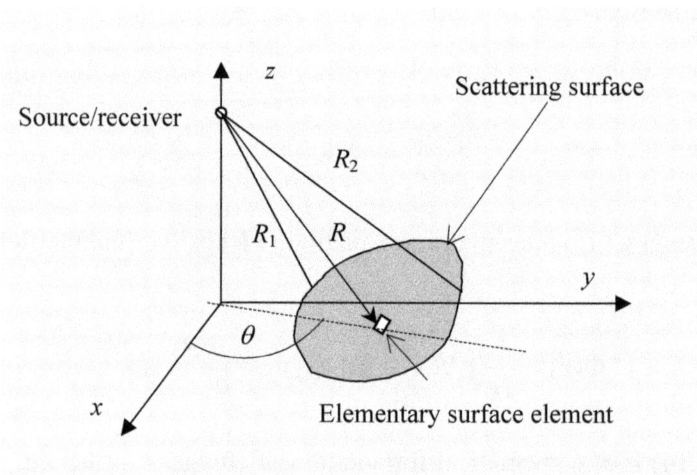


Figure 14.2. Geometry for the determination of reflection from a plane surface of finite size and irregular shape, where $R_1$ and $R_2$ are the minimum and maximum distances to the scattering surface from the source-receiver position, and $R$ is the distance to an elementary surface element.

The source, which has strength $Q$, produces an incident field $\Phi_{inc}$ at a point on the surface that lies at a distance $R$ from the source. This incident field can be expressed as

$$\Phi_{inc} = \frac{Q}{4\pi R}\exp(i\kappa R) \tag{14.15}$$

We assume that the field reflected from the element is equal to the incident field $\Phi_{inc}$ multiplied by the surface's plane wave reflection coefficient $\Re$. The derivative of the field with respect to the normal surface of the relevant element is expressed as

$$\frac{\partial}{\partial n} = \frac{\partial}{\partial z} = \frac{-z}{R}\frac{\partial}{\partial R} \tag{14.16}$$

When equations (14.15) and (14.16) are inserted into equation (14.14), we obtain an expression for the unknown field

$$\Phi(\mathbf{r}) = \frac{-zQ\Re}{(4\pi)^2}\int_S \frac{\partial}{R\partial R}\left(\frac{\exp(2i\kappa R)}{R^2}\right)RdRd\theta \tag{14.17}$$

By evaluating the differential in the integrand, we first get

$$\Phi(\mathbf{r}) = \frac{-zQ\Re}{(4\pi)^2}\int_S \frac{2i\kappa R^2 - 2R}{R^4}\exp(2i\kappa R)dRd\theta \tag{14.18}$$

The integration over $R$ in equation (14.18) is evaluated by partial integration, which produces

$$\int_S \left(\frac{i\kappa}{R^2} - \frac{1}{R^3}\right)\exp(2i\kappa R)dR = -\frac{\exp(2i\kappa R)}{2R^2} \tag{14.19}$$

so that

$$\Phi(\mathbf{r}) = \frac{Q\Re}{16\pi^2}z\int_{\theta_1}^{\theta_2}\left[\frac{\exp(2i\kappa R_1)}{R_1^2} - \frac{\exp(2i\kappa R_2)}{R_2^2}\right]d\theta \tag{14.20}$$

Equation (14.20) is a general expression for the backscattered or reflected field from a surface that is intersected by a wave transmitted from a point source in the same location as the receiver.

This expression is difficult to calculate in most instances, so we consider the particularly simple example of scattering from a circular plate, which has the radius $a$ (see figure 14.3).

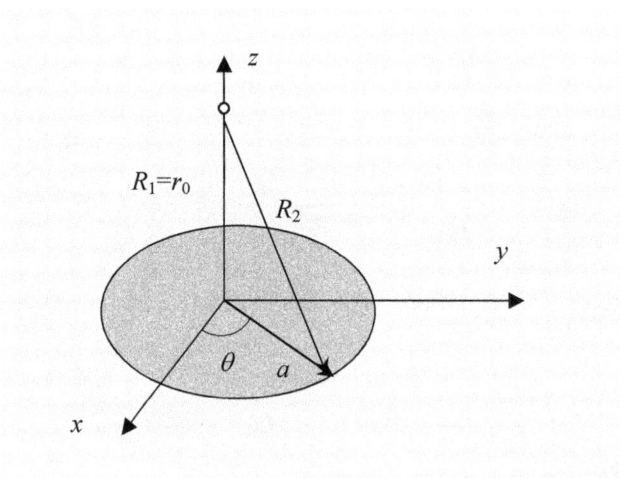


Figure 14.3. Reflection from a flat circular surface with both source and receiver located on the $z$-axis at a height $r_0$ above the center of the plate.

The source and the receiver are collocated on the $z$-axis at a height $r_0$ over the plate's center in such a way that $z = r_0$. In this case, the minimum and maximum distances are, respectively

$$\begin{aligned} R_1 &= r_0 \\ R_2 &= \sqrt{r_0^2 + a_0^2} \end{aligned} \tag{14.21}$$

Since now $R_1$ and $R_2$ are independent of the angle $\theta$, the integral in equation (14.20) equals $2\pi$.

For a surface that is totally reflective, that is, where $\mathfrak{R} = 1$, the reflected field becomes

$$\Phi(\mathbf{r}_0) = \frac{Q}{8\pi r_0}\left\{\exp\left(2i\kappa r_0\right) - \frac{r_0^2}{r_0^2 + a^2}\exp\left[2i\kappa\sqrt{r_0^2 + a^2}\right]\right\} \tag{14.22}$$

The first term in the curly brackets is the specular reflection, while the second term represents diffraction from the edges of the plate.

The scattering cross section is defined as the ratio of the reflected power to the incident intensity, that is, as

$$\sigma = 4\pi r_0^2 \left| \frac{\Phi(r_0)}{\Phi_{inc}(r_0)} \right|^2 \tag{14.23}$$

Consequently, the plate's scattering cross section $\sigma$ is defined as

$$\sigma = \pi r_0^2 \left| 1 - \frac{r_0^2}{r_0^2 + a^2} \exp\left( -2i\kappa r_0 (1 + \sqrt{1 + (\frac{a}{r_0})^2} \right) \right|^2 \tag{14.24}$$

Figure 14.4 contains an example of the normalized scattering cross section $\tilde{\sigma}$, defined as

$$\tilde{\sigma} = \frac{\sigma}{\pi r_0^2} = \left| 1 - \frac{r_0^2}{r_0^2 + a^2} \exp\left( -2i\kappa r_0 (1 + \sqrt{1 + (\frac{a}{r_0})^2} \right) \right|^2 \tag{14.25}$$

plotted as a function of the normalized frequency *ka* for the case where $a = 1$ m, $r_0 = 100$ m, and c = 1500 m/s. The value $ka = 100$ corresponds approximately to a 24 kHz frequency in this case. For small *ka* values, that is, for low frequencies and for long distances when $r_0 >> a$, the scattering cross section approaches

$$\sigma \approx \pi r_0^2 \left( \kappa \frac{a^2}{r_0} \right)^2 \tag{14.26}$$

For low frequencies, the scattering cross section increases proportionally with the square of the frequency, which can be seen from figure 14.4 where also the result of equation (14.26) is plotted as the dotted line.

For large *ka* values, $\sigma$ oscillates around a mean value of $\pi r_0^2$, with maximum and minimum values, respectively, of

$$\sigma_{\max} = \pi r_0^2 \left( \frac{2r_0^2 + a^2}{r_0^2 + a^2} \right)^2$$
$$\sigma_{\min} = \pi r_0^2 \left( \frac{a^2}{r_0^2 + a^2} \right)^2 \qquad (14.27)$$

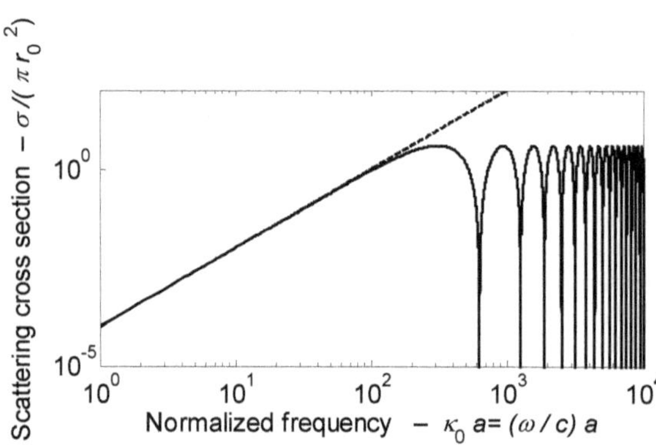


Figure 14.4. The normalized scattering cross section of a circular surface with radius $a$, plotted as a function of $ka$ and calculated using equations (14.25) and (14.26), where $a = 1$ m and $r_0 = 100$ m.

The difference between the minimum and maximum values indicates that the pressure of a signal reflected from the plate may be nearly twice the mean reflected pressure. For an infinitely large and totally reflecting surface, the scattering cross section and target strength *TS* are dependent on the distance from the source to the surface, and are given by

$$\sigma = \pi r_0^2$$
$$TS = 10 \log \left( \frac{r_0^2}{4} \right) \qquad (14.28)$$

We now return to equation (14.22), and we recall that the first term in brackets is the specular reflection, while the second term represents diffraction from the edges of the plate. For an infinitely large plate, the specular component dominates, so that the received reflected signal $\Phi(\mathbf{r}_0)$ is expressed as

$$\Phi(\mathbf{r}_0) = \frac{Q}{8\pi r_0} \exp\left(2i\kappa r_0\right) \tag{14.29}$$

Comparing this with the equation describing the field from a point source, equation (14.15), we understand that equation (14.29) gives us the signal from a point source of strength $Q$ at a distance of $2r_0$, so that the reflected signal from an infinitely large surface is identical to the signal that would be produced by the source's mirror image.

Equation (14.22) is the solution for the reflected signal in the frequency domain. In the time domain, this solution corresponds to two pulses that have the same shape as the source signal. The first is the specular component that arrives after some time, $t_{spec}$, which is equal to the travel time from the source location to the plate and back to the receiver

$$t_{spec} = \frac{2r_0}{c} \tag{14.30}$$

The second pulse comes after a time that equals the travel time to the edge of the plate and back; this is the diffracted component, and travel time, $t_{diff}$, is given by

$$t_{diff} = \frac{2\sqrt{\left(r_0^2 + a^2\right)}}{c} \tag{14.31}$$

Because the second pulse travels a longer distance to the edge of the plate, its amplitude is lower than the amplitude of the specular component, and from equation 14.22, the sign of the diffracted pulse is negative. For a small plate, the edge diffraction interferes with the specular component, so that the scattering cross section varies strongly with the frequency, as we saw in figure 14.4.

Equation (14.22) shows that the size of the scattering cross section increases proportionally with the square of the frequency up to a certain frequency; thereafter, the scattering cross section's size levels out and oscillates between the maximum and minimum values given in equation (14.27). This result is of practical interest when sonar is used to search for a platelike object in the sea.

Remember that the source is located above the plate's center in the case we are discussing. If the sonar's location is outside the plate's circumference, the specular component disappears and the echo is produced only by the edge diffraction. This is also valid in the more general case where the main contributions to an object's echo derive from the discontinuities of the geometric cross section (Freedman 1962a, 1962b).

### *14.3.1 Diffraction from a straight edge*

Next, we consider a plane strip with width $L_x$. This strip is parallel to the $y$-axis and its end edges in the $y$-direction are at an infinite distance. A source-receiver is located at height $z_0$ above the origin of the $x$-$y$ plane, as shown in figure 14.5.

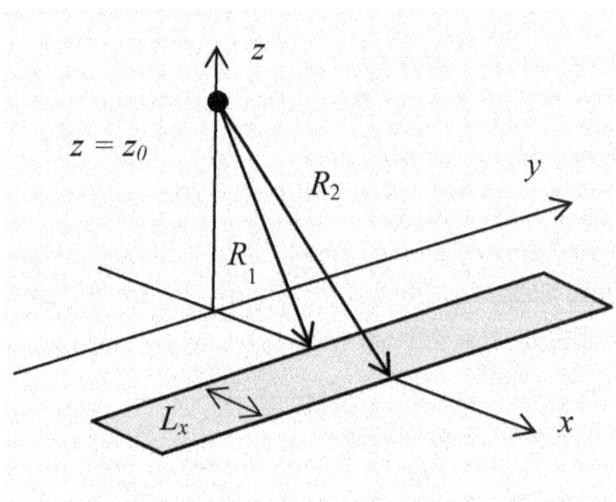


Figure 14.5. A plane strip of width $L_x$ in the $x$-direction; its extent is infinite in the $y$-direction. The collocated source-receiver is at a height $z_0$ above the $x$-$y$ plane. The distance from the source-receiver to the first edge of the strip is $R_1$; the distance to the second edge is $R_2$.

We start with equation (14.20), which is the general expression for the backscattered or reflected field for the case where the source and the receiver are in the same location. The expression of equation (14.20) has two terms of the same type.

In the following paragraphs, we consider a single term of that type, thereby determining the diffracted wave from an edge. We therefore start with the general boundary diffraction term of

$$\Phi(\mathbf{r}) = \frac{Q}{4\pi}\frac{z_0\Re}{4\pi}\int_{\theta_1}^{\theta_2}\left[\frac{\exp(2i\kappa R)}{R^2}\right]d\theta \tag{14.32}$$

The integral over $\theta$ can be converted to an integral over travel time by substituting $t = 2R/c$ and using the expression

$$d\theta = \left(\frac{d\theta}{dt}\right)dt \tag{14.33}$$

which produces

$$\Phi(\mathbf{r}) = \frac{Q}{4\pi}\frac{z_0\Re}{\pi c^2}\int_{t_1}^{t_2}\left[\frac{1}{t^2}\frac{d\theta}{dt}\right]\exp(i\omega t)\,dt \tag{14.34}$$

Because the integral of equation (14.34) has the form of an inverse Fourier transform, the time signal of the boundary diffraction term, $\phi(\mathbf{r}, t)$, can be expressed as

$$\phi(\mathbf{r},t) = \frac{Q}{4\pi}\frac{z_0\Re}{\pi c^2}w(t) \tag{14.35}$$

where the signal $w(t)$ is defined as

$$w(t) = \frac{1}{t^2}\frac{d\theta}{dt} \quad \text{for } t_1 < t < t_2$$
$$= 0 \quad \text{otherwise} \tag{14.36}$$

We illustrate the significance and use of equations (14.35) and (14.36) by considering the scattering from a straight edge, shown in figure 14.6. The edge lies in the $y$-direction, at a distance $x_1$ from the $x$-axis. The source, with source strength $Q$, and the receiver are collocated in a position at a height $z = z_0$ above the origin.

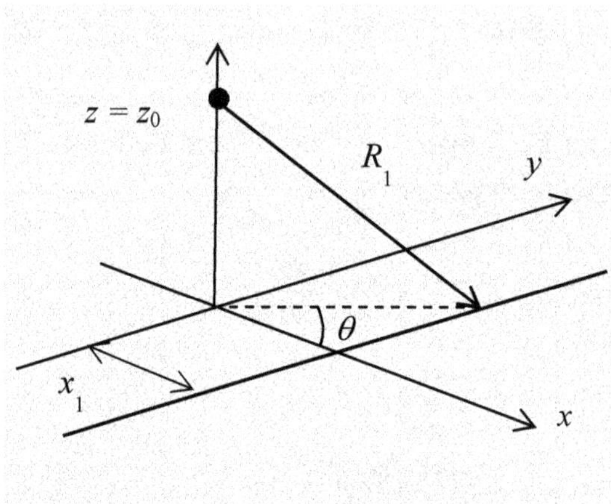


Figure 14.6. Signal scattering from an infinitely long straight edge. The source, with source strength $Q$, and the receiver are collocated at a height $z = z_0$ above the origin. The edge is at the distance $x_1$ from the $y$-axis.

We consider a running point on the edge at an angle $\theta$ and a distance $R_1$. The two-way propagation time $t$ from the source-receiver position to this point is

$$t = \frac{2R_1}{c} = \frac{2}{c}\sqrt{z^2 + \frac{x_1^2}{\cos^2\theta}} \tag{14.37}$$

Furthermore, we define the following two-way travel times

$$\begin{aligned} T_0 &= \frac{2z_0}{c} \\ T_1 &= \frac{2x_1}{c} \\ T &= \sqrt{T_0^2 + T_1^2} \end{aligned} \tag{14.38}$$

The derivative of the travel time in equation (14.37) with respect to the angle $\theta$ is

$$\left(\frac{d\theta}{dt}\right) = \frac{t\cos^2\theta}{T_1^2\tan\theta} \tag{14.39}$$

From equations (14.38) and figure 14.6, we find that

$$\cos^2\theta = \frac{T_1^2}{t^2 - T_0^2}$$
$$\tan\theta = \frac{\sqrt{t^2 - T^2}}{T_1} \qquad (14.40)$$

Then, using equation (14.36), we find that the time response of the diffraction from the straight edge is

$$w(t) = \frac{T_1}{t\left(t^2 + T_1^2 - T^2\right)\sqrt{t^2 - T^2}} \qquad t > T$$
$$w(t) = 0 \qquad t < T \qquad (14.41)$$

Figure 14.7 shows an example of the diffracted signal from an infinite straight edge, calculated by using equation (14.41) with the values $z_0 = 100$ m, $x_1 = 100$ m, and $c = 1500$ m/s.

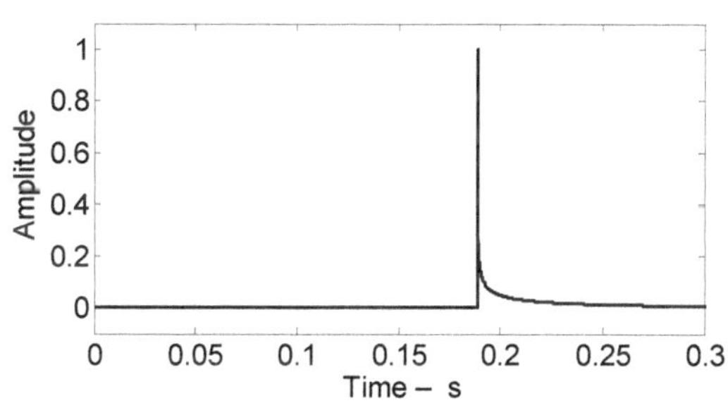


Figure 14.7. Diffraction signal, calculated according to equation (14.41) using the values $z_0 = 100$ m, $x_1 = 100$ m, and $c = 1500$ m/s.

The leading edge of the time response arrives at the time $T = 0.189$ s; afterward, the response decays to zero. From the plane strip shown in figure 14.5, we have two diffraction contributions, one from the first edge and one from the second edge. In addition to these, a specular reflection is created when the source-receiver is in a position straight above the strip. In that case, $R_1 = z_0$ and $R_1$ does not depend on the angle $\theta$.

The integral over $\theta$ yields the factor $2\pi$, so that the specular term becomes

$$\Phi_1(\mathbf{r}) = \frac{Q}{4\pi}\frac{\Re}{2z_0}\exp\left(2i\frac{\omega}{c}z_0\right) \tag{14.42}$$

In the time domain, this is a delta pulse delayed by the two-way travel time $t = 2z_0 / c$, that is, the time response is

$$\phi_{spec}(t) = \frac{Q}{4\pi}\frac{\Re}{2z_0}\delta\left(t - \frac{2z_0}{c}\right) \tag{14.43}$$

Figure 14.8 shows a simulation of the received signal when the source-receiver units are moving in a straight path and located at $z_0 = 100$ m over a plane strip 20 m wide. The minimum two-way travel time is $2z_0 / c = 0.133$ s, which occurs when the source-receiver is directly above the strip at positions from −10 m to 10 m.

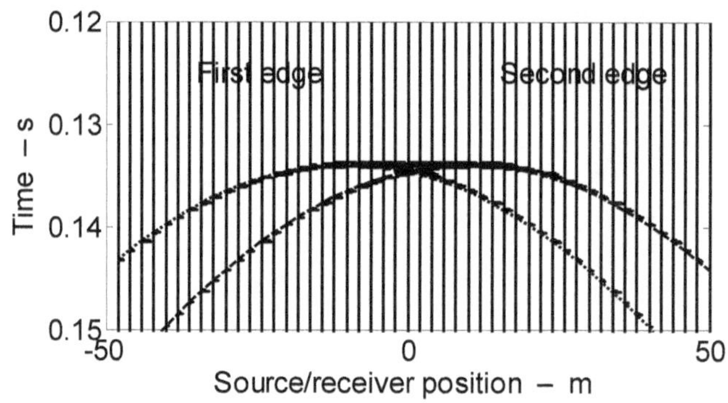


Figure 14.8. Diffraction from a plane strip as a function of source-receiver position. The width of the strip is 20 m, and the source-receiver is 100 m above the strip.

It is important to note that the diffraction theory we have presented here is based on equation (14.14) and the approximations of equations (14.12) and (14.13). In some important applications these approximations are not valid. Our results could be relatively poor approximations as far as the amplitudes of the diffractions are concerned. We shall not discuss these problems further here, but refer to the literature on the subject, in particular to Medwin, Childs, and Jebsen (1982), Jebsen and Medwin (1982), and Svensson and Fred (1999).

## 14.4 Scattering from rough surfaces

Scattering from a rough seabed or a rough sea surface is of considerable importance in underwater acoustics. Wherever the surface of the sea or seabed is involved in the transmission path, the signal may experience additional losses because of diffuse scattering from rough surfaces, and some of the scattered signal may propagate back to the receiver as noiselike reverberation, thus masking the desired signal.

In rough surface scattering, it is important to recognize that the roughness of a surface depends on the acoustic wavelength and on the directions of the incident wave. In section 4.5, this variability was discussed briefly, and the argument was made that the roughness of a given surface is obtained from the Rayleigh parameter $\chi = 2\kappa\sigma_h \sin\theta$, where $\theta$ is the grazing angle at which the acoustic wave intersects the surface, $\kappa$ is the wave number, and $\sigma_h$ is the root mean square of the surface roughness. The Rayleigh criterion states that the surface is acoustically smooth when $\chi$ is very small ie. when $\chi << 1$.

Over the years, a large number of studies have been conducted on scattering from rough surfaces, and many theories have been developed for use in modeling such scattering. Experimental and measured results are summarized by Urick (1983), and Ogilvy (1991) gives a summary of the theoretical models. Other valuable and general references on the subject are Tolstoy and Clay (1966), Medwin and Clay (1998), Clay and Medwin (1972), and Ishimaru (1978a, 1978b). Acoustic scattering from a rough surface is not very different from the scattering of electromagnetic waves. The standard reference on electromagnetic scattering is *The Scattering of Electromagnetic Waves from Rough Surfaces* by Beckmann and Spizzichino (1963).

In this section, we discuss two of the most common theories about scattering from rough surfaces, and apply the theories to scattering from the sea surface and the seabed. First we develop the Helmholtz-Kirchhoff method, following the original work by Eckart (1953), who developed an expression for surface scattering strength as a function of frequency and angle. For low frequencies, he found the scattering strength to be given by the wave number spectrum of the surface roughness. This theory was originally intended for application with the sea surface or a very hard sea bottom; in both cases, the boundary conditions are simple, with total reflection.

Thereafter, we discuss the perturbation technique, also called the Rayleigh-Rice method, which is more appropriate to applications with small-scale roughness. This method has also some advantages for application to seabed scattering, since the correct boundary conditions are applied. In the discussion of this method, we follow the approach by Essen (1994), but we limit our treatment to the fluid boundary case and omit discussion of scattering at an interface with a solid that can support shear waves.

### *14.4.1 The Helmholtz-Kirchhoff model*

Figure 14.9 presents the geometry of the problem we want to solve, where there is a source in position $\mathbf{r}_1$ and a receiver in position $\mathbf{r}_2$. The signal from the source strikes part of the rough surface source, which scatters sound in all directions; we want to determine how much of the scattered field arrives at the receiver. The orientation of the axes is chosen so that the source lies in the *x-z* plane, where $\varphi_1 = 0$. The rough surface is mainly in the *x-y* plane where $z = 0$, but there are deviations in surface heights, which are given by a function $\zeta(x,y)$ having a mean value of zero, that is, $<\zeta(x,y)> = 0$.

At a point on the surface where $\mathbf{r}_e = [x,y,z]$, the incident field $\Phi_{inc}(\mathbf{r}_e)$ is defined as

$$\Phi_{inc}(\mathbf{r}_e) = \frac{Q}{4\pi|\mathbf{r}_1 - \mathbf{r}_e|} B(x,y) \exp\left[-i\mathbf{k}_1 \cdot (\mathbf{r}_1 - \mathbf{r}_e)\right] \tag{14.44}$$

where $Q$ is the source strength; $B(x,y)$ represents the source's directivity and gives the incident field's relative strength at the position $(x,y)$ of the scattering element. The directivity is normalized so that $B(0,0) = 1$, at the center of the insonified surface.

In this discussion, while we mainly follow the original development of Eckart (1953), but we also add modifications found in Tolstoy and Clay (1966). Our starting point is the Helmholtz-Kirchhoff integral developed in the previous section, which says that the scattered field $\Phi(\mathbf{r}_2)$ is expressed by the integral

$$\Phi(\mathbf{r}_2) = \frac{1}{4\pi}\int_S \left\{ \Phi(\mathbf{r}_e)\frac{\partial G(\mathbf{r}_2,\mathbf{r}_e)}{\partial n} - G(\mathbf{r}_2,\mathbf{r}_e)\frac{\partial \Phi(\mathbf{r}_e)}{\partial n} \right\} dS \tag{14.45}$$

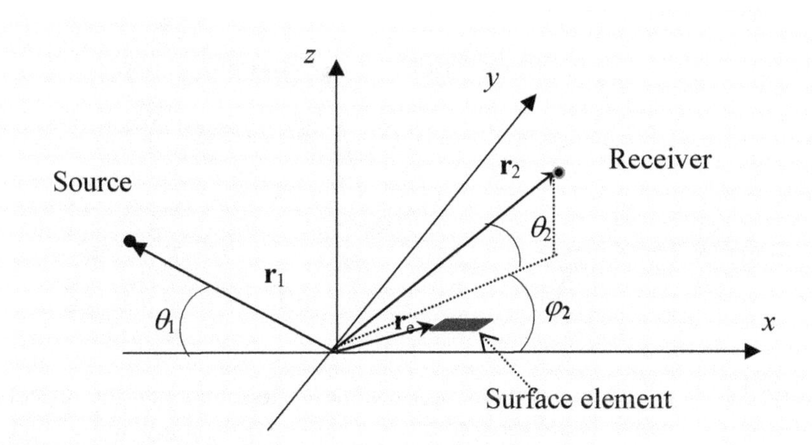


Figure 14.9. Geometry for the calculation of scattering from a rough surface when the source and receiver are in different locations.

The function $G(\mathbf{r}_2,\mathbf{r}_e)$ is the Green's function for propagation from the small surface element at $\mathbf{r}_e$ to the position $\mathbf{r}_2$, and is defined as

$$G(\mathbf{r}_2,\mathbf{r}_e)=\frac{\exp\left[i\mathbf{k}_2\cdot(\mathbf{r}_2-\mathbf{r}_e)\right]}{|\mathbf{r}_2-\mathbf{r}_e|} \tag{14.46}$$

Equations (14.45) and (14.46) determine the scattered field at position $\mathbf{r}_2$. The problem, as stated before, is that the unknown field, which is either $\Phi(\mathbf{r}_2)$ or $\Phi(\mathbf{r}_e)$, occurs on both sides of the equation, so that this equation defines the scattered field only implicitly. Therefore, in order to proceed we are forced to introduce some simplifying assumptions. The first and most significant one is that the scattered field on the surface is equal to the incoming field multiplied by the plane wave reflection coefficient $\Re$; that is, we assume that

$$\Phi(\mathbf{r}_e)=\Re\Phi_{inc}(\mathbf{r}_e) \tag{14.47}$$

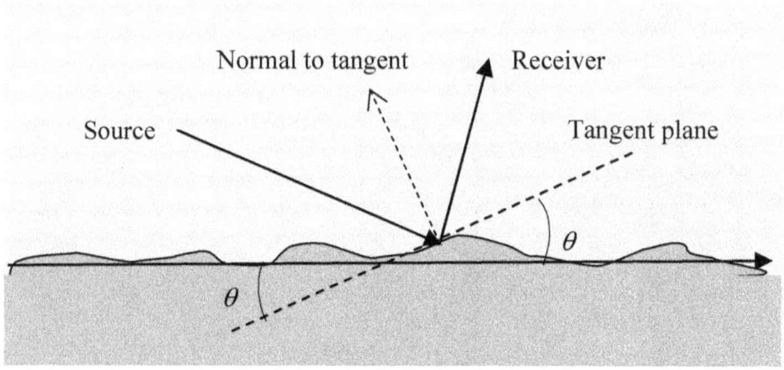


Figure 14.10. The tangent plane approximation assumes that the reflected wave is identical to a wave reflected from a plane surface with an inclination $\theta$ that is equal to the inclination of the tangent plane.

The normal derivative of the scattered field is by the same assumption

$$\frac{\partial \Phi(\mathbf{r}_e)}{\partial \mathbf{n}} = -\Re \frac{\partial \Phi_{inc}(\mathbf{r}_e)}{\partial \mathbf{n}} \tag{14.48}$$

This statement, which is called the ***tangent plane approximation***, is illustrated in figure 14.10.

The assumption is that the rough surface can be considered locally plane, so that the reflected field is identical to the field that would result if the surface were plane, with the inclination angle being equal to the angle of the tangent plane at the position $\mathbf{r}_e$. Note that for this statement to be valid, the radius of the surface's curvature has to be much larger than the acoustic wavelength.

Using equations (14.47) and (14.48), the Helmholtz-Kirchhoff integral now is rewritten as

$$\Phi(\mathbf{r}_2) = \frac{1}{4\pi} \int_S \Re \frac{\partial}{\partial n} \{ \Phi_{inc}(\mathbf{r}_e) G(\mathbf{r}_2, \mathbf{r}_e) \} \, dS \tag{14.49}$$

The second type of approximation or assumption required to solve this problem concerns simplifications in the geometry of figure 14.9. Specifically, we are assuming that both the source and the receiver are at far distances from the scattering surface. Therefore, the incoming field, defined in equation (14.44), is approximated by

$$\Phi_{inc}(\mathbf{r}_e) \approx \frac{QB(x,y)}{4\pi R_1} \exp(i\kappa R_1 + i\mathbf{k}_1 \cdot \mathbf{r}_e) \tag{14.50}$$

In equation (14.50), $R_1 = |\mathbf{r}_1|$ and $B(x,y)$ accounts for the directivity of the source.

In the same way, the Green's function is approximated by

$$G(\mathbf{r}_2, \mathbf{r}_e) \approx \frac{\exp(i\kappa R_2 - i\mathbf{k}_2 \cdot \mathbf{r}_e)}{R_2} \tag{14.51}$$

where $R_2 = |\mathbf{r}_2|$.

In the denominator, we achieve sufficient accuracy by using $R_2$, which is the distance to the scattering surface's center. However, in the exponent, we have to be more accurate, which means also retaining the second-order term. We thereby arrive at an approximation that is a valid assumption in the farfield, and at several wavelengths from the scattering surface. With these approximations, the scattered field becomes

$$\begin{aligned}\Phi(\mathbf{r}_2) = & \frac{Q\exp[i\kappa(R_1 + R_2)]}{(4\pi)^2 R_1 R_2} \\ & \times \int_S \Re B(x,y) i(\mathbf{k}_1 - \mathbf{k}_2) \cdot \mathbf{n} \exp[i(\mathbf{k}_1 - \mathbf{k}_2) \cdot \mathbf{r}_e] dS\end{aligned} \tag{14.52}$$

In the source and receiver directions, the wave numbers have the components $\mathbf{k}_1$ and $\mathbf{k}_2$, with the angles illustrated in figure 14.9.

$$\begin{aligned}\mathbf{k}_1 &= \kappa[\cos\theta_1 \quad 0 \quad -\sin\theta_1] \\ \mathbf{k}_2 &= \kappa[\cos\theta_2 \cos\varphi_2 \quad \cos\theta_2 \sin\varphi_2 \quad \sin\theta_2]\end{aligned} \tag{14.53}$$

and, consequently

$$\mathbf{k}_1 - \mathbf{k}_2 = \kappa[(\cos\theta_1 - \cos\theta_2\cos\varphi_2) \quad -\cos\theta_2\sin\varphi_2 \quad -(\sin\theta_1 + \sin\theta_2)] \tag{14.54}$$

We define the differences between the horizontal and vertical components of the incident and scattered wave numbers as

$$\begin{aligned} 2\alpha &= (\mathbf{k}_1 - \mathbf{k}_2)_x = \kappa(\cos\theta_1 - \cos\theta_2\cos\varphi_2) \\ 2\beta &= (\mathbf{k}_1 - \mathbf{k}_2)_y = -\kappa(\cos\theta_2\sin\varphi_2) \\ 2\gamma &= (\mathbf{k}_1 - \mathbf{k}_2)_z = -\kappa(\sin\theta_1 + \sin\theta_2) \end{aligned} \tag{14.55}$$

where $2\alpha$, $2\beta$, and $2\gamma$ represent the $x$, $y$, and $z$ components of the difference wave number $\mathbf{k}_1 - \mathbf{k}_2$. In the specular direction, $\varphi_2 = 0$ and $\theta_2 = \theta_1$ and, therefore, $\alpha = \beta = 0$ and $\gamma = -\kappa \sin\theta_1$. In the backscattered direction, that is, the direction of scattering returning to the source position, $\varphi_2 = \pi$ and $\theta_2 = \theta_1$, so that $\alpha = \kappa \cos\theta_1$, $\beta = 0$, and $\gamma = -\kappa \sin\theta_1$.

Let $\zeta$ be the surface elevation. The vector $\mathbf{n}$ is the unity vector normal to the small surface element, with components that are written as

$$\mathbf{n} = \begin{bmatrix} -n_x \dfrac{\partial\zeta}{\partial x} & -n_y \dfrac{\partial\zeta}{\partial y} & n_z \end{bmatrix} \frac{1}{\sqrt{1 + \left(\dfrac{\partial\zeta}{\partial x}\right)^2 + \left(\dfrac{\partial\zeta}{\partial y}\right)^2}} \tag{14.56}$$

The size of the scattering element is

$$dS = \frac{dxdy}{\sqrt{1 + \left(\dfrac{\partial\zeta}{\partial x}\right)^2 + \left(\dfrac{\partial\zeta}{\partial y}\right)^2}} \tag{14.57}$$

Therefore, the scattered field defined by equation (14.52) can now be rewritten as

$$\begin{aligned} \Phi(\mathbf{r}_2) = Q &\frac{-2i\exp\left[i\kappa(R_1 + R_2)\right]}{(4\pi)^2 R_1 R_2} \\ &\times \int_x\int_y B(x,y)\Re\left(\alpha\frac{\partial\zeta}{\partial x} + \beta\frac{\partial\zeta}{\partial y} - \gamma\right)\exp\left\{2i\left[\alpha x + \beta y + \gamma\zeta(x,y)\right]\right\} dxdy \end{aligned} \tag{14.58}$$

In order to solve the double integral in equation (14.58), we first assume that the reflection coefficient $\Re$ is constant over the area of integration; therefore, it can be brought outside the integral. Now the integrals in equation (14.58) can be solved by partial integration. To illustrate this, we take the integration over $x$

$$
\begin{aligned}
II_x &= \int_{-\infty}^{\infty} \{\alpha B(x,y)\exp(2i\alpha x)\}\left\{\frac{\partial\zeta}{\partial x}\exp\left[2i\gamma\zeta(x,y)\right]dx\right\} \\
&= \left|\frac{\alpha}{2i\gamma}B(x,y)\exp\left[2i\alpha x + 2i\gamma\zeta(x,y)\right]\right|_{-\infty}^{\infty} - \int_{\infty}^{\infty}\frac{\alpha^2}{\gamma}B(x,y)\exp\left[2i\left(\alpha x + \gamma\zeta(x,y)\right)\right]dx \\
&\quad - \int_{-\infty}^{\infty}\frac{\alpha}{2i\gamma}\exp\left[2i\left(\alpha x + \gamma\zeta(x,y)\right)\right]\frac{\partial B(x,y)}{\partial x}dx
\end{aligned}
\tag{14.59}
$$

The first term in equation (14.59) is a diffraction, or an edge-effect term, since it is given by the value of the source directivity function $B(x,y)$ at the edges of the surface insonified by the source. This term is often ignored when $B(x,y)$ is a rather smooth function that goes to zero for large values of $x$ and $y$. The third term can also be ignored because it is much smaller than the second term, which is the term we retain. When we include the integration over $y$, the solution of equation (14.58) can be now expressed as

$$
\begin{aligned}
\Phi(\mathbf{r}_2) &= \frac{iQ\exp\left[i\kappa(R_1 + R_2)\right]}{8\pi^2 R_1 R_2} \\
&\quad \times \Re\int_x\int_y\left(\frac{\alpha^2+\beta^2+\gamma^2}{\gamma}\right)B(x,y)\exp\left\{2i\left[\alpha x + \beta y + \gamma\zeta(x,y)\right]\right\}dxdy
\end{aligned}
\tag{14.60}
$$

The wave number components $\alpha$, $\beta$, and $\gamma$ are functions of the incident angle. We can use average values for $\alpha$, $\beta$, and $\gamma$ in many practical situations, when the incident radiation is limited to a small range of angles by a directional source. When this is the case, we can move $\alpha$, $\beta$, and $\gamma$ outside the integral so that equation (14.60) becomes

$$\Phi(\mathbf{r}_2)=\frac{i\kappa Q\exp\left[i\kappa(R_1+R_2)\right]}{8\pi^2 R_1 R_2}$$
$$\times\Re f(\theta_1,\theta_2,\varphi_2)\int_x\int_y B(x,y)\exp\left\{2i\left[\alpha x+\beta y+\gamma\varsigma(x,y)\right]\right\}dxdy \tag{14.61}$$

The function $f(\theta_1,\theta_2,\varphi_2)$ is defined as

$$\begin{aligned} f(\theta_1,\theta_2,\varphi_2)&=\frac{\alpha^2+\beta^2+\gamma^2}{\kappa\gamma}\\ &=\frac{1+\sin\theta_1\sin\theta_2-\cos\theta_1\cos\theta_2\cos\varphi_2}{(\sin\theta_1+\sin\theta_2)}\end{aligned} \tag{14.62}$$

The function $f(\theta_1,\theta_2,\varphi_2)$ is equal to sin $\theta_1$ in the specular direction and to $1/\sin\theta_1$ in the backscattering direction. Other expressions are also found in the literature for this function. In Eckart's original work, equation (14.62) is replaced by

$$f(\theta_1,\theta_2,\varphi_2)=\frac{\sin\theta_1+\sin\theta_2}{2} \tag{14.63}$$

which is the result obtained by ignoring the contributions of $\alpha^2$ and $\beta^2$ in equation (14.62). In the following, we use equation (14.63) rather than equation (14.62). The former is better behaved at low grazing angles where equation (14.62) goes to infinity.

Alternatively, we can express this result in terms of the incident field $\Phi_{inc}(x,y)$ on the plane, which is

$$\Phi_{inc}(x,y)=\frac{Q}{4\pi R_1}B(x,y)\exp(i\kappa R_1) \tag{14.64}$$

At this point, equation (14.61) takes the form

$$\begin{aligned}\Phi(\mathbf{r}_2)&=\frac{i\kappa\exp(i\kappa R_2)}{2\pi R_2}\Re f(\theta_1,\theta_2,\varphi_2)\\ &\times\int_x\int_y \Phi_{inc}(x,y)\exp\left\{2i\left[\alpha x+\beta y+\gamma\zeta(x,y)\right]\right\}dxdy\end{aligned} \tag{14.65}$$

Equations (14.61) and (14.65) have a simple interpretation. The scattered field is given as an integral over the surface, and is weighted by the source directivity function $D(x,y)$. The contribution from a surface element has a phase angle that is produced by the directions to the source and receiver, the position of the element $(x,y)$, and the local surface elevation $\zeta(x,y)$.

So far, nothing has been said about the shape of the corrugated surface represented by the term $\zeta(x,y)$, except that it has to satisfy certain criteria because of the approximations we have assumed. In fact, equations (14.61) and (14.65) only reflect the geometry of the problem, that is, the location of the scattering element in relation to the source and receiver positions. In the following section, we consider the effects of a corrugated surface on reflection and scattering, first examining a case where the surface corrugation can be described as a periodic function. Then we look at an example of corrugation as a statistically random surface.

### *14.4.2 Scattering from a periodic corrugated surface*

We can use the result of equation (14.65) to determine scattering from a surface with elevations that can be described as sinusoidal functions in the $x$-direction,

$$\zeta = h\cos(Kx) \tag{14.66}$$

The amplitude of the sinusoidal corrugation is $h$, and the spatial wave number is $K = 2\pi / \Lambda$, where $\Lambda$ is the spatial wavelength.

The relevant exponential function, which involves $\zeta$ in equation (14.60), can be expanded in a power series with Bessel functions (Abramowitz and Stegun 1965).

$$\begin{aligned}\exp(2i\gamma h\cos Kx) &= J_0(2\gamma h) + 2\sum_{n=1}^{\infty}(i)^n J_n(2\gamma h)\cos(nKx)\\ &= J_0(2\gamma h) + \sum_{n=1}^{\infty}(i)^n J_n(2\gamma h)\left[\exp(inKx) + \exp(-inKx)\right]\end{aligned} \tag{14.67}$$

When this result is inserted into equation (14.65), the integral, which now is only over $x$, takes the form

$$\int_x \exp\{2i[\alpha x+\zeta(x)]\}dx =$$

$$\int_x \left\{ J_0(2\gamma h)\exp(2i\alpha x)+\sum_{n=1}^{\infty}(i)^n J_n(2\gamma h)[\exp(2i\alpha x+inKx)+\exp(2i\alpha x-inKx)] \right\} dx \tag{14.68}$$

In the expression above we have also assumed that the insonification is uniform, with $B(x,y) = 1$.

Equation (14.68) shows there is scattering in many directions. Some scattered signals interact resulting in constructive interference, and some interfere destructively. However, when the surface is large, in the sense that the integration over $x$ covers many periods of corrugations, the constructive interferences dominate so that we get scattering only in the direction where the exponential terms in equation (14.68) are constant and equal to unity. We recall equation (14.55), in which $\alpha$ and $\gamma$ are defined as the difference wave numbers in the $x$- and $z$-directions, respectively. In the present example, $\beta = 0$ since we are studying only scattering in the $(x,z)$ plane where $\varphi_2 = 0$.

The first direction in which constructive interference takes place is described by

$$2\alpha = \kappa(\cos\theta_1 - \cos\theta_2) = 0 \tag{14.69}$$

which gives the reflected signal in the ***specular direction*** for $\theta_2 = \theta_1$.

The other directions of constructive interference are given by

$$\exp(2i\alpha x \pm inKx) = 1 \quad n=1, 2, 3... \tag{14.70}$$

The condition $2\alpha = \pm nK$ specifies a set of directions $\cos\theta_2^{(n)}$, which are given by the real solutions from

$$\cos\theta_2^{(n)} = \cos\theta_1 \pm n\frac{\lambda}{\Lambda} \tag{14.71}$$

where the acoustic wavelength is represented by $\lambda$, and the wavelength of the sinusoidal corrugations is $\Lambda$.

The amplitude of the $n^{\text{th}}$ component is

$$A_n = J_n\left(2\gamma^{(n)}h\right) \tag{14.72}$$

with

$$2\gamma^{(n)} = -\frac{2\pi}{\lambda}\left(\sin\theta_1 + \sin\theta_2^{(n)}\right) \tag{14.73}$$

When the roughness amplitude $h$ is smaller than the acoustic wavelength, that is, when $2\pi(h/\lambda) < 1$, the first-order term with $n = 1$ dominates over the higher-order terms and the main scattered signal is the direction given by equation (14.71)

$$\cos\theta_2 = \cos\theta_1 \pm \frac{\lambda}{\Lambda} \tag{14.74}$$

This is an example of selective scattering or the ***Bragg scattering*** of a periodic rough surface where the direction of the main contribution of the scattered acoustic wave $\theta_2$ is given by the direction of the incident wave $\theta_1$ and the wavelengths of the acoustic wave $\lambda$ and the wavelength of the sinusoidal corrugation $\Lambda$, as specified by equation (14.74). This type of scattering is illustrated in figure 14.11.

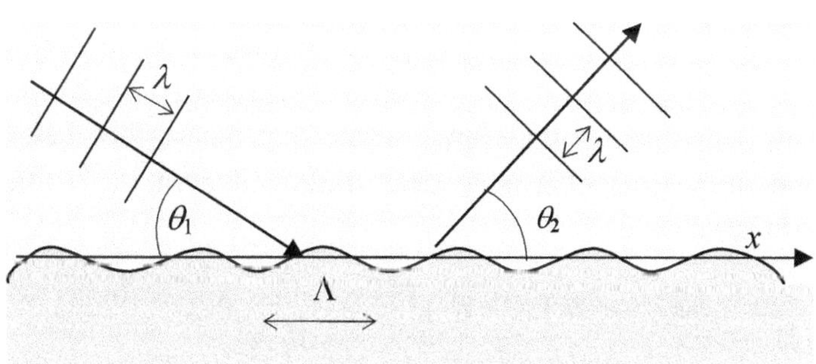


Figure 14.11. A plane wave with angle $\theta_1$ and wavelength $\lambda$ strikes a periodic surface with wavelength $\Lambda$. Scattering contributions from the facets of the rough surface reinforce one another and add constructively at the angle $\theta_2$, which was defined in equation (14.71).

For a source that transmits an acoustic wave with wavelength $\lambda$ in the direction $\theta_1$ toward a periodic corrugated surface with wavelength $\Lambda$, the main component of the scattered field then lies in a direction $\theta_2$ that satisfies equation (14.74).

It follows that when the rough surface is more complicated but can be represented by a sum of several different spatial wave numbers, the scattered field arriving at a receiver in a particular direction $\theta_2$ has its main contribution from the particular spatial wave number that satisfies the Bragg condition, expressed by equation (14.74). This is called selective or resonance scattering and we shall see it again later in this chapter.

### *14.4.3 Scattering from surfaces with random roughness*

We now address scattering from randomly irregular surfaces. A given source-receiver configuration and a given corrugated surface will determine the configuration of the scattered field, as described by either equation (14.61) or equation (14.65). However, any change, however small, in the source-receiver configuration or in the surface properties leads to different results. Since changes this small are beyond control, in many cases it is natural to consider scattering as a random process and to describe the scattered field statistically rather than deterministically.

In the following equation we therefore assume that the surface height $\zeta(x,y)$ is a zero-mean random process. The statistical mean value of the scattered field of equation (14.61) is

$$
\begin{aligned}
<\Phi(\mathbf{r}_2)> &= \frac{i\kappa Q \exp\left[i\kappa(R_1+R_2)\right]}{8\pi^2 R_1 R_2}\Re f(\theta_1,\theta_2,\varphi_2) \\
&\times \int_x\int_y B(x,y)\exp\{2i[\alpha x+\beta y]\}\{<\exp\left[2i\gamma\zeta(x,y)\right]>\}dxdy
\end{aligned}
\tag{14.75}
$$

where < > means ensemble averaging.

The scattered field's ensemble average, $m_0$, is expressed as

$$m_0 = <\exp\left[2i\gamma\zeta(x,y)\right]> = \int \exp\left[2i\gamma\zeta(x,y)\right]w\left[\zeta(x,y)\right]d\zeta \qquad (14.76)$$

with $w(\zeta)$ being the probability density distribution for the surface heights.

If, for instance, we assume a Gaussian distribution with standard deviation $\sigma_h$ and zero mean, then

$$w(h) = \frac{1}{\sqrt{2\pi\sigma_h^2}}\exp\left(-\frac{h^2}{2\sigma_h^2}\right) \qquad (14.77)$$

produces this definition for $m_0$

$$m_0 = \exp\left[-2\left(\gamma^2\sigma_h^2\right)\right] \qquad (14.78)$$

When equation (14.78) is inserted into equation(14.75), we find that the mean value of the specular reflected field is

$$<\Phi(\mathbf{r}_2)> = \Phi_{flat}(\mathbf{r}_2)\exp\left(-2\gamma^2\sigma_h^2\right) \qquad (14.79)$$

This is the same result as the one we obtained and discussed in section 4.6, with the exponential term defined as the coherent reflection coefficient

$$R_{coh} = \exp\left(-2\gamma^2\sigma_h^2\right) \qquad (14.80)$$

We notice that, according to this model, the coherent reflection coefficient of a rough surface is dependent on the statistics of the surface height variations and independent of the shape of the surface.

In order to calculate the mean scattered intensity, we assume that the acoustic field's wavelength is much larger than the largest value of $\zeta(x,y)$. We can then approximate the exponential term containing $\zeta(x,y)$ with

$$\exp\left[2i\gamma\zeta(x,y)\right] = 1 + i2\gamma\zeta(x,y) \qquad (14.81)$$

Setting this approximation into equation (14.65) yields two terms, the first being the specularly reflected field and the second expressing the diffusely scattered field $\Phi_s(\mathbf{r}_2)$ as

$$\Phi_s(\mathbf{r}_2) = \frac{-\kappa\gamma \exp(i\kappa R_2)}{\pi R_2} \Re f(\theta_1, \theta_2, \varphi_2) \int_x \int_y \Phi_{inc}(x, y)\zeta(x, y)\exp\{2i[\alpha x + \beta y]\}dxdy$$

(14.82)

The mean intensity $I_s(\mathbf{r}_2)$ of the scattered field is proportional to

$$I_s(\mathbf{r}_2) = \langle |\Phi_s(\mathbf{r}_2)|^2 \rangle = \left[\frac{-\kappa\gamma}{\pi R_2} \Re f(\theta_1, \theta_2, \varphi_2)\right]^2$$

$$\times \iint_{x,y} \iint_{x',y'} \{\Phi_{inc}(x, y)\Phi^*_{inc}(x', y')\}\{\langle \zeta(x, y)\zeta(x', y')\rangle\}\exp\{2i[\alpha(x - x') + \beta(y - y')]\}\, dxdydx'dy'$$

(14.83)

In equation (14.83) the function

$$\psi(x, x', y, y') = \langle \zeta(x, y)\zeta(x', y')\rangle \tag{14.84}$$

is the spatial correlation function for the surface irregularities. Generally speaking, this function is difficult to estimate. In some cases, we may appropriately assume that the correlation function is dependent only on the differences $\xi = x - x'$ and $\eta = y - y'$, so that the surface irregularities can be modeled as a spatially homogenous and stationary process, that is

$$\psi(\xi, \eta) = \langle \zeta(x, y)\zeta(x - \xi, y - \eta)\rangle \tag{14.85}$$

With this assumption made, the mean scattered intensity can be expressed as

$$I_s(\mathbf{r}_2) = \left[\frac{\kappa\gamma}{\pi R_2} \Re f(\theta_1, \theta_2, \varphi_2)\right]^2 \xi \int_\xi \int_\eta \Gamma(\xi, \eta)\psi(\xi, \eta)\exp\{2i[\alpha\xi + \beta\eta]\}\, d\xi d\eta \tag{14.86}$$

We have introduced the function $\Gamma(\xi, \eta)$, which is entirely determined by the incident field

$$\Gamma(\xi, \eta) = \int_x \int_y \Phi_{inc}(x, y)\Phi^*_{inc}(x - \xi, y - \eta)dxdy \tag{14.87}$$

Furthermore, we assume that the width of the correlation function in $\xi$ and $\eta$ is much smaller than the size of the insonified surface area. With this assumption, we can replace the function $\Gamma(\xi, \eta)$ with its value at the origin, or $\Gamma(0,0)$, which is expressed as

$$\Gamma(0,0) = \int_x \int_y \left|\Phi_{inc}(x,y)\right|^2 dxdy \tag{14.88}$$

This gives the scattered intensity as

$$\begin{aligned} I_s(\mathbf{r}_2) = {} & \left[\frac{\kappa\gamma}{\pi R_2}\Re f(\theta_1,\theta_2,\varphi_2)\right]^2 \int_x \int_y \left|\Phi_{inc}(x,y)\right|^2 dxdy \\ & \times \int_\xi \int_\eta \psi(\xi,\eta)\left\{\exp\left[2i(\alpha\xi+\beta\eta)\right]\right\} d\xi d\eta \end{aligned} \tag{14.89}$$

This is an integral over the area of the surface insonified by the beam of incident sound. Assuming that the insonification is constant over the area, we can take $\Phi_{inc}$ outside the integral so that we can rewrite equation (14.89) as

$$\begin{aligned} I_s(\mathbf{r}_2) = {} & \left|\Phi_{inc}(x,y)\right|^2 A\left[\frac{\kappa\gamma}{\pi R_2}\Re f(\theta_1,\theta_2,\varphi_2)\right]^2 \\ & \times \int_\xi \int_\eta \psi(\xi,\eta)\left\{\exp\left[2i(\alpha\xi+\beta\eta)\right]\right\} d\xi d\eta \end{aligned} \tag{14.90}$$

### *14.4.4 The wave number spectrum*

Instead of using the spatial correlation function, it is often more convenient to use the wave number spectrum of the surface roughness, $\Psi(K_x, K_y)$, which is the two-dimensional Fourier transform of the correlation function, and which can be defined as

$$\Psi(K_x,K_y) = \frac{1}{(2\pi)^2}\int_{-\infty}^{\infty}\int_{-\infty}^{\infty} \psi(\xi,\eta)\exp(i\xi K_x + i\eta K_y)\, d\xi d\eta \tag{14.91}$$

Inserting equation (14.91) into equation (14.89) gives us the diffusely scattered intensity $I_s(\mathbf{r}_2)$ as

$$I_s(\mathbf{r}_2) = \left[\frac{\kappa\gamma}{\pi R_2}\Re f(\theta_1,\theta_2,\varphi_2)\right]^2 \left[\int_x\int_y |\Phi_{inc}(x,y)| dxdy\right](2\pi)^2 \Psi(2\alpha,2\beta)$$

(14.92)

The arguments $\alpha$ and $\beta$ in equations (14.92) were defined in equation (14.55).

The surface scattering coefficient $\sigma_a$, that is, the scattering cross section per unit area of the surface, is

$$\sigma_a = \frac{I_s(\mathbf{r}_2)R_2^2}{\int_x\int_y |\Phi_{inc}(x,y)|^2 dxdy} = \left[\frac{\kappa\gamma}{\pi}\Re f(\theta_1,\theta_2,\varphi_2)\right]^2 (2\pi)^2 \Psi(2\alpha,2\beta) \quad (14.93)$$

Equation (14.93) states that the scattering coefficient of a rough surface is proportional to the two-dimensional wave number spectrum of the surface roughness, the reflection coefficient, and the angular positions of source and receiver. The last term of equation (14.93) singles out a specific part of the two-dimensional wave number spectrum, that is, the components with

$$\begin{aligned} K_x &= 2\alpha \\ K_y &= 2\beta \end{aligned} \quad (14.94)$$

These components are the ***Bragg wave numbers.*** Their values depend on both the frequency of the acoustic signal and the angular positions of the source and receiver. Of the whole spectrum of spatial wave numbers, only the Bragg wave numbers contribute to the scattering in a particular direction. Therefore, in principle it is feasible to determine the roughness spectrum of a surface by measuring the scattering strength as a function of frequency and angle. If the roughness spectrum is constant, the scattering increases with the fourth power of the signal frequency because of the effect of the factor $(\kappa\gamma)^2$ in equation (14.93).

If we use the function $f(\theta_1,\theta_2,\varphi_2)$ given by equation (14.63), equation (14.93) then becomes

$$\sigma_a = 4\gamma^4 |\Re|^2 \Psi(2\alpha,2\beta) \quad (14.95)$$

When $|\Re| = 1$ this is the result obtained by Eckart, except for a factor of $(2\pi)^2$; the reason for that is that our definition of the wave number spectrum in equation (14.91) differs by the factor of $2\pi$.

The scattered intensity at the receiver at a distance $R_2$ is

$$I_s(\mathbf{r}_2) = |\Phi_{inc}|^2 \frac{\sigma_a}{R_2^2} A \tag{14.96}$$

where $A$ is the insonified area of the surface. The surface scattering strength parameter $S_a$, used in chapter 11 for the calculation of the reverberation level, is defined as

$$S_a = 10\log(\sigma_a) \tag{14.97}$$

For an isotropic and homogenous random surface having the same type of roughness in all directions, the roughness spectrum is only a function of a scalar wave number $K$, which is expressed as

$$K = \sqrt{K_x^2 + K_y^2} \tag{14.98}$$

so that the wave number spectrum can be written as

$$\Psi(K_x, K_y) = \frac{1}{2\pi K} G(K) \tag{14.99}$$

Equation (14.99) results after transforming the rectangular coordinates $K_x$ and $K_y$ into cylindrical coordinates $K^2 = (K_x^2 + K_y^2)$.

The height variance of the surface roughness is obtained by integration over the whole wave number spectrum

$$\langle \zeta^2 \rangle = \int G(K)dK \tag{14.100}$$

### *14.4.5 Application to sea surface scattering*

For scattering from the sea surface, the problem of establishing the roughness spectrum may at first glance appear complicated, as the surface of the sea is constantly changing. However, it is quite straightforward to measure the temporal frequency spectrum of the wave motion. Then, provided we know the relationship between the temporal spectrum and the spatial wave number, we can convert the temporal frequency spectrum to a spatial wave number spectrum.

Our example is a so-called fully developed sea with gravity waves. Ocean surface gravity waves are the large swell-type waves that oscillate in response to their own weights. The relationship between the temporal frequency $\varpi$ and the spatial wave number $K$ of the ocean gravity waves is

$$\varpi^2 = gK \tag{14.101}$$

where $g$ represents the acceleration of gravity. The expression for the temporal frequency spectrum $S(\varpi)$ is

$$S(\varpi) = Cg^2\varpi^{-5} \tag{14.102}$$

where $g = 9.81$ m/s$^2$ and $C$, a dimensionless constant, has the value $C = 7.4 \cdot 10^{-3}$. The wave number spectrum of the ocean waves $G(K)$ is related to the temporal frequency spectrum by

$$G(K) = S(\varpi)\frac{d\varpi}{dK} \tag{14.103}$$

Consequently, the ocean surface wave number spectrum for gravity waves is

$$G(K) = \frac{C}{2}K^{-3} \tag{14.104}$$

The sea surface roughness spectrum of equation (14.104) is a reasonable choice for low-frequency acoustics, where the scattering from long-wavelength surface waves is most important. At higher acoustic frequencies, however, scattering from surface waves with shorter wavelengths may be more important and the roughness spectrum given in equation (14.104) is no longer appropriate.

Using the roughness spectrum from equation (14.104), and with $\alpha = \kappa\cos\theta_1$, $\beta = 0$, and $\gamma = -\kappa\sin\theta_1$ from equation (14.95), we find the backscattering cross section $\sigma_a$ for the sea surface when $|\Re| = 1$ to be

$$\sigma_a = \frac{C}{16\pi}\tan^4\theta \qquad (14.105)$$

Figure 14.12 features a plot developed by using equation (14.105) to calculate the backscattering strength, $S_a$ (equation (14.97)), of the sea surface as a function of the grazing angle.

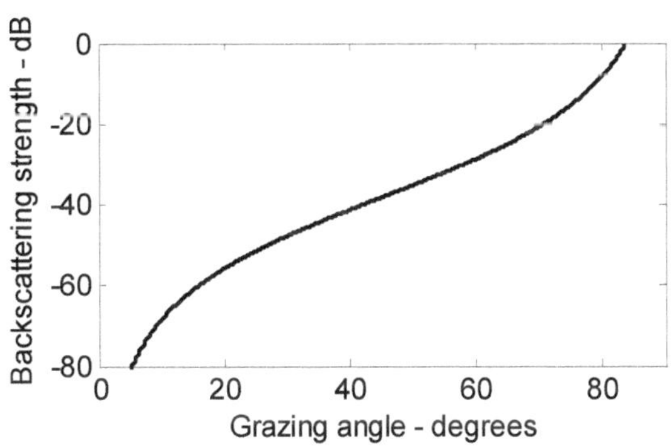


Figure 14.12. Backscattering strength, $S_a$, of the sea surface, plotted as a function of grazing angle. Equation (14.105) was used to calculate the results.

Equation (14.105) is essentially the same as the results obtained by Marsh (1963) and by Bachmann (1973). These findings are most interesting: equation (14.105) shows that scattering from the sea surface is independent of frequency when the wave action is dominated by gravity waves. Note that, according to the results of this calculation, the scattering does not depend on the wind speed. However, note also that these scattering models compare well with some measurements, but not with others; see *Principles of Underwater Sound* by Urick (1983) for a comparison of modeled and measured results.

### 14.4.6 *First-order perturbation theory*

We now consider the second model, which is a first-order perturbation approach frequently called the Rayleigh-Rice model. We consider the case of a rough interface between two acoustic media, which we assume to be water and the seabed, that have sound speeds and densities $c_w$, $\rho_w$, and $c_b$, $\rho_b$, respectively. The geometry of the problem is the same as was considered in the previous section, but since we now have to consider the reflected and transmitted waves in both media, we must expand our notation as shown in figure 14.13.

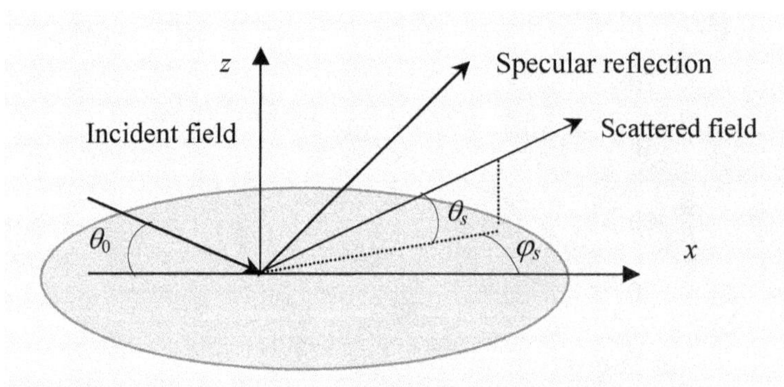


Figure 14.13. Geometry for determining scattering coefficient of a rough surface. The incident wave has the direction given by $\theta_0$. There is a reflected wave in the specular direction and a scattered wave in the direction given by $\theta_s$ and $\varphi_s$.

There is an incident plane wave with the direction given by the horizontal wave number $\mathbf{k}_0$ and a specular reflected wave with the same horizontal wave number. There is also a scattered wave in the direction $\mathbf{k}_s$ because of the rough surface. In addition, there are two other waves, a transmitted wave and a scattered wave within the seabed that are not shown in figure 14.13. Here the specular direction corresponds to $\theta_s = \theta_0$ and $\varphi_s = 0$, and the backscattering direction to $\theta_s = \theta_0$ and $\varphi_s = \pi$.

We assume an incident plane wave $\Phi_0$, which we define by the displacement potential

$$\Phi_0 = A_0 \exp[i(\mathbf{k}_0 \cdot \mathbf{x} - \gamma_{w0} z - \omega t)] \tag{14.106}$$

The horizontal wave number $\mathbf{k}_0$ has components in the $x$- and $y$-directions, so that

$$\mathbf{k}_0 = \frac{\omega}{c_w}\left[\cos\theta_0 \cos\varphi_0\ ,\ \cos\theta_0 \sin\varphi_0\right] \tag{14.107}$$

We assume that, without loss of generality, the incident wave vector is in the $(x,z)$ plane. Therefore, $\varphi_0 = 0$ and $|\mathbf{k}_0| = k0_0 = (\omega / c_w)\ \cos\theta_0$. The vertical wave number in the water is $\gamma_{w0} = (\omega / c_w)\ \sin\theta_0$.

The incident wave generates reflected waves that return into the upper medium, the water, and produces transmitted waves that travel into the seabed. In a case where the surface is completely plane, the reflected wave $\Phi_r$ in the specular direction is given by

$$\Phi_r = A_r \exp\left[i(\mathbf{k}_0 \cdot \mathbf{x} + \gamma_{w0} z)\right] \tag{14.108}$$

The transmitted wave in the seabed, $\Phi_t$, is expressed as

$$\Phi_t = A_t \exp\left[i(\mathbf{k}_0 \cdot \mathbf{x} - \gamma_{b0} z)\right] \tag{14.109}$$

In equations (14.106), (14.108), and (14.109) $A_0$ represents the incident wave's amplitude, and $A_r$ and $A_t$ are the reflected and transmitted waves' amplitudes,

The vertical wave number in the sea bottom is $\gamma_{b0}$, defined as

$$\gamma_{b0} - \sqrt{\left(\frac{\omega}{c_b}\right)^2 - k_0^2} \tag{14.110}$$

The surface of the seabed is basically plane, but there is a small variation in the surface heights; this variation is given by the two-dimensional function $\zeta(x,y)$. We expand the acoustic fields in the water and within the seabed in a power series of the height deviation $(x, y)$, in the general form

$$\Phi = \Phi^{(0)} + \frac{\partial \Phi^{(0)}}{\partial z}\zeta + \frac{1}{2}\frac{\partial^2 \Phi^{(0)}}{\partial z^2}\zeta^2 + \cdots \tag{14.111}$$

In accordance with first-order perturbation theory, the first two terms of equation (14.111) are retained and all terms higher than $\zeta$ are ignored.

The zero-order terms $\Phi^{(0)}$ are the only terms that exist when the surface is completely flat, that is, when $\zeta = 0$. In such a case, the acoustic field in the water $\Phi_w^{(0)}$ is the sum of an incident wave and a reflected wave in the specular direction

$$\Phi_w^{(0)} = A_0 \exp\left[ i\left(\mathbf{k}_0 \cdot \mathbf{x} - \gamma_{w0} z\right)\right] + A_r \exp\left[ i\left(\mathbf{k}_0 \cdot \mathbf{x} + \gamma_{w0} z\right)\right] \tag{14.112}$$

Within the seabed, a transmitted wave $\Phi_b^{(0)}$ is defined as

$$\Phi_b^{(0)} = A_t \exp\left[ i\left(\mathbf{k}_0 \cdot \mathbf{x} - \gamma_{b0} z\right)\right] \tag{14.113}$$

The first-order perturbed (or scattered) wave in the water, $\Phi_w^{(1)}$, has the amplitude $B_r$ and is given in the form

$$\Phi_w^{(1)} = B_r \exp\left[ i\left(\mathbf{k}_s \cdot \mathbf{x} + \gamma_{ws} z\right)\right] \zeta \tag{14.114}$$

Similarly, the first-order field scattered at the seabed surface, $\Phi_b^{(1)}$, is expressed as

$$\Phi_b^{(1)} = B_t \exp\left[ i\left(\mathbf{k}_s \cdot \mathbf{x} - \gamma_{bs} z\right)\right] \zeta \tag{14.115}$$

The horizontal wave number of the scattered wave is $k_s = (\omega / c_w)\cos\,\theta_s$. The vertical wave numbers for the scattered waves in the water and bottom, $\gamma_{ws}$ and $\gamma_{bs}$, respectively, are

$$\begin{aligned} \gamma_{ws} &= \frac{\omega}{c_w}\sin\theta_s \\ \gamma_{bs} &= \sqrt{\left(\frac{\omega}{c_b}\right)^2 - k_s^2} \end{aligned} \tag{14.116}$$

The task now is to determine the amplitudes of the reflected and transmitted waves for both the zero and first orders. To do that, we apply the normal boundary conditions, requiring that the sound pressure and the vertical component of the particle velocity or displacement must be the same on both sides of the boundary for $z = 0$. Furthermore, these boundary conditions must be satisfied separately by the zero- and first-order waves.

For the zero-order waves, we readily obtain

$$\begin{aligned} \rho_w \left(A_0 + A_r\right) &= \rho_b A_t \\ \gamma_{w0} \left(A_0 - A_r\right) &= \gamma_{b0} A_t \end{aligned} \tag{14.117}$$

where $A_0$ represents the incident wave's amplitude, and $A_r$ and $A_t$ are the reflected and transmitted waves' amplitudes, respectively. Solved with respect to $A_r$ and $A_t$, equations (14.117) give us the zero-order reflection and transmission coefficients as

$$\begin{aligned} \Re^{(0)} &= \frac{A_r}{A_0} = \frac{\rho_b \gamma_{w0} - \gamma_{b0} \rho_w}{\rho_b \gamma_{w0} + \gamma_{b0} \rho_w} \\ T^{(0)} &= \frac{A_t}{A_0} = \frac{2\gamma_{w0} \rho_w}{\rho_b \gamma_{w0} + \gamma_{b0} \rho_w} \end{aligned} \tag{14.118}$$

These are the same coefficients for reflection and transmission that we found in section 4.2.

And, as we showed in chapter 4, the reflected and transmitted fields given by equations (14.118) satisfy the energy conservation principle: the sum of the reflected and transmitted intensities is equal to the incident intensity. Note, however, that with additional scattered waves caused by roughness, the requirement for energy conservation is no longer satisfied.

Returning to the case of the plane surface, recall that according to equation (14.111) the first-order term for the sound pressure $p^{(1)}$ is

$$p^{(1)} = \rho \omega^2 \left( \Phi^{(1)} + \frac{\partial \Phi^{(0)}}{\partial z} \zeta \right) \tag{14.119}$$

To determine the first-order term of the particle displacement normal to the interface, we must take into consideration the height displacement $\zeta$ and the inclination of the surface, which is given by the gradient $\nabla_{\mathbf{x}} \zeta$ in the $x$-$y$ plane and defined as

$$\nabla_{\mathbf{x}} \zeta = \left[ \frac{\partial \zeta}{\partial x} \;\; \frac{\partial \zeta}{\partial y} \right] \tag{14.120}$$

Therefore, the first-order normal particle displacement is

$$u_z^{(1)} = \frac{\partial \Phi^{(1)}}{\partial z} + \zeta\left(\frac{\partial^2}{\partial z^2}\Phi^{(0)}\right) - \mathbf{u}_x^{(0)} \cdot \nabla_{\mathbf{x}}\zeta \tag{14.121}$$

In equation (14.121), $\mathbf{u}_x^{(0)}$ is the zero-order particle displacement in the *x-y* plane, and the gradient operator is $\nabla_x$. This is illustrated in figure 14.14.

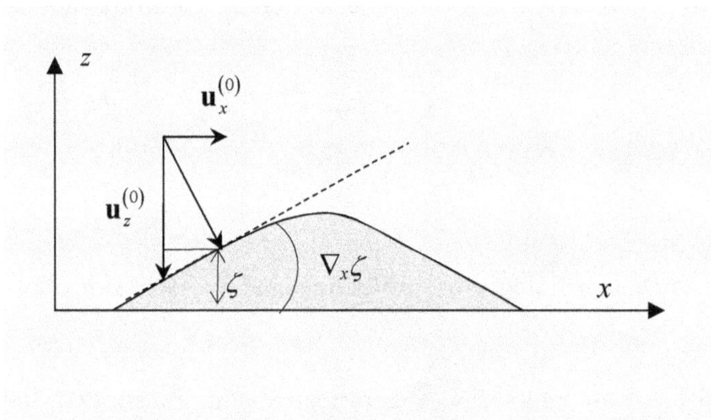


Figure 14.14. Reflection from a small facet of a plane surface; the facet has surface elevation $\zeta$ and inclination $\nabla_{\mathbf{x}}\zeta$.

The next step, which is to express the surface roughness with a Fourier integration over wave numbers, is based on the assumption that surface height deviations can be expressed by

$$\zeta(\mathbf{K}) = \exp(i\mathbf{K} \cdot \mathbf{x}) \tag{14.122}$$

In equation (14.122), **x** is the position vector and **K** is the wave number of the surface roughness spectrum, both with components in the *x-y* plane. Using equation (14.122) means that first we consider the sinusoidal height deviation of a single component (**K**) of the surface roughness spectrum, and later we sum up the scattered contributions from all the components, thereby obtaining a solution for the total scattered field.

At the boundary $z = 0$, the water and seabed's first-order pressures $p_w^{(1)}$, $p_b^{(1)}$ and vertical displacements $u_w^{(1)}$, $u_b^{(1)}$, respectively, become

$$p_w^{(1)} = \rho_w \omega^2 \left[ B_r \zeta - i\gamma_{w0} \left( A_0 - A_r \right) \zeta \right]$$
$$p_b^{(1)} = \rho_b \omega^2 \left[ B_t \zeta - i\gamma_{b0} A_t \zeta \right] \tag{14.123}$$

and

$$u_w^{(1)} = i\gamma_{ws} B_r \zeta - \gamma_{w0}^2 \left( A_0 + A_r \right) \zeta + \mathbf{k}_0 \cdot \mathbf{K} \zeta \left( A_0 + A_r \right)$$
$$u_b^{(1)} = -i\gamma_{bs} B_t \zeta - \gamma_{b0}^2 A_t \zeta + \mathbf{k}_0 \cdot \mathbf{K} \zeta A_t \tag{14.124}$$

where $B_r$ is the amplitude of the first-order scattered field in water and $B_t$ is the amplitude of the first-order scattered field in the lower medium.

Equations (14.123) and (14.124) now can be solved to produce the amplitude of the scattered field in water, $B_r$

$$B_r = 2A_0 i\gamma_{w0} \Re^{(1)} \tag{14.125}$$

where the function $\Re^{(1)}$ is defined by equation (14.125) and found after some algebra to be

$$\Re^{(1)} = \frac{\left(\rho_b - \rho_w\right)\left(\rho_b \mathbf{k}_0 \cdot \mathbf{K} - \rho_w \gamma_{b0} \gamma_{bs}\right) - \rho_b \left(\rho_b \gamma_{w0}^2 - \rho_w \gamma_{b0}^2\right)}{\left(\rho_b \gamma_{w0} + \rho_w \gamma_{b0}\right)\left(\rho_b \gamma_{ws} + \rho_w \gamma_{bs}\right)} \tag{14.126}$$

Equation (14.125), after the insertion of equation (14.126), now defines $B_r$ for a given wave number component **K** of the surface roughness. Since we are going to discuss only $B_r$, the amplitude of the reflected scattered field in the upper medium, here we will not calculate $B_t$, the transmitted field in the seabed.

The energy is scattered in all directions but results in constructive interference in certain directions, as we have seen before in our discussion of scattering from a sinusoidal surface in section 14.4.2. As we discussed in section 14.4.4, the direction with constructive interference is the direction given by the horizontal wave number $\mathbf{k}_s$ given as

$$\mathbf{k}_s = \mathbf{k}_0 - \mathbf{K} \tag{14.127}$$

This means that for a given source and receiver configuration, only the roughness wave number **K** that satisfies equation (14.127) contributes to the roughness scattering. With this value for **K**, equation (14.126) can be expressed as

$$\Re^{(1)} = \frac{(\rho_b - \rho_w)\left(\rho_b \mathbf{k}_0 \cdot \mathbf{k}_s - \rho_b k_0^2 - \rho_w \gamma_{b0}\gamma_{bs}\right) - \rho_b\left(\rho_b \gamma_{w0}^2 - \rho_w \gamma_{b0}^2\right)}{\left(\rho_b \gamma_{w0} + \rho_w \gamma_{b0}\right)\left(\rho_b \gamma_{ws} + \rho_w \gamma_{bs}\right)} \quad (14.128)$$

In order to calculate the total scattering from all the wave number components, we must integrate over all wave numbers, using a relative weight given by the wave number spectrum $\psi(\mathbf{K})$, and thereafter take the ensemble average to get the mean scattered intensity. Thus, we arrive at the scattered intensity in the water

$$I_s = \left\langle \left|\Phi_w^{(1)}\right|^2 \right\rangle = \int_0^{\infty} |B_r|^2\, \Psi(\mathbf{K})\, d\mathbf{K} \quad (14.129)$$

Since $\mathbf{K} = \mathbf{k}_s - \mathbf{k}_0$ we find that the differential $d\mathbf{K}$ can be expressed by angles of the receiver position as

$$d\mathbf{K} = d\left(\mathbf{k}_s - \mathbf{k}_0\right) = d\mathbf{k}_s = \left(\frac{\omega}{c_w}\right)^2 \cos\theta_s \sin\theta_s d\theta_s d\varphi_s \quad (14.130)$$

Therefore, the scattered intensity $I_s$ can be written in terms of an integral over angles by

$$I_s = \iint |B_r|^2 \Psi\left(\mathbf{k}_s - \mathbf{k}_0\right)\left(\frac{\omega}{c_w}\right)^2 \cos\theta_s \sin\theta_s d\theta_s d\varphi_s \quad (14.131)$$

In general, a definition of the surface scattering coefficient $\sigma$ can be developed by beginning with the expression

$$I_s = I_i \int_S \sigma \frac{ds}{r^2} \quad (14.132)$$

Equation (14.132) says that the scattering intensity $I_s$ is proportional to the incident intensity $I_i$ and obtained by integration over the whole area $S$ that contributes to the scattering. Therefore the scattering strength coefficient $\sigma$ is a measure of the scattering strength of the surface per unit surface. Furthermore, the integral of equation (14.132) can be changed into a double integral over the angles $\theta_s$ and $\varphi_s$. This transformation is illustrated in figure 14.15. In the *x*-direction, the area *ds* has the length $(r d\theta_s) / \sin\theta_s$, as shown in the figure. In the *y*-direction, perpendicular to the plane, the length is $r d\varphi_s \cos\theta_s$, therefore, the surface element is

$$ds = r^2 \cot\theta_s d\theta_s d\varphi_s \tag{14.133}$$

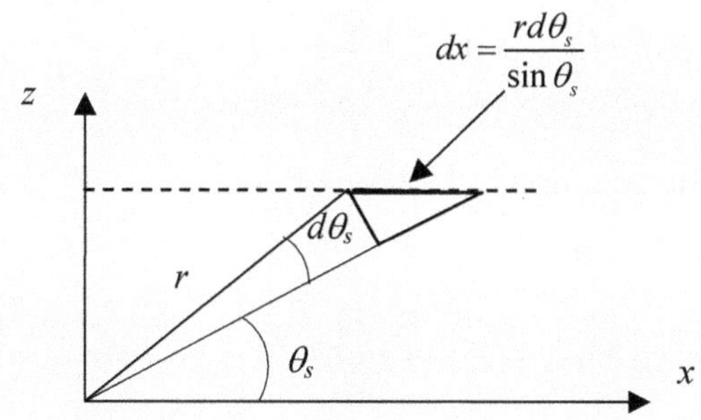


Figure 14.15. Calculation of the surface area for a ray bundle with angular width of $d\theta_s$ and $d\varphi_s$.

By applying equation (14.133), we can express the scattered intensity as

$$I_s = I_i \iint \sigma \frac{r^2 \cot(\theta_s) d\theta_s d\varphi_s}{r^2} \tag{14.134}$$

By comparing equation (14.131) with equation (14.134) we find the value of the surface scattering coefficient $\sigma$ to be

$$\sigma_a = \left(\frac{\omega}{c_w}\right)^2 \left|\frac{B_r}{A_0}\right|^2 \Psi(\mathbf{k}_s - \mathbf{k}_0)\sin^2\theta_s \tag{14.135}$$

Finally, by inserting equation (14.125) into equation (14.135), we find that the expression for the scattering strength can be written as

$$\sigma_a = 4\left(\frac{\omega}{c_w}\right)^4 \sin^2\theta_s \sin^2\theta_0 \left|\Re^{(1)}\right|^2 \Psi(\mathbf{k}_s - \mathbf{k}_0) \tag{14.136}$$

The $x$- and $y$-components of $\mathbf{k}_s - \mathbf{k}_0$ are the same as those given in equation (14.55); with the current notation, they are given as

$$2\alpha = (k_0 - k_s)_x = \frac{\omega}{c_w}(\cos\theta_0 - \cos\theta_s \cos\varphi_s)$$
$$2\beta = (k_0 - k_s)_y = -\frac{\omega}{c_w}\cos\theta_s \sin\varphi_s \tag{14.137}$$

We can therefore write equation (14.136) as

$$\sigma_a = 4\left(\frac{\omega}{c_w}\right)^4 \sin^2\theta_s \sin^2\theta_0 \left|\Re^{(1)}\right|^2 \Psi(2\alpha, 2\beta) \tag{14.138}$$

Equation (14.128) for the reflection coefficient can be written as

$$\Re^{(1)} = \frac{(\rho_b - \rho_w)(\rho_b \mathbf{k}_0 \cdot \mathbf{k}_s - \rho_b k_0^2 - \rho_w \gamma_{b0}\gamma_{bs}) - \rho_b(\rho_b \gamma_{w0}^2 - \rho_w \gamma_{b0}^2)}{(\rho_b \gamma_{w0} + \rho_w \gamma_{b0})(\rho_b \gamma_{ws} + \rho_w \gamma_{bs})} \tag{14.139}$$

where the horizontal wave numbers are

$$k_0 = \frac{\omega}{c_w}\cos\theta_0$$
$$k_s = \frac{\omega}{c_w}\cos\theta_s \cos\varphi_s \tag{14.140}$$

The vertical wave numbers for the zero-order waves in the water and seabed are

$$\gamma_{w0} = \sqrt{\left(\frac{\omega}{c_w}\right)^2 - k_0^2}$$
$$\gamma_{b0} = \sqrt{\left(\frac{\omega}{c_b}\right)^2 - k_0^2} \tag{14.141}$$

and the vertical wave numbers for the scattered waves are

$$\gamma_{ws} = \sqrt{\left(\frac{\omega}{c_w}\right)^2 - k_s^2}$$
$$\gamma_{bs} = \sqrt{\left(\frac{\omega}{c_b}\right)^2 - k_s^2} \qquad (14.142)$$

Equation (14.138), with equations (14.139), (14.140), (14.141), and (14.142), represents the general result for the scattering towards the direction defined by the angles $\theta_s$ and $\varphi_s$ of a plane wave coming from the direction $\theta_0$ in the *x-y* plane as illustrated in figure 14.13. The result is thus valid for both monostatic and bistatic situations.

For ***forward scattering*** in the same plane in which the incident wave travels $\varphi_s = 0$ and consequently

$$k_s = \frac{\omega}{c_w}\cos\theta_s \qquad (14.143)$$

For ***backscattering*** toward the source, $\theta_s = \theta_0$ and $\varphi_s = \pi$. The horizontal wave number of the scattered wave becomes

$$k_s = -k_0 = -\frac{\omega}{c_w}\cos\theta_0 \qquad (14.144)$$

The vertical wave numbers for the backscattered waves are the same as the vertical wave numbers for the zero-order terms, that is, $\gamma_{ws} = \gamma_{w0}$ and $\gamma_{bs} = \gamma_{b0}$. We can then simplify the expression for $|\Re^{(1)}|^2$ in equation (14.139) as

$$\left|\Re^{(1)}\right|^2 = \frac{\left[\rho_b^2\left(2k_0^2 + \gamma_{w0}^2\right) - 2\rho_b\rho_w k_0^2 - \rho_w^2\gamma_{b0}^2\right]^2}{\left|\rho_b\gamma_{w0} + \rho_w\gamma_{b0}\right|^4} \qquad (14.145)$$

and then furthermore to

$$\left|\Re^{(1)}\right|^2 = \frac{\left[(1-\rho)^2\cos^2\theta_0 - \eta^{-2} + \rho^2\right]^2}{\left|\rho\sin\theta_0 + \sqrt{\eta^{-2} - \cos^2\theta_0}\right|^4} \qquad (14.146)$$

In equation (14.146), we have introduced the density ratio $\rho = \rho_b / \rho_w$ and the sound speed ratio $\eta = c_b / c_w$. Note that with an incident angle lower than the critical angle, $\cos\theta_0 > \eta^{-1}$, and the square root term in equation (14.146) becomes imaginary.

For back scattering we therefore obtain

$$\sigma_a = 4\left(\frac{\omega}{c_w}\right)^4 \sin^4\theta_0 \left|\Re^{(1)}\right|^2 \Psi(2\alpha,0) \tag{14.147}$$

We can compare this result with that produced by applying the Eckart theory expressed by equation (14.95). With $\gamma = (\omega / c_w)\sin\theta_0$ we obtain for the Eckart theory

$$\sigma_a = 4\left(\frac{\omega}{c_w}\right)^4 \sin^4\theta_0 \left|\Re^{(0)}\right|^2 \Psi(2\alpha,0) \tag{14.148}$$

with $\Re^{(0)}$ being the ordinary reflection coefficient and thus

$$\left|\Re^{(0)}\right|^2 = \left|\frac{\rho\sin\theta_0 - \sqrt{\eta^{-2} - \cos^2\theta_0}}{\rho\sin\theta_0 + \sqrt{\eta^{-2} - \cos^2\theta_0}}\right|^2 \tag{14.149}$$

For scattering from a rough sea surface we replace $c_b$ and $\rho_b$ with the sound speed and density of air; consequently, $\rho \ll 1$ and $\eta \ll 1$ and $|\Re^{(1)}|^2 = |\Re^{(0)}|^2 \approx 1$. With these values used in equations (14.138) and (14.148), we obtain the same results for the two theories

$$\sigma_a = 4\left(\frac{\omega}{c_w}\right)^4 \sin^4\theta_0 \Psi(2\alpha,0) \tag{14.150}$$

### *14.4.7 Application to bottom scattering*

In principle, for bottom scattering we should be able to determine the roughness spectrum experimentally by making very detailed and accurate depth measurements in order to obtain samples of the height function $\zeta(x,y)$. Then the roughness spectrum $\Psi(\mathbf{K})$ could be found with the help of a two-dimensional Fourier transform, followed by averaging all measurements. In reality, determining the roughness spectrum with this method is very difficult; in the literature, we find only rare cases where it has been done. The roughness spectrum in most cases must be estimated, either on the basis of the information available about the seabed or by inverse modeling of measured scattering.

In the absence of detailed information, often the assumption is made that the roughness spectrum is isotropic, meaning that the roughness has no preferred directions. In such an instance, the roughness wave number spectrum can be expressed by a single scalar wave number $K$, which was introduced in equation (14.99).

In some cases we are justified in assuming isotropic roughness with the wave number spectrum given by equation (14.99). We then find the scattering strength according to the perturbation theory to be

$$\sigma_a = 4\left(\frac{\omega}{c_w}\right)^4 \sin^2\theta_s \sin^2\theta_0 \left|\Re^{(1)}\right|^2 \frac{G(2\alpha)}{4\pi\alpha} \tag{14.151}$$

Note that the function $|\Re^{(1)}|^2$ is dependent on density and sound speed as well as on the angles $\theta_s$ and $\theta_0$. In the following section, we consider some cases of practical interest.

One possible form of the roughness spectrum follows the Gaussian curve, but more commonly we assume that the spectrum follows a power law, which is expressed as

$$\begin{aligned} G(K) &= G_0 K^{-n} \qquad K \geq K_u \\ G(K) &= 0 \qquad\quad\;\; K > K_u \end{aligned} \tag{14.152}$$

where $G_0$ is a scaling constant and $n$ is the exponent.

The lower boundary condition $K_u$ is necessary to ensure that the variance in surface roughness, which was introduced in equation (14.100), is finite; a finite variance prevents the singularity that otherwise occur for $K = 0$. This singularity is caused by normal incidence in the case of backscattering, and by the specular direction $\theta_s = \pi - \theta_0$ in the case of forward scattering. With the limitation that $K > K_u$, this theory's validity is limited to angles that satisfy the condition

$$\left|\frac{\omega}{c_w}\left(\cos\theta_0 + \cos\theta_s\right)\right| \geq K_u \tag{14.153}$$

In acoustics literature, the assumption is often made that when $n$ of equation (14.152) is between 2 and 4, we obtain the best fit to measured values of backscattering. Particularly, the value $n = 3$ is interesting, since this value makes the scattering strength independent of the frequency, as can be seen from solutions to equation (14.151).

We now consider an example of backscattering and forward scattering using the theory developed in this section for two types of bottom sediments, sand and silts, with sound speeds and densities given in table 14.1.

Table 14.1. Sound speeds and densities used in the examples.

| *Medium* | *Density (kg/m³)* | *Sound speed (m/s)* |
|---|---|---|
| Water | 1000 | 1500 |
| Silt bottom | 1500 | 1650 |
| Sand bottom | 1800 | 1800 |

We also assume that the roughness wave number spectrum has the form of equation (14.152) with the scaling constant's value being $G_0 = 0.025$ and the exponent $n = 3$. These values are close to the ones recommended by Jackson, Winebrenner, and Ishimaru (1986) as giving a reasonably good fit to measured values. The value of $n = 3$ for the exponent makes the scattering strengths independent of frequency, as we have noted above.

Figure 14.16 shows the backscattering strength $S_a$ as a function of the incident angle and figure 14.17 shows the forward-scattering strength as a function of the scattering angle for an incident wave with a grazing angle of 30°. Both graphs show the results for sand and silt bottoms. For comparison, we have also calculated the backscattering strength for a pressure release surface, that is, the sea surface, and for a rigid bottom with infinite density and sound speed in addition to the results for the two sediment types.

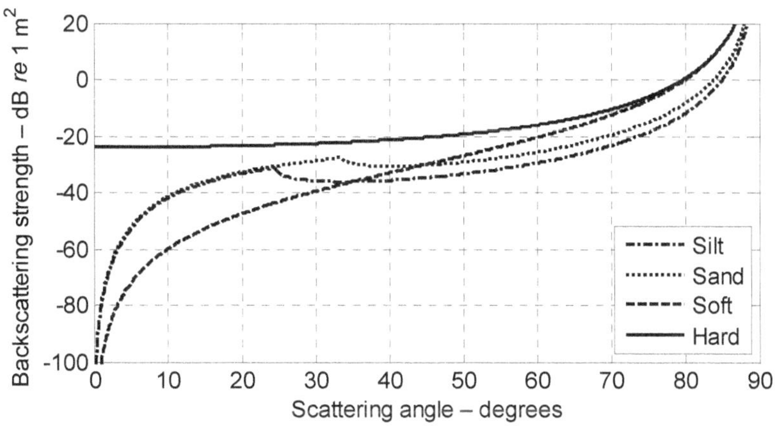


Figure 14.16. Backscattering strength, $S_a$, according to the first-order perturbation theory as a function of scattering angle. The results for sand and silt bottoms are calculated with parameters given in table 14.1. For comparison, the figure also shows the backscattering strength for soft surface and for an infinitely hard surface.

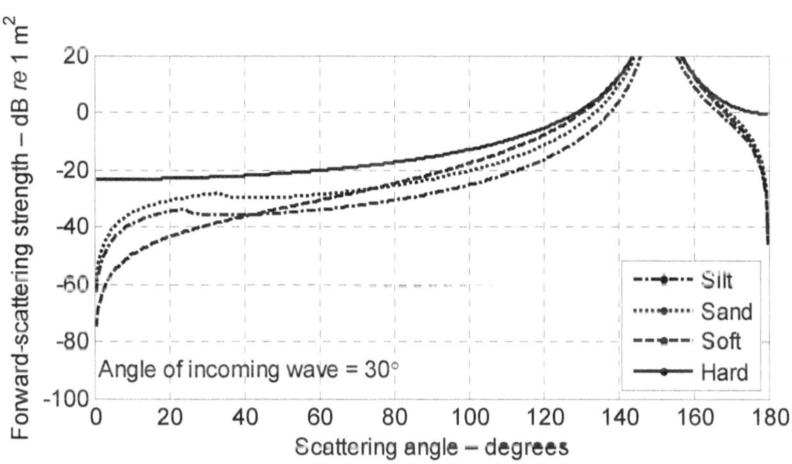


Figure 14.17. Forward-scattering strengths as functions of grazing angle for ocean bottoms composed of sand and silt, and for infinite hard and soft bottoms. The direction of the incident wave is 30°.

We notice a small local maximum in the scattering strength at an incident angle that is equal to the critical angle of total reflection; in this case the critical angle is 33.6° for the sand bottom and 24.6° for the silt bottom. Theoretically, this maximum value could be used to establish the sound speeds within the seabed, but in reality, maxima of this kind are very difficult to measure. The forward scattering strength in figure 14.17 goes towards infinity in the direction of the specular direction, which in this case is at 150° since the incident wave is 30°.

### *14.4.8 Comparison and discussion*

During development of the models, mathematical necessities have forced us to make assumptions and simplifications that are rarely possible in practice. Thus, the models should be used carefully.

The scattering cross sections we have developed in this chapter are used in the sonar equations, and to estimate reverberation from the seabed and the sea surface. At the distance $r$ from the receiver, the intensity of the reverberation is expressed as

$$I_B(r) = I_0 \frac{A}{r^2} \sigma(\theta_s, \theta_0) \qquad (14.154)$$

On the right-hand side of the equation, the incident intensity is represented by $I_0$, and $A$ is the area of the surface being radiated by the sonar.

In this chapter, we have presented the two most often cited mathematical models for scattering from rough surfaces. We now discuss the similarities of the two theories by comparing the backscattering strengths as a function of the grazing angle for a sand and silt bottom with sound speeds and densities given in table 14.1. The results are shown in figures 14.18 and 14.19 for sand and silt, respectively. In addition to the two model results we also show the scattering strength of Lambert's law, which we used in chapter 11 and is given as

$$S_a = 10 \log \mu + 20 \log(\sin \theta_0) \qquad (14.155)$$

In this example we use $\mu = 10^{-3}$.

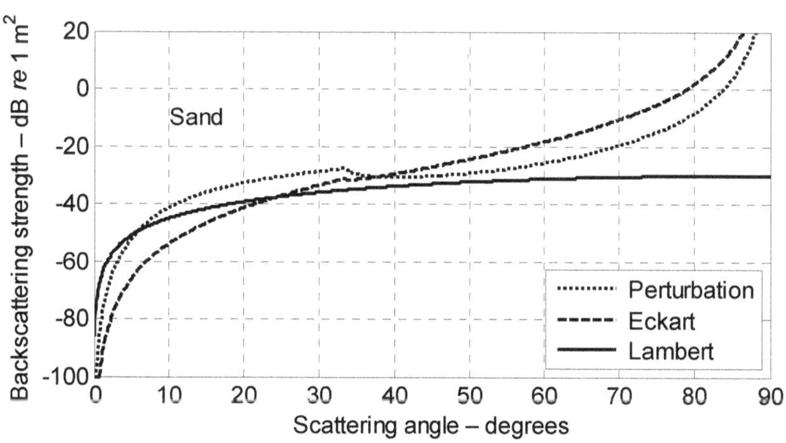


Figure 14.18. Comparing the backscattering strength $S_a$ calculated with the Eckart model and the perturbation theory for a sand bottom. Also shown is the scattering strength according to Lambert's law.

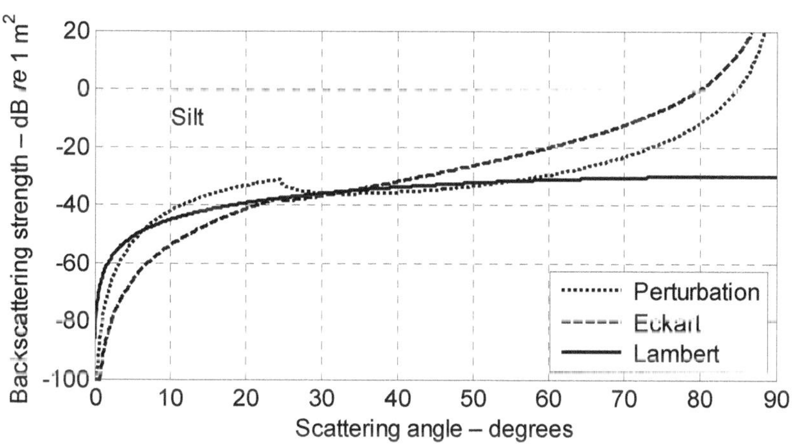


Figure 14.19. Comparing the backscattering strength $S_a$ calculated with the Eckart model and the perturbation theory for a silt bottom. Also shown is the scattering strength according to Lambert's law.

The results of the two models are similar, with both exhibiting a cusp at the critical angles of 33.6° for sand and 24.6° for silt, but there are differences in backscattering levels. At low grazing angles the backscattering strength predicated by the perturbation models is significantly higher than the level of the Eckart model. The Lambert's law values, which are independent of the physical properties of the bottom, are in this case close to the values of the perturbation model for very small grazing angles.

The two models are based on different assumptions and therefore, to some extent, the models are complementary as to region of validity. The perturbation model is likely to be more accurate at low grazing angles and with surface height deviations that are small compared with the acoustic wavelength. The Eckart model is more accurate for angles closer to normal incidence and with large-scale roughness, that is, where the angle of curvature is large compared with the acoustic wavelength. This dichotomy has led to development of ***composite scattering models***, where the two models we have developed here are combined in attempts to provide one model that is valid for a larger range of frequencies and roughness (Brekhovskikh and Lysanov 2003; Bachmann 1973; Jackson, Winebrenner, and Ishimaru 1986).

# Chapter 15

# Elastic waves in solids

## 15.1 Introduction

In chapter 2 we developed a wave equation that describes acoustic waves traveling in gases and fluids; this wave equation has been the basis for all our discussions so far. In the present chapter, we expand the discussion to include acoustic and elastic waves in solid media. The main difference between fluid and solid media is that, in the former, only ***longitudinal waves*** can propagate, but the latter can support both longitudinal and ***transverse waves***. In a ***longitudinal***, or ***compressional*** wave, particles of the medium are displaced only in the direction of the wave propagation, but in a solid medium, ***shear waves*** with particle displacement perpendicular to the direction of propagation can also exist.

Understanding how waves propagate in elastic media is essential in seismic applications and in dealing with wave propagation in rock. In marine acoustics, we normally need to be concerned with shear waves in applications where waterborne acoustic waves interact with the seabed. Such interactions typically take place during long-range propagation in shallow waters, where the propagation paths include many reflections at the seabed boundary. For long-range and low-frequency sonars in shallow water, the conversion of waterborne acoustic waves into shear waves in the sea bottom may be the predominant energy-loss mechanism affecting the sonar's performance.

In this chapter we first develop the wave equation for elastic waves. Then we look at some properties and characteristics of such solid-media waves, limiting our discussion to ***homogeneous isotropic solids,*** where the physical properties are the same everywhere and independent of direction. Lastly, we consider reflection and transmission of compressional and shear waves at an interface between a fluid and a solid.

## 15.2 Stresses and strains in isotropic solids

Stress is the ratio of the average internal forces within a deformable body to the area over which that force is distributed. It is expressed in units of newtons/m$^2$, dynes/cm$^2$, and pounds/ft$^2$. Strain is the relative change in dimensions or shape of a body when subjected to stress. It is expressed as a ratio of the change in dimension to the original dimension. There are several types of strain. For example, tensile strain is expressed as the ratio of the change in length to its original length when subjected to stress. Volume strain is defined as the ratio of change in volume to the original volume when subjected to stress. Shear strain is defined as a change in angular deformation measured in radians.

First we consider the strains resulting from a stress, or a force per unit of surface area, on a small cubical volume element $dV$ with sides $\Delta x$, $\Delta y$, and $\Delta z$ (figure 15.1). In rectangular coordinates, the particle displacement **u** has the components $u_x$, $u_y$, and $u_z$. The volume's relative expansions or compressions are expressed as

$$e_{xx} = \frac{\partial u_x}{\partial x}$$
$$e_{yy} = \frac{\partial u_y}{\partial y} \tag{15.1}$$
$$e_{zz} = \frac{\partial u_z}{\partial z}$$

The relative change of volume, called dilation, is denoted as $\Delta$ and defined as

$$\Delta = \frac{dV}{V} = e_{xx} + e_{yy} + e_{zz} \tag{15.2}$$

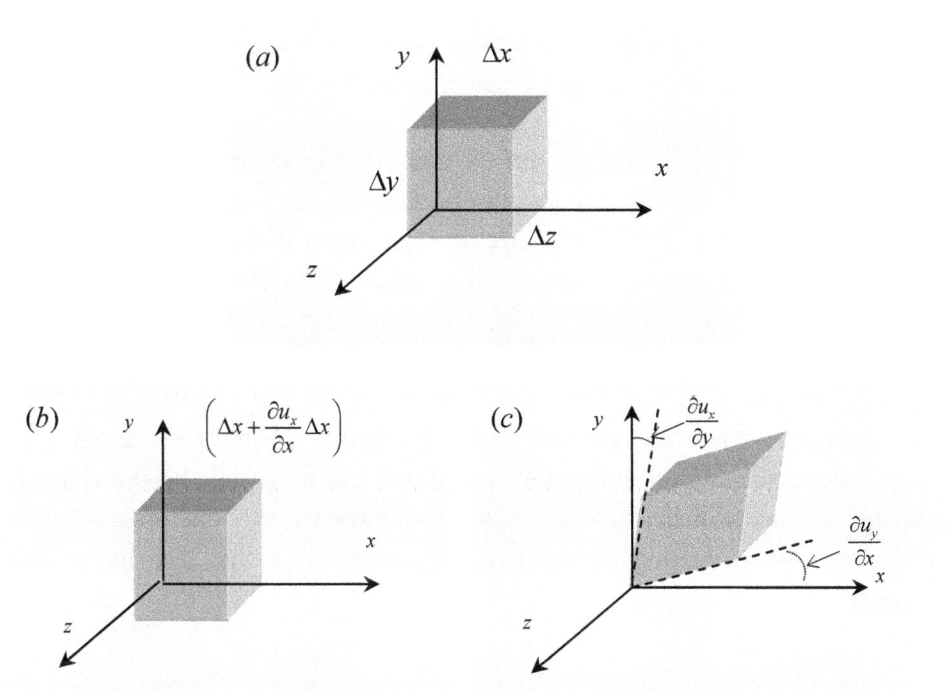


Figure 15.1. Strains, or the deforming effects of stress, in a solid medium. (*a*) the unperturbed cubical element; (*b*) volume strain caused by compression in the *x*-direction; (*c*) an example of shear strains without volume change.

The shear strains, which are also called shear deformations, are defined as

$$
\begin{aligned}
e_{xy} &= \frac{1}{2}\left(\frac{\partial u_x}{\partial y} + \frac{\partial u_y}{\partial x}\right) \\
e_{yz} &= \frac{1}{2}\left(\frac{\partial u_y}{\partial z} + \frac{\partial u_z}{\partial y}\right) \\
e_{xz} &= \frac{1}{2}\left(\frac{\partial u_x}{\partial z} + \frac{\partial u_z}{\partial x}\right)
\end{aligned}
\tag{15.3}
$$

The shear strain $e_{xy}$ is the rotation of the cubic element in the $x$-$y$ plane; $e_{yz}$ and $e_{xz}$ are the rotations in the $y$-$z$ plane and the $x$-$z$ plane, respectively.

Stress is defined as force per unit of surface area and is a vector with components in the three directions $x$, $y$, and $z$. Furthermore, stresses can act in all three planes; therefore, we have nine different components of stress acting on an elementary element. These components are illustrated in figure 15.2. The first index of each component indicates the direction in which the stress acts, while the second index gives the orientation of the plane in which the stress acts.

We now want to relate the stresses $\sigma_{ij}$ to the strains $\varepsilon_{ij}$ and we assume linear elasticity. Because equilibrium requires that the shear strains have symmetric properties, we know that $\sigma_{xy} = \sigma_{yx}$, $\sigma_{zy} = \sigma_{yz}$, and $\sigma_{xz} = \sigma_{zx}$. Therefore, we can conclude that for a homogeneous, isotropic solid, stress has six independent components.

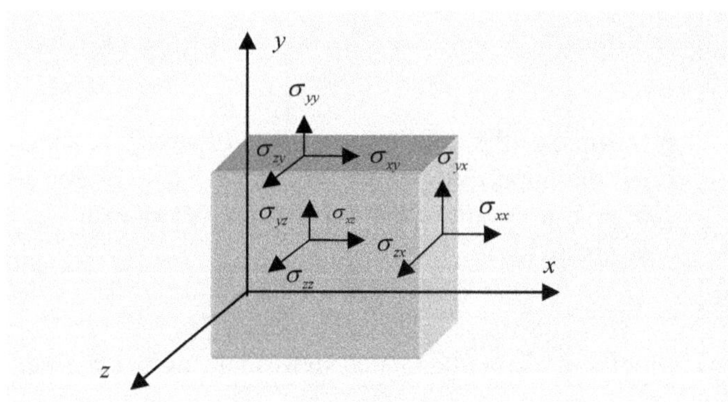


Figure 15.2. Stresses acting on a cubical volume element.

In elastic solids, stresses have linear relationships to strains. If, in addition, the material is also isotropic (that is, the material has the same elastic properties in all directions), the elastic properties are completely described by just two independent elastic coefficients that relate a stress to the resultant strains. Commonly used elasticity coefficients are the Lamé coefficients $\lambda$ and $\mu$, the Young's modulus $E$, the bulk modulus $K$, and the Poisson number $\nu$. If we use the Lamé coefficients to describe these relations, the stress and its resultant strains may be defined as

$$\begin{aligned}
\sigma_{xx} &= (\lambda + 2\mu)e_{xx} + \lambda e_{yy} + \lambda e_{zz} \\
\sigma_{yy} &= \lambda e_{xx} + (\lambda + 2\mu)e_{yy} + \lambda e_{zz} \\
\sigma_{zz} &= \lambda e_{xx} + \lambda e_{yy} + (\lambda + 2\mu)e_{zz} \\
\sigma_{xy} &= 2\mu e_{xy} \\
\sigma_{yz} &= 2\mu e_{yz} \\
\sigma_{zx} &= 2\mu e_{zx}
\end{aligned} \tag{15.4}$$

Table 15.1. Relationships among elastic parameters used to describe isotropic media

| *Elastic parameter* | *Expressed in terms of the parameters* | | | |
|---|---|---|---|---|
| | $\lambda, \mu$ | $H, \mu$ | $K, \mu$ | $E, \nu$ |
| $H =$ | $\lambda + 2\mu$ | $H$ | $K + \frac{4}{3}\mu$ | $E\frac{(1-\nu)}{(1-\nu-2\nu^2)}$ |
| $\mu =$ | $\mu$ | $\mu$ | $\mu$ | $E\frac{1}{2(1+\nu)}$ |
| $K =$ | $\lambda + \frac{2}{3}\mu$ | $H - \frac{4}{3}\mu$ | $K$ | $E\frac{1}{3(1-2\nu)}$ |
| $E =$ | $\mu\frac{(3\lambda+2\mu)}{(\lambda+\mu)}$ | $\mu\frac{(3H-4\mu)}{(H-\mu)}$ | $\frac{9\mu K}{(\mu+3K)}$ | $E$ |

Other elastic coefficients are presented in Table 1 and will be discussed subsequently. To understand the connections between the different elasticity coefficients, we consider a thin, long, square rod subjected to an axial stress $\sigma_{xx}$. When all other stresses are equal to zero, we can conclude from equation (15.4) that

$$\begin{aligned}\sigma_{xx} &= (\lambda + 2\mu)e_{xx} + \lambda e_{yy} + \lambda e_{zz} \\ 0 &= \lambda e_{xx} + (\lambda + 2\mu)e_{yy} + \lambda e_{zz} \\ 0 &= \lambda e_{xx} + \lambda e_{yy} + (\lambda + 2\mu)e_{zz}\end{aligned} \tag{15.5}$$

Young's modulus $E$ is defined as the ratio between the axial stress and the relative axial extension. After solving equation (15.5), we can express Young's modulus as

$$E = \frac{\sigma_{xx}}{e_{xx}} = \frac{\mu(3\lambda + 2\mu)}{(\lambda + \mu)} \tag{15.6}$$

The Poisson ratio $\nu$ is defined as the ratio between the transverse strains, where $e_{yy} = e_{zz}$, and the axial strain $e_{xx}$, so that

$$\nu = -\frac{e_{yy}}{e_{xx}} = \frac{\lambda}{2(\lambda + \mu)} \tag{15.7}$$

By definition, a fluid has $\mu = 0$. Consequently, the Poisson ratio for a fluid is $\nu = 0.5$, which can be seen from equation (15.7).

The bulk modulus $K$, or the volume stiffness, is defined by the relative change of volume when a material is exposed to hydrostatic stresses, which means that $\sigma_{xx} = \sigma_{yy} = \sigma_{zz} = \sigma$. When these values are inserted into equation (15.4), we obtain

$$\sigma = \left(\lambda + \frac{2}{3}\mu\right)\left(e_{xx} + e_{yy} + e_{zz}\right) \tag{15.8}$$

It then follows that

$$\sigma = \left(\lambda + \frac{2}{3}\mu\right)\Delta \tag{15.9}$$

Consequently, the bulk modulus $K$ may be expressed as

$$K = \frac{\sigma}{\Delta} = \lambda + \frac{2}{3}\mu \qquad (15.10)$$

The plane wave modulus $H$ may be defined either as

$$H = \lambda + 2\mu \qquad (15.11)$$

or as

$$H = K + \frac{4}{3}\mu \qquad (15.12)$$

We shall later see that the wave speed of a compressional wave in an elastic medium is defined in terms of the plane wave modulus and the density. Table 15.1 gives the relationships among the elastic parameters defined in the preceding paragraphs.

## 15.3 The wave equation

We shall now derive the wave equation for elastic media. This derivation is similar to the development in section 2.2, but we ignore the higher-order effects and directly develop the linear wave equation.

Figure 15.3 shows the stresses that act on an elementary cubical volume element in the direction of the $x$-axis, and how those stresses vary over the dimensions $\Delta x$, $\Delta y$ and $\Delta z$ of the element. By adding the forces that act in the $x$-direction and applying Newton's law, we obtain an equation that describes acceleration in the $x$-direction as

$$\rho \frac{\partial^2 u_x}{\partial t^2}(\Delta x \Delta y \Delta z) = \left( \frac{\partial \sigma_{xx}}{\partial x} + \frac{\partial \sigma_{xy}}{\partial y} + \frac{\partial \sigma_{xz}}{\partial z} \right)(\Delta x \Delta y \Delta z) \qquad (15.13)$$

For the other two directions we find similar expressions. Therefore, we have the following equations relating stress to the second time derivatives (accelerations) of the particle displacements in all three directions

$$\rho\frac{\partial^2 u_x}{\partial t^2}=\frac{\partial\sigma_{xx}}{\partial x}+\frac{\partial\sigma_{xy}}{\partial y}+\frac{\partial\sigma_{xz}}{\partial z}$$
$$\rho\frac{\partial^2 u_y}{\partial t^2}=\frac{\partial\sigma_{yx}}{\partial x}+\frac{\partial\sigma_{yy}}{\partial y}+\frac{\partial\sigma_{yz}}{\partial z} \qquad (15.14)$$
$$\rho\frac{\partial^2 u_z}{\partial t^2}=\frac{\partial\sigma_{zx}}{\partial x}+\frac{\partial\sigma_{zy}}{\partial y}+\frac{\partial\sigma_{zz}}{\partial z}$$

The next step is to replace the stresses in equation (15.14) with the strains, first using equation (15.4) and then equations (15.1) and (15.3). After some algebra, we obtain the following differential equation for the displacement along the $x$-axis

$$\rho\frac{\partial^2 u_x}{\partial t^2}=(\lambda+\mu)\frac{\partial}{\partial x}\left(\frac{\partial u_x}{\partial x}+\frac{\partial u_y}{\partial y}+\frac{\partial u_z}{\partial z}\right)+\mu\nabla^2 u_x \qquad (15.15)$$

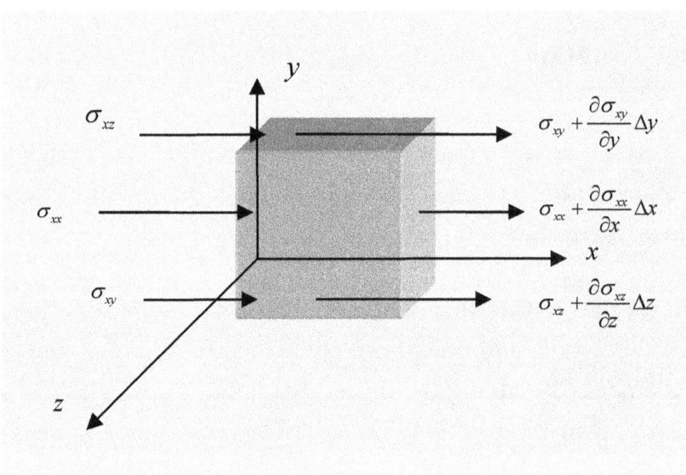


Figure 15.3. The stresses in the direction of the $x$-axis on an elementary cubical volume.

Similar equations may be derived for the other two axial directions. When they are put together, we obtain the vector equation

$$\rho\frac{\partial^2\mathbf{u}}{\partial^2 t}=(\lambda+\mu)\nabla(\nabla\cdot\mathbf{u})+\mu\nabla^2\mathbf{u} \tag{15.16}$$

In this wave equation, $\lambda$ and $\mu$ are the Lamé elasticity coefficients, $\rho$ is the density of the medium, and **u** is the displacement vector with components $u_x$, $u_y$, and $u_z$.

It is often convenient to recast equation (15.16) by expressing the particle displacement vector as two potential functions, a scalar potential $\phi$ and a vector potential $\psi$**,** so that the particle displacement vector is expressed as

$$\mathbf{u}=\nabla\phi+\nabla\times\mathbf{\Psi} \tag{15.17}$$

Then, by inserting equation (15.17) into equation (15.16), we obtain

$$\rho\left(\nabla\frac{\partial\phi^2}{\partial t^2}+\nabla\times\frac{\partial^2\mathbf{\Psi}}{\partial t^2}\right)=(\lambda+\mu)\nabla\left[\nabla\cdot\nabla\phi+\nabla\cdot(\nabla\times\mathbf{\Psi})\right]+\mu\nabla^2(\nabla\phi+\nabla\times\mathbf{\Psi}) \tag{15.18}$$

By definition, $\nabla\cdot(\nabla\times\Psi) = 0$. The terms containing $\phi$ and $\psi$ in equation (15.18) may be separated, which results in the following two wave equations

$$\rho\frac{\partial^2\phi}{\partial t^2}=(\lambda+2\mu)\nabla^2\phi \tag{15.19}$$

$$\rho\frac{\partial^2\mathbf{\Psi}}{\partial t^2}=\mu\nabla^2\mathbf{\Psi} \tag{15.20}$$

From equation (15.19), we conclude that the scalar potential $\phi$ propagates at the compressional wave speed $c_p$ defined as

$$c_p=\sqrt{\frac{(\lambda+2\mu)}{\rho}}=\sqrt{\frac{H}{\rho}} \tag{15.21}$$

and from equation (15.20) we conclude that the vector potential $\mathbf{\Psi}$ propagates with the shear wave speed $c_s$ defined as

$$c_s = \sqrt{\frac{\mu}{\rho}} \tag{15.22}$$

Using the Poisson relationship, the ratio between the two wave speeds defined in equations (15.21) and (15.22) can be expressed as

$$\frac{c_s}{c_p} = \sqrt{\frac{1-2\nu}{2(1-\nu)}} \tag{15.23}$$

Attenuation has been ignored in our development, but can be included in an approximate manner by adding imaginary components to the wave speeds as described in section 2.4.

After inserting the two wave speeds into equations (15.19) and (15.20), respectively, the two wave equations are rewritten as

$$\nabla^2 \phi = \frac{1}{c_p^2} \frac{\partial^2 \phi}{\partial t^2} \tag{15.24}$$

$$\nabla^2 \mathbf{\Psi} = \frac{1}{c_s^2} \frac{\partial^2 \mathbf{\Psi}}{\partial t^2} \tag{15.25}$$

Equations (15.24) and (15.25) are the two wave equations relevant to acousto-elastic wave propagation in an isotropic elastic medium. They may be solved in the same way as the acoustic wave equation was solved in chapter 2. We note that the equations, and consequently their solutions, are mutually independent—compressional waves do not affect the shear waves, and *vice versa.* However, this independence pertains only to waves in homogeneous media. In an inhomogeneous medium with space-dependent parameters (at the interface between two different media, for instance) conversions between compressional waves and shear waves take place. We will see examples of such conversions later in this chapter.

### *15.3.1 Vertically polarized waves*

Often we are only interested in a two-dimensional case in which the particle motions are in the *x-z* plane and where there is no *y*-plane dependence. Shear waves that are polarized so that the particle motion is in the *x-z* plane are called ***vertically polarized shear waves*** or *SV* waves.

When $u_y = 0$ and the derivatives of all quantities with respect to *y* are zero (which we denote by shorthand $\partial / \partial y = 0$), the vector potential $\boldsymbol{\Psi}$ has only a *y*-plane component $\Psi_y$, which, for notational simplicity, we will denote as $\Psi$. The two components of the particle displacement in equation (15.17) then become

$$u_x = \frac{\partial \phi}{\partial x} - \frac{\partial \psi}{\partial z}$$
$$u_z = \frac{\partial \phi}{\partial z} + \frac{\partial \psi}{\partial x} \tag{15.26}$$

Equation (15.4) defines the stress. When we apply equations (15.26), (15.1), and (15.3) to equation (15.4), we find, after some algebra, that the stresses can be expressed by the potentials

$$\sigma_{xx} = (\lambda + 2\mu)\left(\frac{\partial^2 \phi}{\partial x^2} + \frac{\partial^2 \phi}{\partial z^2}\right) - 2\mu\left(\frac{\partial^2 \phi}{\partial z^2} + \frac{\partial^2 \psi}{\partial x \partial z}\right)$$
$$\sigma_{zz} - (\lambda + 2\mu)\left(\frac{\partial^2 \phi}{\partial x^2} + \frac{\partial^2 \phi}{\partial z^2}\right) - 2\mu\left(\frac{\partial^2 \phi}{\partial x^2} - \frac{\partial^2 \psi}{\partial x \partial z}\right) \tag{15.27}$$
$$\sigma_{xz} = \mu\left(2\frac{\partial^2 \phi}{\partial x \partial z} + \frac{\partial^2 \psi}{\partial x^2} - \frac{\partial^2 \psi}{\partial z^2}\right)$$

Equations (15.27) may be simplified further by the application of equations (15.19) and (15.20)

$$\begin{aligned}
\sigma_{xx} &= \rho\frac{\partial^2\phi}{\partial t^2} - 2\mu\left(\frac{\partial^2\phi}{\partial z^2} + \frac{\partial^2\psi}{\partial x\partial z}\right) \\
\sigma_{zz} &= \rho\frac{\partial^2\phi}{\partial t^2} - 2\mu\left(\frac{\partial^2\phi}{\partial x^2} - \frac{\partial^2\psi}{\partial x\partial z}\right) \\
\sigma_{xz} &= \rho\frac{\partial^2\psi}{\partial t^2} + 2\mu\left(\frac{\partial^2\phi}{\partial x\partial z} - \frac{\partial^2\psi}{\partial z^2}\right) \\
&= -\rho\frac{\partial^2\psi}{\partial t^2} + 2\mu\left(\frac{\partial^2\phi}{\partial x\partial z} + \frac{\partial^2\psi}{\partial x^2}\right)
\end{aligned} \tag{15.28}$$

Later in the chapter we will use equations (15.26) for the displacements and equations (15.28) for the stress in terms of the two potentials, but in the next section we will present an alternative formulation also found in the literature.

### *15.3.2 Formulation by differential equations*

We have chosen to express the wave equations in terms of the vector potential $\mathbf{\Psi}$ and the scalar potential $\phi$. The use of these potentials is quite common in the literature (Jensen et al. 1994, Aki and Richards 2002), but it is not really necessary since we can also work directly with stresses and strains by setting up a matrix differential wave equation for the relevant variables. In the case of vertical polarization, the relevant variables are $u_x$, $u_z$, $\sigma_{zz}$, and $\sigma_{zx}$. We start with equation (15.4) and replace the strains with the displacement differentials by using equations (15.1) and (15.3). Because $\partial/\partial y = 0$ and $u_y = 0$ for vertically polarized waves, we find that

$$\begin{aligned}
\sigma_{xx} &= (\lambda + 2\mu)\frac{\partial u_x}{\partial x} + \lambda\frac{\partial u_z}{\partial z} \\
\sigma_{zz} &= \lambda\frac{\partial u_x}{\partial x} + (\lambda + 2\mu)\frac{\partial u_z}{\partial z} \\
\sigma_{zx} &= \mu\left(\frac{\partial u_z}{\partial x} + \frac{\partial u_x}{\partial z}\right)
\end{aligned} \tag{15.29}$$

Using equation (15.14), we get for the same conditions

$$\begin{aligned} \rho\frac{\partial^2 u_x}{\partial t^2} &= \frac{\partial \sigma_{xx}}{\partial x} + \frac{\partial \sigma_{xz}}{\partial z} \\ \rho\frac{\partial^2 u_z}{\partial t^2} &= \frac{\partial \sigma_{zx}}{\partial x} + \frac{\partial \sigma_{zz}}{\partial z} \end{aligned} \tag{15.30}$$

We assume plane harmonic waves and substitute $\partial/\partial t = (-i\omega)$ and $\partial/\partial x = (ik)$. Next, we eliminate $\sigma_{xx}$ and collect all the differentials with respect to $z$ on the left-hand side. We then get the vector differential equation

$$\frac{d}{dz}\begin{bmatrix} u_z \\ \sigma_{zz} \\ \sigma_{zx} \\ u_x \end{bmatrix} = \begin{bmatrix} 0 & \dfrac{1}{(\lambda+2\mu)} & 0 & -ik\dfrac{\lambda}{(\lambda+2\mu)} \\ -\rho\omega^2 & 0 & -ik & 0 \\ 0 & -ik\dfrac{\lambda}{(\lambda+2\mu)} & 0 & 4k^2\dfrac{\mu(\lambda+\mu)}{(\lambda+2\mu)} - \rho\omega^2 \\ -ik & 0 & \dfrac{1}{\mu} & 0 \end{bmatrix} \begin{bmatrix} u_z \\ \sigma_{zz} \\ \sigma_{zx} \\ u_x \end{bmatrix} \tag{15.31}$$

Introducing the wave speeds $c_p$ and $c_s$ from equations (15.21) and (15.22), we obtain

$$\frac{d}{dz}\begin{bmatrix} u_z \\ \sigma_{zz} \\ \sigma_{zx} \\ u_x \end{bmatrix} = \begin{bmatrix} 0 & \dfrac{1}{\rho c_p^2} & 0 & -ik\left(1-2\dfrac{c_s^2}{c_p^2}\right) \\ -\rho\omega^2 & 0 & -ik & 0 \\ 0 & -ik\left(1-2\dfrac{c_s^2}{c_p^2}\right) & 0 & 4k^2 c_s^2\rho\left(1-\dfrac{c_s^2}{c_p^2}\right) - \rho\omega^2 \\ -ik & 0 & \dfrac{1}{\rho c_s^2} & 0 \end{bmatrix} \begin{bmatrix} u_z \\ \sigma_{zz} \\ \sigma_{zx} \\ u_x \end{bmatrix} \tag{15.32}$$

When $c_p$, $c_s$, and $\rho$ are known functions of the depth $z$, we can use standard methods to integrate the differential equations. Note that the integrated solution must satisfy the boundary conditions at both $z = 0$ and $z = \infty$.

## 15.4 Reflection and transmission of plane waves

The reflection and transmission coefficients at interfaces between solid media can be determined in the same way as those for interfaces between fluid media, but we need to include shear waves in our calculations. In general, shear waves are both vertically and horizontally polarized, but in most marine acoustic applications, we only need to consider vertically polarized shear waves. It then follows that we must consider four waves (down/up, pressure/shear) and therefore work with $4 \times 4$ matrices. We will now see how the general problem can be solved with the use of matrices and vectors, and thereafter consider two specific simple examples.

In the general case, where we have compressional and vertically polarized shear waves that propagate as plane waves in the $x$-$z$ plane, we can define the scalar potential $\phi$ and the vector potential $\psi$ as

$$\begin{aligned}\phi &= \left[D_p \exp\left(i\gamma_p z\right) + U_p \exp\left(-i\gamma_p z\right)\right]\exp\left[i\left(kx - \omega t\right)\right] \\ \psi &= \left[D_s \exp\left(i\gamma_s z\right) + U_s \exp\left(-i\gamma_s z\right)\right]\exp\left[i\left(kx - \omega t\right)\right]\end{aligned} \tag{15.33}$$

Waves $D_p$ and $D_s$ are downward-moving (that is, traveling in the positive $z$-direction) compressional and shear waves, and $U_p$ and $U_s$ are upward-moving compressional and shear waves. The vertical wave numbers are $\gamma_p$ and $\gamma_s$ for the compressional and the shear waves, respectively

$$\begin{aligned}\gamma_{p,1} &= \frac{\omega}{c_{p,1}}\sin\theta_{p,1} \\ \gamma_{p,2} &= \frac{\omega}{c_{p,2}}\sin\theta_{p,2} \\ \gamma_{s,2} &= \frac{\omega}{c_{s,2}}\sin\theta_{s,2}\end{aligned} \tag{15.34}$$

The angles are given by Snell's law, which requires that the horizontal wave number $k$ must be the same for all waves. Consequently, $k$ is defined as

$$k = \frac{\omega}{c_{p,2}} \cos\theta_{p,2} = \frac{\omega}{c_{s,2}} \cos\theta_{s,2} = \frac{\omega}{c_{p,1}} \cos\theta_{p,1} \tag{15.35}$$

At an interface between solid media the continuity conditions require that the normal stress $\sigma_{zz}$, normal displacement $u_z$, and tangential stress $\sigma_{zx}$ be continuous and that the tangential stress in the fluid be zero. Using equations (15.26) and the last two equations in equations (15.28), the relevant stresses and displacements at the interface $z = 0$, may be written

$$\begin{bmatrix} \sigma_{zz} \\ \sigma_{zx} \\ u_z \\ u_x \end{bmatrix} = \begin{bmatrix} -\rho\left(\omega^2 - 2k^2c_s^2\right) & -\rho\left(\omega^2 - 2k^2c_s^2\right) & -2\rho k\gamma_s c_s^2 & 2\rho k\gamma_s c_s^2 \\ -2\rho k\gamma_p c_s^2 & 2\rho k\gamma_p c_s^2 & \rho\left(\omega^2 - 2k^2c_s^2\right) & \rho\left(\omega^2 - 2k^2c_s^2\right) \\ i\gamma_p & -i\gamma_p & ik & ik \\ ik & ik & -i\gamma_s & i\gamma_s \end{bmatrix} \begin{bmatrix} D_p \\ U_p \\ D_s \\ U_s \end{bmatrix}$$

(15.36)

Now we consider the simplest case, where waves arrive at a plane interface between a fluid and a solid medium, as depicted in figure 15.4.

In the fluid medium, the shear speed is zero, and we see from figure 15.4 that there is the incident compressional wave $D_{p,1}$, which has the grazing angle $\theta_{p,1}$, plus a reflected compressional wave $U_{p,1}$. Therefore, equation (15.36) may be simplified to become

$$\begin{bmatrix} \sigma_{zz} \\ u_z \end{bmatrix} = \begin{bmatrix} -\rho_1\omega^2 & -\rho_1\omega^2 \\ i\gamma_{p,1} & -i\gamma_{p,1} \end{bmatrix} \begin{bmatrix} D_{p,1} \\ U_{p,1} \end{bmatrix} \tag{15.37}$$

This result is identical to equation (5.7), except that now we use $\sigma$ to represent stress rather than the sound pressure $p$ where $p = -\sigma$.

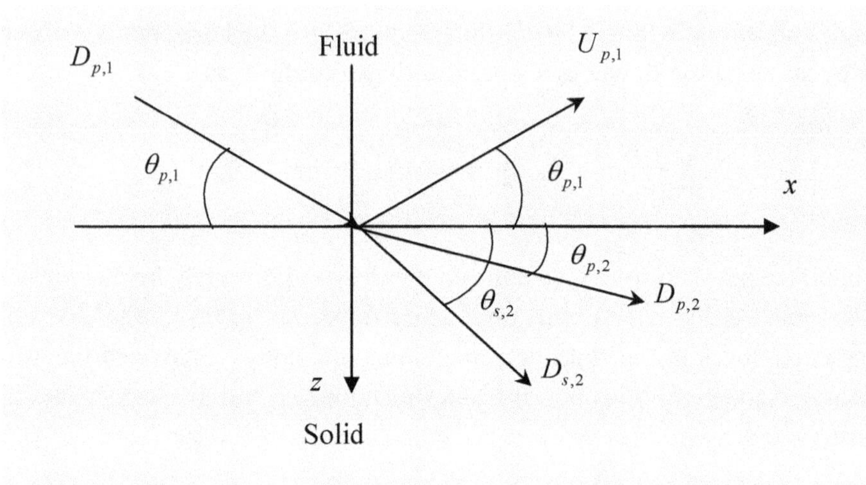


Figure 15.4. Incident wave ($D_{p,1}$), reflected wave ($U_{p,1}$), transmitted compressional wave ($D_{p,2}$), and transmitted shear wave ($D_{s,2}$) at the interface between fluid and solid media.

In our present example, the lower medium contains only downward-moving pressure and shear waves. Since there are no conditions placed on the $x$-component of the particle displacement, we need only consider the first three equations of equations (15.36)

$$\begin{bmatrix} \sigma_{zz} \\ u_z \\ \sigma_{zx} \end{bmatrix} = \begin{bmatrix} -\rho_2\left(\omega^2 - 2k^2c_{s,2}^2\right) & -2\rho_2 k\gamma_{s,2}c_{s,2}^2 \\ i\gamma_{p2} & ik \\ -2\rho_2 k\gamma_{p,2}c_{s2}^2 & \rho_2\left(\omega^2 - 2k^2c_{s,2}^2\right) \end{bmatrix} \begin{bmatrix} D_{p,2} \\ D_{s,2} \end{bmatrix} \tag{15.38}$$

The boundary conditions require continuity of normal stress $\sigma_{zz}$ and of vertical particle displacement $u_z$ at the interface, and therefore require that the stress $\sigma_{zz}$ and the vertical displacements $u_z$ of equation (15.37) equal those of equation (15.38). This gives the following equations to solve for the unknown waves

$$\begin{bmatrix} D_{p,1} \\ U_{p,1} \end{bmatrix} = \begin{bmatrix} -\rho_1\omega^2 & -\rho_1\omega^2 \\ i\gamma_{p,1} & -i\gamma_{p,1} \end{bmatrix}^{-1} \begin{bmatrix} -\rho_2\left(\omega^2 - 2k^2c_{s,2}^2\right) & -2\rho_2 k\gamma_{s,2}c_{s,2}^2 \\ i\gamma_{p,2} & ik \end{bmatrix} \begin{bmatrix} D_{p,2} \\ D_{s,2} \end{bmatrix} \tag{15.39}$$

From equation (15.35), it follows that

$$\begin{aligned} \cos\left(2\theta_{s,2}\right) &= 2\cos^2\theta_{s,2} - 1 = 2\frac{k^2}{\omega^2}c_{s,2}^2 - 1 \\ \sin\left(2\theta_{s,2}\right) &= 2\sin\theta_{s,2}\cos\theta_{s2} = 2k\gamma_{s,2}\frac{c_{s,2}^2}{\omega^2} \end{aligned} \tag{15.40}$$

We must also take into consideration that $\sigma_{zx} = 0$ at the interface, which, from the last row of equation (15.38), gives us

$$D_{p,2} = D_{s,2}\frac{\left(\omega^2 - 2k^2c_{s,2}^2\right)}{2k\gamma_{p,2}c_{s,2}^2} = -\frac{\gamma_{s,2}}{\gamma_{p,2}}\frac{\cos\left(2\theta_{s,2}\right)}{\sin\left(2\theta_{s,2}\right)}D_{s,2} \tag{15.41}$$

When we know the amplitude of the incident wave $D_{p,1}$ in the water, equations (15.39) and (15.41) determine the amplitudes of the reflected wave $U_{p,1}$ and the two transmitted waves $D_{p,2}$ and $D_{s,2}$,. After some algebra, we find that the reflection coefficient $r_{12}$ is

$$r_{12} = \frac{U_{p,1}}{D_{p,1}} = \frac{Z_{p,2}\cos^2\left(2\theta_{s,2}\right) + Z_{s,2}\sin^2\left(2\theta_{s,2}\right) - Z_{p,1}}{Z_{p,2}\cos^2\left(2\theta_{s,2}\right) + Z_{s,2}\sin^2\left(2\theta_{s,2}\right) + Z_{p,1}} \tag{15.42}$$

and the transmission coefficients $t_{p,12}$ and $t_{s,12}$ for compressional and shear waves, respectively, are

$$t_{p,12} = \frac{D_{p,2}}{D_{p,1}} = \frac{2\frac{\rho_1}{\rho_2}Z_{p,2}\cos\left(2\theta_{s,2}\right)}{Z_{p,2}\cos^2\left(2\theta_{s,2}\right) + Z_{s,2}\sin^2\left(2\theta_{s,2}\right) + Z_{p,1}} \tag{15.43}$$

$$t_{s,12} = \frac{D_{s,2}}{D_{p,1}} = \frac{2\frac{\rho_1}{\rho_2} Z_{s,2} \sin\left(2\theta_{s,2}\right)}{Z_{p,2}\cos^2\left(2\theta_{s,2}\right) + Z_{s,2}\sin^2\left(2\theta_{s,2}\right) + Z_{p,1}} \tag{15.44}$$

where the impedances $Z_{p,1}$, $Z_{p,2}$, and $Z_{s,2}$ are defined as

$$\begin{aligned} Z_{p,1} &= \omega \frac{\rho_1}{\gamma_{p,1}} \\ Z_{p,2} &= \omega \frac{\rho_2}{\gamma_{p,2}} \\ Z_{s,2} &= \omega \frac{\rho_2}{\gamma_{s,2}} \end{aligned} \tag{15.45}$$

The reflection and transmission coefficients we have developed above are generalizations of the corresponding equations discussed in section 4.2. Note that when the shear wave speed in the lower medium approaches zero, it leads to $\sin(2\theta_{s,2}) = 0$ and $\cos(2\theta_{s,2}) = -1$, according to equation (15.40), and therefore $t_{s,12} = 0$. The reflection and transmission coefficients for compressional waves then become, as we found earlier in equations (4.14) and (4.15)

$$r_{12} = \frac{U_{p,1}}{D_{p,1}} = \frac{\gamma_{p,1}\rho_2 - \gamma_{p,2}\rho_1}{\gamma_{p,1}\rho_2 + \gamma_{p,2}\rho_1} \tag{15.46}$$

$$t_{p,12} = \frac{D_{p,2}}{D_{p,1}} = \frac{2\gamma_{p,1}\rho_1}{\gamma_{p,1}\rho_2 + \gamma_{p,2}\rho_1} \tag{15.47}$$

### *15.4.1 Intensities*

Now we determine the intensities of the reflected and transmitted waves. Intensity is measured in power (watt) per square meter. The general expression for intensity $I$ is the magnitude of the real part of the product of the stress and the complex conjugated particle velocity normal to the boundary

$$I = \left| \text{Re} \left\{ \sigma_{zz} \dot{u}_z^* \right\} \right| \tag{15.48}$$

For the incident intensity $I_i$ we obtain

$$I_i = \left| \text{Re} \left\{ \omega^3 \rho_1 \gamma_{p,1} D_{p,1}^2 \right\} \right| \tag{15.49}$$

and for the reflected intensity $I_r$

$$I_r = \left| \text{Re} \left\{ \omega^3 \rho_1 \gamma_{p,1} U_{p,1}^2 \right\} \right| = I_i \left| r_{12} \right|^2 \tag{15.50}$$

The intensity of the transmitted compressional and shear waves can be expressed as (omitting, for the moment, the Re{ } notation)

$$I_t = \left| \left[ -\rho_2 \left( \omega^2 - 2k^2 c_{s,2}^2 \right) D_{p,2} - 2\rho_2 k \gamma_{s,2} c_{s,2}^2 D_{s,2} \right] \left[ \omega \gamma_{p,2} D_{p,2}^* + \omega k D_{s,2}^* \right] \right| \tag{15.51}$$

After multiplying the two terms and using equations (15.40), we find

$$I_t = \rho_2 \omega^3 \left| \begin{matrix} \gamma_{p,2} \cos\left(2\theta_{s,2}\right) D_{p,2} D_{p,2}^* + k \cos\left(2\theta_{s,2}\right) D_{p,2} D_{s,2}^* - \\ \gamma_{p,2} \sin\left(2\theta_{s,2}\right) D_{s,2} D_{p,2}^* - k \sin\left(2\theta_{s,2}\right) D_{s,2} D_{s,2}^* \end{matrix} \right| \tag{15.52}$$

We now use equation (15.41), but substitute for $D^*_{s,2}$ in the second term of equation (15.52) and for $D^*_{p,2}$ in the third term, thereby obtaining

$$I_t = \rho_2 \omega^3 \left| \begin{matrix} \left[ \gamma_{p,2} \cos\left(2\theta_{s,2}\right) - k \dfrac{\gamma_{p,2}}{\gamma_{s,2}} \sin\left(2\theta_{s,2}\right) \right] \left| D_{p,2} \right|^2 \\ \left[ -k \sin\left(2\theta_{s,2}\right) + \gamma_{p,2} \dfrac{\gamma_{s,2}}{\gamma_{p,2}} \cos(2\theta_{s,2}) \right] \left| D_{s,2} \right|^2 \end{matrix} \right| \tag{15.53}$$

finding $I_t$ to be

$$I_t = \omega^3 \rho_2 \gamma_{p,2} \left| D_{p,2} \right|^2 + \omega^3 \rho_2 \gamma_{s,2} \left| D_{s,2} \right|^2 \tag{15.54}$$

The transmitted intensity can therefore be expressed as the sum of two terms, one given by the intensity $I_{tp}$ of the transmitted compressional wave and one by the intensity $I_{ts}$ of the shear wave

$$I_{tp} = \mathrm{Re}\left\{ \frac{\rho_2 \gamma_{p,2}}{\rho_1 \gamma_{p,1}} \right\} I_i \left| t_{p,12} \right|^2 \tag{15.55}$$

and

$$I_{ts} = I_i \,\mathrm{Re}\left\{ \frac{\rho_2 \gamma_{s,2}}{\rho_1 \gamma_{p,1}} \right\} \left| t_{s,12} \right|^2 \tag{15.56}$$

If we substitute these for the reflection and transmission coefficients used in equations (15.42), (15.43), and (15.44), we find that $I_i = I_r + I_{tp} + I_{ts}$, which demonstrates that energy is conserved as the plane wave interacts with the interface.

Now we have found a general expression that defines the reflection and transmission coefficients at an interface between a fluid and a solid medium. Next we consider two examples that address reflection within a sedimentary seabed having shear rigidity.

### *15.4.2 Reflection from a seabed with shear rigidity*

In this section we examine the reflection coefficient in a seabed having rigidity, and we study the importance of the conversion of compressional waves to shear waves within the seabed. In water, we assume $c_{p,1}$ = 1500 m/s and $\rho_1$ = 1000 kg/m$^3$. Values of these parameters for seabed materials are quite variable, but in marine sediments the compressional speed is usually higher than the sound speed in water, and the shear speed is lower. The seabed material's density is always greater than that of the water. Therefore, we can normally assume that

$$
\begin{aligned}
c_{p,2} &> c_{p,1} \\
c_{s,2} &< c_{p,1} \\
\rho_2 &> \rho_1
\end{aligned}
\tag{15.57}
$$

In general, there may be two critical angles: one for the compressional wave and one for the shear wave

$$
\begin{aligned}
\theta_{p,crit} &= \cos^{-1}\left(\frac{c_{p,1}}{c_{p,2}}\right) \\
\theta_{s,crit} &= \cos^{-1}\left(\frac{c_{p,1}}{c_{s,2}}\right)
\end{aligned}
\tag{15.58}
$$

With the wave speed relationships given in equation (15.57) only $\theta_{p,crit}$ exists.

Figure 15.5 contains plots of the reflected and transmitted intensities for the pressure waves and the shear wave, normalized with respect to the intensity of the incident wave and calculated with equations (15.50), (15.55), and (15.56).

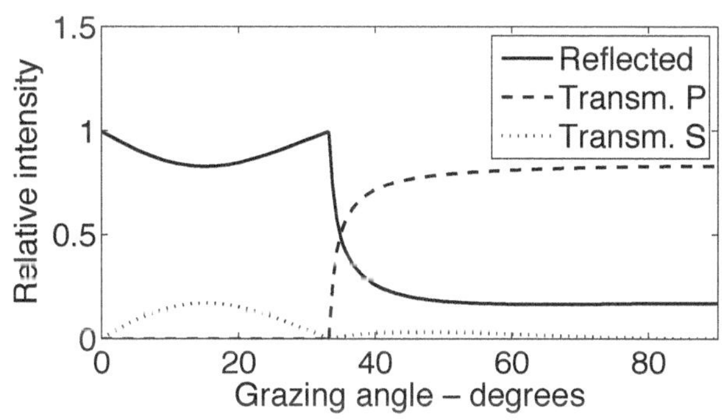


Figure 15.5. Reflected and transmitted intensities, relative to the incident intensity, for shear and compressional waves, plotted as functions of the grazing angle for a sedimentary seabed with low shear speed.

In these calculations, we have used the numerical values $c_{p,1} = 1500$ m/s, $c_{p,2} = 1800$ m/s, $c_{s,2} = 500$ m/s, $\rho_1 = 1000$ kg/m$^3$, and $\rho_2 = 2000$ kg/m$^3$. Since $c_{p,1} > c_{s,2}$, there is no critical angle for the shear wave, but the compressional wave has a critical angle of 33.5°.

For incident waves with grazing angles smaller than this angle, no compressional wave is transmitted into the seabed. Because the transmitted shear wave reduces the reflected intensity, the magnitude of the reflection coefficient is less than unity for an incident wave with an angle lower than the critical angle, as can be seen from figure 15.5. This demonstrates that shear wave conversion can be important for the seabed reflection loss at angles shallower than the critical angle.

The bottom reflection loss, *RL*, is defined as

$$RL = -20\log|R| \tag{15.59}$$

where *R* is the reflection coefficient of the bottom.

Figure 15.6 illustrates this using the same parameters as used to generate figure 15.5. We also see that there is no conversion to shear waves at the critical angle $\theta_{p,crit}$ and at zero angle; for these cases, the reflection is total and the reflection loss is zero. At all other angles, some of the acoustic intensity is converted to shear intensity.

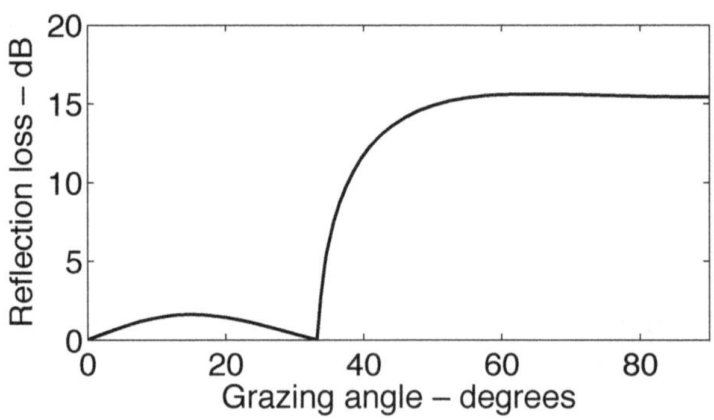


Figure 15.6. Reflection loss as a function of the grazing angle for a sedimentary seabed with low shear speed.

Figure 15.6 should also be compared with figure 4.7, which shows the reflection loss for a fluid sediment based on the same numerical values but with compressional wave attenuation included as a parameter. We see that shear wave conversion in the solid sediment and absorption in a fluid sediment may have similar effects on the bottom reflection loss, and it may be impossible to tell the difference.

Now we will consider an example of a highly rigid seabed with the values $c_{p,2} = 4000$ m/s, $\rho_2 = 2200$ kg/m$^3$, and $c_{s,2} = 1800$ m/s. The critical angle for the compressional wave is approximately 70° and the critical angle for the shear wave is 33.5°. This is an example of multiple critical angles, defined by equation (15.58). The reflection and the transmission coefficients are plotted in figure 15.7, and the reflection loss is plotted in figure 15.8. Reflection is total at angles less than the shear critical angle. In general cases of multiple critical angles, reflection is total at angles less than the smallest critical angle. At angles between the two critical angles, the reflection coefficient is less than unity because the incident compressional wave is converted to a transmitted shear wave within the seabed. When the angle of incidence is greater than the larger critical angle, both shear waves and pressure waves are transmitted into the seabed.

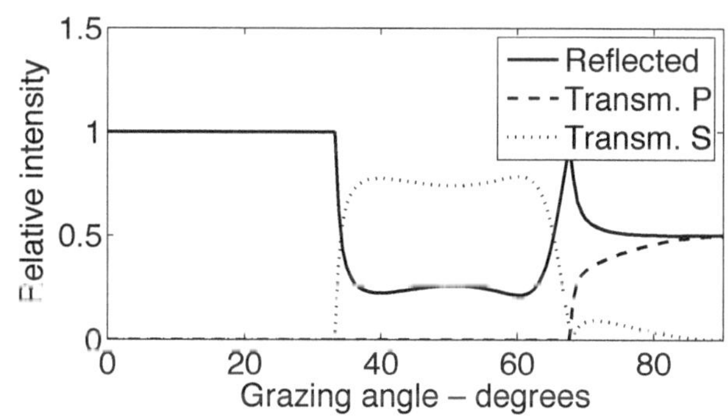


Figure 15.7. Reflected and transmitted intensities for shear and pressure waves, plotted as a function of the grazing angle for a seabed with high shear speed.

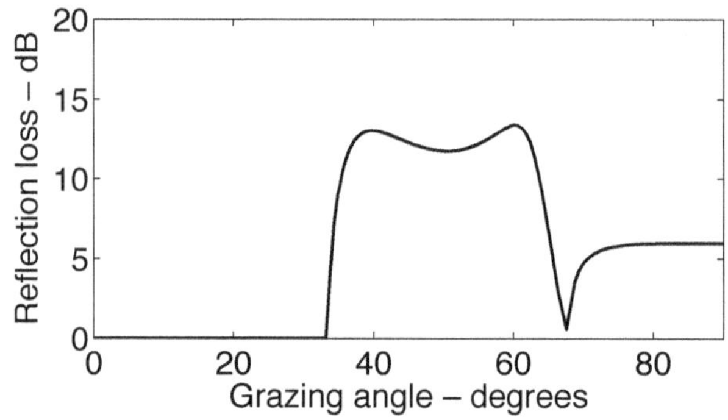


Figure 15.8. Reflection loss (dB) as function of grazing angle for a seabed with high shear speed.

From these two examples, we conclude that conversion of compressional waves to shear waves generally increases the reflection loss in the seabed. This effect is particularly important for the quality of acoustic transmission in shallow waters, where the sound propagation is bottom limited. In such instances, even a small loss upon each reflection can play a major role because of the large number of reflections.

## 15.5 Elastic multilayer media

When we examined acoustic waves in stratified fluid media, we were able to find some relatively simple recursions for calculating transmission and reflection coefficients for any number of layers. In this section, we will advance to the case of multiple elastic layers in a solid medium. A most useful description of the multilayer solution for elastic media is found in the article by Westwood, Tindle, and Chapman (1996).

In principle, the elastic case with shear waves is solved in the same way as the fluid case. In the elastic case, we have four waves (down/up, pressure/shear) and $4 \times 4$ matrices. The general problem can be solved by the use of matrices and vectors, which we will discuss here, and later we will look at some simple examples.

The waves in layer number $n$ of a solid medium having multiple elastic layers can be defined for $z_{n-1} < z < z_n$ as

$$\begin{aligned}\phi &= \left[D_{p,n}\exp\left(i\gamma_{p,n}(z-z_n)\right)+U_{p,n}\exp\left(-i\gamma_{p,n}(z-z_n)\right)\right]\exp\left[i(kx-\omega t)\right]\\ \psi &= \left[D_{s,n}\exp\left(i\gamma_{s,n}(z-z_n)\right)+U_{s,n}\exp\left(-i\gamma_{s,n}(z-z_n)\right)\right]\exp\left[i(kx-\omega t)\right]\end{aligned} \tag{15.60}$$

Here, $D_{p,n}$ and $D_{s,n}$ are downward-moving (that is, moving in the positive $z$-direction) compressional and shear waves, and $U_{p,n}$ and $U_{s,n}$ are upward-moving compressional and shear waves at the interface $z = z_n$ as shown in figure 15.9. The corresponding waves at the upper interface $z = z_{n-1}$ are given by a transport matrix in the form

$$\begin{bmatrix} D'_{p,n} \\ U'_{p,n} \\ D'_{s,n} \\ U'_{s,n} \end{bmatrix} = \begin{bmatrix} \exp\left(-i\gamma_{p,n}d_n\right) & 0 & 0 & 0 \\ 0 & \exp\left(i\gamma_{p,n}d_n\right) & 0 & 0 \\ 0 & 0 & \exp\left(-i\gamma_{s,n}d_n\right) & 0 \\ 0 & 0 & 0 & \exp\left(i\gamma_{s,n}d_n\right) \end{bmatrix} \begin{bmatrix} D_{p,n} \\ U_{p,n} \\ D_{s,n} \\ U_{s,n} \end{bmatrix} \tag{15.61}$$

where $d_n = z_n - z_{n-1}$ is the thickness of layer $n$.

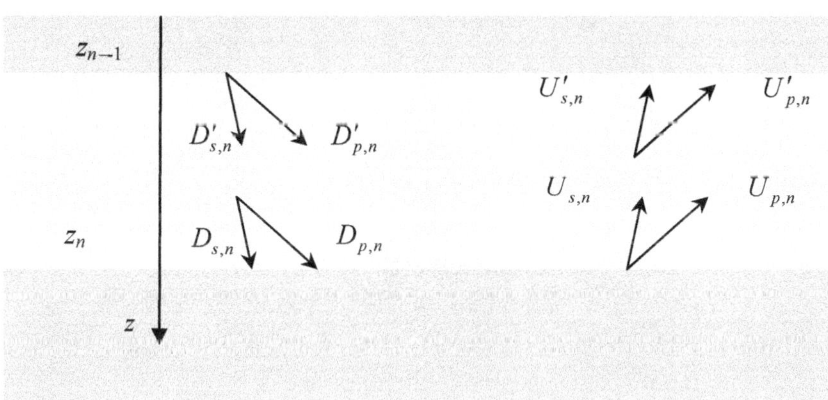


Figure 15.9. Waves in an elastic layer $n$ of a multilayer solid medium.

When we define the wave vector $\mathbf{B}_n = \left[D_{p,n}\ U_{p,n}\ D_{s,n}\ U_{s,n}\right]$, where indicates transposition, and the wave vector $\mathbf{B'}_n$ as $\mathbf{B}'_n = \left[D'_{p,n}\ U'_{p,n}\ D_{s,n}\ U'_{s,n}\right]$, we can rewrite equation (15.61) as

$$\mathbf{B}'_n = \mathbf{T}_n \mathbf{B}_n \qquad (15.62)$$

with $\mathbf{T}_n$ being the square matrix defined by equation (15.61). The boundary conditions at all interfaces include stresses $\sigma_{zz}$ and $\sigma_{xz}$, as well as the particle displacements $u_x$ and $u_z$, which are given by equation (15.36). At the interface $z = z_n$, the stresses and displacements may be written in terms of wave amplitudes as

$$\begin{bmatrix} \sigma_{zz} \\ \sigma_{zx} \\ u_z \\ u_x \end{bmatrix} = \begin{bmatrix} -\rho_n\left(\omega^2 - 2k^2c_{s,n}^2\right) & -\rho_n\left(\omega^2 - 2k^2c_{s,n}^2\right) & -2\rho_n k\gamma_{s,n}c_{s,n}^2 & 2\rho_n k\gamma_{s,n}c_{s,n}^2 \\ -2\rho_n k\gamma_{p,n}c_{sn}^2 & 2\rho_n k\gamma_{p,n}c_{s,n}^2 & \rho_n\left(\omega^2 - 2k^2c_{s,n}^2\right) & \rho_n\left(\omega^2 - 2k^2c_{s,n}^2\right) \\ i\gamma_{p,n} & -i\gamma_{p,n} & ik & ik \\ ik & ik & -i\gamma_{s,n} & i\gamma_{s,n} \end{bmatrix} \begin{bmatrix} D_{p,n} \\ U_{p,n} \\ D_{s,n} \\ U_{s,n} \end{bmatrix}$$

(15.63)

When we introduce the stress-displacement vector **W**

$$\mathbf{W}(z) = \left[\sigma_{zz}\quad \sigma_{zx}\quad u_z\quad u_x\right]^{\dagger} \qquad (15.64)$$

Equation (15.63) may be written in matrix form as

$$\mathbf{W}_n(z) = \mathbf{Q}_n \mathbf{B}_n \qquad (15.65)$$

where $\mathbf{Q}_n$ is the square matrix in equation (15.63).

Next, we consider the interface between layer $n-1$ and $n$ at the depth $z = z_{n-1}$. Then the stress–displacement vector just above the interface is

$$\mathbf{W}_{n-1}(z_{n-1}) = \mathbf{Q}_{n-1}\mathbf{B}_{n-1} \qquad (15.66)$$

and the stress-displacement vector just below the interface is

$$\mathbf{W}_n(z_{n-1}) = \mathbf{Q}_n \mathbf{B}'_n = \mathbf{Q}_n \mathbf{T}_n \mathbf{B}_n \tag{15.67}$$

The boundary conditions require that the stresses and displacements should be continuous at the interface. Therefore, the right-hand sides of equations (15.66) and (15.67) must be equal. This leads to the relationship

$$\mathbf{B}_{n-1} = \mathbf{Q}_{n-1}^{-1} \mathbf{Q}_n \mathbf{T}_n \mathbf{B}_n \tag{15.68}$$

Repeatedly using equation (15.68) gives us an expression for the waves in the upper layer $n = 0$. The initial values in the lower semi-infinite layer $N$ are

$$\mathbf{B}_N = \begin{bmatrix} D_{p,N} & 0 & D_{s,N} & 0 \end{bmatrix}^{\dagger} \tag{15.69}$$

While this recursive formulation can be used to find the reflected and incident fields in all layers, we are likely to be most interested in the relationship between the reflected and incident fields in the upper layer $n = 0$. When this layer is a fluid, the reflection coefficient $R$ is found to be

$$R = \frac{U_{p,0}}{D_{p,0}} \tag{15.70}$$

where $U_{p0}$ represents a reflected compressional wave in the layer $n = 0$. Such situations are encountered in marine seismic exploration where the transmitter and receiver are in water. The reflection coefficient $R$ in equation (15.70) includes all simple and multiple reflections as well as all conversions between compressional waves and shear waves.

In principle, this equation can be used to find all the waves; in reality, this approach involves a large number of calculations, which means numerical methods have to be applied. As well, we should be aware of a problem that may occur when one or more of the layers are of fluid material. In a viscidity-free medium, shear stress cannot exist and therefore $\sigma_{zx} = 0$, which means that the movement in the $x$-direction is free, that is, there are no conditions attached to the horizontal displacement $u_x$. This unconditional horizontal displacement may create problems in our numerical solutions for calculating the reflection coefficient.

### *15.5.1 Unconsolidated layer over a solid subsurface*

We will now consider a relatively simple example where we have seawater as the $n = 0$ layer; a fluid sediment layer 1, and a solid lower half-space as layer 2. This situation is illustrated in figure 15.10.

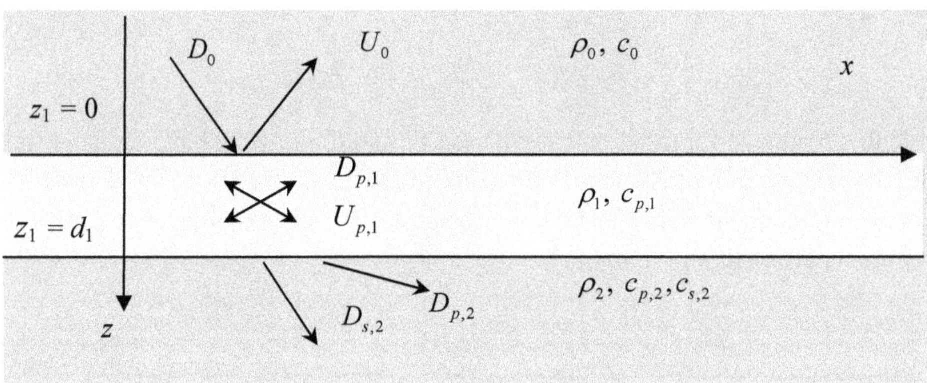


Figure 15.10. Three-layer model with shear rigidity only in the lower half-space.

Since the layer $n = 1$ is a fluid layer, we obtain the expression

$$\begin{aligned}\begin{bmatrix} D_0 \\ U_0 \end{bmatrix} &= \frac{1}{t_{01}} \begin{bmatrix} 1 & r_{01} \\ r_{01} & 1 \end{bmatrix} \begin{bmatrix} D'_{p,1} \\ U'_{p,1} \end{bmatrix} \\ &= \frac{1}{t_{01}} \begin{bmatrix} 1 & r_{01} \\ r_{01} & 1 \end{bmatrix} \begin{bmatrix} \exp(-i\gamma_{p,1} d_1) & 0 \\ 0 & \exp(i\gamma_{p,1} d_1) \end{bmatrix} \begin{bmatrix} D_{p,1} \\ U_{p,1} \end{bmatrix}\end{aligned} \tag{15.71}$$

Here, $r_{01}$ and $t_{01}$ are the reflection and transmission coefficients between the two fluid media. Further, the relationship between the reflected pressure wave $U_{p,1}$ and the incoming pressure wave $D_{p1}$ is the ordinary reflection coefficient for a fluid/solid interface, $r_{12} = U_{p,1} / D_{p,1}$, which we obtained from equation (15.42). Thus, we find that the total reflection coefficient is

$$R = \frac{U_0}{D_0} = \frac{r_{01} + r_{12} \exp\left(2i\gamma_{p,1} d_1\right)}{1 + r_{01} r_{12} \exp\left(2i\gamma_{p,1} d_1\right)} \tag{15.72}$$

We use equation (15.72) to calculate the reflection loss $RL = -20\log|R|$ as a function of the grazing angle and the frequency. The thickness of the layer is $d_1 = 5$ m. The values used for the densities, sound speeds, and attenuations are given in table 15.2. The attenuations are introduced by adding imaginary components to the wave speeds as described in section 2.4. The result of the calculation is shown in figure 15.11.

Table 15.2. Values of the geoacoustic parameters used to obtain figure 15.11

| *Sounds speeds* (m/s) | *Attenuations* (dB/λ) | *Densities* (kg/m$^3$) |
|---|---|---|
| $c_0 = 1500$ | 0 | $\rho_0 = 1000$ |
| $c_{p,1} = 1700$ | 0.5 | $\rho_0 = 1500$ |
| $c_{p,2} = 4700$ | 0.5 | $\rho_0 = 2000$ |
| $c_{s,2} = 2000$ | 0.5 | – |

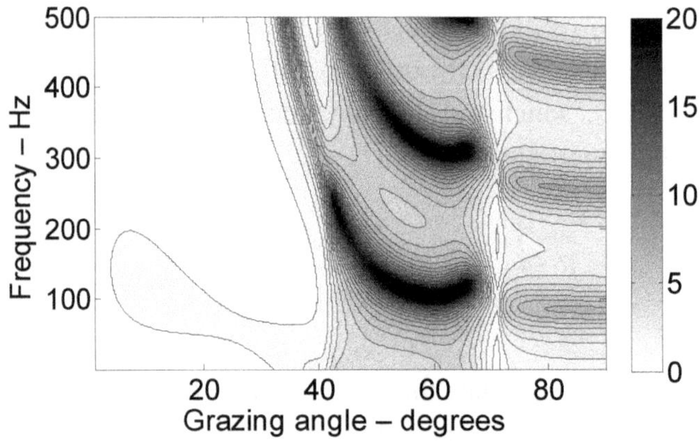


Figure 15.11. Reflection loss, plotted as a function of the grazing angle and frequency, for a seabed with a thin fluid sediment layer over a solid half-space. The geoacoustic parameter values are given in table 15.2.

Since all sound speeds within the seabed are higher than the sound speed in water, there are three critical angles. For the compressional wave arrival at the water-sediment interface the critical angle is $\theta_{p,1} = \cos^{-1}(1500 / 1700) = 28°$. At the half-space interface, the critical angles for the shear and compressional waves are $\theta_{s,2} = \cos^{-1}(1500 / 2000) = 41°$ and $\theta_{p,2} = \cos^{-1}(1500 / 4700) = 71°$, respectively.

We see from figure 15.11 that, at low frequencies, the reflection loss is essentially given by the geoacoustic properties of the solid half-space with very low reflection loss for angles up to the shear wave critical angle of 41°. At high frequencies, the reflection loss is determined by the conditions at the upper interface, and the lower half-space has little effect on the reflected field. At such frequencies there is total reflection at incident angles up to the critical angle of 28°. The transition between low and high frequencies occurs at the frequency where the layer is approximately one wavelength thick. In this case, the thickness of the sediment layer is 5 m and the transition frequency is 340 Hz, which agrees well with the behavior observed in figure 15.11.

# Chapter 16

# Sound in marine sediments

## 16.1 Introduction

Generally, marine sediments, soils, and sedimentary rocks are porous media. Such media are composed of a framework made up of interlocking solid particles. The voids, or pores, between the particles may be filled with a fluid or a gas. As a wave propagates through such a medium, the pore fluid or gas may move in relation to the solid particles. Sound propagation in such materials is of great importance in underwater acoustics and seismic exploration and also important for many other acoustical applications, such as in material science and fabrication of acoustic damping elements for buildings.

In porous media, sound speed and attenuation of a propagating wave are functions of the physical properties of the fluid and solid constituents—that is, functions of both porosity and framework rigidity. Because these functional dependencies are complex, and since the acoustic properties of marine sediments are difficult to determine theoretically, sound speeds and attenuations in marine sediments and soils are generally considered empirical quantities whose values can be found only by measurement. Hamilton (1980, 1987) provides very useful references about established values of sound speeds and attenuations within composite porous media. In order to make use of such information, it is necessary to have a broad understanding of the physical processes that are involved when sound is propagating in composite porous media. In this chapter, we discuss the common physical models, with emphasis on models for sound speed and attenuation in marine sediments. First we derive the Gassmann model of propagation in porous media, which can be considered an extension of the Wood's equation that we derived in chapter 3; the extension is to include the frame rigidity. In addition, we present some of the conventional viscous models, also known as "dashpot models," which are commonly used to describe attenuation of sound in sediments and rocks.

However, the major part of this chapter is spent on discussion of the Biot theory as it relates to sound propagation in porous media. This theory, which is a further extension of the Gassmann model, also takes into consideration the movement of pore fluid in relation to the framework. The Biot theory's importance lies in its prediction of three kinds of body waves in porous material: two types of pressure or compressional waves, plus a shear wave. The Biot model's shear wave and first type of pressure wave are very similar to those we already know from ordinary elastic wave theory; however, Biot's second pressure wave is a highly attenuated wave, which is significant mainly because it pulls energy away from the first, or ordinary, compressional wave, thereby increasing absorption significantly.

## 16.2 The Gassmann model for a porous medium with a rigid framework

In section 3.3 we derived Wood's equation for sound speed in a two-component material, and then applied the model to suspensions of solid particles in water. We concluded that one weakness in this model is that it does not take framework rigidity into consideration. This limitation makes the model ineffective for treating porous solids, where frame stiffness is a significant factor. Wood's equation does not deal adequately with both low-porosity sediments and porous rocks in which the solid particles are interconnected to form a rigid framework.

To begin, we present the Gassmann model, which factors frame rigidity into Wood's equation. We begin by considering a fluid-saturated porous medium, where the solid particles are cemented together in a framework that has its own elasticity modulus $K_r$ and shear modulus $\mu_r$. We assume that the fluid does not react with the solid framework in any way and, consequently, has no effect on the rigidity of the frame.

Let **u** be a vector that represents the displacement of particles connected to the framework, and let **U** be a vector for the fluid displacement. The relative volume change of the elements connected to the framework $e$ is defined as

$$e = \nabla \cdot \mathbf{u} \tag{16.1}$$

and the relative volume of fluid flowing through the framework $\zeta$ is defined as

$$\zeta = \phi \nabla \cdot (\mathbf{u} - \mathbf{U}) \tag{16.2}$$

where $\nabla$ represents the gradient operator and $\phi$ is the porosity.

We assume linear relations between the six independent components of a stress $\sigma$, which are denoted as $\sigma_{xx}$, $\sigma_{yy}$, $\sigma_{zz}$, $\sigma_{xy}$, $\sigma_{yz}$, and $\sigma_{zx}$, and the deformation caused by each. We express these relations as

$$\begin{aligned}
\sigma_{xx} &= He - 2\mu\left(e_{yy} + e_{zz}\right) - C\zeta \\
\sigma_{yy} &= He - 2\mu\left(e_{xx} + e_{zz}\right) - C\zeta \\
\sigma_{zz} &= He - 2\mu\left(e_{xx} + e_{yy}\right) - C\zeta \\
\sigma_{xy} &= 2\mu e_{xy} \\
\sigma_{yz} &= 2\mu e_{yz} \\
\sigma_{zx} &= 2\mu e_{zx}
\end{aligned} \tag{16.3}$$

where the ***plane wave modulus*** $H$ is equal to $(\lambda + 2\mu)$ and $C$ represents a new elastic modulus that presently is unknown, but which will be determined as we proceed.

We first define the mean normal stress $\sigma$ as

$$\sigma = \frac{1}{3}\left(\sigma_{xx} + \sigma_{yy} + \sigma_{zz}\right) \tag{16.4}$$

and add up equations (16.1), (16.2), and (16.3), which gives that

$$\sigma = \left(H - \frac{4}{3}\mu\right)e - C\zeta \tag{16.5}$$

We introduce the ***bulk modulus*** $K$, where $K = H - 4/3\ \mu$, and find that

$$\sigma = Ke - C\zeta \tag{16.6}$$

As pressure is the negative of stress, for the pore pressure $p_f$ we assume that

$$-p_f = Ce - M\zeta \tag{16.7}$$

where $M$ is a general elasticity modulus that we determine shortly.

Thus, equations (16.6) and (16.7) represent stress and pore pressure as functions of the elastic parameters $H$, $K$, $M$, and $C$. The values for these parameters are, for the time being, unknown, but we relate them to the properties of the framework and the particulate and fluid media. This can be done by two simple mental experiments, the jacketed and the unjacketed tests.

### Jacketed test

Here we imagine that a sample of a porous material is in a tight bag with an opening that allows the pore fluid to stream out when the bag is exposed to an external pressure, as shown in the left-hand drawing in figure 16.1.

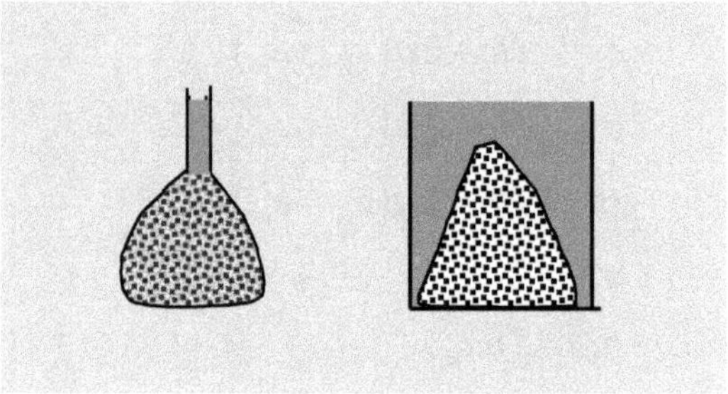

Figure 16.1. Left, the jacketed and right, the unjacketed tests of a composite medium having solid particles and a pore fluid.

From the result of the jacketed test, it follows that the medium's pore pressure is not affected by the external pressure; consequently, $p_f = 0$. By inserting this value into equations (16.6) and (16.7), we find that $\zeta = (C / M)e$ and the definition for $\sigma$ becomes

$$\sigma = \left(K - C^2/M\right)e \tag{16.8}$$

The rigidity, or the stiffness, of the test sample under the conditions shown on the left in figure 16.1 is essentially only the rigidity of the framework, $K_r$. Consequently

$$K_r = K - C^2/M \tag{16.9}$$

### Unjacketed test

In our next imagined experiment, presented on the right-hand side of figure 16.1, the sample is placed in a container filled with the same fluid as that in the pores. Thus, the ambient pressure and the pore pressure are identical: $\sigma = -p_f$. Using this value in equations (16.6) and (16.7) we find that

$$\sigma = \frac{K - C^2/M}{1 - C/M}e \tag{16.10}$$

Now the entire stress rests on the fixed particles; therefore, it follows that $e = \sigma / K_s$, so that

$$K_s = \frac{K - C^2/M}{1 - C/M} \tag{16.11}$$

where $K_s$ is the elasticity modulus for the material of the solid particles.

We further find that

$$p_f = \frac{KM - C^2}{K - C}\zeta \tag{16.12}$$

Equation (16.12) gives the displacement of the fluid $\zeta$, which is caused by a change in the pore pressure $p_f$. The fluid displacement can also be expressed as the difference between the compressibility of the fluid ($1 / K_f$) and the compressibility of the solid material ($1 / K_s$) multiplied by the porosity $\phi$, that is

$$\zeta = \phi\left(\frac{1}{K_f} - \frac{1}{K_s}\right)p_f \tag{16.13}$$

and, consequently

$$\phi\left(\frac{1}{K_f} - \frac{1}{K_s}\right) = \frac{K - C}{KM - C^2} \tag{16.14}$$

Equations (16.11) and (16.14) can be solved with regard to the unknown quantities of the elastic parameters $K$, $C$, and $M$. To do so we introduce a new auxiliary elastic constant $L$ defined as

$$L = \frac{K_f(K_s - K_r)}{\phi(K_s - K_f)} \tag{16.15}$$

Now we can define $C$, $M$, and $K$

$$C = \frac{LK_s}{K_s + L} \tag{16.16}$$

$$M = \frac{CK_s}{K_s - K_r} \tag{16.17}$$

$$\begin{aligned} K &= K_s \frac{K_r + L}{K_s + L} = K_r + C^2/M \\ &= K_r + \frac{(K_s - K_r)^2}{\phi K_s (K_s - K_f)/K_f + (K_s - K_r)} \end{aligned} \tag{16.18}$$

At this point we have theoretically determined the elastic qualities of the composite material, and have linked these qualities to those of the solid material, the pore fluid, and the framework. We now consider the dynamic conditions that result from time-varying stresses and strains caused by the acoustic waves.

The sound speed of the pressure or compressional wave $c_p$ is expressed as

$$c_p = \sqrt{\frac{H}{\rho}} \tag{16.19}$$

where the plane wave modulus $H$ is defined as $H = K + 4/3\ \mu$, with $K$ representing the bulk modulus, and the density $\rho$ is the average density given as

$$\rho = \phi \rho_f + (1-\phi)\rho_s \tag{16.20}$$

The shear modulus of the composite material is equal to the shear modulus of the framework, so that $\mu = \mu_r$ and the shear wave speed $c_s$ is defined as

$$c_s = \sqrt{\frac{\mu}{\rho}} \tag{16.21}$$

The suspensions to which we applied Wood's equation in section 3.3 correspond to $K_r = 0$ and $\mu_r = 0$. It is easy to verify these values with equation (16.18). This verification leads to Wood's equation, which we can consider to be a special case of the Gassmann model.

Some special limiting cases for weak and strong frameworks are useful to understand. For a composite medium where the cementation is weak, both $K_r$ and $\mu_r$ are very small. In equation (16.18), if we neglect $K_r$ in the terms with $(K_s - K_r)$, we find the effective bulk modulus $K$ of the composite medium to be approximately

$$K = K_r + \frac{K_s K_f}{\phi K_s + (1-\phi) K_f} \tag{16.22}$$

The second term on the right-hand side of equation (16.22) gives us the bulk modulus from Wood's equation for a suspension of solid particles in a fluid. We see that when the particles are in contact, as would be the case when the porosity is low, there is an additional term that can be attributed to frame stiffness. Equation (16.22), which we have already seen in section 3.3, accounts for the increase in sound speed in a suspension having a high concentration of solid particles.

The Gassmann model yields sound speeds as functions of the elastic parameters for porous materials. Unfortunately, determining these parameters by independent measurements is very difficult, especially for the parameters of the framework, that is, the bulk modulus of the frame $K_r$ and the shear modulus of the frame $\mu_r$. Because of this difficulty, the model can hardly be used to estimate sound speeds, which have to be measured. However, an advantage of the Gassmann model lies in its insight into how the values of the constituents' densities and elastic properties affect the sound speed in the composite material. For instance, we can use the model to estimate changes in sound speeds within a framework medium having a fluid in its pores when that fluid is replaced by another fluid or a gas. We consider one such example in the next section.

## 16.3 Acoustics of sediments with gas bubbles

Geophysical surveys have demonstrated that the presence of entrained gas in shallow sediments is widespread in many parts of the world. Bubbles of air can be introduced into sediments by raindrops, breaking waves, and ships. Bubbles of other gases can be introduced into sediments by various biological activities in water and sediment. Gas in sediment, even in small quantities, will strongly modify the medium's acoustic properties, influencing sound propagation and scattering. In this section we describe a heuristic model for sound propagation in gassy sediments based on applying the effective density and compressibility of a gassy fluid. The model assumes single scattering and that the bubbles are far from each other and from any boundaries, so that the bubbles may behave as if they are in an unbounded medium.

We assume that the pore space, or the collective volume of the pores, is partially filled with gas bubbles in addition to the water present in the pores. We develop the models by extending the Gassmann theory presented in section 16.2, the extension being that now we are dealing with a three-phase medium consisting of solid particles, pore fluid, and gas bubbles. Part of the total pore volume $\phi$ consists of the volume of gas $\phi n_g$ and the rest of the pore volume $\phi$ $(1 - n_g)$ is water. Therefore, we will modify the pore fluid's density and bulk modulus to account for the relative volume fraction of gas bubbles $n_g$. This model is essentially the same as that developed by Anderson and Hampton (1980a, 1980b), but may differ in some details.

For the density of the gaseous pore water $\rho_f$ we have the expression

$$\rho_f = n_g \rho_g + \left(1 - n_g\right) \rho_w \tag{16.23}$$

where $\rho_g$ and $\rho_w$ are, respectively, the densities of pore gas and pore water.

In the same way, we can express the compressibility of water containing gas bubbles $K_w^{-1}$ as

$$K_f^{-1} = n_g K_g^{-1} + \left(1 - n_g\right) K_w^{-1} \tag{16.24}$$

where $K_w$ is the bulk modulus of the pore water and $K_g$ is the bulk modulus of a gas bubble.

In section 13.5, we found that the oscillation in the radius of a spherical bubble is given by equation (13.89); consequently, we can define the compressibility of the bubble as

$$K_g^{-1} = \frac{3}{\rho_w a^2 \omega^2} \frac{1}{\left[\left(\dfrac{\omega_r}{\omega}\right)^2 - 1\right] - i\delta} \tag{16.25}$$

In equation (16.25), we see the damping constant of the bubble oscillations, $\delta$, which was discussed in section 13.5. The (angular) resonance frequency of the gas bubbles $\omega_r$ is defined as

$$\omega_r = \frac{1}{a}\sqrt{\frac{3\gamma p_a}{\rho_w}} \tag{16.26}$$

where the value of the gas constant $\gamma$ is approximately 1.4 for an ideal gas and $p_a$ represents the ambient pressure. For very low frequencies, that is, where $\omega \ll \omega_r$, equations (16.25) and (16.26) produce

$$K_g = \frac{\rho_w a^2 \omega_r^2}{3} = \gamma p_a \tag{16.27}$$

Since the atmospheric pressure is approximately $10^5$ Pa, the bulk modulus for a gas bubble at a depth $d$ in water is approximately

$$K_g = 1.4 \cdot 10^5 \left(1 + \frac{d}{10}\right) \tag{16.28}$$

where $d$ is measured in meters. Using the same numerical values, next we find the resonance frequency $f_r$ of a bubble with radius $a$ that is located at a water depth $d$ as

$$f_r = \frac{\omega_r}{2\pi} = \frac{3.2}{a}\sqrt{1 + \frac{d}{10}} \tag{16.29}$$

where $f_r$ is measured in hertz and the radius $a$ and depth $d$ are expressed in meters.

After these substitutions, the sound speed for the compressional wave $c_p$ is expressed as

$$c_p = \sqrt{\frac{K + 4/3\mu}{\rho}} \tag{16.30}$$

and the shear wave sound speed $c_s$ is written as

$$c_s = \sqrt{\frac{\mu}{\rho}} \tag{16.31}$$

The expressions defining the effective density $\rho$ and bulk modulus $K$, which are to be substituted into equations (16.30) and (16.31), are

$$\rho = (1 - \phi)\rho_s + \phi\rho_f \tag{16.32}$$

$$K = K_r + \frac{(K_s - K_r)^2}{\phi K_s (K_s - K_f)/K_f + (K_s - K_r)} \tag{16.33}$$

In equations (16.32) and (16.33) we use the density and the bulk modulus of the fluid as given in equations (16.23) and (16.24).

Now we are ready to illustrate this model's use by considering an example with the physical parameters given in table 16.1.

Table 16.1. Parameters used with the Gassmann model

| *Parameter* | *Value* | *Parameter* | *Value* |
|---|---|---|---|
| $\mu$ | $0.1 \cdot 10^9$ Pa | $\rho_s$ | $1.8 \cdot 10^3$ kg/m$^3$ |
| $K_r$ | $0.5 \cdot 10^9$ Pa | $\rho_w$ | $1.0 \cdot 10^3$ kg/m$^3$ |
| $K_s$ | $4.0 \cdot 10^{10}$ Pa | $\rho_g$ | 1.3 kg/m$^3$ |
| $K_w$ | $2.25 \cdot 10^9$ Pa | $\phi$ | 0.35 |
| $K_g$ | $4.6 \cdot 10^5$ Pa | $\delta$ | 0.1 |
| $a$ | 0.0002 m | Depth | 100 m |

The first example considers the low-frequency case, where the frequency of the acoustic wave is much lower than the resonance frequency of the gas bubbles. Figure 16.2 shows the low-frequency wave speed as a function of relative gas content, using the values given in table 16.1. The value of $K_g$ is obtained from equation (16.28) for the water depth of 100 m and the resonance frequency is 53.0 kHz according to equation (16.29).

We see that the pressure wave's speed is highly dependent on the medium's gas content; even a very small amount of gas will reduce the speed dramatically as compared to wave speed in a fluid-saturated formation. The shear wave's speed, on the other hand, is scarcely affected by the presence of either gas or fluid in the pores; in fact, the presence of gas affects only the medium's density, not the shear modulus.

In the next example, we consider the high-frequency case where the acoustic frequency is of the same order of magnitude as the resonance frequency of the gas bubbles. When this is the situation, the full expression for the bulk modulus of the gas bubbles from equation (16.25) is used. In figure 16.3, the compressional wave's sound speed is plotted as function of frequency for a porosity of 35% and a relative pore-space gas content of $10^{-4}$, or 0.01%.

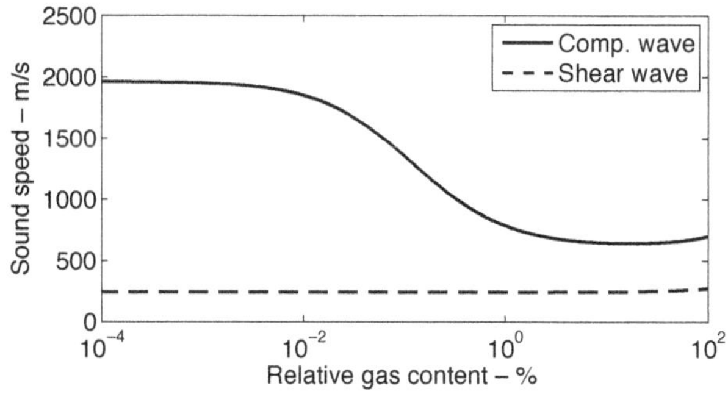


Figure 16.2. Sound speeds for compressional and shear waves for frequencies lower than the resonance frequency given by equation (16.26), plotted as a function of the relative gas content. The physical parameters are provided in table 16.1.

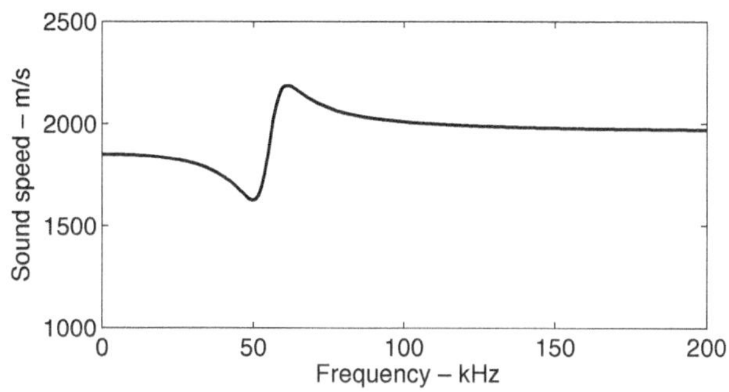


Figure 16.3. Sound speed of gassy sediment, plotted as function of frequency. The relative gas content is $10^{-4}$, and the porosity is 35%. The other physical parameters appear in table 16.1.

Each bubble has a 0.2 mm radius. Assuming the sediment to be at a 100 m water depth, the bubble resonance frequency is approximately 53.1 kHz. We see that the sound speed is strongly dependent on the frequency, particularly in the vicinity of the resonance frequency. At lower and higher signal frequencies, where respectively $n_g = 0.01\%$ and $n_g = 0$, the sound speed approaches the values shown in figure 16.3.

## 16.4 Absorption models

In all media, acoustic waves suffer losses when some of their acoustic energy is absorbed and eventually transformed into heat. For elastic waves in soils and marine sediments, there are likely to be a number of possible loss mechanisms, such as viscoelastic losses, scattering losses, and friction losses.

In the field of seismic applications, ***friction losses*** are traditionally considered to be the most important mechanism of loss. This loss mechanism is a nonlinear process, and therefore very difficult to model and describe quantitatively, but the general assumption is that friction losses are proportional to the propagating wave's frequency. This theory is quite plausible, considering that each period in a harmonic wave leads to the same relative loss.

Another loss mechanism, ***scattering losses***, can be caused by reflection and scattering from the particles and inhomogeneities in a composite material. The theory explaining such losses is, to a certain extent, described in chapter 13, which introduced scattering and reflection by particles. According to this theory, scattering losses usually are important only when particles match the wavelength size of the seismo-acoustic wave. At the frequencies normally used in underwater acoustics and seismic applications, this condition usually is not relevant, so losses caused by scattering can be ignored for most such applications.

The third loss mechanism discussed here is ***viscoelastic loss***. Viscoelastic losses are caused by the movement of viscous pore fluids in relation to the fixed particles and other matter in a porous medium. Experience with this acoustic phenomenon indicates that this loss mechanism may be quite important at the frequencies relevant to marine acoustics and seismic exploration. Therefore, a number of theories have been developed that try to quantify the losses resulting from such relative movement. Most of these models are phenomenological—that is, they describe the effect, without trying to give a detailed account of what takes place. We now discuss some of the most popular models of this type.

Losses in the medium can be modeled if we assume that the elasticity moduli expressed in the frequency domain are complex functions of the frequency. We let $M(\omega)$, with a real part $M_r(\omega)$ and an imaginary part $M_i(\omega)$, be a general frequency-dependent elasticity modulus, which either can represent the pressure wave or the shear wave, and write

$$M(\omega) = M_r(\omega) - iM_i(\omega) \tag{16.34}$$

From the wave equation, we know that a plane wave $S(\omega,x)$, propagating along the $x$-axis, is defined as

$$S(\omega,x) = A(\omega)\exp\left[i(\kappa x - \omega t)\right] \tag{16.35}$$

At $x = 0$, $A(\omega)$ is the frequency function of the transmitted pulse, and $\kappa$ is the wave number. Because of the complex elastic modulus $M(\omega)$ of equation (16.34), we know $\kappa$ to be complex and can express $\kappa$ as

$$\kappa = \frac{\omega}{c} + i\alpha \tag{16.36}$$

Here, $c$ is the wave speed and $\alpha$ is the absorption coefficient. The sign for $\alpha$ is chosen in order to obtain a decaying amplitude for the wave as it propagates along the positive $x$-axis. In this form, $\alpha$ is measured in nepers per meter. To get the absorption loss in decibels per meter, we must multiply $\alpha$ by 20log e = 8.69.

The wave speed $c$ and absorption coefficient $\alpha$ are found by solving

$$\left(\frac{1}{c} + i\frac{\alpha}{\omega}\right)^2 = \frac{\rho}{M(\omega)} \tag{16.37}$$

By splitting both sides of this equation into their real and imaginary parts, after some calculation we find these expressions for $c$ and $\alpha$

$$\begin{aligned} c &= \sqrt{\frac{2|M(\omega)|^2}{\rho\left[|M(\omega)| + M_r(\omega)\right]}} \\ \alpha &= \omega\sqrt{\frac{\rho\left[|M(\omega)| - M_r(\omega)\right]}{2|M(\omega)|^2}} \end{aligned} \tag{16.38}$$

When the losses are small, that is, when $M_i(\omega) \ll M_r(\omega)$, a good approximation of equations (16.38) would be

$$c = \sqrt{\frac{M_r(\omega)}{\rho}}$$
$$\alpha = \frac{\omega M_i(\omega)}{2cM_r(\omega)} \tag{16.39}$$

Usually, absorption into the medium is characterized by the absorption coefficient $\alpha$, but another measure of absorption is the logarithmic decrement $\delta$, for which absorption loss is determined in nepers per wavelength. The log decrement is defined as

$$\delta = \alpha\lambda = \alpha\frac{c}{f} = 2\pi\alpha\frac{c}{\omega} \tag{16.40}$$

A third measurement of absorption is the quality factor $Q$, or $Q$-value, which in this context is defined as the ratio between the real and the imaginary part of the elastic modulus $M(\omega)$

$$Q = \frac{M_r(\omega)}{M_i(\omega)} \tag{16.41}$$

By using equations (16.39) and equation (16.40), we find that

$$Q = \frac{\pi}{\delta} = \frac{\omega}{2\alpha c} = \frac{\pi}{\alpha\lambda} \tag{16.42}$$

Generally, both the absorption coefficient and the phase speed depend on the wave frequency, but the frequency dependency is normally very weak, and even difficult to determine by measurements. It is usual to assume that the sound speed is constant, and that $\alpha$ increases linearly with the frequency, which also means that both $Q$ and $\delta$ are constant.

We now examine the effects of these assumptions on the shape of a pulse transmitted through a medium with absorption. The wave $S(x,\omega)$, defined in equation (16.35), is an attenuated wave, which after propagating a distance $x$ can be described as

$$S(x,\omega) = A(\omega)\exp\left[-\frac{|\omega|x}{2cQ}\right]\exp\left[-i\omega\left(t - \frac{x}{c}\right)\right] \tag{16.43}$$

To simplify the calculations, we let $A(\omega)$ be constant, thereby assuming that the transmitted pulse is a $\delta$-pulse in time. After propagating a distance $x$, such a pulse is defined as a time signal $s(t,x)$, taking the form

$$s(t,x)=\frac{1}{\pi}\frac{(x/2cQ)}{(x/2cQ)^2+(x/c-t)^2} \tag{16.44}$$

Equation (16.44) is the inverse Fourier transform of equation (16.43). We see that the amplitude decreases and the pulse width increases proportionally with the distance $x$ because of absorption loss. The effect of absorption is illustrated in figure 16.4, where we have an example of a time response calculated using equation (16.44). In figure 16.5, we have extended this example by calculating the time responses at several distances $x$.

We see that the pulse becomes wider and the amplitude lower at greater distances. Further, we observe that the time response is symmetrical at the nominal time of arrival $t$; therefore $t = x/c$ and, consequently, the $\delta$-pulse starts before $t = 0$. Therefore, the pulse is noncausal and hence, nonphysical. The condition for causality, that is, the demand that the time response be zero for $t < 0$, means that there must be a connection between the real and imaginary parts of the frequency function; this is a fundamental and general condition for Fourier transforms. The connection between the frequency function's real and imaginary parts is given by the Kramers-Kronig relation (Aki and Richards 1980). In the present context, the significance of the Kramers-Kronig relation is that wave propagation in media with absorption is always dispersive, with a frequency-dependent sound speed. However, the dispersion may be quite small and very difficult to measure in most cases of interest in underwater acoustics. Thus, for most practical purposes, we may assume that the sound speed is constant with frequency.

Friction as a loss mechanism can also be modeled in the same way by adding a small imaginary term to the elasticity moduli. We can do this, for instance, by assuming that the ordinary Lamé parameters $\lambda$ and $\mu$ each have a complex part and thus may be expressed as

$$\begin{aligned}\lambda &\Rightarrow \lambda - i\lambda' \\ \mu &\Rightarrow \mu - i\mu'\end{aligned} \tag{16.45}$$

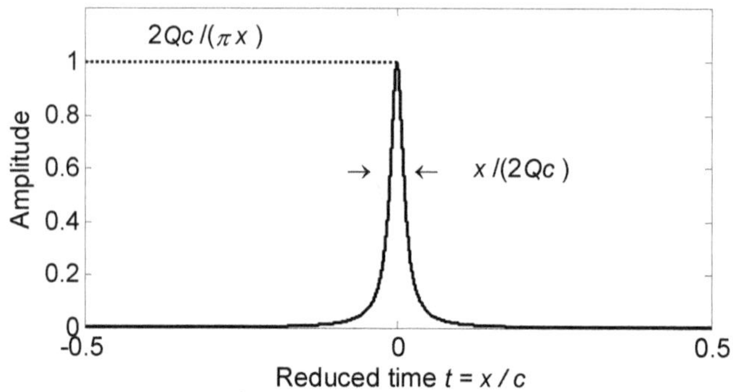


Figure 16.4. The impulse response of a $\delta$-pulse in time after propagating a distance $x$ in a medium with an absorption coefficient proportional to frequency, corresponding to a constant $Q$-value.

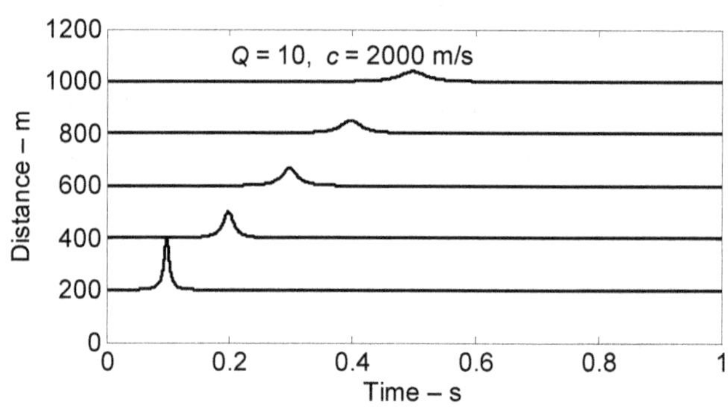


Figure 16.5. Time responses as a function of the propagation distance in a constant $Q$ medium, in which the amount of absorption is proportional with wave frequency.

If the elasticity moduli are constant and independent of the frequency, absorption will increase linearly with the frequency and, strictly speaking, the system becomes noncausal. In practice, as mentioned earlier, in most cases this is not a significant problem.

The other main group of absorption models, the viscoelastic moduli, assume that the relation between stress and strains is not only dependent on instantaneous values, but also on time history. Such relationships between stress and deformations are called ***creep functions***. We have earlier touched upon such functions in section 4.2, in connection with relaxation and absorption. A number of different creep functions can be imagined; one of them is represented by the Voigt model, shown in the left-hand third of figure 16.6, which in the time domain can be expressed as

$$\sigma(t) = K_1 e(t) + N\dot{e}(t) \tag{16.46}$$

In the frequency domain, equation (16.46) defines the elastic modulus $M(\omega)$ as

$$M(\omega) = K_1 - i\omega N \tag{16.47}$$

Another viscous model, the Maxwell model, is featured as the center drawing of figure 16.6.

Here, the same stress is applied to both elements. In this instance, the strain has two components, and the relative volume change of the element is given as

$$e = \sigma\left(\frac{1}{K_2} - \frac{1}{i\omega N}\right) \tag{16.48}$$

Therefore, the elastic modulus of the Maxwell element is

$$M(\omega) = \frac{-i\omega N K_2}{K_2 - i\omega N} \tag{16.49}$$

The Voigt model is elastic at low frequencies and viscous at high frequencies. At low frequencies, absorption in the Voigt model is proportional to the square of the frequency. The Maxwell model is elastic at high frequencies and viscous at low frequencies. These two models are special cases of a more general model, the three-parameter model, which appears on the right in figure 16.6. For the three-parameter model, the complex modulus $M(\omega)$ is expressed as

$$M(\omega) = \frac{\sigma}{e} = \frac{1}{\dfrac{1}{K_2} + \dfrac{1}{K_1 - i\omega N}} = \frac{K_2(K_1 - i\omega N)}{(K_1 + K_2) - i\omega N} \tag{16.50}$$

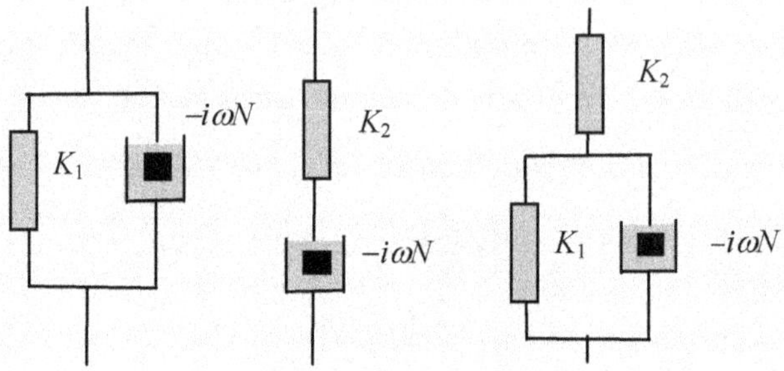


Figure 16.6. Viscoelastic absorption models: left, the Voigt model; center, the Maxwell model; and right, a general three-component model. The elastic elements denoted as $K_i$ and $N$ are viscous damping elements that represent a kind of "shock absorber."

The Maxwell model represents a special case of $M(\omega)$ of equation (16.49), in which $K_1 = 0$, and the Voigt model is the special case where $K_2$ approaches infinity. For the general three-parameter model, equation (16.49), absorption loss increases with the square of the frequency at low frequencies, but at higher frequencies, absorption is constant. The sound speed and the log decrement of the attenuation are plotted as a function of frequency and shown in figure 16.7. Note that both the attenuation (log decrement) and the sound speed are shown as relative values with respect to the values at $\omega = \omega_r$. The frequency $\omega_r$ is defined as the frequency where $M_i(\omega)$ has its maximum value.

It is also instructive to look at the time response of the strain $e(t)$ after a change in stress. As an example, we look at the three-parameter model with the frequency domain function given by equation (16.50). After having been subjected to a stress increase $\sigma_0$ at $t = 0$, the time response $e(t)$ is

$$e(t) = \frac{\sigma_0}{K_2} + \frac{\sigma_0}{K_1}\left[1 - \exp\left(-K_1 t/N\right)\right] \tag{16.51}$$

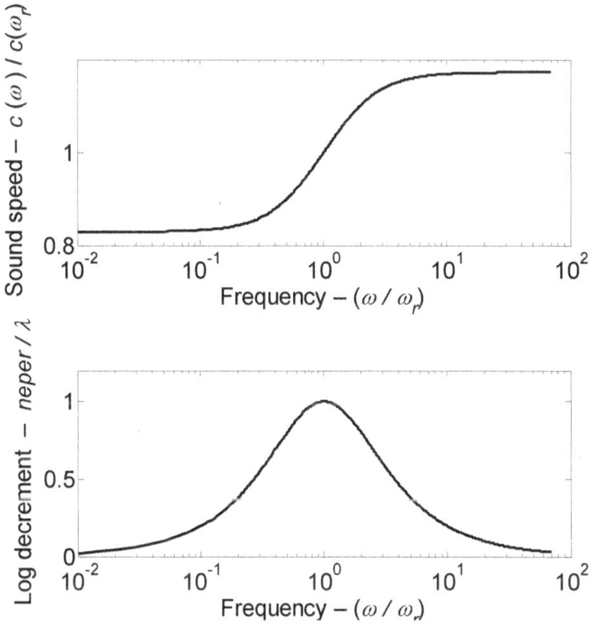


Figure 16.7. Using the general three-parameter model, relative values for (top) sound speed and (bottom) log decrement are plotted; both plots are normalized with the values at $\omega = \omega_0$.

This tells us that some period of time must elapse before the stress-generated deformation is stabilized; in the interim, the instability leads to losses, since dissipation is given as the product of the stress and the deformation rate. All real materials have a finite response time that, among other things, leads to the energy loss from elastic waves. In fluid-saturated soil and minerals, we can assume that an important viscous loss mechanism is related to fluid movement relative to the framework and its fixed particles.

At this point, we have discussed and modeled some of the most probable physical processes responsible for the absorption of elastic waves propagating in soil and minerals. However, we still do not know enough about the parameters used in the models; in order to obtain realistic values for them, we have to carry out measurements. The absorption coefficients can be measured either in laboratories or by field experiments. The advantage of field experiments is that they measure the medium *in situ*, but such experiments are expensive, and gaining full control of the geometry and the wave propagation environment may be difficult.

Measurements made in the laboratory are easier to implement since conditions are controlled. Laboratory measurements of absorption and sound speed are usually made by transmission experiments, in which an elastic wave is transmitted through a test piece of the medium, which could be acquired through core sampling, for example. An important stipulation, which is sometimes difficult to meet, is that the sampling must not change the medium's structure. Another is that the test piece must be preserved under the right conditions with regard to pressure and temperature.

Still another difficulty with laboratory measurements is related to the size of the sample and the frequency used in the measurement. The sample must be large compared to the wavelength used to take the measurement, which usually leads to the frequency used in laboratory measurements being considerably higher than the frequencies applied in practical application of seismics and sonar, the domains in which we wish to use the values. Use of higher frequencies means that we have to extrapolate from high to low values of frequencies, which adds uncertainty to the laboratory assessment.

In the next section, the Biot model for sound propagation in porous media is discussed; this model includes viscous absorption caused by the flow of fluid in the pore space of the medium.

## 16.5 The Biot theory for acoustic waves in porous media

In the 1950s Maurice Biot developed a comprehensive theory about sound propagation in porous media; this theory really did not become well known for underwater acoustic application until the 1970s when Robert Stoll used it to describe sound propagation in marine sediments. Because of Stoll's work, the theory is sometimes referred to as the Biot-Stoll theory. The Biot theory was originally developed in relation to fluid-saturated marine sediments, but the theory can also be used to problems in air acoustics, particularly in connection with porous acoustic absorbents (Allard 1993).

In this section, we first develop the wave equations for acoustic waves in porous media. To a large extent, this discussion is based on Biot's original work (1956a, 1956b, 1962) and on Stoll's book *Sediment Acoustics* (1989). Next, we demonstrate examples of such waves, using the theory to calculate sound speed and attenuations in fluid and air-saturated sands. To close this section, we study what happens when one of the media at an interface is a porous medium, and we develop the relevant reflection and transmission coefficients.

The Biot theory can be considered an extension of Gassmann's theory, discussed earlier in this chapter; Biot's extension takes into consideration the movement of pore fluid in relation to a framework-type medium. A most important feature of the Biot theory is that it predicts three kinds of body waves within a porous medium: two types of pressure waves and a shear wave. The first type of compressional wave and the shear wave are very similar to the ordinary waves we already know from elastic wave theory. With the propagation of these waves, the solid material and the pore fluid move almost in phase. However, with the propagation of the second pressure wave, the pore fluid's movement is almost in opposite phase to the movement of the solid framework; consequently, the second wave becomes strongly attenuated because of viscous losses, which may explain why the second wave has never been observed directly in marine sediments.

We consider a porous medium that consists of a solid elastic material with coupled, or connected, pores having a gas or fluid inside. The solid material is characterized by the density $\rho_s$ and the elasticity modulus $K_s$. The solid particles are cemented into a framework that has its own elasticity modulus $K_r$ and shear modulus $\mu_r$. The pores may contain gas or fluid with the density $\rho_f$ and bulk modulus $K_f$. The relative pore volume, or porosity, is $\phi$.

As we did in section 16.2, we define a vector $\mathbf{u}$ to represent the displacement of particles attached to the framework, and $\mathbf{U}$ as a vector signifying the displacement of the fluid. For the elements connected to the framework, the relative volume change $e$ is expressed as

$$e = \nabla \cdot \mathbf{u} \tag{16.52}$$

The fluid flowing through the framework, $\zeta$, is described as

$$\zeta = \phi \nabla \cdot (\mathbf{u} - \mathbf{U}) \tag{16.53}$$

By assuming linear elasticity, we found earlier in equation (16.3) that

$$\begin{aligned}
\sigma_{xx} &= He - 2\mu\left(e_{yy} + e_{zz}\right) - C\zeta \\
\sigma_{yy} &= He - 2\mu\left(e_{xx} + e_{zz}\right) - C\zeta \\
\sigma_{zz} &= He - 2\mu\left(e_{xx} + e_{yy}\right) - C\zeta \\
\sigma_{xy} &= 2\mu e_{xy} \\
\sigma_{yz} &= 2\mu e_{yz} \\
\sigma_{zx} &= 2\mu e_{zx}
\end{aligned} \tag{16.54}$$

For the pore pressure $p_f$, we have from equation (16.7) that

$$-p_f = Ce - M\zeta \tag{16.55}$$

Equations (16.54) and (16.55) define stresses and pore pressure as functions of the elastic parameters $H$, $\mu$, $K$, $M$, and $C$. In section 16.2, we found the relations between these parameters and those of the fluid, the solid material, and the framework by developing equations (16.16), (16.17), and (16.18). Now we are ready to construct wave equations describing the movement of fluid relative to the framework.

### *16.5.1 The wave equations*

We first consider the compressional waves by focusing on the movement of a volume element connected to the framework. For movement in the $x$-direction, Newton's second law tells us that

$$\begin{aligned}
\frac{\partial \sigma_{xx}}{\partial x} &= \frac{\partial^2}{\partial t^2}\left[\phi \rho_f U_x + (1-\phi)\rho_s u_x\right] \\
&= \frac{\partial^2}{\partial t^2}\left[\phi \rho_f u_x + (1-\phi)\rho_s u_x - \phi \rho_f \left(u_x - U_x\right)\right]
\end{aligned} \tag{16.56}$$

We now take the differential of the $x$-component of the stress in equation (16.56) and substitute for $\sigma_{xx}$ from equation (16.54), thereby obtaining a one-dimensional differential equation. After repeating these steps for the $y$- and the $z$-directions, we get the wave equation

$$\nabla^2\left(He - C\zeta\right) = \frac{\partial^2}{\partial t^2}\left(\rho e - \rho_f \zeta\right) \tag{16.57}$$

where the density of the composite material is $\rho = \phi\rho_f + (1-\phi)\rho_s$.

A second equation, which describes the movement of fluid relative to the frame, is for the $x$-direction

$$-\frac{\partial p_f}{\partial x} = \frac{\partial^2}{\partial t^2}\left[\rho_f u_x - \frac{\rho_f}{\phi}\left(\phi\left(u_x - U_x\right)\right)\right] - \frac{\eta}{B}\frac{\partial}{\partial t}\left[\phi\left(u_x - U_x\right)\right] \qquad (16.58)$$

The last term on the right-hand side of the equation represents the viscous flow resistance, where $\eta$ is the viscosity of the fluid and $B$ is the ***permeability*** of the porous medium. We now assume harmonic time variation and differentiate equation (16.58) with respect to $x$ and $t$, and use equation (16.55) to insert for $p_f$. After having done the same for the other directions, we get another wave equation

$$\nabla^2\left(Ce - M\zeta\right) = \left(\rho_f e - \frac{\rho_f}{\phi}\zeta\right) + \frac{i\omega\eta}{B}\varsigma \qquad (16.59)$$

In equation (16.59), we have assumed that the flow of fluid is parallel to the pressure gradient, something that generally cannot be the case since the fluid has to follow the pores, which are twisted or tortuous, and thus can be oriented in all directions. In order to take this factor into consideration, we must replace $\rho = \rho_f / \phi$ with the mass coupling density $\rho_c$, for which $\rho_c \geq \rho_f / \phi$, and which is defined as

$$\rho_c = \alpha\frac{\rho_f}{\phi} \qquad (16.60)$$

The factor $\alpha$ in equation (16.60) is called a structure constant, and describes the tortuousity of the pores, which is given by the geometry of the pores. A value $\alpha \geq 1$ for the structure constant indicates that less fluid flows in curved conduits than in conduits that are straight and parallel to the pressure gradient. In a uniform medium having all pores parallel to the pressure gradient, it follows that $\alpha = 1$, while for a medium with pores in all directions the theoretical value is $\alpha = 3$.

With this additional modification equation (16.59) can be rewritten as

$$\nabla^2\left(Ce - M\zeta\right) = \left(\rho_f e - \rho_c\zeta\right) + \frac{i\omega\eta}{B}\varsigma \qquad (16.61)$$

Equations (16.57) and (16.61) are two coupled wave equations; later we see that they define two different compressional waves. Next we must introduce another modification that will make our treatment more realistic.

The last term in equation (16.61) represents the viscous flow resistance. Actually, as that term is expressed now, it is valid only for low-frequency signals. For higher-frequency signals we must introduce corrections that take into account the nonuniformity of pore flow over the pores' cross section. This deviation can be modeled by replacing the viscosity $\eta$ with a dynamic viscosity, $\eta \to F\eta$, where $F$ is a dimensionless function dependent on both the frequency and the pore geometry.

The function $F$ can be calculated for a few simple cases of different pore geometries. For pores with a circular cross section and a radius $a_p$, the function $F$ is given in Biot (1956b) as a complex function of frequency that can be expressed as

$$F(\chi) = F_r(\chi) + iF_i(\chi) = \frac{1}{4}\frac{\chi T(\chi)}{1 - 2T(\chi)/i\chi} \tag{16.62}$$

where the function $T(\chi)$ is given by the Kelvin functions $ber(\kappa)$ and $bei(\kappa)$ and by their derivatives $ber'(\chi)$ and $bei'(\chi)$, or by the Bessel functions $J_0(z)$ and $J_1(z)$.

$$\begin{aligned} T(\chi) &= \frac{ber'(\chi) + ibei'(\chi)}{ber(\chi) + ibei(\chi)} \\ &= -\left[\exp(i3\pi/4)\right]\frac{J_1\left[(\chi \exp(i3\pi/4)\right]}{J_0\left[(\chi \exp(i3\pi/4)\right]} \end{aligned} \tag{16.63}$$

The argument $\chi$ is defined as

$$\chi = a_p\sqrt{\frac{\omega \rho_f}{\eta}} \tag{16.64}$$

In figure 16.8, the function $F(\chi)$ is plotted as a function of the argument $\chi$. The transition from low to high frequencies occurs at about the frequency where $\chi \approx 1$, which corresponds to a transition frequency $\omega_0$

$$\omega_0 = \frac{\eta}{\rho_f a_p^2} \tag{16.65}$$

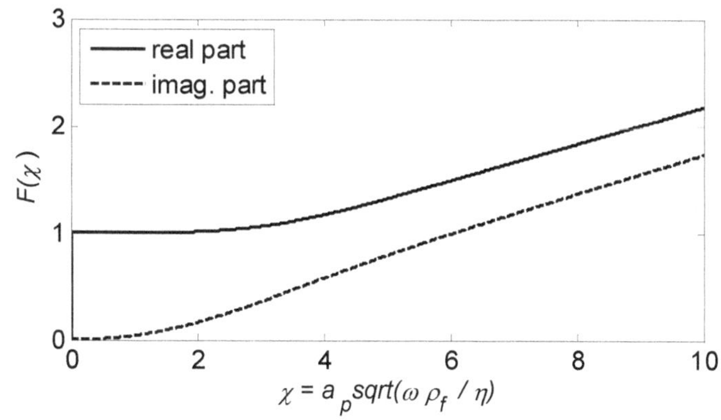


Figure 16.8. The frequency function $F(\chi)$, plotted as a function of the definition of the argument $\chi$, which is expressed as $\chi = a_p(\omega\rho_f / \eta)^{1/2}$.

With this modification the wave equations for harmonic compressional waves in a porous medium become

$$\nabla^2\left(He - C\zeta\right) = -\omega^2\left(\rho e - \rho_f\zeta\right)$$
$$\nabla^2\left(Ce - M\zeta\right) = -\omega^2\left(\rho_f e - \rho_c\zeta\right) + F(\chi)\frac{i\omega\eta}{B}\zeta \tag{16.66}$$

Thus, we have found that flow resistance as a function of the frequency is given by the function $F(\chi)$, which for circular pores is given by equation (16.63). In reality, the pores are never circular; nevertheless, it is common to use the functional relation given by equation (16.63) but with $a_p$ interpreted as a characteristic ***pore-size parameter.*** Obviously, the pore-size parameter must be related to permeability but the relationship is difficult to determine. For well-sorted sands with spherical grains having the mean grain diameter $d_m$, one may use the following expressions for the permeability $B$ and the pore-size parameter $a_p$ (Hovem and Ingram 1979)

$$a_p = \frac{d_m}{3}\frac{\phi}{1-\phi}$$
$$B = \frac{d_m^2}{36k}\frac{\phi^3}{(1-\phi)^2} \tag{16.67}$$

Here $k$ is a new free parameter that, in principle, can be determined from measurements. The relationships summarized in equations (16.67) are found to be quite accurate if the porous material consists of spherical grains that are relatively large, in which case $k \approx 5$.

For shear deformations, we can carry out similar derivations. First we introduce the rotations $\mathbf{\Omega}$ and $\mathbf{\Theta}$, for which

$$\begin{aligned} \mathbf{\Omega} &= \nabla \times \mathbf{u} \\ \mathbf{\Theta} &= \phi \nabla \times (\mathbf{u} - \mathbf{U}) = \nabla \times \mathbf{w} \end{aligned} \tag{16.68}$$

now we are able to express the wave equations for shear strains as

$$\begin{aligned} \mu \nabla^2 \mathbf{\Omega} &= \frac{\partial^2}{\partial t^2}\left(\rho \mathbf{\Omega} - \rho_f \mathbf{\Theta}\right) \\ \frac{\eta}{B}\frac{\partial \mathbf{\Theta}}{\partial t} &= \frac{\partial^2}{\partial t^2}\left(\rho_f \mathbf{\Omega} - \rho_c \mathbf{\Theta}\right) \end{aligned} \tag{16.69}$$

After we again make the modification regarding frequency dependence of the fluid flow in the pores, the equations for harmonic shear waves traveling in a porous medium become

$$\begin{aligned} \mu \nabla^2 \mathbf{\Omega} &= -\omega^2 \left(\rho \mathbf{\Omega} - \rho_f \mathbf{\Theta}\right) \\ -\frac{i\omega F(\chi)\eta}{B}\mathbf{\Theta} &= \omega^2 \left(\rho_f \mathbf{\Omega} - \rho_c \mathbf{\Theta}\right) \end{aligned} \tag{16.70}$$

From this derivation of Biot's theory, we can see that the Biot theory has a large number of parameters that are very difficult to determine by independent measurements. Next we present some examples of applications of the theory but note two important points: first, the selected parameters are not necessarily correct, and second, the results may be quite different when other values are used for the parameters.

### *16.5.2 Wave speeds and attenuations*

Now we discuss the waves described by these equations, beginning with the compressional waves described in equation (16.66). To begin, we assume that these are plane and harmonic waves that propagate in the $x$-direction, and consequently can be written in the form

$$\begin{aligned} e &= A_1 \exp\left[i\left(\kappa x - \omega t\right)\right] \\ \zeta &= A_2 \exp\left[i\left(\kappa x - \omega t\right)\right] \end{aligned} \tag{16.71}$$

After the wave number $\kappa$ is determined by inserting equations (16.71) into equations (16.66), we see that the condition for finding a solution different from the trivial solution $A_1 = A_2 = 0$ is that the determinant must be zero

$$\begin{vmatrix} H\kappa^2 - \rho\omega^2 & \rho_f\omega^2 - C\kappa^2 \\ C\kappa^2 - \rho_f\omega^2 & \rho_c\omega^2 - M\kappa^2 + \dfrac{i\omega\eta F\left(\chi\right)}{B} \end{vmatrix} = 0 \tag{16.72}$$

This is an equation of the fourth degree, and it has four solutions. Two of the solutions, which are basically different from each other, represent what are called the ***first*** and ***second compressional waves***. For each of these wave types, there is one solution that goes in the positive direction and another that goes in the negative direction. The first and second compressional wave types have the complex wave numbers $\kappa_{p1}$ and $\kappa_{p2}$, which respectively yield the wave speeds $c_{p1}$ and $c_{p2}$ and the attenuations $\alpha_{p1}$ and $\alpha_{p2}$. Therefore, we can write

$$\kappa_{pi} = \frac{\omega}{c_{pi}} + i\alpha_{pi} \tag{16.73}$$

For the shear wave we can determine the wave speed and attenuation in a similar fashion. We assume that this is a vertically polarized wave so that the displacement in the $y$-direction is zero. We also assume that all derivatives with respect to $y$ are zero, which means that $\mathbf{u}_y = 0$ and $\partial / \partial y = 0$. Thus, the rotations $\mathbf{\Omega}$ and $\mathbf{\Theta}$ only yield components in the $x$-direction, so that now we can define $\mathbf{\Omega}$ and $\mathbf{\Theta}$ as scalars with solutions of the form

$$\begin{aligned} \mathbf{\Omega} &= C_1 \exp\left[i\left(\kappa_s x - \omega t\right)\right] \\ \mathbf{\Theta} &= C_2 \exp\left[i\left(\kappa_s x - \omega t\right)\right] \end{aligned} \tag{16.74}$$

The determinant must satisfy the condition

$$\begin{vmatrix} \mu\kappa_s^2 - \rho\omega^2 & \rho_f\omega^2 \\ \rho_f\omega^2 & -\left[\rho_c\omega^2 + \dfrac{i\omega\eta F(\chi)}{B}\right] \end{vmatrix} = 0 \qquad (16.75)$$

producing

$$\left(\mu\kappa_s^2 - \rho\omega^2\right)\left(\rho_c\omega^2 + i\omega\frac{F(\chi)\eta}{B}\right) + \omega^4\rho_f^2 = 0 \qquad (16.76)$$

which has the solution

$$\left(\frac{\kappa_s}{\omega}\right)^2 = \frac{\rho}{\mu}\frac{\left(\rho_c - \dfrac{\rho_f^2}{\rho}\right) + i\dfrac{\eta F(\chi)}{\omega B}}{\rho_c + i\dfrac{\eta F(\chi)}{\omega B}} \qquad (16.77)$$

Equation (16.75) has two solutions, one for shear waves traveling in the positive direction and the other for shear waves going in the negative direction. Therefore, we find that there is only one kind of shear wave in a homogeneous porous material.

### *16.5.3 Examples of sound speeds and attenuations*

The sound speeds and attenuations for the two compressional waves are found by solving equation (16.72), and the sound speed and attenuation of the shear wave by calculating equation (16.77). We demonstrate the use of these equations by calculating the sound speeds and attenuations for water- and air-saturated sands, using the parameters given in table 16.2.

We first consider water-saturated sand with the pores completely filled with water. Figure 16.9 shows the resulting sound speeds of the two types of compressional waves and the shear wave as functions of frequency. We notice that the speeds of the first compressional wave and the shear wave are nearly constant with frequency; therefore, these two waves have very little dispersion. The second compressional wave, on the other hand, is highly dispersive. Although we show results up to a very high frequency, it is important to note that the modeled results are incorrect when the wavelengths start to approach the grain size.

Table 16.2. Parameter values for water- and air-saturated sands

| *Parameter* | *Water-saturated sand* | *Air-saturated sand* |
|---|---|---|
| $K_s$ | $3.8 \cdot 10^{10}$ Pa | $3.8 \cdot 10^{10}$ Pa |
| $K_f$ | $2.15 \cdot 10^{9}$ Pa | $0.15 \cdot 10^{6}$ Pa |
| $K_r$ | $0.6 \cdot 10^{9}$ Pa | $0.6 \cdot 10^{9}$ Pa |
| $\mu_r$ | $0.6 \cdot 10^{9}$ Pa | $0.6 \cdot 10^{9}$ Pa |
| $\rho_s$ | 2500 kg/m$^3$ | 2500 kg/m$^3$ |
| $\rho_f$ | 1000 kg/m$^3$ | 1.4 kg/m$^3$ |
| $\phi$ | 0.37 | 0.37 |
| $a_p$ | $3.5 \cdot 10^{-5}$ m | $3.5 \cdot 10^{-5}$ m |
| $\eta$ | $1.0 \cdot 10^{-3}$ Pa | $1.7 \cdot 10^{-5}$ Pa |
| $B$ | $0.23 \cdot 10^{-10}$ m$^2$ | $0.23 \cdot 10^{-10}$ m$^2$ |
| $\alpha$ | 1.25 | 1.25 |

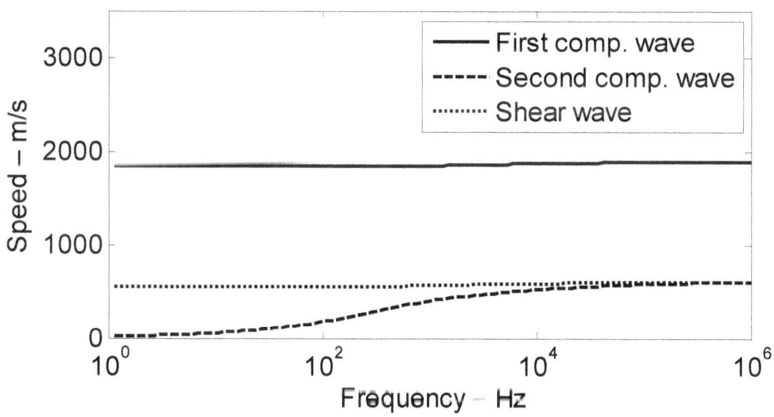


Figure 16.9. The speeds of the first and second compressional waves and the shear wave, plotted as functions of frequency. The values are calculated for the parameters of water-saturated sand given in table 16.2.

Figure 16.10 shows attenuations of the first and second compressional waves and the shear wave, calculated for water-saturated sand using the same values used to create figure 16.9. The attenuations are plotted in decibels per meter, that is, equal to

$20\log(e) \cdot \alpha = 8.686 \cdot \alpha$, where $\alpha$ is the imaginary part of the wave number for each of waves from the calculations of equations (16.72) and (16.73) for the compressional waves and equation (16.77) for the shear wave. The attenuations of the first compressional wave and the shear wave at low frequencies are proportional with the frequency squared, but at high frequencies, attenuations are proportional with the square root of the frequency. The transition from low to high frequency occurs approximately around the frequency $f_0 = \omega_0 / 2\pi$, which was obtained in equation (16.65). With the current parameter values, the frequency $f_0$ is about 1000 Hz. We also notice in figure 16.10 that the attenuation of the second compressional wave is very high. A more thorough analysis will show that the log decrement approaches the value of $2\pi$, which corresponds to 55 dB/$\lambda$ at frequencies lower than $f_0$; this tells us that the second wave is not a propagating wave at all, but instead is a diffusion process.

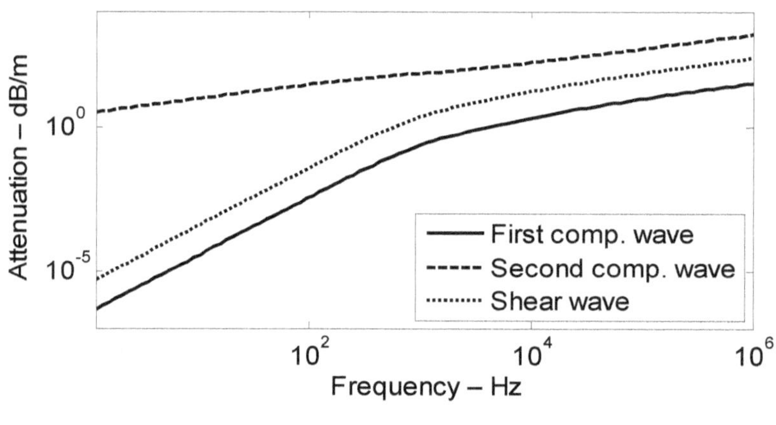


Figure 16.10. Attenuation, as measured in decibels per meter, for the two compressional waves and the shear wave, plotted as a function of frequency. The values are calculated for the parameters of water-saturated sand given in table 16.2.

Figures 16.11 and 16.12 use the same values as figures 16.9 and 16.10; the only difference between the two groups of figures is that now the pores contain air instead of water. In figure 16.11, we again see that the first pressure wave and the shear wave have sound speeds in the air-saturated sand that are nearly constant with frequency, and that the second pressure wave is very dispersive. In figure 16.12, the attenuations as a function of frequency behave much like those for the water-saturated case, but the values are much lower for the first compressional wave and the shear wave.

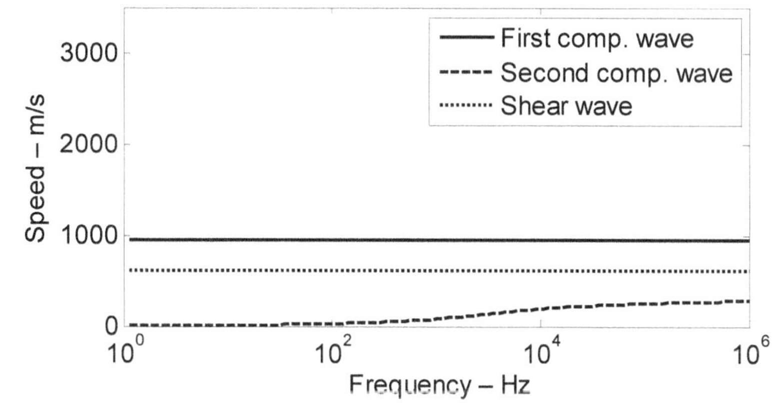


Figure 16.11. Sound speeds in air-saturated sand, plotted as a function of frequency. The values are calculated for the parameters of air-saturated sand given in table 16.2.

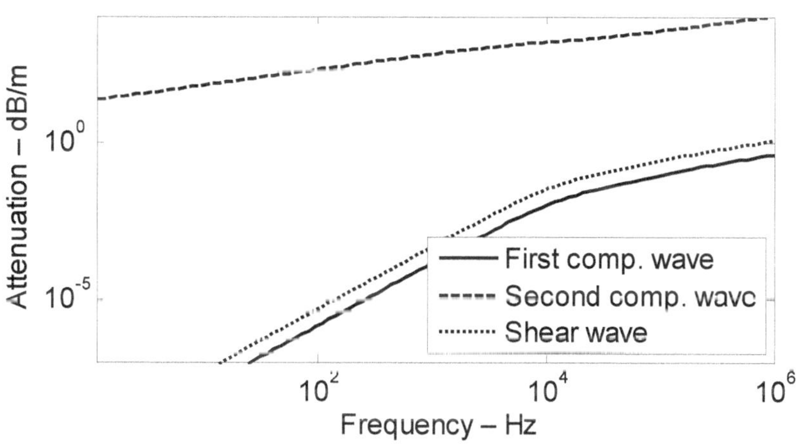


Figure 16.12. Attenuations, in decibels per meter, for waves traveling in air-saturated sand, showing that absorption loss is a function of wave frequency. The values are calculated for the parameters of air-saturated sand given in table 16.2.

### *16.5.4 Approximate expressions of sound speed and attenuation*

In the following paragraphs, we develop limiting values for sound speed and attenuation at both low and high frequencies. Such limiting values can be useful in assigning values to the Biot model parameters.

The first wave is an ordinary pressure wave having a relatively small loss and a small dispersion. At very low frequencies we find that the wave speed $c_{p1}$ becomes approximately

$$\omega \to 0 \qquad c_{p1} \to \sqrt{\frac{H}{\rho}} \tag{16.78}$$

At higher frequencies the ordinary wave's sound speed increases slightly. This happens because the effective mass is reduced at higher frequencies, where only part of the fluid can follow the wave's rapid oscillations.

At low frequencies, that is, at frequencies for which $\omega \to 0$, viscous attenuation for the first or ordinary pressure wave is approximately

$$\alpha_{p1} \to \frac{\omega^2}{2c_{p1}^3} \frac{B}{\eta} \frac{HM - C^2}{H} \tag{16.79}$$

From this expression, we see that, at low frequencies, viscous attenuation increases both with the square of the frequency and in proportion to the medium's permeability. At high frequencies, that is, at frequencies for which $\omega \to \infty$, viscous attenuation increases with the frequency's square root, so that

$$\alpha_{p1} \to \frac{1}{c_{p1}} \sqrt{\frac{\omega \eta}{\rho}} \tag{16.80}$$

At low frequencies, the speed of the shear wave $c_s$ becomes

$$c_s \to \sqrt{\frac{\mu}{\rho}} \tag{16.81}$$

At high frequencies the speed $c_s$ increases because the effective mass is reduced at higher frequency

$$c_s \to \frac{\sqrt{\dfrac{\mu}{\rho}}}{\sqrt{1-\dfrac{\rho_f^2}{\rho\rho_c}}} \tag{16.82}$$

At low frequencies the shear wave's viscous attenuation also increases with the square of the frequency; and at high frequencies shear wave attenuation increases with the square root, just as it does for the pressure waves. We see from figure 16.9 that the second compressional wave is strongly dispersive, with its sound speed approaching zero at low frequencies. This wave's attenuation, shown in figure 16.10, is also very large, and at all frequencies increases proportionally with the square root of the frequency.

At low frequencies the second wave is actually a diffusion process rather than a wave. Such a process is characterized by having a wave number with equally large real and imaginary parts, so that it follows from equation (16.73) that the attenuation becomes $2\pi$, measured in nepers per wavelength. We can observe this change in figure 16.10. In water-saturated sands and marine sediments, the second compressional wave usually cannot be observed and has little or no relevance except that it results in a viscous loss.

### *16.5.5 The Biot model with a lossy framework*

In figure 16.10, we saw plots of viscous attenuation resulting from the fluid's movement in relation to the framework, as predicted by the Biot theory; in addition, Biot's theory also allows losses in the framework to be taken into consideration by assuming that $K_r$ and $\mu_r$ are complex, so that respectively they are defined as

$$\begin{aligned} K_r' &= K_r(1-i\delta_c) \\ \mu_r' &= \mu_r(1-i\delta_s) \end{aligned} \tag{16.83}$$

where $\delta_c$ is the log decrement for pressure waves and $\delta_s$ is the decrement for shear waves traveling in the framework. Figure 16.13 contains an example of attenuation changing as a function of signal frequency for a case that includes losses in the framework as well as in the pore fluid. While the values are the same as those used in figure 16.10, here $\delta_c = \delta_s = 0.1$, so that in this example the attenuation increases almost linearly with increasing frequency.

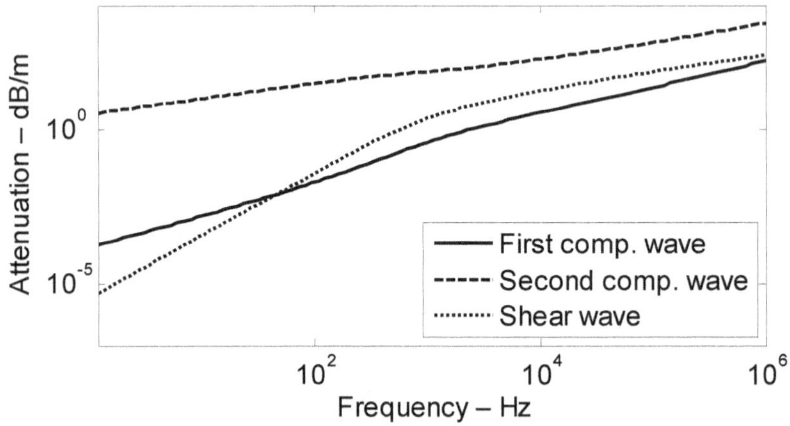


Figure 16.13. Wave attenuations, measured in decibels per meter, when losses in the framework are included for the two compressional waves and the shear wave; here attenuation is plotted as a function of frequency. The values are calculated for the parameters of water-saturated sand, given in table 16.2.

### *16.5.6 Reflection and transmission at a porous interface*

In this section, we discuss reflection and transmission of acoustic waves at an interface between a fluid medium and a porous medium. The problem of reflection and transmission at an interface with a porous medium has been treated in reference works such as *Sediment Acoustics* (Stoll 1989) and in articles by Hovem (1981), Stoll and Kan (1981), and Dutta and Odé (1983). Here we begin our discussion by assuming that the substance in the permeable medium's pores is identical to the adjacent fluid. For the sake of simplicity, we also ignore any shear waves, which limits this case's validity to materials with a relatively weak framework or a normal incidence.

We look at our combination of media in three different situations, which we designate as situations A, B, and C.

***Situation A.*** An incident wave in a fluid medium generates a reflected wave, plus two transmitted waves of the first and second compressional types, in the porous medium.

***Situation B.*** An incident wave of the first compressional type in the porous medium generates a transmitted wave in the fluid medium, plus two reflected waves of the first and second compressional types in the porous medium.

***Situation C.*** An incident wave of the second type in the porous medium generates a transmitted wave in the fluid medium, plus two reflected waves of the first and second types in the porous medium.

In order to calculate the amplitudes and the intensities of the transmitted and reflected waves, we need to know the boundary conditions at the interface between the fluid and the porous medium. While there are a number of possibilities, in our discussion we assume an open exchange of pore fluid at the interface. Alternatively, we could assume that the interface is closed, with no exchange of fluid; this requires the existence of a thin impermeable film between the porous medium and the ambient fluid.

We also make use of Snell's law, which states that horizontal wave number components must be equal, which means that

$$\kappa_0 \cos\theta_0 = \kappa_1 \cos\theta_1 = \kappa_2 \cos\theta_2 \tag{16.84}$$

Here, the angles $\theta_i$, where $i = 0,1,2$, are the grazing angles. The index 0 refers to the wave in the fluid medium that borders the porous medium; the two types of compressional waves in the porous medium are indicated by the indices 1 and 2. Note that the wave numbers $\kappa_0$, $\kappa_1$, and $\kappa_2$ are generally complex, as they also contain the loss terms of the respective waves.

However, here we assume that the waves in the ambient fluid are free from loss, so that both $\kappa_0$ and the angle $\theta_0$ are real. Wave numbers $\kappa_1$ and $\kappa_2$ are still complex, which means that the angles $\theta_1$ and $\theta_2$ become complex. Consequently, we cannot use the usual methods for calculation of trigonometric functions, but must calculate sine and cosine as follows.

$$\cos\theta_1 = \frac{\kappa_0}{\kappa_1}\cos\theta_0$$
$$\cos\theta_2 = \frac{\kappa_0}{\kappa_2}\cos\theta_0 \qquad (16.85)$$

and

$$\sin\theta_1 = \sqrt{\left[1-\left(\frac{\kappa_0}{\kappa_1}\cos\theta_0\right)^2\right]}$$
$$\sin\theta_2 = \sqrt{\left[1-\left(\frac{\kappa_0}{\kappa_2}\cos\theta_0\right)^2\right]} \qquad (16.86)$$

We now consider situations A, B, and C, stipulating that the porous medium has open pores.

We first consider situation A, which is shown in figure 16.14.

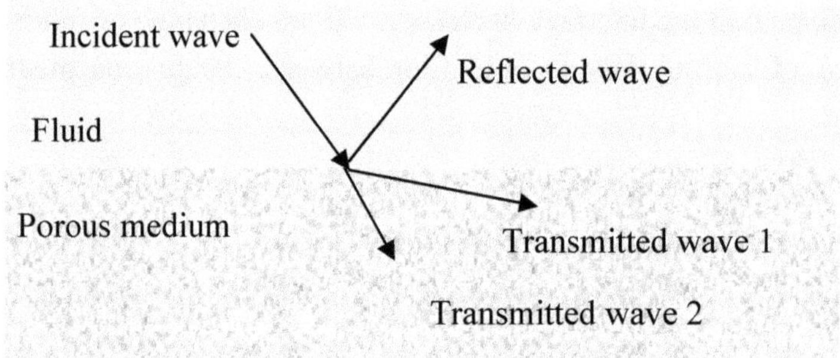


Figure 16.14. An incident wave in the fluid generates a reflected wave and two transmitted compressional waves in the porous medium. Shear waves are ignored in this analysis.

In the fluid medium, there are an incident wave with amplitude $A_i$ and a reflected wave with amplitude $A_r$. In the porous medium, there are two transmitted compressional waves of the first and second types with respective amplitudes $B_{t1}$ and $B_{t2}$.

The field in the fluid, $\Phi_0$, can be expressed as

$$\begin{aligned}\Phi_0 &= A_i \exp(i\kappa_0 \cos\theta_0 x + i\kappa_0 \sin\theta_0 z) \\ &\quad + A_r \exp(i\kappa_0 \cos\theta_0 x - i\kappa_0 \sin\theta_0 z)\end{aligned} \tag{16.87}$$

where the time dependency $\exp(-i\omega t)$ is implied. In the porous medium, we introduce the vector **u** to represent the displacement of the solid material and the displacement vector **U** to represent that of the pore fluid. The relative displacement of the pore fluid in relation to the solid material, **w**, is defined as $\mathbf{w} = \phi(\mathbf{u} - \mathbf{U})$, where $\phi$ is the porosity. Next we express the displacements as the gradients of two scalar potentials

$$\begin{aligned}\mathbf{u} &= \nabla\Phi_u \\ \mathbf{w} &= \nabla\Phi_w\end{aligned} \tag{16.88}$$

The volume change in the solid $e$ and that in the pore fluid $\xi$ are defined as

$$\begin{aligned}e &= \nabla\cdot\mathbf{u} = \nabla^2\Phi_u \\ \zeta &= \nabla\cdot\mathbf{w} = \nabla^2\Phi_w\end{aligned} \tag{16.89}$$

The two potentials have contributions from both the first and second compressional waves. For plane harmonic waves, it follows that

$$\begin{aligned}\Phi_u &= \left[A_{t1}\exp(i\kappa_1\cos\theta_1 x + i\kappa_1\sin\theta_1 z) + A_{t2}\exp(i\kappa_2\cos\theta_2 x + i\kappa_2\sin\theta_2 z)\right] \\ \Phi_w &= \left[B_{t1}\exp(i\kappa_1\cos\theta_1 x + i\kappa_1\sin\theta_1 z) + B_{t2}\exp(i\kappa_2\cos\theta_2 x + i\kappa_2\sin\theta_2 z)\right]\end{aligned} \tag{16.90}$$

In equation (16.90) $A_{t1}$ and $A_{t2}$ are the amplitudes of the displacements of the porous solid material for the first and second waves, respectively, and $B_{t1}$ and $B_{t2}$ respectively represent the amplitudes of the relative displacement of the pore fluid relative to the solid frame for the two waves.

For harmonic waves, the strain normal to the interface, $\sigma_{zz}$, and the pore pressure $p$ are given by the following expressions; note that we have assumed that $\mu = 0$.

$$\begin{aligned}\sigma_{zz} &= Ke - C\zeta = -\left[A_{t1}\kappa_1^2(K - m_1 C) + A_{t2}\kappa_2^2(K - m_2 C)\right] \\ p &= -(Ce - M\zeta) = A_{t1}\kappa_1^2(C - m_1 M) + A_{t2}\kappa_2^2(C - m_2 M)\end{aligned} \tag{16.91}$$

From the first of the two wave equations, equation (16.57), we see that the two waves are mutually dependent. Therefore, it follows that for $j = 1,2$ we obtain this expression for the ratios $m_j$

$$m_j = \frac{B_{tj}}{A_{tj}} = \frac{\kappa_j^2 K - \omega^2 \rho}{\kappa_j^2 C - \omega^2 \rho_f} \tag{16.92}$$

In the fluid medium, the sound pressure $p_f$ of the incident wave is expressed as

$$p_f = K_f \kappa_0^2 A_i \tag{16.93}$$

and the vertical component $u_z$ of the particle displacement normal to the interface is defined as

$$u_z = i\kappa_0 \sin\theta_0 A_i \tag{16.94}$$

We have three boundary conditions to use for the determination of the reflected wave amplitude $A_r$ and the two transmitted wave amplitudes $A_{t1}$ and $A_{t2,}$ assuming that the incident wave amplitude $A_0$ is known.

The boundary conditions are

***Boundary condition A***. Continuity of total stress, $\sigma_{zz} = -p_f$:

$$\kappa_0^2 K_f \left(A_i + A_r\right) = A_{t1}\kappa_1^2 (K - m_1 C) + A_{t2}\kappa_2^2 (K - m_2 C) \tag{16.95}$$

***Boundary condition B***. Continuity of total fluid pressure, $p = p_f$:

$$\kappa_0^2 K_f \left(A_i + A_r\right) = A_{t1}\kappa_1^2 (C - m_1 M) + A_{t2}\kappa_2^2 (C - m_2 M) \tag{16.96}$$

***Boundary condition C***. Continuity of total displacement normal to the interface, that is, in the $z$-direction

$$\left(U_i + U_r\right)_z = \left[(1-\phi)u_t + \phi U_t\right]_z = \left(u_t - w_t\right)_z \tag{16.97}$$

which gives us

$$\begin{aligned} i\kappa_0 \sin\theta_0 \left(A_i - A_r\right) = i\kappa_1 \sin\theta_1 \left(1 - m_1\right) A_{t1} \\ + i\kappa_2 \sin\theta_2 \left(1 - m_2\right) A_{t2} \end{aligned} \tag{16.98}$$

Equations (16.95), (16.96), and (16.98) can be solved for the three unknown amplitudes $A_r$, $A_{t1}$, and $A_{t2}$. The solutions, expressed as transmission and reflection coefficients, are

$$R_{01} = \frac{\kappa_0 A_r}{\kappa_0 A_i} = \frac{1 - s_1 - s_2}{1 + s_1 + s_2} \tag{16.99}$$

$$T_{01} = \frac{\kappa_1 A_{t1}}{\kappa_0 A_i} = \frac{2s_1}{1 + s_1 + s_2} \frac{\sin\theta_0}{(1 - m_1)\sin\theta_1} \tag{16.100}$$

$$T_{02} = \frac{\kappa_2 A_{t2}}{\kappa_0 A_i} = \frac{2s_2}{1 + s_1 + s_2} \frac{\sin\theta_0}{(1 - m_2)\sin\theta_2} \tag{16.101}$$

where

$$s_1 = \frac{\kappa_0 \sin\theta_1}{\kappa_1 \sin\theta_0} \frac{1 - m_1}{m_2 - m_1} \frac{K_f\left(K - C - m_2 C + m_2 M\right)}{KM - C^2} \tag{16.102}$$

$$s_2 = \frac{\kappa_0 \sin\theta_2}{\kappa_2 \sin\theta_0} \frac{1 - m_2}{m_1 - m_2} \frac{K_f\left(K - C - m_1 C + m_1 M\right)}{KM - C^2} \tag{16.103}$$

The coefficients $s_1$ and $s_2$ can be thought of as normalized impedances for the first and second waves, respectively, where the normalization is with regard to the impedance of the wave in the ambient fluid. This normalization can be illustrated by replacing the porous medium with a nonporous fluid medium, which is achieved by setting the porosity equal to zero and by assuming a nonrigid frame. In this case

$$\begin{aligned} \phi &= 0 \\ K_r &= 0 \\ \mu_r &= 0 \end{aligned} \tag{16.104}$$

For the elastic Biot parameters, the values of equation (16.104) produce

$$K = M = C = K_s \tag{16.105}$$

giving the wave speeds of the two compressional waves

$$\begin{aligned} c_{p1} &= \frac{\omega}{\kappa_1} = \sqrt{\frac{K_s}{\rho_s}} \\ c_{p2} &= \frac{\omega}{\kappa_2} = 0 \end{aligned} \tag{16.106}$$

Furthermore, from equation (16.92) we see that $m_1 = 0$ and $m_2 = 1$, and consequently the normalized impedance coefficients $s_1$ and $s_2$ are expressed as

$$s_1 = \frac{\kappa_0 \sin\theta_1}{\kappa_1 \sin\theta_0} \frac{K_f}{K_s} = \frac{c_0 \rho_f \sin\theta_1}{c_{p1} \rho_s \sin\theta_0} \tag{16.107}$$

$$s_2 = 0$$

These results demonstrate that the new reflection and transmission coefficients we have found with equations (16.99) and (16.100) reduce to the conventional reflection and transmission coefficients for an interface between two fluid media.

We now calculate the reflected and transmitted intensities. Intensity is given as the magnitude of the real part of the complex product of the vertical particle velocity and the stress. The incident intensity $I_i$ is therefore

$$I_i = \left|\mathrm{Re}\left(\sigma_i \cdot \dot{u}_{iz}^*\right)\right| = \left|\mathrm{Re}\left(\omega \kappa_0^3 K_f \sin\theta_0 \left|A_i\right|^2\right)\right| \tag{16.108}$$

and the reflected intensity is

$$I_r = \left|\mathrm{Re}\left(\sigma_r \cdot \dot{u}_{rz}^*\right)\right| = \left|\mathrm{Re}\left(\omega \kappa_0^3 K_f \sin\theta_0 \left|A_r\right|^2\right)\right| \tag{16.109}$$

so that the relative reflected intensity $I_{R0}$ is

$$I_{R0} = \left|R_{01}\right|^2 = \left|\frac{1 - s_1 - s_2}{1 + s_1 + s_2}\right|^2 \tag{16.110}$$

The total transmitted intensity in the porous medium is expressed as

$$I_t = \left|\mathrm{Re}\left[(1-\phi)\sigma_{zz} \cdot \dot{u}_z^* - p_f \phi \dot{U}_z^*\right] = \mathrm{Re}\left[\sigma(\dot{u}_z^* - \dot{w}_z^*)\right]\right| \tag{16.111}$$

Equation (16.111) provides four contributions to the transmitted intensity $I_t$

$$\begin{aligned} I_t = & \left|A_{t1}\right|^2 \omega \kappa_1^2 \kappa_1^* \left(K - m_1 C\right)\left(1 - m_1^*\right) \sin\theta_1^* \\ & + \left|A_{t2}\right|^2 \omega \kappa_2^2 \kappa_2^* \left(K - m_2 C\right)\left(1 - m_2^*\right) \sin\theta_2^* \\ & + A_{t1} A_{t2}^* \omega \kappa_1^2 \kappa_2^* \left(K - m_1 C\right)\left(1 - m_2^*\right) \sin\theta_2^* \\ & + A_{t2} A_{t1}^* \omega \kappa_2^2 \kappa_1^* \left(K - m_2 C\right)\left(1 - m_1^*\right) \sin\theta_1^* \end{aligned} \tag{16.112}$$

We then normalize equation (16.112) with regard to the incident intensity, and get

$$
\begin{aligned}
I_t = & \left|T_{01}\right|^2 \frac{\kappa_1}{\kappa_0} \frac{\left(K - m_1 C\right)}{K_f}\left(1 - m_1^*\right) \frac{\sin\theta_1^*}{\sin\theta_0} \\
& + \left|T_{02}\right|^2 \frac{\kappa_2}{\kappa_0} \frac{\left(K - m_2 C\right)}{K_f}\left(1 - m_2^*\right) \frac{\sin\theta_2^*}{\sin\theta_0} \\
& + T_{01} T_{02}^* \frac{\kappa_1}{\kappa_0} \frac{\left(K - m_1 C\right)}{K_f}\left(1 - m_2^*\right) \frac{\sin\theta_2^*}{\sin\theta_0} \\
& + T_{02} T_{01}^* \frac{\kappa_2}{\kappa_0} \frac{\left(K - m_2 C\right)}{K_f}\left(1 - m_1^*\right) \frac{\sin\theta_1^*}{\sin\theta_0}
\end{aligned}
\tag{16.113}
$$

From equations (16.100), (16.101), (16.102), and (16.103), we see that the two transmission coefficients $T_{01}$ and $T_{02}$ are related in such a way that

$$
T_{02} = -T_{01} \frac{\kappa_1}{\kappa_2} \frac{K - C - m_1 C + m_1 M}{K - C - m_2 C + m_2 M} \tag{16.114}
$$

We insert equation (16.114) for $T_{02}$ in the fourth term and for $T_{01}$ in the third term of equation (16.113), and get two new terms representing the intensity of the transmitted waves. After some algebraic calculations, we find that the first and second pressure waves' relative transmitted intensities, designated as $I_{T1}$ and $I_{T2}$, respectively, can be expressed as

$$
I_{T1} = \frac{4\,\mathrm{Re}\{s_1\}}{\left|1 + s_1 + s_2\right|^2} \tag{16.115}
$$

$$
I_{T2} = \frac{4\,\mathrm{Re}\{s_2\}}{\left|1 + s_1 + s_2\right|^2} \tag{16.116}
$$

It is easy to verify that the energy balance has been maintained since

$$
I_{R0} + I_{T1} + I_{T2} = 1 \tag{16.117}
$$

We next consider case *B* where there is an incident wave of either the first or second type in the porous medium. This wave generates two reflected waves in the porous medium, one wave of the first type and one of the second type, and a transmitted wave in the fluid, as shown in figure 16.15.

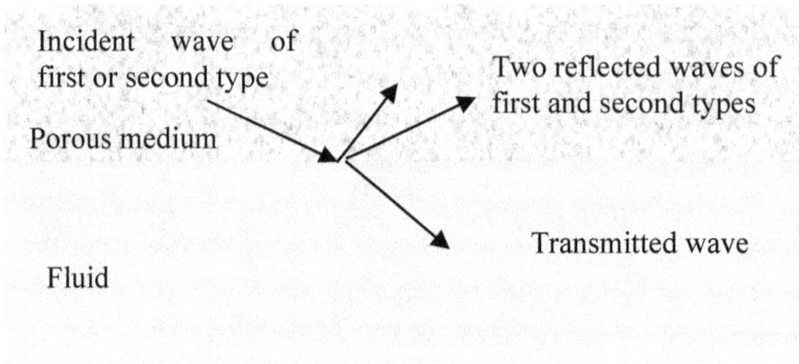


Figure 16.15. An incident wave in the porous medium generates a transmitted wave in the fluid and two reflected compressional waves, one being of the first type and the other being the second type, in the porous medium. Shear waves are ignored in this analysis.

In this case, the incident wave is of the first type and has the amplitude $A_{i1.}$ In the porous medium, the incident wave creates two reflected pressure waves of the first and second types with the amplitudes $A_{r1}$ and $A_{r2}$, respectively, and in the fluid medium, a transmitted wave we designate as $A_{t0}$.

Using the same boundary conditions as before, we find that the amplitudes of the reflected and transmitted waves, normalized with respect to the amplitude of the incident wave, can be expressed as

$$R_{11} = \frac{\kappa_1 A_{r1}}{\kappa_1 A_{i1}} = -\frac{1 - s_1 + s_2}{1 + s_1 + s_2} \tag{16.118}$$

$$R_{12} = \frac{\kappa_2 A_{r2}}{\kappa_1 A_{i1}} = \frac{2 s_2}{1 + s_1 + s_2} \frac{(1 - m_1)\sin\theta_1}{(1 - m_2)\sin\theta_2} \tag{16.119}$$

$$T_{10} = \frac{\kappa_0 A_{t0}}{\kappa_1 A_{i1}} = \frac{2(1 - m_1)}{1 + s_1 + s_2} \frac{\sin\theta_1}{\sin\theta_0} \tag{16.120}$$

Again using the procedure by which we obtained equations (16.113) and (16.114), we now can express the transmitted intensity as

$$I_{T10} = \frac{4\,\mathrm{Re}\{s_1\}}{\left|1 + s_1 + s_2\right|^2} \tag{16.121}$$

Next, we find that the reflected intensity in the first pressure wave is expressed as

$$I_{R1} = \frac{\left|1 - s_1 + s_2\right|^2}{\left|1 + s_1 + s_2\right|^2} \tag{16.122}$$

and the reflected intensity in the second wave is

$$I_{R2} = \frac{4\,\mathrm{Re}\left\{s_1 s_2^*\right\}}{\left|1 + s_1 + s_2\right|^2} \tag{16.123}$$

As we did in equation (16.115), we again verify our results by checking the energy balance

$$I_{T10} + I_{R1} + I_{R2} = 1 \tag{16.124}$$

Finally, let us consider case C, where the incident wave in the porous medium is a compressional wave of the second type. We have an incident wave with the amplitude $A_{i2}$ traveling in a porous medium adjacent to a fluid layer. In the porous medium, this incident wave produces two reflected waves of the first and second compressional types, $R_{21}$ and $R_{22}$, respectively; these waves have the corresponding amplitudes $A_{r1}$ and $A_{r2}$. In the fluid medium, the incident wave also creates a transmitted wave $A_{t0}$.

By the same method used for cases A and B, we find the reflection and transmission coefficients to be

$$R_{22} = \frac{\kappa_2 A_{r2}}{\kappa_2 A_{i2}} = -\frac{1 + s_1 - s_2}{1 + s_1 + s_2} \tag{16.125}$$

$$R_{21} = \frac{\kappa_1 A_{r1}}{\kappa_2 A_{i2}} = \frac{2 s_1}{1 + s_1 + s_2} \frac{(1 - m_2)\sin\theta_2}{(1 - m_1)\sin\theta_1} \tag{16.126}$$

$$T_{20} = \frac{\kappa_0 A_{t0}}{\kappa_1 A_{i1}} = \frac{2(1 - m_2)}{1 + s_1 + s_2} \frac{\sin\theta_2}{\sin\theta_0} \tag{16.127}$$

We can define the transmitted intensity of the wave traveling in the fluid medium as

$$I_{T20} = \frac{4\,\mathrm{Re}\left\{s_2\right\}}{\left|1 + s_1 + s_2\right|^2} \tag{16.128}$$

The reflected intensity for the first compressional wave, $I_{R1}$ is expressed as

$$I_{R1} = \frac{4\,\mathrm{Re}\left\{s_1^* s_2\right\}}{\left|1 + s_1 + s_2\right|^2} \tag{16.129}$$

and the reflected intensity for the second compressional wave, $I_{R2}$, as

$$I_{R2} = \frac{\left|1 + s_1 - s_2\right|^2}{\left|1 + s_1 + s_2\right|^2} \tag{16.130}$$

Again, as we did with equations (16.115) and (16.122), we verify our results by checking to make sure the energy balance has been maintained

$$I_{T20} + I_{R1} + I_{R2} = 1 \tag{16.131}$$

### *16.5.7 Examples*

Our first example deals with acoustic reflection from water-saturated sand. The parameters for this case appear in table 16.2, and the speeds and attenuations are the same as those provided in figures 16.9 and 16.10.

In figures 16.16 and 16.17 we have plots of the reflected and transmitted intensities for waves with 100 Hz and 1000 Hz frequencies, respectively; the wave intensities are given as functions of the grazing angles. In figures 16.18 and 16.19, we have the reflected and transmitted intensities for the incident wave in air, moving toward the air-saturated sand, for waves of 100 and 1000 Hz frequencies, respectively.

From figure 16.9, we know that in the porous material, the first pressure wave has a speed $c_{p1}$ of approximately 1835 m/s, while the second pressure wave has a wave speed $c_{p2}$ = 180 m/s at 100 Hz and a speed $c_{p2}$ = 527 m/s at 1000 Hz. In the water, the wave's sound speed is 1466 m/s, which for the first wave gives us a critical grazing angle of about 37.4°.

From figures 16.16 and 16.17 we can see that for the reflected wave and the transmitted first compressional wave, the conditions are not very different from those we would predict using ordinary acoustic theory. This is particularly true at the 100 Hz frequency, where the conversion of energy to the second compressional wave is very small.

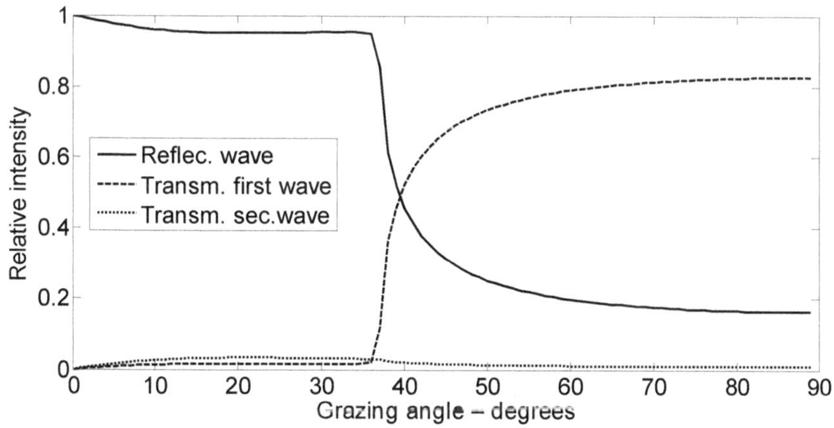


Figure 16.16. Reflected and transmitted intensities at an interface between water and water-saturated sand. The incident wave and the reflected wave are in the water and there are two transmitted waves, one of each the first and second kinds, in the porous medium. The frequency is 100 Hz.

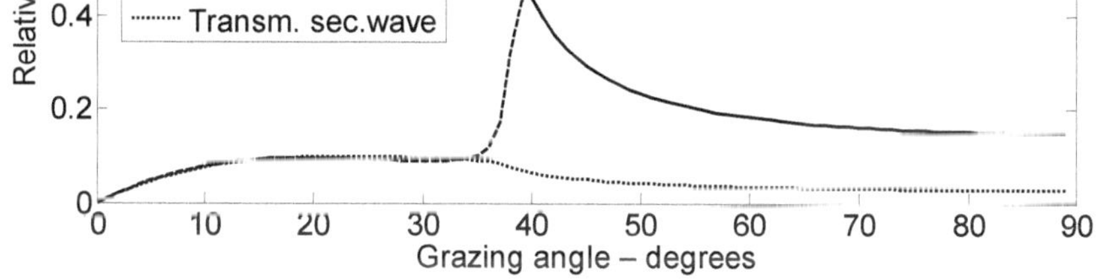


Figure 16.17. Reflected and transmitted intensities at an interface between water and water saturated sand. The incident wave and the reflected wave are traveling in the water. There are two transmitted waves, one of each the first and second kinds, in the porous medium. The frequency is 1000 Hz.

At the 1000 Hz frequency the energy conversion to the second wave is slightly greater, which we can see especially affects the transmission at incident angles lower than the critical angle, where otherwise we would have expected total reflection.

The next example deals with sand identical to that in our first case, but it now has air in the pores rather than water. The parameters appear in table 16.2. Sound speeds are provided in figure 16.11; attenuations in air-saturated sands are given in figure 16.12. The sound speed of the incident wave in air is $c_0$ = 327 m/s. In the air-filled sand, the speed of the first pressure wave, $c_{p1}$, was 943 m/s; for the second wave, we find the wave speed $c_{p2}$ to be 26 m/s at 100 Hz and 82 m/s at 1000 Hz. In air-saturated sand, the second of the two transmitted pressure waves dominates, instead of the first pressure wave. In fact the amplitude of the first wave is so low that it is difficult to see in the plots of figures 16.18 and 16.19. The low amplitude of the transmitted first wave compared with the amplitude of the transmitted second wave is understandable, since the second compressional wave is better matched to the impedance of the incident wave than is the first wave.

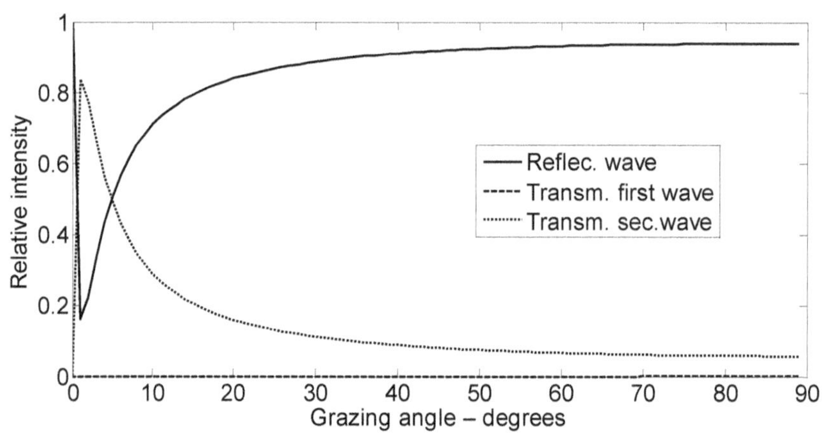


Figure 16.18. Reflected and transmitted intensities at an interface between air and air-saturated sand. The incident wave and the reflected wave are in the air and there are two transmitted waves, one each of the first and the second kind, in the porous medium. The frequency is 100 Hz.

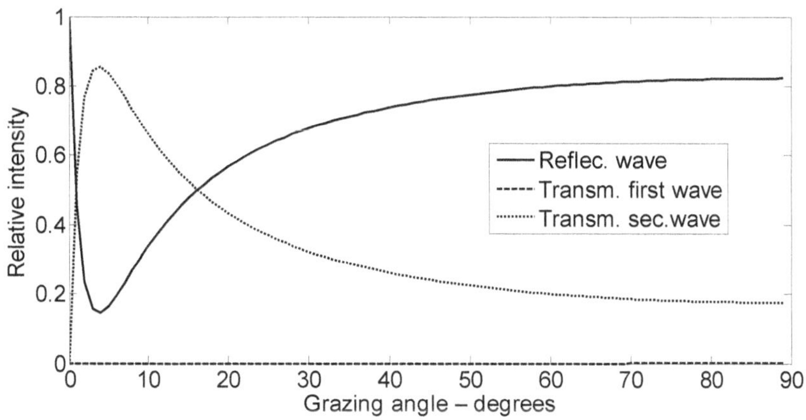


Figure 16.19. Reflected and transmitted intensities at an interface between air and air-saturated sand. The incident wave and the reflected wave are in the air and there are two transmitted waves, one each of the first and the second kind, in the porous medium. The frequency is 1000 Hz.

### *16.5.8 Comments on the Biot model's validity*

The Biot theory's relevance and importance as it is applied to marine sediments is still debated in acoustics literature. In particular, much discussion concerns whether the second compressional wave actually exists. A major problem with the application of the Biot theory is the large number of parameters used and the fact that most of them cannot be determined by independent measurements. In the examples presented here, we have chosen input parameters that frequently appear in the literature, and have found that the second compressional wave is highly attenuated in water-saturated sand. At low frequencies, by which we mean lower than about 1000 Hz, the second wave is a diffusion-process wave rather than a propagating wave. The second wave's main importance is that it drains power from the first or ordinary compressional wave, and thus causes the attenuation that follows the wave's relative movement between air or water and a solid.

# Chapter 17

# Waves in layered media

## 17.1 Introduction

In this chapter, we return to the problem of wave propagation in layered media. We treated this subject earlier in chapter 7, where we discussed the particular methodology of normal modes. This chapter is an expansion of that discussion. This topic is relevant to long-range acoustic propagation, where the sea bottom and its layered structure strongly affect propagation. Our treatment is short and mainly aims to provide physical insight into the problem of propagation and its solution. We shall also limit our discussion to ***range-independent*** environments, in which all layers are horizontal.

Among the books recommended for additional reading are the classic acoustics volumes by Ewing, Jardetzky, and Press (1957) and Brekhovskikh (1980). More recent publications include the volumes by Brekhovskikh and Godin (1999, 2002), Brekhovskikh and Lysanov (2003), and Frisk (1994). The text by Jensen et al. (1994) provides much valuable material on mathematical and numerical methods.

We begin our discussion by demonstrating that an acoustic wave in layered media can be expressed as an integral over horizontal wave numbers, where the integrand contains the solution of the depth-separated wave equation. In general, the integral must be solved numerically, and in this chapter we examine two methods of obtaining the solution. The first is to use the residue technique, in which the acoustic field is expressed as the sum of the residues of the integrand's poles. We will see that this sum is identical to the modal expansion we derived in chapter 7. The second method uses direct numerical calculation, commonly referred to as the ***wave number integration technique*** or ***fast field technique***. This technique is used in several acoustic propagation models, including the OASES (SAFARI) model (Schmidt 1988).

Our starting point is the Helmholtz equation, which gives us the wave equation in the frequency domain. By means of a spatial Fourier transform, this equation is transformed from the spatial domain, designated as either $x$ or $r$, to the wave number domain $k$, which results in an ordinary differential equation in the frequency-wave number domain $f$-$k$. Once we impose boundary conditions, the differential equation can be integrated and solved by common analytical or numerical techniques. Then the equation's solution is transformed twice to return it from the frequency-wave number domain to the time-spatial domain, which allows us to determine the acoustic field's impulse response at any point in the waveguide. One condition for transforming the equation from the spatial domain to the wave number domain is range independence—that is, there can be no range variation in the medium's geometry or characteristics. Although the requirement for range independence limits the technique's applicability, the ocean environment approximates this criterion in many cases.

## 17.2 The two-dimensional case in *x-z* coordinates

In the next section we look at an example of wave propagation in a case that requires use of cylindrical coordinates, but first we examine propagation in rectangular coordinates. Although this situation is less relevant to long-range propagation in the ocean than the problem involving cylindrical coordinates, the rectangular coordinates example provides insights that will be useful when we deal with propagation in a setting with cylindrical coordinates.

The wave equation, with a source term included, was developed in chapter 2 and given by equations (2.46) and (2.127). For the two-dimensional case, the frequency-domain wave equation with a point source at $x = 0$ and $z = z_s$ may be written

$$\left[\frac{\partial^2}{\partial x^2}+\frac{\partial^2}{\partial z^2}+\kappa^2(z)\right]\phi(x,z,\omega)=S(\omega)\delta(x)\delta(z-z_s) \tag{17.1}$$

Note that $\phi(x,z,\omega)$ is now the displacement potential in the space-frequency domain. $S(\omega)$ is the frequency spectrum of the source, and $\kappa = \omega/c$ is the wave number.

The Fourier transform of $\phi(x,z,\omega)$ in the spatial domain $x$ to $\Phi(k,z,\omega)$ in the horizontal wave number domain $k$ is given by

$$\Phi(k,z,\omega) = \frac{1}{2\pi}\int_{-\infty}^{\infty} \phi(x,z,\omega)\exp(-ikx)\,dx \tag{17.2}$$

The inverse transform is written

$$\phi(x,z,\omega) = \int_{-\infty}^{\infty} \Phi(k,z,\omega)\exp(ikx)\,dk \tag{17.3}$$

Applying the Fourier transform to equation (17.1) produces

$$\left[\frac{d^2}{dz^2} + \left(\kappa^2(z) - k^2\right)\right]\Phi(k,z,\omega) = \frac{S(\omega)}{2\pi}\delta(z - z_s) \tag{17.4}$$

Equation (17.4) is the depth-separated wave equation and is a function of the depth coordinate $z$, the horizontal wave number $k$, and the frequency $\omega$. It is an ordinary differential equation whose solution is the sum of the solution of the homogeneous equation plus a particular solution with the source term included.

For the special case in which the sound speed is constant with depth $z$, we have shown in chapter 7 that the solution of the homogeneous equation has the form

$$\Phi(k,z,\omega) = A_+ \exp(i\gamma z) + A_- \exp(-i\gamma z) \tag{17.5}$$

The solution in equation (17.5) represents a downward ($A_+$) and an upward ($A_-$) propagating plane wave. The amplitudes $A_+$ and $A_-$ are given by the boundary conditions, and the vertical wave number $\gamma$ is given by

$$\begin{aligned} \gamma &= \sqrt{\kappa^2 - k^2} \qquad & k \le \kappa \\ \gamma &= i\sqrt{k^2 - \kappa^2} \qquad & k > \kappa \end{aligned} \tag{17.6}$$

Note that where the wave number is imaginary, the sign is selected to make $\Phi(k,z,\omega)$ decay with increasing depth.

We have a certain freedom to select a particular solution of equation (17.4), and here we choose the solution to be the acoustic field from a sound source in free space; this particular solution is called the Green's function $G(k,z,\omega)$. In our present case, we can write $G(k,z,\omega)$ as

$$\begin{aligned} G(k,z,\omega) &= A(\gamma)\exp[i\gamma(z - z_s)] \qquad & z > z_s \\ G(k,z,\omega) &= A(\gamma)\exp[-i\gamma(z - z_s)] \qquad & z < z_s \end{aligned} \tag{17.7}$$

The amplitude $A(\gamma)$ is a function of the source strength. To determine $A(\gamma)$, we integrate equation (17.4) from $z_s - \varepsilon$ to $z_s + \varepsilon$ and then let $\varepsilon \to 0$. This produces

$$\frac{d}{dz}G(k,z,\omega)\Bigg|_{z_S-\varepsilon}^{z_S+\varepsilon} = \frac{S(\omega)}{2\pi} \tag{17.8}$$

The left side of equation (17.8) is determined from equation (17.7) to be

$$\left[\frac{d}{dz}G(k,z,\omega)\right]_{z_S-\varepsilon}^{z_S+\varepsilon} = 2i\gamma A(\gamma) \tag{17.9}$$

The unknown amplitude $A(\gamma)$ is therefore

$$A(\gamma) = \frac{S(\omega)}{4\pi i\gamma} \tag{17.10}$$

and function $G(k,z,\omega)$ becomes

$$G(k,z,\omega) = S(\omega)\frac{\exp(i\gamma|z-z_s|)}{4\pi i\gamma} \tag{17.11}$$

Note that the $k$ dependence of $G(k,z,\omega)$ appears in $\gamma(k)$, defined in equations (17.6).

In figure 17.1, we have an example of the wave number spectrum $G(k,z,\omega)$ for a point source in a medium of infinite extent. The results were calculated by means of equation (17.11), using the values $S(\omega) = 1$, a frequency of 150 Hz, and a sound speed of 1500 m/s. The figure is calculated for $|z - z_s| = 1$ m, which here corresponds to one-tenth of a wavelength of the sound signal.

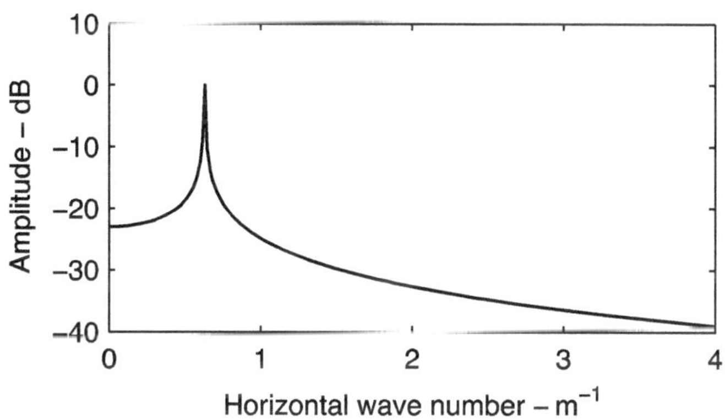


Figure 17.1. Example of the wave number spectrum $G(k,z,\omega)$ as a function of horizontal wave number $k$ for a point source in a medium of infinite extent for fixed values of $\omega$ and $z$.

From equation (17.11), we see that $G(k,z,\omega)$ has a singularity when $\gamma = 0$ or $k = \kappa$. In the example shown in figure 17.1 the spectrum has a singularity at $k = 0.628$.

Next, we consider an example in which a point source is located at depth $z_s$ above an interface between two semi-infinite fluid media that have sound speeds $c_0$ and $c_1$ and densities $\rho_0$ and $\rho_1$. We wish to determine the amplitude and phase of the reflected signal received at the position $(x,z)$ when the point source is located at $(0,z_s)$ as depicted in figure 17.2. The upper medium contains both a direct wave from the source and an upward-propagating reflected wave with amplitude $A_0$. The lower medium contains only the downward-propagating transmitted wave with amplitude $B_1$. Therefore, the fields in the upper and lower media may be written

$$\begin{aligned}\Phi_0(k,z,\omega) &= \frac{S(\omega)\exp\left(i\gamma_0|z-z_s|\right)}{4\pi i\gamma_0} + A_0(k)\exp(-i\gamma_0 z)\\ \Phi_1(k,z,\omega) &= B_1(k)\exp(i\gamma_1 z)\end{aligned} \tag{17.12}$$

The vertical wave number in the upper medium is designated $\gamma_0$. The two unknown amplitudes $A_0$ and $B_1$ are determined by the usual boundary conditions of continuity of pressure and vertical particle displacement at the interface $z = 0$, and are therefore given by the reflection and transmission coefficients at the interface.

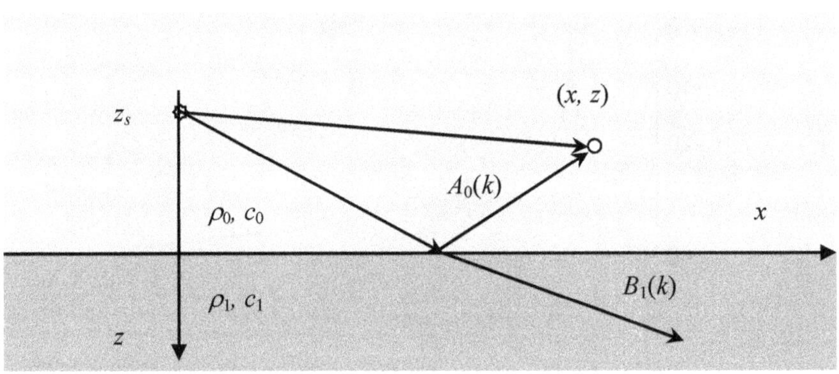


Figure 17.2. Direct, reflected, and transmitted signals from a point source located above the boundary between two semi-infinite fluid media.

This gives us the results

$$A_0(k) = \frac{S(\omega)\exp\left(i\gamma_0|z_s|\right)}{4\pi i\gamma_0} R_{01}(k) \tag{17.13}$$

$$B_1(k) = \frac{S(\omega)\exp\left(i\gamma_0 |z_s|\right)}{4\pi i\gamma_0} T_{01}(k) \tag{17.14}$$

The reflection and transmission coefficients $R_{01}(k)$ and $T_{01}(k)$ are the same as we determined in section 4.2.

Using equations (17.13) and (17.14) in equation (17.12) we now transform equation (17.12) from the wave number domain back to the spatial domain by taking the inverse transform of equation (17.3). Then the total field $\phi_0(x,z,\omega)$ in the upper medium and the transmitted field $\phi_1(x,z,\omega)$ in the lower medium are given as

$$\phi_0(x,z,\omega) = \int_{-\infty}^{\infty} \frac{S(\omega)}{4\pi i\gamma_0} \left\{\exp\left[i\gamma_0 |z - z_s|\right] + R_{01}(k)\exp\left[-i\gamma_0 (z + z_s)\right]\right\} \exp(ikx)dk \tag{17.15}$$

$$\phi_1(x,z,\omega) = \int_{-\infty}^{\infty} \frac{S(\omega)}{4\pi i\gamma_0} T_{01} \exp\left[-i\left(\gamma_0 z_s - \gamma_1 z\right)\right] \exp(ikx)dk \tag{17.16}$$

The vertical wave numbers $\gamma_0$ and $\gamma_1$ are found by using equation (17.6) for the upper and lower media, respectively. Note that the source depth $z_s$, as defined in figure 17.2, is negative. The horizontal wave number $k$ is the same in both media. Note that in most cases, the integrals in equation (17.15) and (17.16) must be calculated numerically; examples will be presented and commented upon later in this chapter.

## 17.3 The solution in cylindrical coordinates

In many cases, the ocean and its environment can appropriately be considered as being transversely isotropic—that is, media properties are dependent only on the depth coordinate $z$ and not on the coordinates $x$ and $y$. In such cases, we may use cylindrical coordinates instead of rectangular coordinates. The coordinate system to be used in transversely isotropic cases is illustrated in figure 17.3. It includes the $r$-coordinate, indicating horizontal range, and the $z$-coordinate, indicating depth, and the distance from the origin $R$.

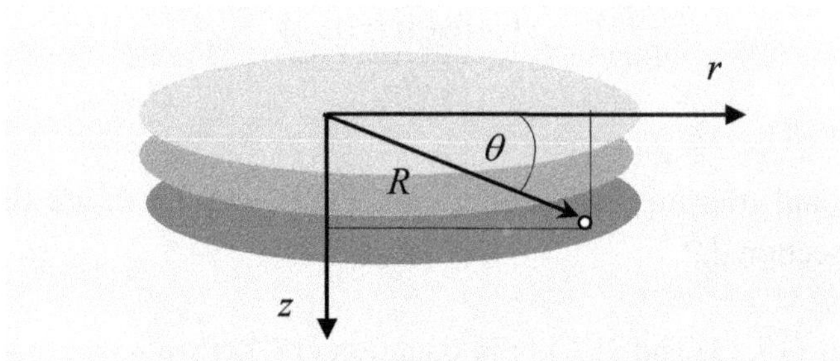


Figure 17.3. Geometry of a cylindrical coordinate system where $r$ is the horizontal range coordinate, $\theta$ is the angle with the horizontal plane, and $z$ is the depth coordinate. The distance from the origin in the figure to the point is $R$.

The wave equation for a cylindrically symmetric environment (see equation (2.85)) is written in cylindrical coordinates as

$$\frac{1}{r}\frac{\partial}{\partial r}\left(r\frac{\partial \Phi}{\partial r}\right)+\frac{\partial^2 \Phi}{\partial z^2}+\kappa^2\Phi=0 \tag{17.17}$$

The solution of this equation can be expressed as a product of two functions: one function that is dependent only on depth, and another function that is dependent only on radial distance. This approach was discussed in section 2.5 with the development of the depth and range equations (2.89) and (2.90).

***The depth equation***

$$\frac{d^2\Phi_z(z)}{dz^2}=-\left(\kappa^2-k^2\right)\Phi_z(z) \tag{17.18}$$

***The range equation***

$$\frac{1}{r}r\frac{d\Phi_r(r)}{dr}=\frac{d^2\Phi_r(r)}{dr^2}+\frac{1}{r}\frac{d\Phi_r(r)}{dr}=-k^2\Phi_r(r) \tag{17.19}$$

Recall that the range equation is a standard type of differential equation and that its solution is a linear combination of the Bessel function $J_0(kr)$ and the Neumann function $Y_0(kr)$ (Abramowitz and Stegun 1965). When the solution we are seeking includes $r = 0$, the part of the solution containing the Neumann function must be omitted since it becomes infinite for $r = 0$. This means that the solution of the range equation in cylindrical coordinates is written as

$$\Phi_r(r) = C_0 J_0(kr) \tag{17.20}$$

The value of the integration constant $C_0$ is determined by the boundary conditions.

For the special case of a point source in an infinite medium, the solution to the depth equation is given by equation (17.11), and the field at depth $z$ and distance $r$ from a point source at depth $z_s$ is

$$\Phi(r,z,\omega) = S(\omega)\int_0^\infty \frac{\exp(i\gamma|z-z_s|)}{4\pi i\gamma} J_0(kr)k\,dk \tag{17.21}$$

In cases involving only one dimension, the transformation from wave number $k$ to distance $x$ is accomplished using the Fourier transform (see, for example, equations (17.15) and (17.16)). When working with cylindrical coordinates, we must use the Bessel transform of equation (17.21), which can be regarded as the cylindrical version of the two-dimensional Fourier transform. To understand this, consider the two-dimensional Fourier transform of an arbitrary function $f(x,y)$ to $F(k_x,k_y)$

$$F(k_x,k_y) = \left(\frac{1}{2\pi}\right)^2 \int_{-\infty}^{\infty}\int_{-\infty}^{\infty} f(x,y)\exp\left[-i(k_x x + k_y y)\right]dx\,dy \tag{17.22}$$

If, as a consequence of cylindrical symmetry, $f(x,y)$ is not dependent on $x$ and $y$ separately, but depends only on the radial distance $r = (x^2 + y^2)^{1/2}$, the two interdependent integrals in equation (17.22) can be transformed into a single integral by making the substitutions

$$\begin{aligned} k_x &= k\cos\varphi \qquad & x &= r\cos\theta \\ k_y &= k\sin\varphi \qquad & y &= r\sin\theta \end{aligned} \tag{17.23}$$

These replacements produce the relationship

$$F(k_x,k_y) = \int_0^\infty f(r)r \int_0^{2\pi} \frac{1}{2\pi} \exp\left[-ikr\cos\left(\varphi-\theta\right)\right] d\theta dr \tag{17.24}$$

The integral over $\theta$ is a standard integral whose solution is the Bessel function

$$J_0(kr) = \frac{1}{2\pi} \int_0^{2\pi} \exp\left[-ikr\cos\left(\varphi-\theta\right)\right] d\theta \tag{17.25}$$

Next, when we define the radial wave number spectrum $F(k)$ as

$$F\left(k\right) = 2\pi F\left(k_x,k_y\right) \tag{17.26}$$

we find that

$$F(k) = \int_0^\infty f(r)J_0(kr)rdr \tag{17.27}$$

The inverse transform may be written

$$f(r) = \int_0^\infty F(k)J_0(kr)kdk \tag{17.28}$$

The Bessel transform of equations (17.27) and (17.28) is the cylindrical version of the two-dimensional Fourier transform, and it is the appropriate transform to use in cases with cylindrical symmetry. Therefore, the wave equation's general solution is expressed as a function of horizontal range $r$ and depth $z$, and is written as

$$\phi(r,z,\omega)=\int_0^\infty \Phi_z(k,z,\omega)J_0(kr)kdk \tag{17.29}$$

In equation (17.29), $\Phi_z(k,z,\omega)$ is the solution for the depth-separated wave equation, that is, equation (17.18).

We may use equation (17.27) to express $\Phi_z(k,z,\omega)$ as

$$\Phi_z(k,z,\omega)=\int_0^\infty \phi(r,z,\omega)J_0(kr)rdr \tag{17.30}$$

The Bessel function $J_0(kr)$ also can be expressed as the sum of two Hankel functions of the first and second kind

$$J_0(kr)=\frac{1}{2}\left[H_0^{(1)}(kr)+H_0^{(2)}(kr)\right] \tag{17.31}$$

The terms $H_0^{(1)}(kr)$ and $H_0^{(2)}(kr)$ represent outgoing and incoming waves, respectively, when the time variation is $\exp(-i\omega t)$. In most cases, we are only interested in the outgoing field so the complete solution for the wave equation in the frequency domain is given by

$$\Phi_z(k,z,\omega)=\frac{1}{2}\int_0^\infty \phi(r,z,\omega)H_0^{(1)}(kr)rdr \tag{17.32}$$

Equation (17.29) then becomes

$$\phi(r,z,\omega)=\frac{1}{2}\int_0^\infty \Phi_z(k,z,\omega)H_0^{(1)}(kr)kdk \tag{17.33}$$

In applications where we are only interested in the acoustic field at such long ranges that the argument $kr >> 1$ holds, we can use the asymptotic expression

$$H_0^{(1)}(kr) \approx \sqrt{\frac{2}{\pi kr}} \exp\left[i(kr - \pi/4)\right] \tag{17.34}$$

Equation (17.33) then becomes

$$\phi(r,z,\omega) = \frac{\exp(-i\pi/4)}{\sqrt{2\pi r}} \int_0^\infty \Phi(k,z,\omega)\sqrt{k}\exp(ikr)\,dk \tag{17.35}$$

Equation (17.35) shows again that we have a Fourier-like transform, but now with the amplitude decreasing as $r^{1/2}$, which is to be expected with cylindrical propagation.

### *17.3.1 Solid media*

So far in our discussion of waves in layered media, we have only examined compressional waves in fluid media, but we can extend this discussion to include shear waves in elastic media, which we examined in chapter 15. With vertically polarized waves and without azimuthal variation, the vector potential has only a $z$ component, which we can treat as if it were a scalar field $\psi$. In a medium with constant parameters, the complete solution for compressional and shear waves becomes

$$\begin{aligned} \phi(r,z,\omega) &= \int_0^\infty \left\{ D_p \exp(i\gamma_p z) + U_p \exp(-i\gamma_p z) \right\} kJ_0(kr)\,dk \\ \psi(r,z,\omega) &= \int_0^\infty \left\{ k^{-1} D_s \exp(i\gamma_s z) + k^{-1} U_s \exp(-i\gamma_s z) \right\} kJ_0(kr)\,dk \end{aligned} \tag{17.36}$$

The factors $k^{-1}$ in front of the downward- and upward-moving shear waves $D_s$ and $U_s$ are introduced for mathematical convenience, as a later derivation will show.

The boundary conditions at an interface with another medium involve the normal particle displacements and the normal and tangential stresses. The displacements in the case of cylindrical symmetry are given as

$$\mathbf{u} = \nabla\phi + \nabla\times\nabla\left(\psi\mathbf{e}_z\right) \tag{17.37}$$

Therefore, the radial and vertical particle displacements $u_r(r,z)$ and $u_z(r,z)$ are

$$\begin{aligned} u_r(r,z,\omega) &= \frac{\partial}{\partial r}\phi(r,z,\omega) + \frac{\partial^2}{\partial r\partial z}\psi(r,z,\omega) \\ u_z(r,z,\omega) &= \frac{\partial}{\partial z}\phi(r,z,\omega) - \frac{1}{r}\frac{\partial}{\partial r}\left[r\frac{\partial}{\partial r}\psi(r,z,\omega)\right] \\ &= \frac{\partial}{\partial z}\phi(r,z,\omega) + k^2\psi(r,z,\omega) \end{aligned} \tag{17.38}$$

The last expression in equation (17.38) is obtained by substituting from the cylindrical wave equation, which is

$$\frac{1}{r}\frac{\partial}{\partial r}r\frac{\partial}{\partial r}\psi(r,z,\omega) = -\left(\frac{\omega^2}{c_s^2} - \gamma_s^2\right)\psi(r,z,\omega) = -k^2\psi(r,z,\omega) \tag{17.39}$$

If we insert equations (17.36) into equations (17.38) and perform some algebra, we can express the vertical and horizontal displacements as

$$u_r(r,z,\omega) = \int_0^\infty \begin{Bmatrix} -k\left[D_p\exp(i\gamma_p z) + U_p\exp(-i\gamma_p z)\right] \\ -i\gamma_s\left[D_s\exp(i\gamma_s z) - U_s\exp(-i\gamma_s z)\right] \end{Bmatrix} kJ_1(kr)dk \tag{17.40}$$

$$u_z(r,z,\omega) = \int_0^\infty \begin{Bmatrix} i\gamma_p\left[D_p\exp(i\gamma_p z) - U_p\exp(-i\gamma_p z)\right] + \ldots.. \\ \ldots..k\left[D_s\exp(i\gamma_s z) + U_s\exp(-i\gamma_s z)\right] \end{Bmatrix} kJ_0(kr)dk \tag{17.41}$$

In the derivation of equation (17.40) we have used the property

$$\frac{dJ_0(kr)}{dr} = -kJ_1(kr) \tag{17.42}$$

As stated before, the boundary conditions also require the horizontal and tangential stresses $\sigma_{zz}(r,z)$ and $\sigma_{rz}(r,z)$ to be continuous. These stresses are given by

$$\sigma_{zz}(r,z) = \lambda\nabla\cdot\mathbf{u} + 2\mu\frac{\partial u_z}{\partial z} = \lambda\nabla^2\phi + 2\mu\left(\frac{\partial^2\phi}{\partial z^2} + k^2\frac{\partial\psi}{\partial z}\right)$$
$$\sigma_{rz}(r,z) = \mu\left(\frac{\partial u_r}{\partial z} + \frac{\partial u_z}{\partial r}\right) = \mu\left(2\frac{\partial^2\phi}{\partial r\partial z} + \frac{\partial}{\partial r}\left(\frac{\partial^2\psi}{\partial z^2} + k^2\psi\right)\right) \quad (17.43)$$

This can be written as

$$\sigma_{zz}(r,z,\omega) = \int_0^\infty \left\{ \begin{matrix} \left(2\mu k^2 - \omega^2\rho\right)\left[D_p\exp(i\gamma_p z) + U_p\exp(-i\gamma_p z)\right] \\ +2i\mu k\gamma_s\left[D_s\exp(i\gamma_s z) - U_s\exp(-i\gamma_s z)\right] \end{matrix} \right\} kJ_0(kr)dk$$
$$+\frac{\lambda}{r}\int_0^\infty \left\{ \begin{matrix} k\left[D_p\exp(i\gamma_p z) + U_p\exp(-i\gamma_p z)\right] \\ +i\gamma_s\left[D_s\exp(i\gamma_s z) - U_s\exp(-i\gamma_s z)\right] \end{matrix} \right\} kJ_1(kr)dk \quad (17.44)$$

$$\sigma_{rz}(r,z,\omega) = \mu\int_0^\infty \left\{ \begin{matrix} -2ik\gamma_p\left[D_p\exp(i\gamma_p z) - U_p\exp(-i\gamma_p z)\right] - \ldots.. \\ \ldots..(2k^2 - \frac{\omega^2}{c_s^2})\left[D_s\exp(i\gamma_s z) + U_s\exp(-i\gamma_s z)\right] \end{matrix} \right\} kJ_1(kr)dk \quad (17.45)$$

The boundary conditions require that the stresses $\sigma_{rz}$ and $\sigma_{zz}$ and the displacements $u_r$ and $u_z$ be the same on both sides of an interface. These requirements give us four equations with which we can determine the amplitudes for the upward- and downward-going compressional and shear waves, $D_p$, $U_p$, $D_s$, and $U_s$, by using the method discussed in section 15.4.

## 17.4 Acoustic waveguide

In chapter 7 we discussed the acoustic waveguide and presented the modal expansion technique. When deriving the modal equations, we assumed that the acoustic field could be expressed in equation (7.9) as a sum of mode functions $\xi_m(z)$, with the mode functions being the solutions of the depth equation for particular horizontal wave numbers $k_m$.

Note that the key assumption was that the normal modes constitute a complete set of orthogonal functions, and therefore we could express the field by a sum over mode functions. While this is correct in the context in which it was developed, that is, for a perfectly reflecting surface and bottom, the assumption is not strictly valid for other boundary conditions. In this section, we discuss the modal expansion technique in more detail. We also outline a more general method for calculating an acoustic field in a waveguide. Our approach follows those described in the articles by Westwood, Tindle, and Chapman (1996) and Bucker (1970), and we also refer to the approaches discussed by Boyle (1984) and Stakgold (1979).

Figure 17.4 shows a range-independent sound channel with a water depth $D$ and an arbitrary sound speed profile $c(z)$. Although in principle the bottom could have any number of fluid or solid layers, we treat only fluid layers in the following discussion.

First, recall that the general solution of the depth equation $\phi(r,z,\omega)$ can be expressed by equation (17.33) as an integral over all horizontal wave numbers. The first step is therefore to find an expression $\Phi(k\, z,\omega)$ for the problem depicted in figure 17.4.

We consider the water column to be divided into two parts at the source depth $z_s$. In the water above this depth, where $z < z_s$, we assume that the solution is given by a function $U(k,z,\omega)$; below the source depth, where $z > z_s$, the solution is given by the function $L(k,z,\omega)$.

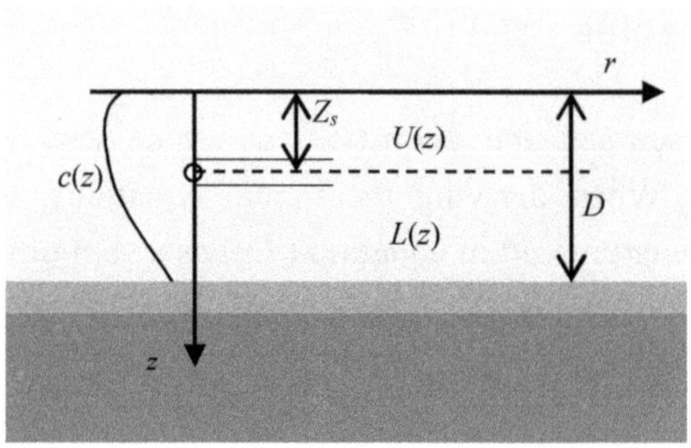


Figure 17.4. A representation of a range-independent sound channel having water depth $D$ and sound speed profile $c(z)$. The waveguide is bounded below by a half space basement. The source is at range $r = 0$ and depth $z = z_S$.

Both solutions must satisfy the depth-separated equation and, therefore, also satisfy the equations

$$U''(k,z,\omega)+\left[\left(\frac{\omega}{c(z)}\right)^2-k^2\right]U(k,z,\omega)=0 \quad 0<z<z_s$$
$$L''(k,z,\omega)+\left[\left(\frac{\omega}{c(z)}\right)^2-k^2\right]L(k,z,\omega)=0 \quad z_s<z<D \tag{17.46}$$

The two functions $U(k, z, \omega)$ and $L(k, z, \omega)$ are determined by the media properties in two different volumes of water, one above and one below the source depth. The functions and their derivatives must satisfy the boundary conditions at the upper surface (for $U$) and the lower interface (for $L$).

At the source depth, the two solutions must be fitted in a way that satisfies the boundary conditions at that depth. To achieve this, we multiply the two solutions by the coefficients $A$ and $B$, respectively, and choose their values to satisfy those boundary conditions. The first boundary condition that must be satisfied is that the pressure, and therefore $\Phi(k,z,\omega)$, must be continuous, which means that

$$AU(k,z_s,\omega) - BL(k,z_s,\omega) = 0 \tag{17.47}$$

We find the second boundary condition by integrating equation (17.4) from $z_s - \varepsilon$ to $z_s + \varepsilon$, and thereafter let $\varepsilon \to 0$

$$\left[\frac{d}{dz}\Phi\left(k,z,\omega\right)\right]_{z_S-\varepsilon}^{z_S+\varepsilon} = -AU'\left(k,z_s,\omega\right) + BL'\left(k,z_s,\omega\right) = \frac{S(\omega)}{2\pi} \tag{17.48}$$

Solving equations (17.47) and (17.48) we determine the two coefficients to be

$$\begin{aligned} A &= \frac{S(\omega)}{2\pi}\frac{L(k,z_s,\omega)}{W(k,z_s,\omega)} \\ B &= \frac{S(\omega)}{2\pi}\frac{U(k,z_s,\omega)}{W(k,z_s,\omega)} \end{aligned} \tag{17.49}$$

where

$$W(k,z_s,\omega) = -U'(k,z_s,\omega)L(k,z_s,\omega) + U(k,z_s,\omega)L'(k,z_s,\omega) \tag{17.50}$$

The function $W(k,z_s,\omega)$ in equations (17.49) is known in mathematics as the Wronskian and is evaluated at the source depth for a given horizontal wave number.

The complete solution to the depth equation is

$$\begin{aligned} \Phi\left(k,z,\omega\right) &= \frac{S(\omega)}{2\pi}\frac{L\left(k,z_s,\omega\right)}{W\left(k,z_s,\omega\right)}U\left(k,z,\omega\right) \qquad 0 < z < z_s \\ \Phi\left(k,z,\omega\right) &= \frac{S(\omega)}{2\pi}\frac{U\left(k,z_s,\omega\right)}{W\left(k,z_s,\omega\right)}L\left(k,z,\omega\right) \qquad z_s < z < D \end{aligned} \tag{17.51}$$

Next, we include in our calculations the boundary conditions above and below the source; we do so by introducing the plane wave reflection coefficients evaluated relative to the source depth. Within a thin depth interval $z_s - \varepsilon < z < z_s + \varepsilon$ around this reference depth, we can consider the sound speed to be constant. Consequently, the two solutions are plane waves in that interval and can be expressed as

$$\begin{aligned} U(k,z,\omega) &= A\left\{\exp\left[-i\gamma_s\left(z-z_s\right)\right]+R_u\exp\left[i\gamma_s\left(z-z_s\right)\right]\right\} \\ L(k,z,\omega) &= B\left\{\exp\left[i\gamma_s\left(z-z_s\right)\right]+R_l\exp\left[-i\gamma_s\left(z-z_s\right)\right]\right\} \end{aligned} \tag{17.52}$$

where $R_u$ and $R_l$ are the reflection coefficients from the surface and bottom, respectively. The term $\gamma_s$ is the vertical wave number at the source depth.

The derivatives of $U$ and $L$ with respect to depth needed to compute $W$ in equation (17.50) are

$$\begin{aligned} U'(k,z,\omega) &= -Ai\gamma_s\left\{\exp\left[-i\gamma_s\left(z-z_s\right)\right]-R_u\exp\left[i\gamma_s\left(z-z_s\right)\right]\right\} \\ L'(k,z,\omega) &= Bi\gamma_s\left\{\exp\left[i\gamma_s\left(z-z_s\right)\right]-R_l\exp\left[-i\gamma_s\left(z-z_s\right)\right]\right\} \end{aligned} \tag{17.53}$$

By inserting equations (17.53), evaluated at the source depth, into equation (17.50), we now find that the Wronskian at the source depth is

$$W\left(k,z_s,\omega\right) = 2iAB\gamma_s\left(1-R_uR_l\right) \tag{17.54}$$

The complete solution of the wave equation for the waveguide is then found by inserting equations (17.51) into equation (17.33) and then integrating over all horizontal wave numbers, by which means we arrive at

$$\begin{aligned} \phi(r,z,\omega) &= \frac{1}{4\pi}\int_0^\infty S(\omega)\frac{U\left(k,z,\omega\right)L\left(k,z_s,\omega\right)}{W\left(k,z_s,\omega\right)}H_0^{(1)}(kr)kdk \qquad z<z_s \\ \phi(r,z,\omega) &= \frac{1}{4\pi}\int_0^\infty S(\omega)\frac{L\left(k,z,\omega\right)U\left(k,z_s,\omega\right)}{W\left(k,z_s,\omega\right)}H_0^{(1)}(kr)kdk \qquad z>z_s \end{aligned} \tag{17.55}$$

Note that the two arbitrary coefficients $A$ and $B$ are no longer present in the solution since they occur in both the numerator and denominator of equations (17.55). Equations (17.55) are a general solution, expressed as an integral over all horizontal wave numbers.

In principle there are three ways of evaluating this integral in an approximate manner:

(1) Expansion using the method of stationary phase, which leads to the ray-tracing approach.
(2) Evaluation by use of branch line integrals and the residue theorem, which leads to the normal mode technique.
(3) Evaluation by numerical integration, which leads to the wave number integration technique.

The ray-tracing approach was treated in chapter 6; the other methods are discussed here.

### *17.4.1 Normal mode expansion*

The integrals of equations (17.55) can be evaluated approximately by the use of contour integrals and the residue theorem. The technique of residue integration will not be treated here, but can be found in advanced textbooks on applied mathematics, such as the book by Kreyszig (1999, chapter 15).

By application of the technique of residue integration, the sum of the residues at the poles yields the normal mode expansion of the acoustic field. The poles of the integrands of equations (17.55) occur for a set of discrete values $k_m$ of the horizontal wave number where

$$W\left(k_m, z_s, \omega\right) = 2i\gamma_s\left(1 - R_u R_l\right) = 0 \tag{17.56}$$

These values of $k_m$ are the wave numbers of the normal modes in the modal expansion.

The theorem of residues states that the pole contributions to the integral are the sum of the residues and are given by

$$\phi(r,z,\omega) = 2\pi i \frac{S(\omega)}{4\pi} \sum_m \left\{ \frac{U(z,k_m,\omega) L(z_s,k_m,\omega)}{\left[\dfrac{\partial W(k,z_s,\omega)}{\partial k}\right]_{k=k_m}} \right\} k_m H_0^{(1)}(k_m r) \qquad z < z_s$$

$$\phi(r,z,\omega) = 2\pi i \frac{S(\omega)}{4\pi} \sum_m \left\{ \frac{L(z,k_m,\omega) U(z_s,k_m,\omega)}{\left[\dfrac{\partial W(k,z_s,\omega)}{\partial k}\right]_{k=k_m}} \right\} k_m H_0^{(1)}(k_m r) \qquad z > z_s$$

(17.57)

The derivative of the Wronskian with respect to the horizontal wave number is

$$\left. \left( \frac{\partial W(k,z_s,\omega)}{\partial k} \right) \right|_{k=k_m} = -2i\gamma_s \left. \frac{\partial (R_u R_l)}{\partial k} \right|_{k=k_m} \qquad (17.58)$$

where we have used the fact that, by equation (17.56), $1 - R_u R_l = 0$ at $k = k_m$.

The terms in curly brackets in equation (17.57) are recognized as the normalized mode functions, valid for all $z$ that we encountered in section 7.2. The steps to calculate the acoustic field are therefore: first, compute $R_u$ and $R_l$ for complex $k$ values. This involves dividing the environment into layers of constant or $1 / c^2$ linear sound speeds and working from the surface to $z_s$ and from the bottom half-space to $z_s$. Then solve equation (17.56) to find the eigenvalues $k_m$ and evaluate the corresponding mode functions, which are required for the modal sum of equation (17.57). For large ranges where $kr >> 1$, the Hankel function may be replaced by the approximation given by equation (17.34).

It should be noted that this normal mode solution is not a complete representation of the field in the waveguide because the contributions from the branch line integrals are not included. In most cases of practical interest, particularly at long ranges, these contributions are not very important and can be neglected. However, in some cases, depending on source-receiver geometry, signal frequency, and mode locations, the contributions from the branch line integrals can be significant, even at long ranges. Because the branch point locations correspond to waves propagating near the critical angle at the lower half-space interface, neglecting the branch line integral results in omitting the contributions of head wave or refracted arrivals. We shall see examples where refracted waves are important in chapter 18. We will not discuss the numerical technique further here, but refer readers to the articles by Bucker (1970), Tindle and Chapman (1994), Levinson et al. (1994), and Westwood, Tindle, and Chapman (1996).

### *17.4.2 Example: The Pekeris waveguide*

We now consider a simple example in order to illustrate the technique we have just outlined. We assume an acoustic waveguide with water depth $D$, a constant sound speed $c_0$ in the water, and water density $\rho_0$; the waveguide lies over a fluid bottom with properties $c_1$ and $\rho_1$. The source is at the depth $z_s$. We want to determine the acoustic field at a depth $z$ and range $r$. In the case of a constant sound speed over the whole water column, the upper and lower fields are defined as

$$\begin{aligned} U(k,z,\omega) &= \left\{\exp\left[-i\gamma\left(z-z_s\right)\right] + R_s \exp\left[i\gamma\left(z+z_s\right)\right]\right\} \\ L(k,z,\omega) &= \left\{\exp\left[i\gamma\left(z-z_s\right)\right] + R_b \exp\left[i\gamma\left(2D-z-z_s\right)\right]\right\} \end{aligned} \tag{17.59}$$

Here $R_b$ and $R_s$ are the ordinary plane wave reflection coefficients for the bottom and the surface, respectively. Since the vertical wave number is now constant, we use $\gamma$ instead of $\gamma_s$. By comparing equations (17.59) to equations (17.52), we see that when the sound speed is constant over the whole water column the reflection coefficients $R_u$ and $R_l$ are

$$\begin{aligned} R_u &= R_s \exp\left(2i\gamma z_s\right) \\ R_l &= R_b \exp\left[2i\gamma\left(D-z_s\right)\right] \end{aligned} \tag{17.60}$$

The Wronskian, evaluated at the source depth, now becomes

$$W(k,z_s,\omega)=(2i\gamma)\left[1-R_sR_b\exp(2i\gamma D)\right] \tag{17.61}$$

and its zeros are given by

$$1-R_sR_b\exp(2i\gamma D)=0 \tag{17.62}$$

which is the same characteristic equation we found in section 7.3.

The mode functions evaluated at the source depth are

$$\begin{aligned} U(k,z_s,\omega)&=\left\{1+R_s\exp\left[2i\gamma z_s\right]\right\}\\ L(k,z_s,\omega)&=\left\{1+R_b\exp\left[2i\gamma(D-z_s)\right]\right\} \end{aligned} \tag{17.63}$$

Therefore, the complete solution of equations (17.46) can be written as

$$\begin{aligned} \Phi(k,z,\omega)&=\frac{S(\omega)}{4\pi i\gamma\left[1-R_sR_b\exp(2i\gamma D)\right]}\left\{1+R_b\exp\left[2i\gamma(D-z_s)\right]\right\}\\ &\quad\times\left\{\exp\left[-i\gamma(z-z_s)\right]+R_s\exp\left[i\gamma(z+z_s)\right]\right\} \qquad z<z_s\\ \Phi(k,z,\omega)&=\frac{S(\omega)}{4\pi i\gamma\left[1-R_sR_b\exp(2i\gamma D)\right]}\left\{1+R_s\exp\left[2i\gamma z_s\right]\right\}\\ &\quad\times\left\{\exp\left[i\gamma(z-z_s)\right]+R_s\exp\left[i\gamma(2D-z-z_s)\right]\right\} \qquad z>z_s \end{aligned} \tag{17.64}$$

The complete solution satisfying all boundary conditions at the surface and the seabed for $z > z_s$ is

$$\Phi(k,z,\omega)=\frac{S(\omega)}{4\pi i}\frac{\left\{\exp(-i\gamma z_s)+R_s\exp(i\gamma z_s)\right\}\left\{\exp(i\gamma z)+R_b\exp\left[i\gamma(2D-z)\right]\right\}}{\gamma\left[1-R_sR_b\exp(2i\gamma D)\right]} \tag{17.65}$$

For $z < z_s$ the solution is the same as equation (17.65) except we interchange $z$ and $z_s$.

We can obtain the wave field at range $r$ and depth $z$ by integrating over all horizontal wave numbers. Using equation (17.32) we obtain

$$\phi(r,z,\omega) = \frac{S(\omega)}{8\pi i}\int_0^\infty \frac{\{\exp(-i\gamma z_s)+R_s\exp(i\gamma z_s)\}\{\exp(i\gamma z)+R_b\exp[i\gamma(2D-z)]\}}{\gamma[1-R_sR_b\exp(2i\gamma D)]} H_0^{(1)}(kr)\,kdk \tag{17.66}$$

We now consider the case of a pressure release surface and an infinitely hard seabed and, consequently, total reflection at both the seabed and the surface. Then it follows that

$$R_s = -1 \qquad R_b = 1 \tag{17.67}$$

The poles of the integrand are found by setting $\exp[(2i\gamma D)] = -1$, which results in the solutions for the vertical wave number

$$\gamma_m D = (2m-1)\frac{\pi}{2} \qquad m = 1,2,... \tag{17.68}$$

The corresponding horizontal wave numbers are

$$k_m = \sqrt{\left(\frac{\omega}{c}\right)^2 - \gamma_m^2} \tag{17.69}$$

When we take the derivative with respect to $k$ of the integrand's denominator in equation (17.66), we obtain

$$\begin{aligned}\frac{\partial}{\partial k}\{\gamma[1+\exp(2i\gamma D)]\}_{\gamma=\gamma_m} &= [1+\exp(2i\gamma D)]\left.\frac{\partial\gamma}{\partial k}\right|_{\gamma=\gamma_m} \\ &\quad + [2iD\gamma\exp(2i\gamma D)]\left.\frac{\partial\gamma}{\partial k}\right|_{\gamma=\gamma_m} \\ &= 2iDk_m\end{aligned} \tag{17.70}$$

By using equations (17.68) and (17.70) we obtain the residue solution of equation (17.66)

$$\phi(r,z,\omega)=\frac{-iS(\omega)}{4}\sum_m H_0^{(1)}(k_m r)\frac{\sin(\gamma_m z_s)}{\sqrt{D/2}}\frac{\sin(\gamma_m z)}{\sqrt{D/2}}\exp(-i\omega t) \qquad (17.71)$$

For large ranges where $k_m r >> 1$, we can apply the asymptotic expression for the Hankel function, thus obtaining

$$\phi(r,z,\omega)=\frac{-iS(\omega)}{2}\exp\left(-i\frac{\pi}{4}\right)\sum_m\sqrt{\frac{1}{2\pi k_m r}}\frac{\sin(\gamma_m z_s)}{\sqrt{D/2}}\frac{\sin(\gamma_m z)}{\sqrt{D/2}}\exp[i(k_m r-\omega t)]$$

(17.72)

The term $(1 / r)^{1/2}$ in equation (17.72) means that each mode undergoes cylindrical spreading with range.

In equations (17.71) and (17.72) the normalized mode functions are

$$\xi_m(z)=\sqrt{\frac{2}{D}}\sin(\gamma_m z) \qquad (17.73)$$

where the normalization ensures that

$$\int_0^D \xi_m(z)\,\xi_n(z)\,dz=\begin{cases}0 & n\neq m\\ 1 & n=m\end{cases} \qquad (17.74)$$

Equations (17.71) and (17.72) show that, at large distances, the field is made up of modes with vertical wave numbers $\gamma_m$ and horizontal wave numbers $k_m$ given by equations (17.68) and (17.69), respectively.

Transmission loss is defined relative to the sound level $\phi_0(r_0)$ at a reference distance $r_0$ from the point source in an infinite medium; $\phi_0(r_0)$ is expressed as

$$\phi_0(r_0) = \frac{S(\omega)}{4\pi r_0} \tag{17.75}$$

Therefore the transmission loss *TL*, measured in dB relative to the reference distance $r_0 = 1$, is defined as

$$TL = -20\log\sqrt{\frac{2\pi}{r}}\left|\sum_{m=0}^{\infty}\xi_m(z_s)\xi_m(z)\frac{\exp(ik_m r)}{\sqrt{k_m}}\right| \tag{17.76}$$

This is identical to the result we found in section 7.2; see equation (7.22).

## 17.5 Ray interpretation of normal modes

Equations (17.64) may also be interpreted from a ray-theory perspective, and this may be clearer if they are rearranged into the following form

$$\begin{aligned}
\Phi(k,z,\omega) &= \frac{S(\omega)}{4\pi i\gamma\left[1 - R_s R_b \exp(2i\gamma D)\right]} \\
&\times\left\{\begin{matrix}\exp\left[i\gamma(z_s - z)\right] + R_s\exp\left[i\gamma(z + z_s)\right] + \\ R_b\exp\left[i\gamma(2D - z_s - z)\right] + R_b R_s\exp\left[i\gamma(2D - z_s + z)\right]\end{matrix}\right\} \qquad z < z_s \\
\Phi(k,z,\omega) &= \frac{S(\omega)}{4\pi i\gamma\left[1 - R_s R_b \exp(2i\gamma D)\right]} \\
&\times\left\{\begin{matrix}\exp\left[i\gamma(z - z_s)\right] + R_s\exp\left[i\gamma(z + z_s)\right] + \\ R_b\exp\left[i\gamma(2D - z_s - z)\right] + R_b R_s\exp\left[i\gamma(2D - z + z_s)\right]\end{matrix}\right\} \qquad z > z_s
\end{aligned} \tag{17.77}$$

The four terms in the curly brackets of equations (17.77) can be identified with the four waves that are transmitted from the source to the receiver without striking the surface and the bottom more than once. For the case of $z > z_s$, these waves are the direct, the surface-reflected, the seabed-reflected, and both the surface- and seabed-reflected waves. These paths are illustrated in the panels of figure 17.5.

Their contributions to the acoustic field are:

Direct wave: $\exp\left[(i\gamma(z - z_s)\right]$
Surface-reflected wave: $R_s \exp\left[(i\gamma(z + z_s)\right]$
Seabed-reflected wave: $R_b \exp\left[(i\gamma(2D - z - z_s)\right]$
Surface- and seabed-reflected wave: $R_s R_b \exp\left[(i\gamma(2D + z_s - z)\right]$

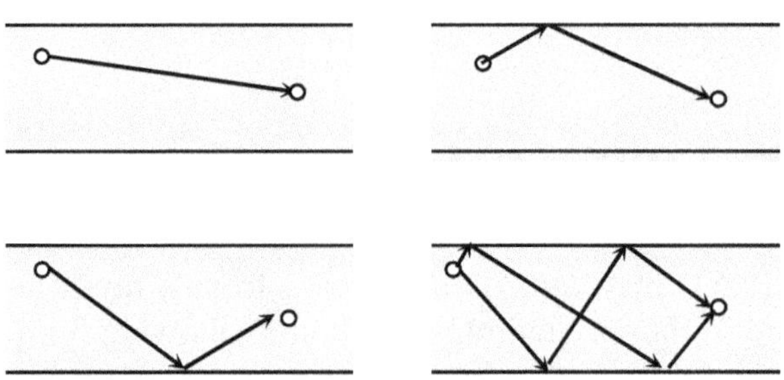

Figure 17.5. The four ray paths between the source and receiver in an isospeed waveguide that involve no more than one surface and one bottom reflection. The four paths are represented by the four terms in braces in equations (17.77).

In addition to these four arrivals, there is an infinitely long train of multiple arrivals that have been reflected at the surface or the bottom more than once. Their contributions are expressed in the form of a geometric series that can be summed as

$$\sum_{p=0}^{\infty} (R_s R_b)^p \exp(2i\gamma pD) = \frac{1}{1 - R_s R_b \exp(2i\gamma D)} \tag{17.78}$$

Note that the right side of equation (17.78) appears as the denominator in equations (17.65) and (17.77). Thus we observe that the mode solution to equation (17.56) for the isospeed waveguide can be interpreted as an infinite sum of rays as well.

## 17.6 The wave number integration technique

We have shown earlier that the acoustic field can be expressed as an integral over horizontal wave numbers, where the integrand contains the solution of the depth-separated equation. Using the farfield approximation of the Hankel functions, the wave number integral $\phi(r,z,\omega)$ can be defined as

$$\phi(r,z,\omega)=\frac{\exp(-i\pi/4)}{\sqrt{2\pi r}}\int_0^\infty \Phi(k,z,\omega)\sqrt{k}\exp(ikr)dk \tag{17.79}$$

The numerical integration of equation (17.79) is not a trivial task and care must be exercised to obtain accurate results. The problem is caused by the characteristics of the integrand kernel, which may have a number of peaks of large magnitude for values of $k$ located at the pole locations. We have seen evidence of this in section 17.4.2 where we found that the poles of the integrand for the Pekeris waveguide are located as given by equation (17.68). In the absence of attenuation losses in the medium the poles are located on the real $k$ axis and at these locations the integrand will go to infinitely high values. With medium losses, that is, sound absorption in the water column or in the bottom materials, the pole positions are complex and the magnitude of the integrand is finite, but the peaks may still be very high and narrow and the integrand may be difficult to integrate accurately.

To integrate equation (17.79) numerically, we must truncate the integrand to a finite interval of the horizontal wave number $k$. In this interval of wave numbers the integrand is sampled at wave numbers $k_q$, defined as

$$k_q = k_{min} + q\Delta k \qquad q = 0,1,\ldots,(N-1) \tag{17.80}$$

where $k_{min}$ is the minimum wave number where the integration starts, $\Delta k$ is the sampling interval, and $N$ is the number of sampling points. Consequently, the wave number where the integration ends is $k_{max} = k_{min} + (N-1)\Delta k$. In selecting the appropriate values for $k_{min}$ and $k_{max}$ we must take into account that the wave number $k$ depends on the frequency. It is therefore advisable to select the integration limits to be dependent on the frequency, for instance as

$$k_{\min} = \frac{\omega_{\min}}{c_{\max}}$$
$$k_{\max} = \frac{\omega_{\max}}{c_{\min}} \tag{17.81}$$

In equations (17.81) $c_{min}$ and $c_{max}$ are the minimum and maximum values of the phase speeds in the problem under investigation, and $\omega_{max}$ and $\omega_{min}$ are the maximum and minimum (angular) frequencies of interest.

The integral in equation (17.79) is now replaced by the sum

$$\begin{aligned}\phi(r,z,\omega) \approx & \frac{\Delta k}{\sqrt{2\pi r}} \exp(irk_{min}) \exp(-i\pi/4) \\ & \times \sum_{q=0}^{N-1} \left[ \Phi(k_q, z, \omega) \sqrt{k_q} \exp(iq\Delta kr) \right]\end{aligned} \tag{17.82}$$

Equation (17.82) is an approximation to the integral of equation (17.79), so there will always be degree of inaccuracy. We can interpret these errors as folding, aliasing, or wraparound errors, well known in all applications of the discrete Fourier transform. These errors occur because the exponential function in the sum is a periodic function of range with period

$$r_{per} = \frac{2\pi}{\Delta k} \tag{17.83}$$

Therefore equation (17.82) does not yield $\phi(r,z,\omega)$, but instead produces

$$\tilde{\phi}(r,z,\omega) = \sum_{j=-\infty}^{\infty} \phi(r + jr_{per}, z, \omega) \tag{17.84}$$

that is, the sum of all signals in all range intervals of width $r_{per}$. However, there will be no errors if the true value of $\phi(r,z,\omega)$ is equal to zero for $r \geq r_{per}$ and $r < 0$.

In cases where the medium attenuation is very low we can improve the accuracy of integration by application of a technique known as ***offset contour integration.*** A formal derivation and justification of this approach may be found in Jensen et al. (1994), chapter 4. By this technique we replace the real wave number $k$ with a complex wave number

$$k \to k - i\varepsilon \tag{17.85}$$

where $\varepsilon$ is a small imaginary component. After the replacement, equation (17.82) becomes

$$\begin{aligned}\phi(r,z,\omega)\exp(-\varepsilon r) \approx & \frac{\Delta k}{\sqrt{2\pi r}}\exp(irk_{min})\exp(-i\pi/4) \\ & \times \sum_{q=0}^{N-1}\left[\Phi(k_q - i\varepsilon, z, \omega)\sqrt{k_q - i\varepsilon}\exp(iq\Delta kr)\right]\end{aligned} \tag{17.86}$$

Equation (17.86) still has the form of a Fourier transform since $\Delta k$ is real and the result is a periodic function of range of $r$ that may be expressed as

$$\tilde{\phi}(r,z,\omega) = \sum_{j=-\infty}^{\infty}\phi(r + jr_{per}, z, \omega)\exp(-j\varepsilon r_{per}) \tag{17.87}$$

A comparison of equations (17.87) and (17.84) shows that the aliasing errors are now reduced by, at least, the factor of $\exp(-\varepsilon r_{per})$. However, the periodic range interval $r_{per}$ has not been changed with the offset integration.

In many applications there is a need to determine the acoustic field for a large number of ranges, in particular for determining the transmission loss as function of range. This can efficiently be done by means of a FFT (fast Fourier transform), an approach leading to the fast field program (FFP) introduced by DiNapoli and Deavenport (1980).

We perform a discretization of the range and choose the same number $N$ of range samples as used for the discretization of wave number. The discrete ranges are therefore defined as

$$r_p = r_{min} + p\Delta r \qquad p = 0,1,\ldots,(N-1) \tag{17.88}$$

where $r_{min}$ is the minimum range and $\Delta r$ is the range increment. The maximum range becomes $r_{max} = r_{min} + (N-1)\Delta r$. With these values for $r$ and $k$, the exponential term of equation (17.79) becomes

$$\begin{aligned}\exp(ikr) &= \exp\left[i(k_{min} + q\Delta k)(r_{min} + p\Delta r)\right] \\ &= \exp\left[ik_{min}(r_{min} + p\Delta r)\right]\exp\left[iq\Delta k r_{min}\right]\exp(iqp\Delta k\Delta r)\end{aligned} \tag{17.89}$$

Next, we select the sampling intervals $\Delta k$ and $\Delta r$ to satisfy

$$\Delta k \cdot \Delta r = \frac{2\pi}{N} \tag{17.90}$$

where the number $N$ is an integer that, for numerical efficiency considerations is chosen to be a power of two (that is, $N = 2^n$). The discrete version of equation (17.79) then becomes

$$\begin{aligned}\phi(r,z,\omega) \approx &\frac{\Delta k}{\sqrt{2\pi r}}\exp(irk_{min})\exp(-i\pi/4) \\ &\times\sum_{q=0}^{N-1}\left[\Phi(k_q,z,\omega)\exp(iq\Delta k r_{min})\sqrt{k_q}\right]\exp\left(2\pi i\frac{pq}{N}\right)\end{aligned} \tag{17.91}$$

We recognize the summation term of equation (17.91) as a discrete Fourier transform (DFT), which can be computed very efficiently by means of the fast Fourier transform or FFT, thereby simultaneously yielding the field at all $N$ ranges This is called the ***fast field method.***

### *17.6.1 Example: The Pekeris waveguide*

We will demonstrate the technique developed in this chapter by again considering the same Pekeris waveguide that we studied earlier in chapter 7 using the normal mode method. To begin, note that the Helmholtz equation's solution for the case of constant sound speed in the water column is given by either equations (17.64) or equation (17.65) depending on the source and receiver depths. We use equation (17.64) for $z > z_s$ and assume that the reflection coefficient at the sea surface is $R_s = -1$. For the sea bottom's reflection coefficient $R_b$, we use the reflection coefficient for an infinite, homogenous, fluid half-space. From equation 4.12, the coefficient is

$$R_b = \frac{\rho_1\gamma_0 - \rho_0\gamma_1}{\rho_0\gamma_1 + \rho_0\gamma_1} \tag{17.92}$$

Note that the index 0 refers to the water's properties, and the index 1 refers to the properties of the bottom.

The function $\Phi(k,z,\omega)$ in equation (17.65) can now be written

$$\Phi(k,z,\omega) = -\frac{S(\omega)}{2\pi}\frac{\sin(\gamma z_s)\exp(i\gamma z)\{1 + R_b\exp[2i\gamma(D-z)]\}}{\gamma\left[1 + R_b\exp(2i\gamma D)\right]} \tag{17.93}$$

The expression for the field at any distance $r$ from the source is given by equation (17.33), which by the use of the approximation of equation (17.34) becomes

$$\phi(r,z,\omega) = \int_0^\infty \frac{\Phi(k,z,\omega)}{\sqrt{2\pi ikr}}\exp(ikr)kdk \tag{17.94}$$

We consider an example where the water has a depth $D$ of 100 m with sound speed 1500 m/s and density 1000 kg/m$^3$. The bottom sound speed is 1700 m/s, the density is 1500 kg/m$^3$ and the attenuation coefficient of sound in the bottom is equivalent to 0.5 dB / wavelength. The source depth is 25 m and the receiver depth is 75 m.

Figure 17.6 shows the magnitude of the integrand kernel of equation (17.94) plotted as a function of the real horizontal wave number *k*. The integrand is calculated for the frequency of 50 Hz and is normalized with respect to its maximum value. Four different representations appear in figure 17.6, plotted as functions of (a) horizontal wave number, (b) grazing angle, (c) phase speed, and (d) group speed. With a constant sound speed $c_0$ in the water, the phase speed $v$ and the group speed $w$ are defined as

$$v = \frac{\omega}{k} = \frac{c_0}{\cos\theta_0}$$
$$w = \frac{d\omega}{dk} = c_0 \frac{k}{\kappa_0} = c_0 \cos\theta_0 \qquad (17.95)$$

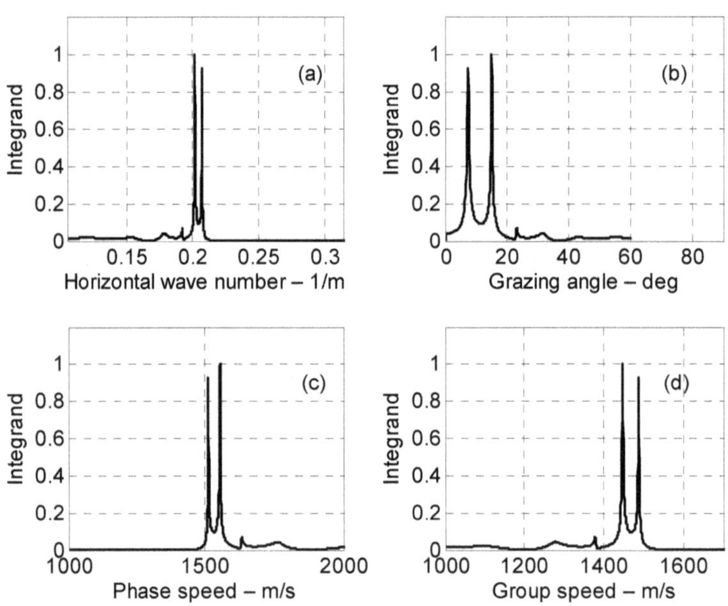


Figure 17.6. Four representations of the integrand for a Pekeris waveguide plotted as functions of (a) wave number; (b) grazing angle; (c) phase speed; and (d) group speed. Source depth is 25 m, receiver depth is 75m, and the frequency is 50 Hz.

We see that the integrand has a number of peaks. In the upper left-hand portion of figure 17.6, the two highest peaks occur when the horizontal wave number is $k = 0.2077$ m$^{-1}$ and $k = 0.2021$ m$^{-1}$ and there is also a much lower peak at $k = 0.1923$ m$^{-1}$. In the upper right-hand part of this figure, we see that the two strongest modes correspond to the grazing angles 7.5° and 15.2°, respectively, and the weaker mode to 23.2°. From the lower plots we see that the two peaks correspond to the phase speeds of 1513 m/s and 1554 m/s, and the group speeds of 1447 m/s and 1487 m/s. These are the same values we found in section 7.3.2 when we considered the same waveguide using the normal mode approach. Thus, the peaks of the integrand can be associated with the normal modes.

We want to calculate that transmission loss as a function of range to a maximum distance of 30 km for the frequencies of 50 Hz and 100 Hz. This means that the $r_{per}$ must be at least 30 km for both frequencies.

As indicated before, the sharpness of the integrand peaks is dependent on the attenuation of sound in the water and the bottom. With little or no attenuation the peaks are extremely sharp and therefore a very low value of $\Delta k$ is required for accurate results. For given integration limits $k_{min}$ and $k_{max}$ this may require a very high value of the block length $N$ and consequently a long computation time.

After some experimenting with parameters, we choose the integration limits according to equations (17.81) with $c_{\text{min}} = 1500$ m/s and $c_{\text{max}} = 1700$ m/s and we use $N = 4096$ for 50 Hz and $N = 8192$ for 100 Hz. With these values the periodic range length $r_{per}$ is 98,304 m and 49,152 m, respectively, for 50 Hz and 100 Hz. For both frequencies we apply the contour offset technique with $\varepsilon = 10^{-4}$. The results will show that these integration parameters yield quite good results, but certainly other integration parameters may be equally good.

Figure 17.7 shows transmission loss versus range for the frequencies of 50 Hz and 100 Hz and the wave number integration results are compared with the results obtained by normal mode theory, that is, implementation of equation (17.76) and with sea bottom absorption included as described in section 7.4.

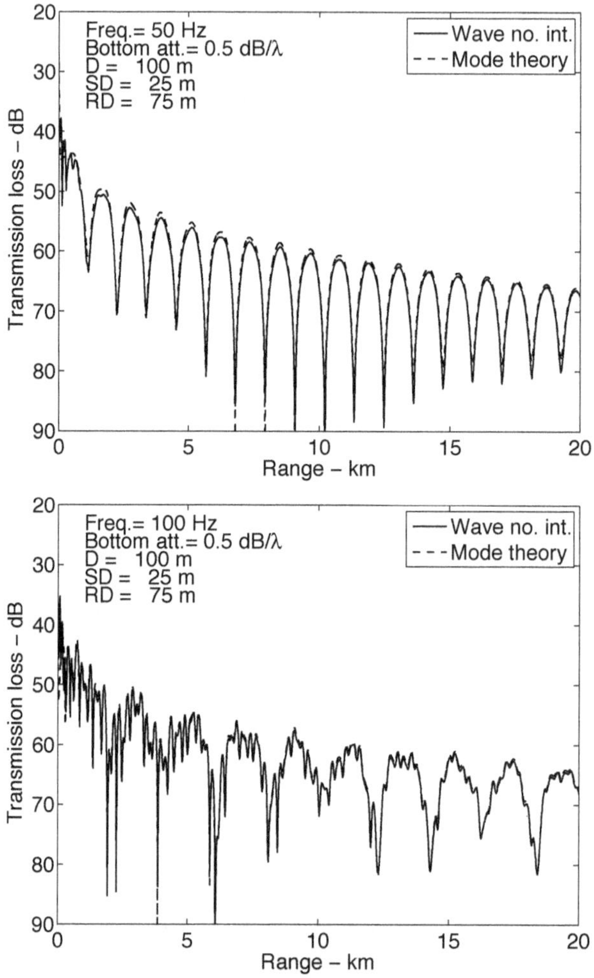


Figure 17.7. Transmission loss versus range for a Pekeris waveguide for the frequencies of 50 Hz and 100 Hz. The wave number integration result (solid line) is compared with the results of a normal mode calculation (dashed line).

## 17.7 Time-domain solutions and discussions

We have so far concentrated on the frequency domain solutions of the wave equations. However, in some applications it may also be necessary to determine the time domain solutions, that is, the time series of acoustic signals, in particular for broadband signals. Studies of time responses are often more instructive and give a better understanding of the physics of wave propagation.

The time-domain solution $s(r,z,t)$ is obtained by a Fourier transform of the frequency solution $\phi(r,z,\omega)$, which gives

$$s(r,z,t) = \int_0^\infty \phi(r,z,\omega)\exp(-i\omega t)\,dt \tag{17.96}$$

where $\phi(r,z,\omega)$ is obtained from equation (17.35) as

$$\phi(r,z,\omega) = \frac{\exp(-i\pi/4)}{\sqrt{2\pi r}}\int_0^\infty \Phi(k,z,\omega)\sqrt{k}\exp(ikr)\,dk \tag{17.97}$$

In practice the Fourier transform is calculated numerically using the fast Fourier transform, which implies sampling of the frequency function. The samplings in the frequency and wave number domains are not independent, so that the sampling theorem must be satisfied in both domains to avoid aliasing errors.

We are not going into details on this, but show some results of pulse transmission in a Pekeris waveguide with the same parameters as used in section 17.6.1, but now the source depth is fixed to 50 m. The source pulse is a short Ricker-shaped pulse with 50 Hz bandwidth centered on 125 Hz.

To perform the calculations we use the OASES (SAFARI) program (Schmidt 1988). This is a general and widely used program for the calculation of acoustic fields in waveguides using the wave number integration technique. We present the results and discuss its interpretation in terms of mode theory.

Figure 17.8 shows the integration kernel, both as a function of wave number *k* and as a function of group speed, plotted for the frequency of 125 Hz with source and receiver at 50 m depths.

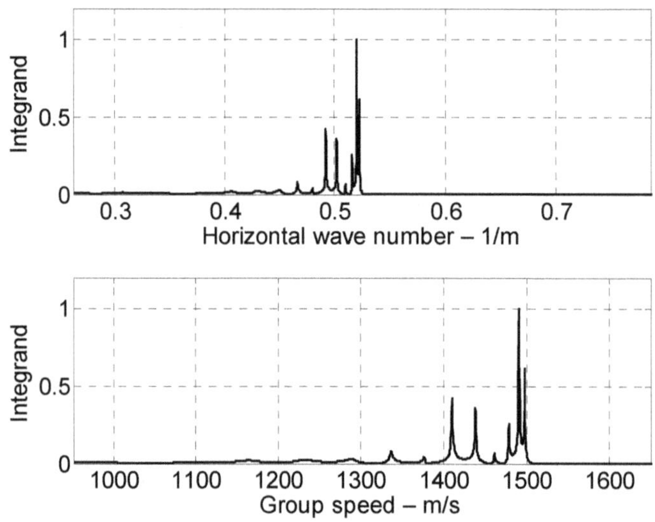


Figure 17.8. The integrand for the frequency of 125 Hz for the case where both the receiver and source are at 50 m depth.The upper panel shows the integrand as a function of wave number, the lower panel as a function of group speed.

From the discussion of the same case in chapter 7 (see equation (7.57)), we find that for the frequency of 125 Hz there are eight propagating modes, and figure 17.9 shows the first five modal functions. From this plot we see that, with the source (and receiver) at 50 m depth, the mode functions of modes 2, 4, and 6 are nearly zero at that depth and thus only weakly excited. The dominating modes are in this case, therefore, the modes 1, 3, and 5. From figure 17.8 we find that the group speeds of these modes are 1498 m/s, 1478 m/s, and 1438 m/s, respectively.

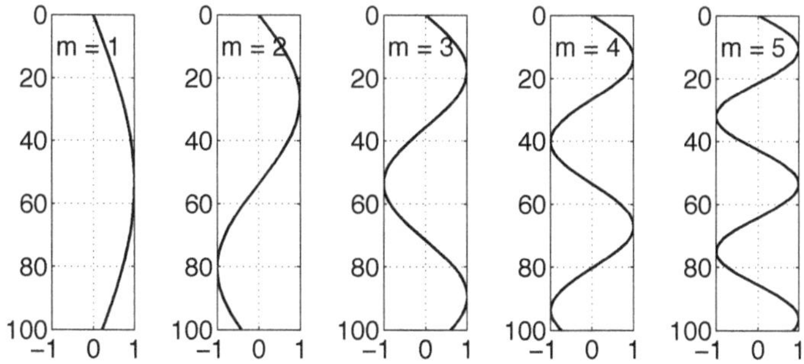


Figure 17.9. Modal functions for the first five modes

The time series as a function of range for a fixed receiver depth of 50 m are shown in figure 17.10, and figure 17.11 shows the time series as a function of depth at a fixed horizontal range of 10 km. In both figures, the responses are shown as a function of reduced time $t_{red}$

$$t_{red} = t - \frac{r}{c_{red}} \tag{17.98}$$

The reduced sound speed is 1500 m/s in this case.

Both plots clearly show the mode structure with the first mode arriving first and the higher modes later. In figure 17.10 the dashed lines indicate the expected arrival times of the first strongest modes 1, 3, and 5, with group speeds of 1498 m/s, 1478 m/s, and 1438 m/s, respectively.

The depth plots of figure 17.11 show that the amplitude of the modes varies with depth, the first strong mode has one maximum at 50 m depth, the second strong mode has three maxima at about 15 m, 50 m, and 90 m. When we compare these observations with the modal functions of figure 17.11 we see that these modes are in reality the modes 1 and 3. A plot of mode amplitude versus depth shows the same information about the modal structure as the mode shape functions in figure 17.9.

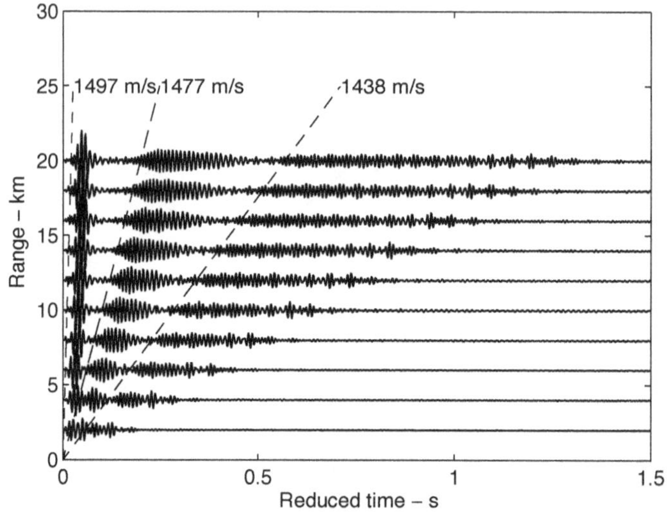


Figure 17.10. Time series of the sound pressure, plotted as a function of range; the receivers are all at 50 m depth. The dashed lines indicate waves arriving with the speeds of 1498 m/s, 1478 m/s, and 1438 m/s, which are the group speeds of modes 1, 3, and 5.

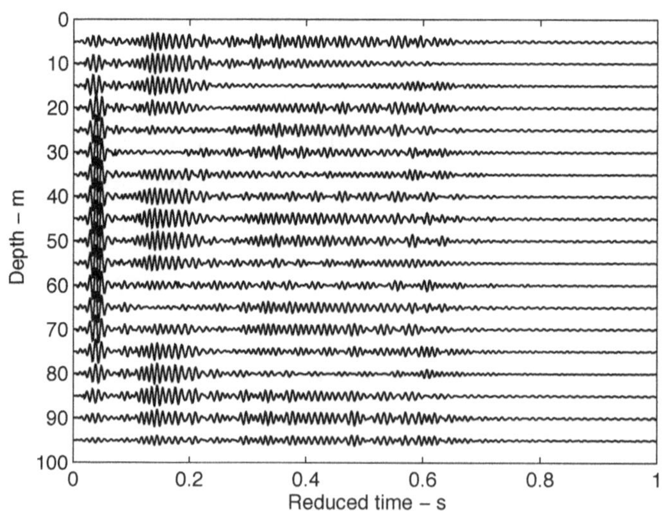


Figure 17.11. Time series of the sound pressure, plotted as a function of depth; the receiver is a range of 10 km

# Chapter 18

# Geoacoustic measurements

## 18.1 Introduction

The structure and composition of the ocean bottom are very important in sonar performance, especially for low-frequency signals and long-range propagation in shallow water. For such applications, therefore, it is essential to have precise information about sound speeds and densities for the upper layers in the seabed in order to assess the acoustic propagation conditions. The bottom properties relevant for sonar operations are referred to as geoacoustic properties.

In some cases, the geoacoustic properties can be acquired by in situ measurement, or by taking samples of the bottom material and subsequently making measurements in the laboratory. In practice, this direct approach is often not sufficient and has to be supplemented by information acquired by remote measurement techniques to obtain the area coverage and the depth resolution required. The indirect approach, which provides the means to estimate parameters of geoacoustic models from measurements of the acoustic field, is generally known as geoacoustic inversion. In this chapter we will not attempt to cover the inversion techniques as such, but we refer to the *IEEE Journal of Oceanic Engineering* for July 2003, which is a special issue on geoacoustic inversion that contains a large number of papers on recent research in that field.

In this chapter we shall describe some of the seismic methods used most frequently to acquire such information. However, we will limit discussion to applications and techniques for finding the structure and composition of the upper portion of the sea bottom just beneath the sea floor, which means that we will not cover the deep seismic techniques used to look for oil and gas.

There are also other relevant applications for geoacoustic measurement, for example in underwater site surveys, which are often conducted in connection with construction on the seafloor. In such work, a good knowledge of the site seabed, and its sediment strength and stability, is crucial. As well, interest has increased recently in marine habitats of various types, often in relation to environmental concerns.

Our starting point for discussion is the situation depicted in figure 18.1. Here we have a point source positioned in water at a height $z_s$ over the interface with the sea bottom, which is modeled as a layered medium with either fluid or solid layers with densities $\rho_n$, sound speeds $c_{pn}$, and shear speeds $c_{sn}$, where $n = 1,2,..., N$ for the layers in the bottom. In the figure, the bottom has $N = 2$ layers.

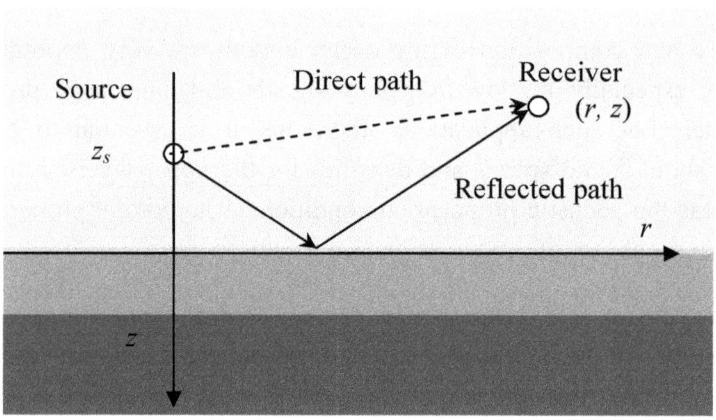


Figure 18.1. A point source, located at height $z_s$ above a layered sea floor, and a receiver at height $z$ and horizontal distance $r$ from the source. The receive signal consists of a direct and a bottom reflected arrival.

In chapter 17 we found that the received signal at depth $z$ and range $r$, $\phi_0(r,z)$, can be expressed by an integral over the horizontal wave numbers $k$

$$\phi_0(r,z,\omega) = \frac{S(\omega)}{4\pi i}\int_0^{\infty}\frac{\exp\left(i\gamma_0\left|z-z_s\right|\right)}{\gamma_0}kJ_0(kr)\,dk$$
$$+\frac{S(\omega)}{8\pi i}\int_0^{\infty}R_b(k)\frac{\exp\left(i\gamma_0\left|z+z_s\right|\right)}{\gamma_0}kH_0^{(1)}(kr)\,dk \tag{18.1}$$

In equation (18.1) $S(\omega)$ represents the frequency function of the source signal, the vertical wave number component is $\gamma_0$, and the horizontal wave number component is $k$. The first term in equation (18.1) describes the direct signal from the source, and the second term describes the signal reflected from the medium below. The reflectivity of the seafloor, defined by the plane wave reflection coefficient $R_b(k)$, is a function of the incident angle expressed by the horizontal wave number component, $k$. The reflection coefficient $R_b(k)$ contains all reflections plus the multiple reflections from underlying layers for both compressional waves and shear waves.

We can further simplify the expression for the reflected signal by replacing the Hankel functions with the asymptotic expression that applies to distances that are much greater than the signal wavelength. By using the result that was developed in section 17.3, we find that the reflected signal $\phi_R(r,z,\omega)$ can be defined with good approximation as

$$\phi_R(r,z,\omega) = \frac{S(\omega)\exp(-i\pi/4)}{4\pi\sqrt{2\pi r}}\int_0^{\infty}R_b(k)\frac{\sqrt{k}}{i\gamma_0}\exp\left(i\gamma_0\left|z+z_s\right|+ikr\right)dk \tag{18.2}$$

## 18.2 Refraction measurements

Refraction measurement, a technique for determining sound speeds and layer thicknesses in the seafloor's upper layers, is much used in site surveys for geotechnical investigation and other purposes. The terms *refraction* and *refracted waves* are commonly used in the seismic literature to describe head wave phenomena; the terms *head wave* and *refracted wave* are used interchangeably.

In order to illustrate what is meant by refraction, we will calculate the reflected signal from a bottom that is a fluid half-space like that shown in figure 18.1, but where $d_1 = \infty$. In such a case, the plane wave reflection coefficient $R_b(k)$ is given as

$$R_b(k) = \frac{\rho_1\gamma_0 - \rho_0\gamma_{p,1}}{\rho_1\gamma_0 + \rho_0\gamma_{p,1}} \tag{18.3}$$

Here, the $\rho_0$ and $\rho_1$ are the densities of the water and the bottom and $\gamma_0$ and $\gamma_{p,1}$ are the vertical wave numbers of the water and sea bottom, respectively.

In figure 18.2, we have an example of the pulse responses from a homogenous seabed as calculated by the OASES (SAFARI) model (Schmidt 1988). The calculation uses equation (18.1) and the reflection coefficient definition from equation (18.3).

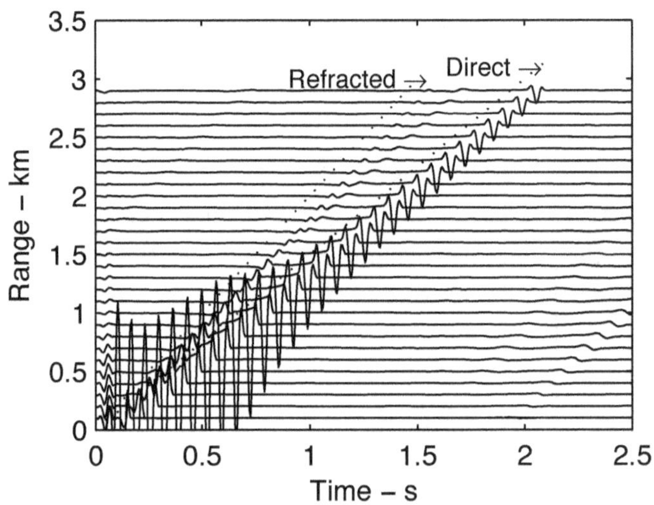


Figure 18.2. The signal from a point source located above a fluid-fluid interface, for receivers on the bottom spaced at 100 m intervals, with a maximum source-receiver range of 2900 m.

The figure shows the signals arriving at receivers that are located on the seabed at z = 0 and spaced at 100 m intervals, with the maximum source-receiver distance being 2900 m. The source is located at $z_s$ = 1 m above the seabed. The water and seabed sound speeds are $c_0$ = 1500 m/s, $c_{p,1}$ = 2000 m/s, respectively, and $\rho_1 / \rho_0 = 2$. The transmitted signal from the source is a short transient pulse with a center frequency of 20 Hz. At each given distance there appear to be two arrivals.

The propagation speeds can be determined from the arrival times and the distance. In this case, we find that the first arrival has a speed of 2000 m/s and the second arrival has a speed of 1500 m/s. The first arrival is the refracted wave and it propagates with the speed of sound in the seabed. The other wave is actually two waves that cannot be separated in the figure: the direct wave from the source and the wave reflected from the specular point at the interface. Both propagate with the sound speed of the water. Note that the refracted arrival is not included in the normal mode solution that we discussed in section 7.4 as we have noted earlier. In this example, observe that the sound speed in the bottom is higher than the sound speed in the water. This condition is required to have a refracted wave.

For signals arriving at the seafloor with incident angles closer to the horizontal than the critical angle, the vertical wave number $\gamma_{p,1}$ in the lower medium is imaginary, so that we more conveniently can express the reflection coefficient of equation (18.3) by amplitude and phase as

$$R_b(k) = |R_b(k)| \exp(-i\beta) \tag{18.4}$$

with the phase angle being defined as

$$\beta - 2\arctan\left(\left|\frac{\gamma_{p,1}\rho_0}{\gamma_0\rho_1}\right|\right) \tag{18.5}$$

Therefore, the reflected and refracted signals $\Phi_R(r,z)$ can be expressed as

$$\phi_R(r,z,\omega) = \frac{S(\omega)\exp(-i\pi/4)}{4\pi\sqrt{2\pi r}} \int_0^\infty |R_b(k)| \frac{\sqrt{k}}{i\gamma_0} \exp\left[i(\gamma_0|z+z_s| + ikr - i\beta\right] dk \tag{18.6}$$

It is instructive to carry out an approximate calculation using the stationary-phase method. In this approach, the main contribution to the integral comes from the $k$-values found where the exponent in the integrand in equation (18.6) has its stationary values, that is, where the derivative with respect to $k$ is equal to zero. With equations (18.4) and (18.5) inserted into equation (18.2), we find that the integral has a stationary value when

$$\frac{\partial}{\partial k}\left[\gamma_0 \left|z+z_s\right| + kr - \beta(k)\right] = 0 \tag{18.7}$$

Since

$$\frac{\partial \gamma_0}{\partial k} = -\frac{k}{\gamma_0} \tag{18.8}$$

the dominant parts of the signal will therefore appear at a distance $r$, which is defined as

$$\begin{aligned} r &= \frac{\partial \beta(k)}{\partial k} + \left|z+z_s\right| \frac{k}{\gamma_0} \\ &= \frac{\partial \beta(k)}{\partial k} + \left|z+z_s\right| \cot\theta \end{aligned} \tag{18.9}$$

The ray's radial displacement $\Delta r$ is expressed as the derivative of the reflection coefficient's phase angle, so that

$$\Delta r = \frac{\partial \beta(k)}{\partial k} \tag{18.10}$$

For angles greater than the critical angle, the phase angle of the reflection coefficient is equal to zero. Therefore, we actually have two situations

$$\begin{aligned} r &= \left|z+z_s\right| \cot\theta & \theta > \theta_{crit} \\ r &= \left|z+z_s\right| \cot\theta + \Delta r & \theta < \theta_{crit} \end{aligned} \tag{18.11}$$

Thus, for angles less than the critical angles, the signal travels partly along a horizontal path, $\Delta r$, that follows the interface with the lower medium. The length of this part of the path is the ray path from source to receiver that follows the interface.

Consider now the situation pictured in figure 18.3, in which there is a source at height $d_s$ above the seabed and a receiver at height $d_r$, and the source and receiver are separated by the horizontal distance $r$. Including the direct wave, we have three possible waves traveling from the source to the receiver in figure 18.3.

The direct wave arrives at a time $\tau_d$ that is defined as

$$\tau_d = \frac{\sqrt{r^2 + (d_r - d_s)^2}}{c_0} \tag{18.12}$$

and the reflected wave arrives at a time $\tau_{refl}$, defined as

$$\tau_{refl} = \frac{1}{c_0}\sqrt{r^2 + (d_s + d_r)^2} \tag{18.13}$$

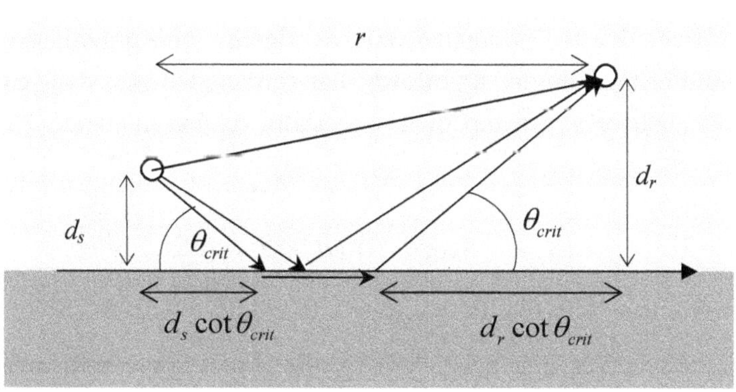


Figure 18.3. Direct, reflected, and refracted wave paths between the transmitter and receiver.

To determine the travel time of the refracted signal, we must take into account that the ray path of the refracted wave has the three segments we see in figure 18.3. In the water, the ray path has two segments, with lengths of $d_s / \sin(\theta_{crit})$ and $d_r / \sin(\theta_{crit})$, respectively, and the signal travels at the wave speed $c_0$. For the signal traveling in the bottom, we have the horizontal segment of the ray path, which has a length $r - d_s \cot(\theta_{crit}) - d_r \cot(\theta_{crit})$; this wave travels at the sound speed in the bottom, $c_{p1}$. The total travel time for the refracted wave $\tau_1$ is equal to the sum of these three contributions and is defined as

$$\tau_1 = \left\{ \frac{d_r + d_s}{c_0 \sin\theta_{crit}} + \frac{1}{c_{p,1}}\left[ r - (d_r + d_s)\cot\theta_{crit} \right] \right\} \tag{18.14}$$

Since the critical angle is

$$\cos\theta_{crit} = \frac{c_0}{c_{p1}} \tag{18.15}$$

equation (18.14) can be transformed to

$$\tau_1 = \frac{r}{c_{p,1}} + \frac{(d_r + d_s)}{c_0}\sin\theta_{crit} \tag{18.16}$$

Because the refracted wave is created only by signals with grazing angles that are smaller than the critical angle, it follows that refracted waves will exist only for source-receiver distances greater than a certain critical distance $r_{crit}$, which is expressed as

$$r_{crit} = (d_r + d_s)\cot\theta_{crit} = (d_r + d_s)\frac{c_0}{\sqrt{c_{p,1}^2 - c_0^2}} \tag{18.17}$$

For short source-receiver distances (SRDs), the direct wave will arrive first; the reflected wave arrives later. With increasing SRD, the arrival time difference will decrease because of the reduction of the difference in path length between the direct and the reflected waves. The refracted wave will be the first wave to arrive when the SRD is so large that the grazing angle becomes smaller than the critical angle. We can see this phenomenon in the plots in figure 18.2. The crossover point $r_{cross}$, where the refracted wave arrives simultaneously with the direct wave, is found to be

$$\begin{aligned} r_{cross} &= (d_r + d_s)\frac{c_{p,1}}{c_{p,1} - c_0}\sin\theta_{crit} \\ &= (d_r + d_s)\sqrt{\frac{c_{p,1} + c_0}{c_{p,1} - c_0}} \end{aligned} \tag{18.18}$$

In figure 18.4, the arrival times of the three wave types (direct, reflected, and refracted) are plotted as a function of the source-receiver distance. This example has the parameters $c_0$ = 1500 m/s and $c_{p,1}$ = 3000 m/s, giving us a critical angle of 60°. The source is suspended 100 m above the seafloor, while the receiver lies 200 m above the seafloor. This geometry gives us $r_{crit}$ = 173 m and $r_{cross}$ = 520 m.

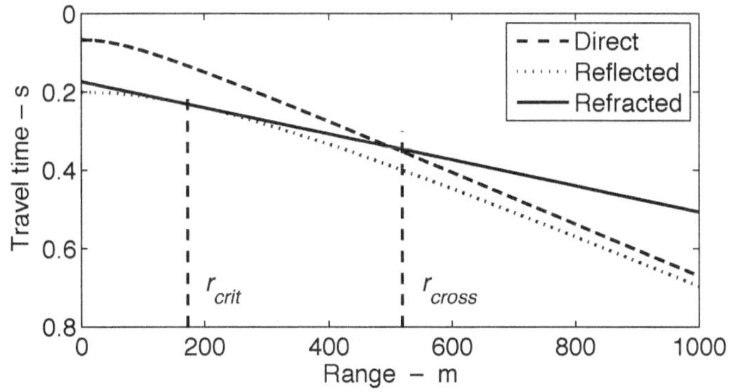


Figure 18.4. Arrival times of the direct, reflected, and refracted wave types.

At this point we can generalize our example by specifying that the refracting medium has an arbitrary number of layers that have different thicknesses and wave speeds. First we will consider the case of a fluid medium overlying a two-layer medium. The medium's upper layer has the thickness $d_1$ with a sound speed $c_1$ and the lower medium has the sound speed $c_2$, as shown in figure 18.5.

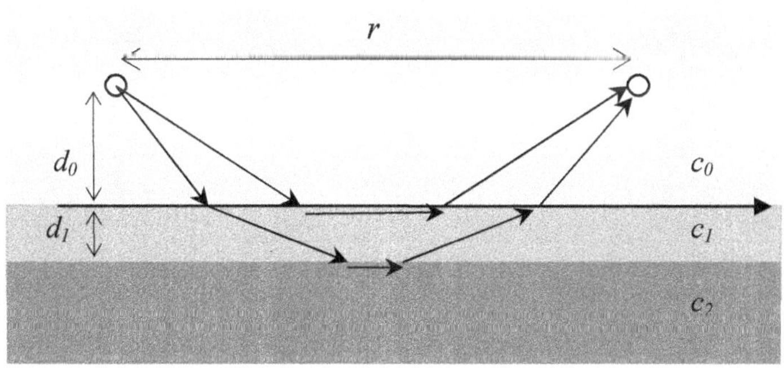


Figure 18.5. Refracted ray paths in a two-layer model. The source and receiver are separated by the distance $r$ and both are a height $d_0$ above the seabed. The thickness of the upper layer is $d_1$; the thickness of the second layer is infinite.

For the sake of simplicity, we assume that the source and receiver are at the same depth, a height $d_0$ above the seabed. In this case, we have two refracted waves, which are shown in the figure.

There are two critical angles, given as

$$\cos\theta_{crit1} = \frac{c_0}{c_1}$$
$$\cos\theta_{crit2} = \frac{c_0}{c_2} \tag{18.19}$$

The first refracted wave arrives after a time $\tau_1$, which was defined in equation (18.16); $\tau_1$ now can be rewritten as

$$\tau_1 = \frac{r}{c_1} + \frac{(d_s + d_r)\sqrt{c_1^2 - c_0^2}}{c_1 c_0} \tag{18.20}$$

By the same argument the second wave's arrival time $\tau_2$ is expressed as

$$\tau_2 = \frac{r}{c_2} + \frac{(d_s + d_r)\sqrt{c_2^2 - c_0^2}}{c_0 c_2} + \frac{2d_1\sqrt{c_2^2 - c_1^2}}{c_1 c_2} \tag{18.21}$$

Figure 18.6 shows the propagation time of the first arrival, plotted as a function of source-receiver separation.

The first arrival is also called the ***first break.*** At short distances, the first arrival will be the direct wave propagating with the sound speed in water $c_0$. At longer ranges, the direct wave will be overtaken by the first refracted wave, which has a speed equal to the sound speed $c_1$ in the upper layer. At even longer ranges, the deeper refracted wave, traveling with the sound speed $c_2$ in the second layer, will arrive first. By extending the segments for the two refracted arrivals back to zero range, we can define the times $\tau_{01}$ and $\tau_{02}$. As indicated in the figure, the slopes of the line segments are given directly as the inverses of the wave speeds.

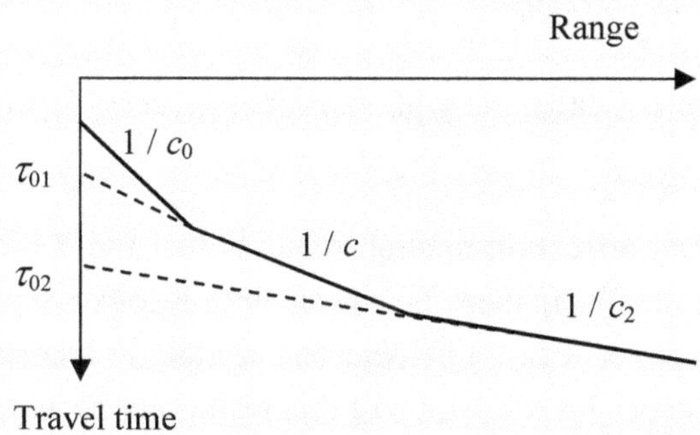


Figure 18.6. Arrival times for the refracted waves in a two-layer model. The arrival times of the first break is a broken line with inclinations given by the sound speed $c_0$ of the water and the sound speeds $c_1$ and $c_2$, respectively, in the first and second layers of the bottom. By extending the segments for the two refracted arrivals back to zero range, we can define the times $\tau_{01}$ and $\tau_{02}$.

After having determined the waves' speeds from the corresponding segment slopes, we can find the travel times $\tau_{01}$ and $\tau_{02}$ by extrapolating the line segments of the two refracted arrivals back to zero range, as indicated by the broken lines in figure 18.6. These starting points will enable us to determine the thicknesses of the refracting layers by using equations (18.20) and (18.21).

After setting $r = 0$ in these two equations, we can find the height of the source and receiver above sea bottom $d_0$ and the layer thicknesses $d_1$

$$d_0 = \frac{\tau_{01}}{2} \frac{c_0 c_1}{\sqrt{c_1^2 - c_0^2}} \tag{18.22}$$

$$d_1 = \frac{1}{2}\left( \tau_{02} - \tau_{01} \frac{c_1}{c_2} \sqrt{\frac{c_2^2 - c_0^2}{c_1^2 - c_0^2}} \right) \frac{c_2 c_1}{\sqrt{c_2^2 - c_1^2}} \tag{18.23}$$

This method can be extended to an arbitrary number of layers, given that the wave speeds in the layers are increasing monotonically. Thus, the arrival time of the wave refracted from layer $n$, $\tau_n$, can be determined to be

$$\tau_n = \frac{r}{c_n} + \sum_{i=0}^{n-1} \frac{2d_i}{c_i} \sin\theta_i = \frac{r}{c_n} + \sum_{i=0}^{n-1} \frac{2d_i}{c_i} \sqrt{1 - \frac{c_i^2}{c_n^2}} \tag{18.24}$$

where the angle $\theta_i$ is the critical angle with respect to layer $n$; $\theta_i$ is obtained from the expression $\cos\theta_i = c_i / c_n$.

As mentioned earlier, the refraction method requires that the wave speeds increase monotonically with the depth. If there is a layer with a lower speed than the layer above it, this low-speed layer will not be detected, leading to incorrect calculation of all thicknesses of the deeper layers. Nor will this technique be accurate where layer differences in wave speed and layer thickness are insufficient to provide detectable refracted arrivals.

Figure 18.7 shows a typical marine measurement setup for collection of refraction data from the seafloor. A vessel tows the source, and one or more hydrophones are suspended from a buoy having a radio link to the vessel. Figure 18.8 contains an example of an original recording from a marine refraction measurement. We can easily observe the direct wave, the reflected wave, and at least two refracted waves, in addition to a number of multiple reflections.

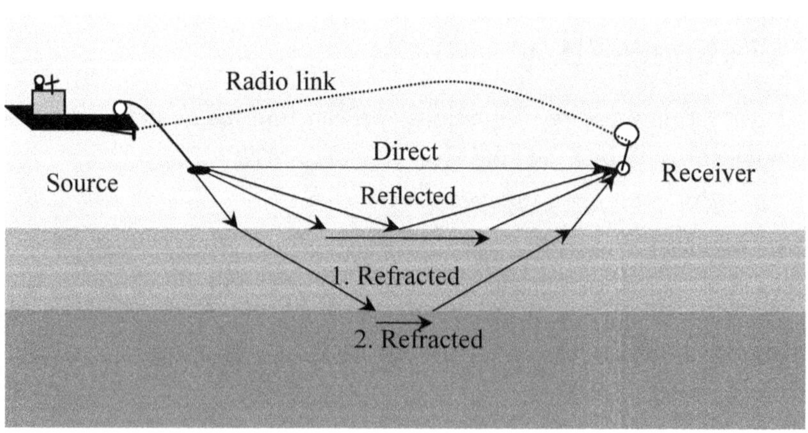


Figure 18.7. Marine refraction measurement.

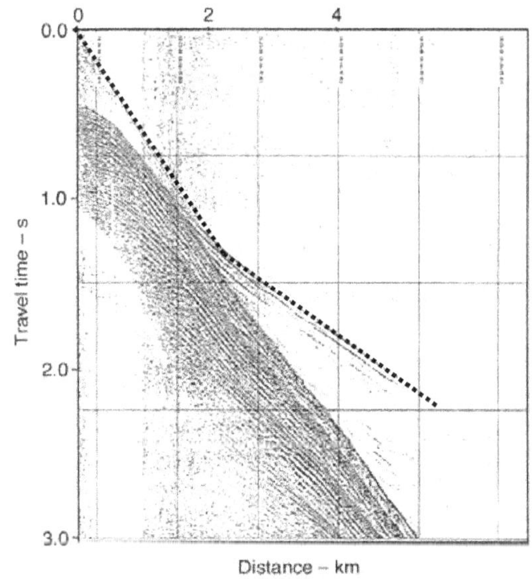


Figure 18.8. Example of refraction measurements taken from an original recording. The dashed lines indicate the wave speeds $c_1 = 1600$ m/s and $c_2 = 3300$ m/s in the first and second layers of the bottom, respectively.

## 18.3 Wide-angle reflection measurements

As mentioned at the beginning of this chapter, in this text we will not attempt to cover the whole field of marine seismic reflection techniques. For an overview of the field and for listings of more literature on this subject, the reader may consult *Exploration Seismology* (Sheriff and Geldhart 1995). Here, however, we shall present a special reflection measurement technique that is especially designed to measure wave speeds and layer thicknesses of thin layers in the seabed's upper part. The equipment and method of data gathering are almost the same as those for obtaining refraction data, which were presented in figure 18.5; the difference between the two applications is that in reflection measurement the source-receiver separation is relatively shorter, so that waves reflected by the layering are excited and received, but refracted waves are not received.

Figure 18.9 shows the trajectory of a particular ray with an initial angle at the source equal to $\theta_0$. This ray generates a reflected wave from the water-bottom interface and from interfaces deeper in the bottom.

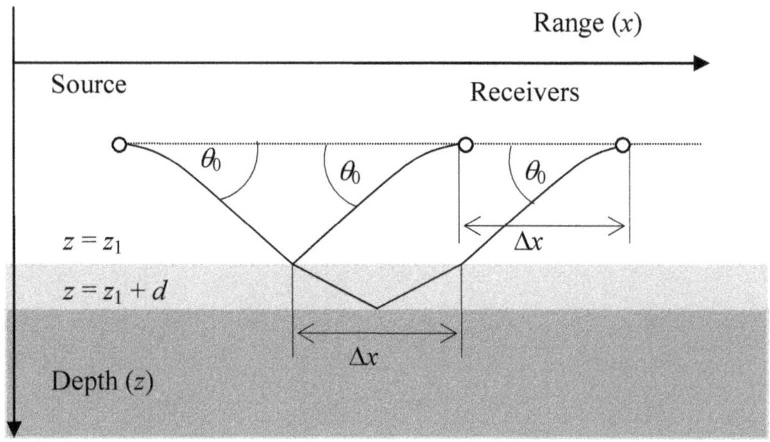


Figure 18.9. A signal with an initial angle at the source of $\theta_0$ is reflected both from the water-bottom interface and from a layer at a depth $d_1$ in the seabed. The reflected rays are received at two receivers spaced a distance $\Delta x$ apart

In this figure we are considering only one layer of a thickness $d_1$ over a semi-infinite half-space. As indicated, the two reflected waves are detected by receivers spaced a distance $\Delta x$ apart. The important point to notice is that the two reflected signals originate from the same ray with an initial angle of $\theta_0$ at the source, where the sound speed is $c_0$. The two received rays at the receivers therefore have the same ray parameter $\xi$, which is given by Snell's law and is defined as (see chapter 6)

$$\frac{\cos\theta_0}{c_0} = \frac{\cos\theta(z)}{c(z)} = \xi \tag{18.25}$$

For a given sound speed profile $c(z)$, the transmission time $\tau_1$ and range $x_1$ for the reflected wave at the depth $z_1$ are found by integration to be

$$\tau_1 = 2\int_0^{z_1} \frac{dz}{c(z)\sin\theta(z)} = 2\int_0^{z_1} \frac{dz}{c(z)\sqrt{1-c(z)^2\,\xi^2}} \tag{18.26}$$

$$x_1 = 2\int_0^{z_1} \cot\theta(z)\,dz = 2\int_0^{z_1} \frac{\xi c(z)\,dz}{\sqrt{1-c(z)^2\,\xi^2}} \tag{18.27}$$

For the wave reflected from a reflector at depth $z_2$, we obtain similar expressions for $\tau_2$ and $x_2$. However, when the thickness of the layer $d_1 = z_2 - z_1$ is small and the speed $c_1$ does not vary much within the layer, we instead have expressions for $\tau_2$ and $x_2$ that are approximately

$$\tau_2 = \tau_1 + \frac{2d_1}{c_1\sqrt{1-c_1^2\xi^2}} \tag{18.28}$$

$$x_2 = x_1 + \frac{2d_1\xi c_1}{\sqrt{1-c_1^2\xi^2}} \tag{18.29}$$

Between $\tau_2$ and $\tau_1$ and between $x_2$ and $x_1$, the respective differences $\Delta\tau$ and $\Delta x$ are expressed as

$$\begin{aligned} \Delta\tau &= \tau_2 - \tau_1 \\ \Delta x &= x_2 - x_1 \end{aligned} \tag{18.30}$$

By inserting equations (18.28) and (18.29) into equations (18.30) we obtain

$$\frac{\Delta x}{\Delta\tau} = \xi c_1^2 \tag{18.31}$$

With the help of figure 18.10, we can use this expression to determine both the sound speed in the layer, and the layer thickness. This figure shows a sketch of travel time as a function of source-receiver separation for the signals reflected from the water-bottom interface and from the deeper interface. We find two points on each of the curves with the same gradient, and read off the difference in travel time $\Delta\tau$ and range $\Delta x$.

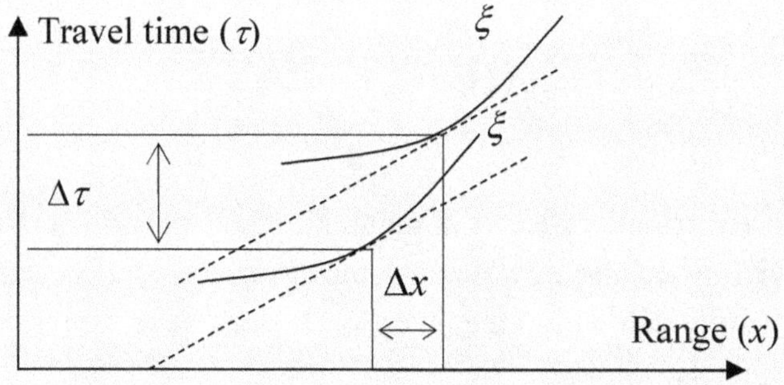


Figure 18.10. Sketch of travel time versus range for the signals reflected from the water-bottom interface and from the deeper interface. The two marked points have the same gradient, given by the ray parameter $\xi$. The sound speed for the sediment layer can be calculated from measurements of $\Delta\tau$ and $\Delta x$.

The sound speed $c_1$ and layer thickness $d_1$ can then be calculated using equation (18.31) to obtain

$$c_1 = \sqrt{\frac{\Delta x}{\xi \Delta \tau}} \tag{18.32}$$

$$d_1 = \frac{1}{2} c_1 \Delta \tau \sqrt{1 - \xi^2 c_1^2} \tag{18.33}$$

This process can be repeated to determine the sound speeds and thicknesses for the deeper layers. Note that the technique does not require knowledge of the sound speed variation in the whole water column. This method is particularly suited for determining sound speed and thickness for thin seabed layers.

## 18.4 Measurements with interface waves

Knowledge of shear rigidity in the seabed is important for many applications in geophysics and marine technology. It is, however, quite difficult to measure the shear speed in the bottom directly in situ, so in most cases the shear speed is determined from measurements on core samples in the laboratory. This method may be unreliable because the mechanical properties of a sample may be changed in the coring and handling processes. Another possible way to measure the seabed's shear wave structure is to use seismo-acoustic interface waves, also known as Rayleigh, Stoneley, or Scholte waves, that may exist at an interface between two media, at least one of which must be a solid. References to this technique are found in the papers by Rauch (1980), Jensen and Schmidt (1986), and Caiti, Stoll, and Akal (1992, 1994). In this section we will discuss this technique in more detail.

To begin, we consider the situation shown in figure 18.9, where the water has the depth *D,* constant sound speed $c_0$, and density $\rho_0$. The sea bottom is modeled as a solid medium with compressional and shear speeds $c_{p,1}$ and $c_{s,1}$, and its density is $\rho_1$.

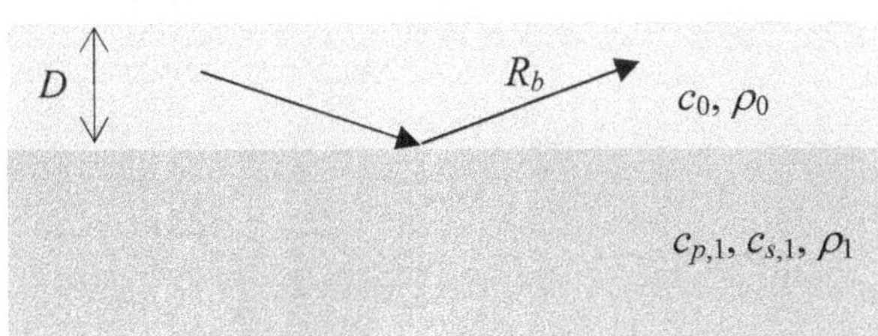


Figure 18.11. An underwater acoustic channel with water depth *D* over a homogenous solid elastic bottom; $R_b$ is the reflection coefficient of the bottom.

The situation depicted in figure 18.11 is the acoustic waveguide that we considered earlier, but unlike the case treated in section 18.2, we are now considering a solid seafloor that can support shear wave propagation. We shall see this implies that a slow-propagating wave can exist at the interface between the water and the solid bottom, and that the speed of this interface wave is directly related to the shear speed of the bottom material. Consequently, a measurement of the speed of the interface wave offers a method for determining the shear speed within the upper strata of the sea bottom.

First, consider the situation from figure 18.11 when the water depth is infinite. In such a case, there are no reflections from the sea surface; the acoustic field is entirely determined by the reflection from the bottom, that is, by the reflection coefficient $R_b$, which is now the reflection coefficient at an interface between a fluid and a solid medium, a situation we discussed in section 15.4.

The reflection coefficient, given by equation 15.42, is

$$R_b = \frac{Z_{p,1}\cos^2\left(2\theta_{s,1}\right) + Z_{s,1}\sin^2\left(2\theta_{s,1}\right) - Z_0}{Z_{p,1}\cos^2\left(2\theta_{s,1}\right) + Z_{s,1}\sin^2\left(2\theta_{s,1}\right) + Z_0} \qquad (18.34)$$

The impedances in equation (18.34) are defined as

$$Z_0 = \omega\frac{\rho_0}{\gamma_0} \qquad Z_{p,1} = \omega\frac{\rho_1}{\gamma_{p,1}} \qquad Z_{s,1} = \omega\frac{\rho_1}{\gamma_{s,1}} \qquad (18.35)$$

and the angles are defined as

$$\begin{aligned} \cos\left(2\theta_{s,1}\right) &= 2\cos^2\left(\theta_{s,1}\right) - 1 = 2\frac{k^2}{\omega^2}c_{s,1}^2 - 1 \\ \sin\left(2\theta_{s,1}\right) &= 2\sin\left(\theta_{s,1}\right)\cos\theta_{s,1} = 2k\gamma_{s,1}\frac{c_{s,1}^2}{\omega^2} \end{aligned} \qquad (18.36)$$

In figure 18.12, we have plotted the reflection coefficient $R_b$, which has been calculated as a function of the horizontal wave number $k$ for the parameters: frequency $f = 5$ Hz, $c_0 = 1500$ m/s, $c_{p,1} = 2500$ m/s, $c_{s,1} = 400$ m/s, $\rho_0 = 1000$ kg/m$^3$, and $\rho_1 = 2000$ kg/m$^3$. The attenuations of the compressional and the shear waves are both set to 0.2 dB per wavelength. Note the sharp maximum value, or pole, located at $k = 0.088$; this is the pole of the interface wave.

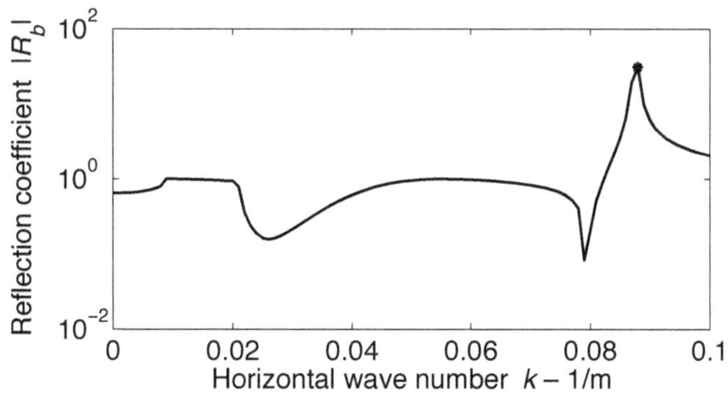


Figure 18.12. The absolute value of the reflection coefficient of a homogeneous elastic medium, calculated as a function of the horizontal wave number $k$.

Since the phase speed $v$ is given by

$$v = \frac{\omega}{k} \tag{18.37}$$

this pole location corresponds to a phase speed of 357 m/s, a value that is about 90% of the shear wave speed of 400 m/s.

The relationship between the grazing angle $\theta_0$ and the horizontal wave number $k$ is

$$k = \frac{\omega}{c_0}\cos\theta_0 \tag{18.38}$$

With a frequency of 5 Hz and water sound speed of 1500 m/s, the $k$ values corresponding to real angles range from $k = 0$ to $k = 0.021$. Therefore, the pole of the interface wave is outside the region of real angles.

We can find the location of this pole more accurately by setting the denominator of the reflection coefficient in equation (18.34) equal to zero

$$Z_{p,1}\cos^2\left(2\theta_{s,1}\right) + Z_{s,1}\sin^2\left(2\theta_{s,1}\right) + Z_0 = 0 \tag{18.39}$$

Using equations (18.35) and (18.36), this expression produces

$$\left(2-\frac{\omega^2}{k^2c_{s,1}^2}\right)^2+4\frac{\gamma_{p,1}\gamma_{s,1}}{k^2}+\frac{\gamma_{p,1}}{\gamma_0}\frac{\rho_0}{\rho_1}\frac{\omega^4}{c_{s,1}^4k^4}=0 \tag{18.40}$$

This equation has a real-value solution for $k$ if the phase speed $v = \omega / k$ is smaller than all three body wave speeds, that is, when $v < (c_0, c_{p,1}, c_{s,1})$. When this is the case, the vertical wave numbers in equation (18.40) are all imaginary and can be expressed as

$$\gamma_0 = ik\sqrt{1-\left(\frac{v}{c_0}\right)^2} \qquad \gamma_{p,1} = ik\sqrt{1-\left(\frac{v}{c_{p,1}}\right)^2} \qquad \gamma_{s,1} = ik\sqrt{1-\left(\frac{v}{c_{s,1}}\right)^2} \tag{18.41}$$

When inserted into equation (18.40), these expressions produce the dispersion equation for the phase speed of the interface wave

$$\left(2-\frac{v^2}{c_{s,1}^2}\right)^2-4\sqrt{1-\frac{v^2}{c_{p,1}^2}}\sqrt{1-\frac{v^2}{c_{s,1}^2}}+\frac{\rho_0}{\rho_1}\frac{v^4}{c_{s,1}^4}\frac{\sqrt{1-\frac{v^2}{c_{p,1}^2}}}{\sqrt{1-\frac{v^2}{c_0^2}}}=0 \tag{18.42}$$

We now turn to the more general situation in which we have finite water depth $D$ as shown in figure 18.11. In this case, we are dealing with the acoustic waveguide that we earlier discussed in section 17.4, where we found that the sound field in an acoustic waveguide can be expressed by equations (17.55). We also found that the poles of these equations are given by the zeroes of the denominator, and thus are given by

$$1-R_bR_s\exp\left(2i\gamma_0D\right)=0 \tag{18.43}$$

This is the same as defined earlier in section 17.4 where the poles of the integrand in equation (17.55), which give the acoustic field as a function of frequency and position, are defined by the equations (17.54) and (17.56).

We assume that the reflection coefficient of the sea surface is $R_s = -1$ and therefore

$$1+R_b\exp\left(2i\gamma_0D\right)=0 \tag{18.44}$$

We find the solutions to equation (18.44) using the same approach as before and we first find that

$$i\frac{\rho_0\gamma_{p,1}}{\rho_1\gamma_0}\tan\left(\gamma_0 D\right)=\left(\frac{kc_{s,1}}{\omega}\right)^4\left\{4\frac{\gamma_{p,1}\gamma_{s,1}}{k^2}+\left[2-\left(\frac{\omega}{kc_{s,1}}\right)^2\right]^2\right\} \tag{18.45}$$

By applying equation (18.41) this expression can be transformed into

$$\left(2-\frac{v^2}{c_{s,1}^2}\right)^2-4\sqrt{1-\frac{v^2}{c_{p,1}^2}}\sqrt{1-\frac{v^2}{c_{s,1}^2}}+\frac{\rho_0}{\rho_1}\frac{v^4}{c_{s,1}^4}\frac{\sqrt{1-\frac{v^2}{c_{p,1}^2}}}{\sqrt{1-\frac{v^2}{c_0^2}}}\tanh\left[\left(\frac{\omega D}{v}\right)\sqrt{1-\frac{v^2}{c_0^2}}\right]=0 \tag{18.46}$$

Equation (18.46) is the dispersion equation for the case with finite water depth and we see that, when $D \rightarrow \infty$, this expression becomes identical to the expression of equation (18.42) for infinite water depth. The dispersion equation gives the phase speed as a function of frequency for given media parameters and layer thicknesses.

The dispersion equation can be solved numerically; figure 18.13 shows an example of such a numerical solution. While the phase speed $v$ is found as the numerical solution of equation (18.46), the group speed $w$ can be found by differentiation, that is, by taking the derivative, and is expressed as

$$w=\frac{d\omega}{dk}=\frac{v}{1-\dfrac{\omega}{v}\dfrac{dv}{d\omega}} \tag{18.47}$$

Figure 18.13 shows an example computed using the geoacoustic parameter values $\rho_0 = 1000$ kg/m$^3$, $\rho_1 = 2000$ kg/m$^3$, $c_0 = 1500$ m/s, $c_{p,1} = 2500$ m/s, and $c_{s,1} = 400$ m/s. Since the frequency appears only in a product with the water depth $D$, the speed must be a function of the product of $f$ and $D$.

Notice that the interface wave's phase speed is slightly lower than the shear wave speed, and that the phase speed decreases slightly with increasing frequency. This means that the interface wave is dispersive on the general case. In the limiting case, where the water depth is infinite, the speed of the interface wave is approximately

90% of the shear speed in the bottom, while with a zero water depth, the speed is somewhat higher, about 95% of the bottom shear speed.

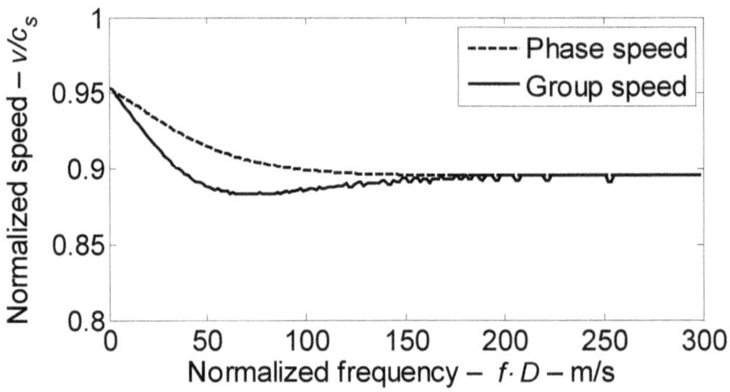


Figure 18.13. The phase and group speeds of an interface wave, relative to the shear speed, expressed as a function of the frequency-thickness product $f \cdot D$ for the numerical values given in the text.

The interface wave at a boundary between a vacuum (or air) and a solid is called the Rayleigh wave, and the equation for the wave speed is obtained from equation (18.46) by setting $D = 0$, resulting in the dispersion equation of the Rayleigh wave

$$\left(2 - \frac{v^2}{c_{s,1}^2}\right)^2 - 4\sqrt{1 - \frac{v^2}{c_{p,1}^2}}\sqrt{1 - \frac{v^2}{c_{s,1}^2}} = 0 \tag{18.48}$$

By solving equation (18.48), we find that the phase speed of the Rayleigh is approximately 95% of the shear wave speed, as can be seen in figure 18.13. Thus, in a solid medium the measurement of the Rayleigh wave speed is a quite accurate measure of shear wave speed.

Since we know from equations (18.41) that all the vertical wave numbers are imaginary, we also know that the signal amplitudes will decrease exponentially with increasing distance from the interface. We should also note that an interface wave cannot be excited by an incident plane wave. This can be most easily understood by considering the grazing angle of the wave in the uppermost medium. This angle is expressed as

$$\cos\theta_0 = \frac{c_0}{\omega}k = \frac{c_0}{v} \tag{18.49}$$

Since the condition for a real solution of equation (18.48) is that $v < c_0$, equation (18.49) tells us that the angle $\theta_0$ must be imaginary and, consequently, cannot be the grazing angle of a propagating plane wave. However, the interface wave can be excited by a point source close to the interface, that is, as a nearfield effect.

The interface waves are confined to a narrow stratum close to the interface, which means that they have cylindrical propagation loss (that is, $1/r$) rather than spherical spreading loss (that is, $1/r^2$), as would be true of waves from a point source located in a medium of infinite extent. Cylindrical spreading loss indicates that, once an interface wave is excited, it is likely to dominate other waves that experience spherical spreading at long distances. This effect is familiar from earthquakes, where exactly this kind of interface wave, the Rayleigh wave, often causes the greatest damage.

### *18.4.1 The Rayleigh wave and particle displacement*

To develop a greater understanding of the Rayleigh wave, we will determine the particle movements or displacements it causes. The analysis developed here applies equally well to the Scholte wave that is observed at interfaces between liquid and solid materials. In the solid medium, we have two waves: one a compressional wave, and the other, a shear wave. We shall see that these waves are evanescent with amplitude decaying with depth in the solid layer.

The displacement potentials for the compressional wave, $\phi(x,z,\omega)$, and the shear wave, $\psi(x,z,\omega)$, are expressed as

$$\begin{aligned}\phi(x,z,\omega) &= D_{p,1}\exp\left(i\gamma_{p,1}z + ikx\right)\exp(-i\omega t)\\ \psi(x,z,\omega) &= D_{s,1}\exp\left(i\gamma_{s,1}z + ikx\right)\exp(-i\omega t)\end{aligned} \tag{18.50}$$

In equations (18.50) $D_{p,1}$ is the amplitude of the compressional wave and $D_{s,1}$ is the amplitude of the shear wave in the solid. The particle displacements in the $x$- and $z$-directions—$u_z$ and $u_x$, respectively—generally are given by equations (15.26); for this case, we obtain

$$u_z = \frac{\partial \phi}{\partial z} + \frac{\partial \psi}{\partial x}$$
$$u_x = \frac{\partial \phi}{\partial x} - \frac{\partial \psi}{\partial z} \tag{18.51}$$

The particle displacements in the *x*- and *z*-directions become

$$\begin{aligned} u_z(x,z,\omega) &= \left[ i\gamma_{p,1} D_{p,1} \exp\left(i\gamma_{p,1} z\right) + ikD_{s,1} \exp\left(i\gamma_{s,1} z\right)\right] \exp\left(ikx - i\omega t\right) \\ u_x(x,z,\omega) &= \left[ ikD_{p,1} \exp\left(i\gamma_{p,1} z\right) - i\gamma_{s,1} D_{s,1} \exp\left(i\gamma_{s1} z\right)\right] \exp\left(ikx - i\omega t\right) \end{aligned} \tag{18.52}$$

In chapter 15 we found the shear stress $\sigma_{zx}$ as part of equation (15.36), which defines the shear stress as

$$\begin{aligned} \sigma_{zx} &= \left[ -2k\rho_1 \gamma_{p,1} c_{s,1}^2 D_{p,1} \exp\left(i\gamma_{p,1} z\right) + \rho_1 \left(\omega^2 - 2k^2 c_{s,1}^2\right) D_{s,1} \exp\left(i\gamma_{s,1} z\right)\right] \\ &\quad \times \exp\left(kx - i\omega t\right) \end{aligned} \tag{18.53}$$

Because the boundary condition requires that the shear stress $\sigma_{zx}$ must be zero at a boundary between vacuum (air) and a solid, the amplitudes of the two waves are dependent

$$D_{p,1} = D_{s,1} \frac{\omega^2 - 2k^2 c_{s,1}^2}{2k\gamma_{p,1} c_{s,1}^2} \tag{18.54}$$

Therefore, when calculated as functions of depth and time, the horizontal and vertical components $u_z(z,t)$ and $u_x(z,t)$ of the particle displacements, respectively, are given as

$$\begin{aligned} u_z(z,t) &= D_{p,1} Re\left\{ \left[ i\gamma_{p,1} \exp\left(i\gamma_{p,1} z\right) + i \frac{2k^2 \gamma_{p,1} c_{s,1}^2}{\omega^2 - 2k^2 c_{s,1}^2} \exp\left(i\gamma_{s,1} z\right)\right] \exp\left(-i\omega t\right)\right\} \\ u_x(z,t) &= D_{p,1} Re\left\{ \left[ ik \exp\left(i\gamma_{p,1} z\right) - i \frac{2k\gamma_{p,1} \gamma_{s,1} c_{s,1}^2}{\omega^2 - 2k^2 c_{s,1}^2} \exp\left(i\gamma_{s,1} z\right)\right] \exp\left(-i\omega t\right)\right\} \end{aligned}$$

(18.55)

Note that the vertical wave numbers $\gamma_{p,1}$ and $\gamma_{s,1}$ are imaginary and were defined in equation (18.41), while the horizontal wave number *k*, which is defined as $k = \omega / v$, was obtained by solving equation (18.48).

In figure 18.14, we have a plot of the horizontal and vertical particle displacements as functions of depth *z*, respectively represented by $u_z(z)$ and $u_x(z)$. We have assumed that the Poisson ratio is 0.25, which corresponds to $c_s / c_p = 0.56$. The displacement amplitudes are plotted as a function of depth relative to a Rayleigh wavelength, that is, relative to $\lambda_R = 2\pi / k$, with the phase speed being given by equation (18.48). The plots in figure 18.15 are examples of particle movements that have been calculated using equations (18.55).

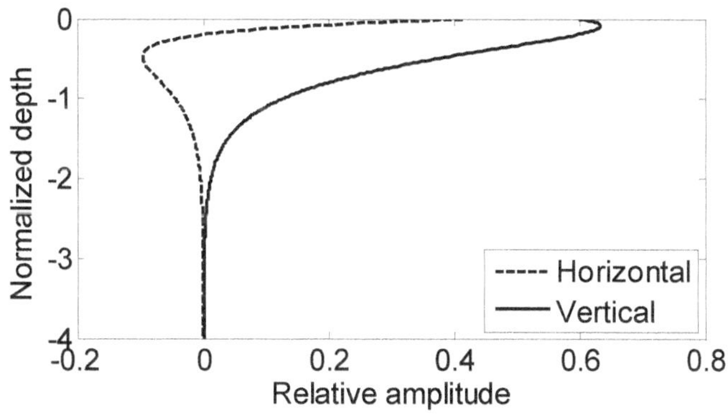


Figure 18.14. Plots of the displacement amplitudes of the horizontal and vertical components of the particle movement caused by a Rayleigh wave, calculated as a function of depth.

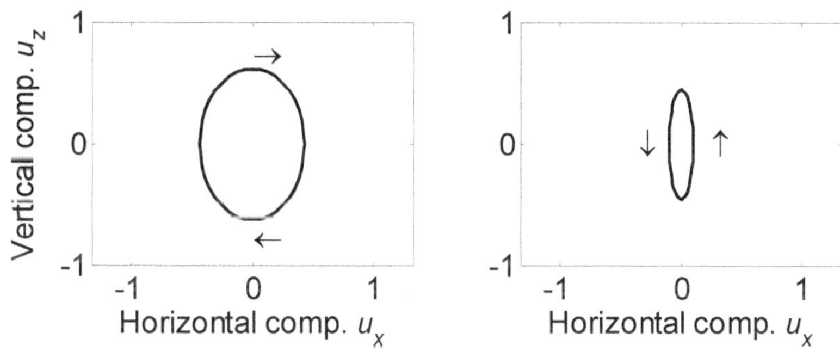


Figure 18.15. Particle movement caused by a Rayleigh wave (left) at the interface $z = 0$ and (right) at a depth $z = 0.4 \cdot \lambda_R$. Arrows show the directions of movement.

At the depth $z = 0$ we find that the amplitudes of the particle movement along the $x$- and $z$-axes, represented respectively by $u_z(z,t)$ and $u_x(z,t)$, are defined as

$$\begin{aligned} u_z(z,t) &= -0.62 \cdot \sin(\omega t) \\ u_x(z,t) &= 0.43 \cdot \cos(\omega t) \end{aligned} \tag{18.56}$$

Equations (18.56) describe a prograde elliptical movement. At the depth $z = 0.4 \cdot \lambda_R$, the particle movement becomes retrograde, so that the particle movement is given as

$$\begin{aligned} u_z(z,t) &= 0.45 \sin(\omega t) \\ u_x(z,t) &= -0.1 \cos(\omega t) \end{aligned} \tag{18.57}$$

### *18.4.2 Pulse responses of interface waves*

To continue our examination of interface waves, we now discuss the time responses obtained from receivers at different distances from the source. The situation we consider is shown in figure 18.16. The water depth is 100 m and we use an impulsive type source with a relatively broad frequency band centered at 5 Hz. The purpose of the experiment is to measure the sound speed in the bottom, but here we use simulated data to illustrate the method. We consider the case where the bottom consists of two elastic layers, layer 1 with thickness $d$ over a semifinite half-space. The geoacoustic parameters are given in table 18.1. The responses are calculated by the computer code OASES (SAFARI) program, which is actually a numerical solution for equation (18.1).

Table 18.1

| | Density $\rho$ kg/m$^3$ | Comp. speed $c_p$ m/s | Shear speed $c_s$ m/s | Attenuation dB per wavelength |
|---|---|---|---|---|
| Water | 1000 | 1500 | 0 | 0 |
| Layer 1 | 2000 | 2500 | 400 | 0.2 |
| Layer 2 | 2200 | 3000 | 600 | 0.2 |

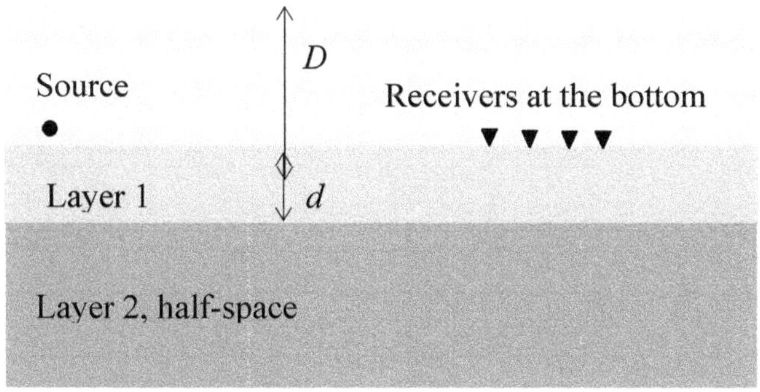


Figure 18.16. Measurement of interface waves with an impulsive point source a short distance above the seabed, and a number of hydrophone receivers on the seafloor.

We consider two cases. In the first case the bottom is homogenous and modeled as a homogeneous semi-infinite half-space with the parameters of layer 1 in table 1; this is the case where the layer thickness $d$ is infinite. The time series responses are shown in figure 18.17. In the second case the bottom is layered with layer 1 having a thickness of $d = 20$ m over the homogenous half-space. The time series responses for this case are shown in figure 18.18. These two examples exhibit a number of interesting features that we now discuss in some detail.

Consider first the homogeneous case shown in figure 18.17. Each trace shows that the responses are composed of a late arrival and a group of earlier arrivals. The respective wave speeds can easily be deduced from the time plots. We find that the slow wave has a speed of approximately 350 m/s, which is the expected speed of an interface wave in accordance with our earlier discussions. The pulse shape of this wave is actually a replica of the source signal; hence the interface wave propagates without distortion in the case of a homogeneous bottom.

The first group of arrivals in each trace is actually the sum of three contributions: a direct, a bottom-reflected, and a refracted arrival. The wave speed of the direct and bottom-reflected signal is equal to the sound speed in the water, which is 1500 m/s. The refracted arrivals are recognized as the first-break arrivals that appear at the longer ranges. The refracted wave has the speed of the compressional sound speed in the bottom, which in this case is 2500 m/s. Note that the surface-reflected signal arrives too late to be observed in the time series traces of figures 18.17 and 18.18.

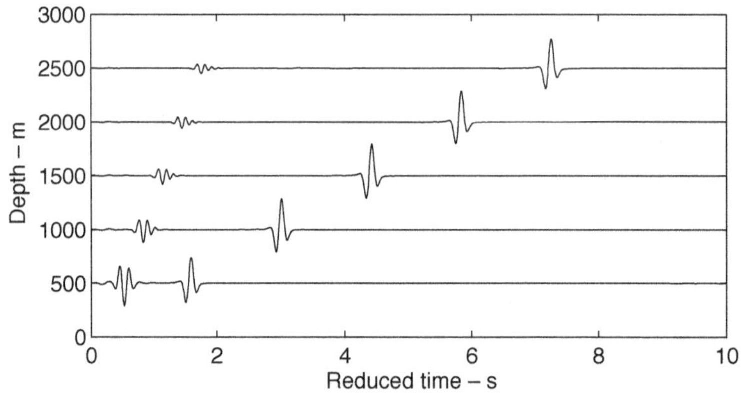


Figure 18.17. Time series of signals received at five depths with spacing of 500 m from an impulsive point source positioned slightly above a homogeneous elastic bottom.

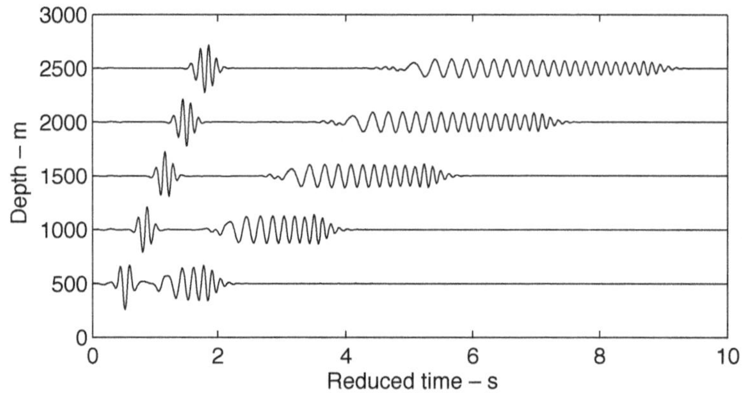


Figure 18.18. The signal received from a point source positioned above a stratified elastic medium.

We now turn to the layered bottom case and the time series traces shown in figure 18.18. In the first group of arrivals the direct and the bottom-reflected arrivals are about the same as in the previous case, but there are now two refracted waves. These can be analyzed as outlined in section 18.2 whereby we find that one propagates with the compressional speed of layer 1 and the other with the compressional speed of the half-space, 2500 m/s and 3000 m/s, respectively.

However, the most striking feature of figure 18.18 is the interface wave arrivals, which are now spread out in time due to time dispersion. This means that the interface wave now is dispersive with the lower frequency components travelling faster than the higher frequency components. From the graph we find the maximum and minimum speeds of the interface wave are approximately 450 m/s and 350 m/s. These two wave speeds are the expected interface wave speeds for the shear speeds in layer 1 and the half space, respectively.

Dispersive signals such as those illustrated in figure 18.18 can be examined more closely through a ***time-frequency analysis,*** in which the signal is passed through a filter bank consisting of a number of contiguous band-pass filters that, when combined, cover the entire frequency band of the transmitted signal. By detecting the signal amplitudes of the filtered outputs as a function of time, we can make a two-dimensional representation of the received dispersive signal as a function of both time and frequency.

Figure 18.19 shows an example of such analysis applied to the signal received at the 2.5 km range in figure 18.18. The shading bar indicates the relative intensity of the time-frequency components. This type of time-frequency analysis is commonly referred to as Gabor analysis. We should note that the speed measured by this method is the group speed $w$. From figure 18.19 we see that the interface wave starting at about 5 seconds is dispersed both in time and frequency.

When we also know the distance between transmitter and receiver, we can convert arrival times to wave speeds and an example is shown in figure 18.20 using the same trace as used for creating figure 18.19. From this plot we find that the speed varies from about 280 m/s for the slowest high-frequency components to about 500 m/s for the fastest low-frequency components.

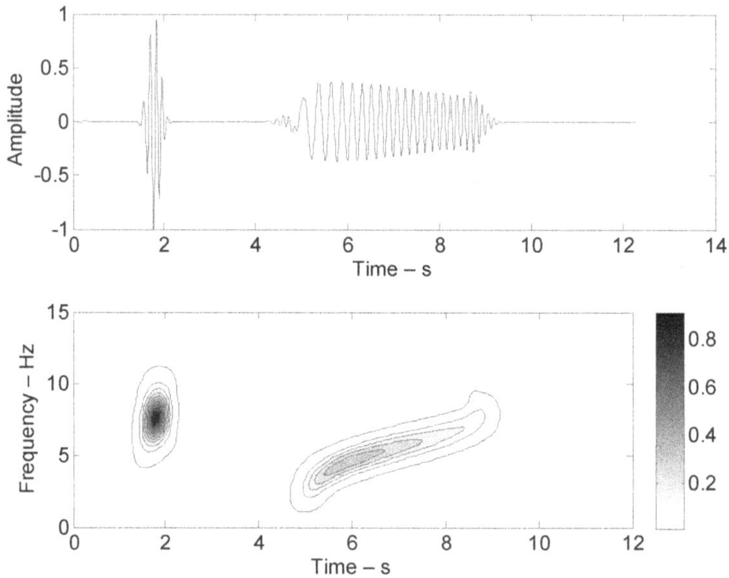


Figure 18.19. Top: The time signal at a receiver 2.5 km from the source (see figure 18.18). Below: Time-frequency analysis of the same signal showing the frequency dispersion of the interface wave arrival.

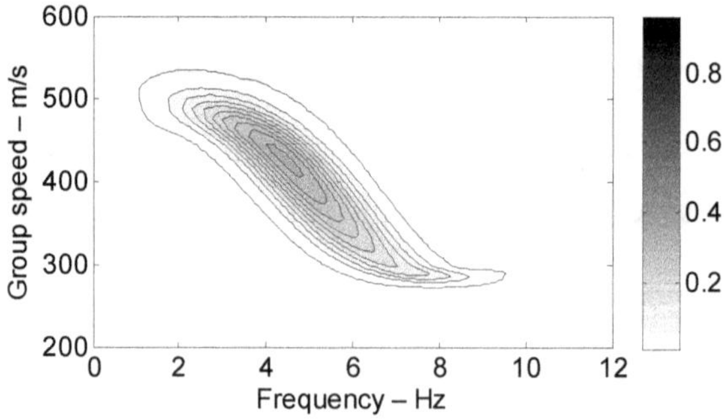


Figure 18.20. The frequency dispersion curve for the interface wave received at a range of 2.5 km from the source.

### *18.4.3 A practical case*

We will end this discussion on interface waves by presenting a result from a field experiment similar to the simulated case of the previous paragraph. Small explosive charges were used as sound sources and the signals were received at a 24-hydrophone array positioned on the seafloor; the hydrophones were spaced 2.5 m apart at a distance of 230–290 m from the source. The signal frequency band was approximately 5–16 Hz. The experimental setup is illustrated in figure 18.21.

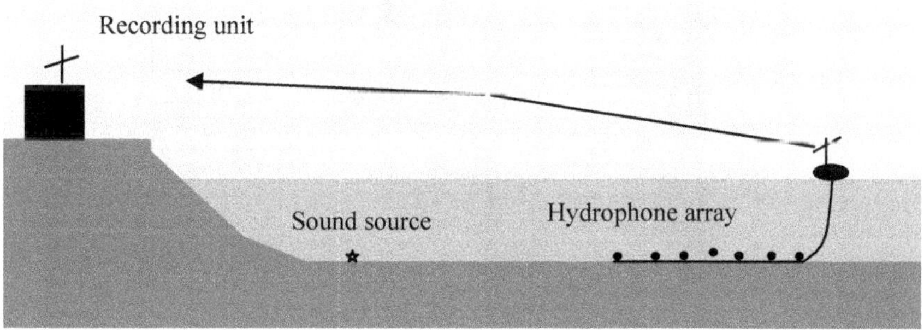


Figure 18.21. Experimental setup for reception of interface waves by a 24-hydrophone array situated on the seabed.

Figure 18.22 contains plots of the 24 signals received by the hydrophone array. As in the modeled numerical example we considered earlier, the first arrivals are a mixture of refracted and direct waves. The interface waves are in the 1.5 s to 3.5 s time interval and are clearly dispersive, with the higher-frequency components arriving later than the lower-frequency components. This indicates that the shear speed varies with depth in the seafloor.

Thus, by inverting a measurement of the frequency dispersion of interface waves, it is possible to estimate how the shear wave speed and modulus will change with increasing depth in the bottom sediment. The first step is to perform a time-frequency analysis of the interface waves, as was done to obtain figure 18.20.

The resulting curve for group speed versus frequency is then inverted to produce an estimate of the shear speed profile in the bottom. Details on the inversion scheme can be found in the papers by Caiti, Stoll, and Akal (1992 and 1994). Figure 18.23

shows the estimate of shear speed as a function of the depth in the bottom; these results are adapted from Frivik and Hovem (1998). The shear speed increases almost linearly from approximately 120 m/s at the surface to 350 m/s at 20 m depth. It is difficult to verify these results by independent measurement in the field. However the results appear reasonable and are likely the best estimate that can be obtained.

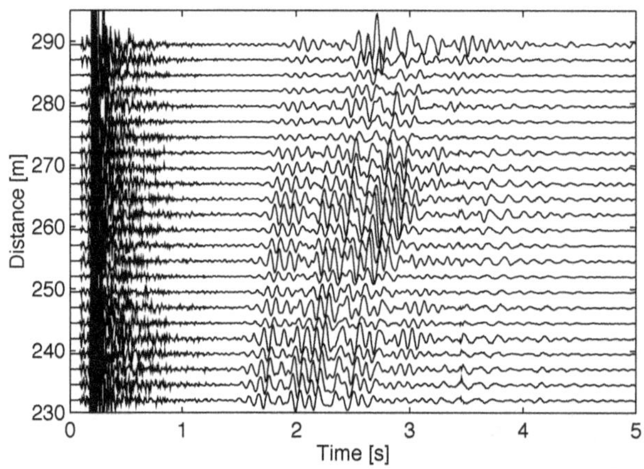


Figure 18.22. Original recording of the signals received by a 24-hydrophone array located on the seabed at a 230–290 m distance from the source.

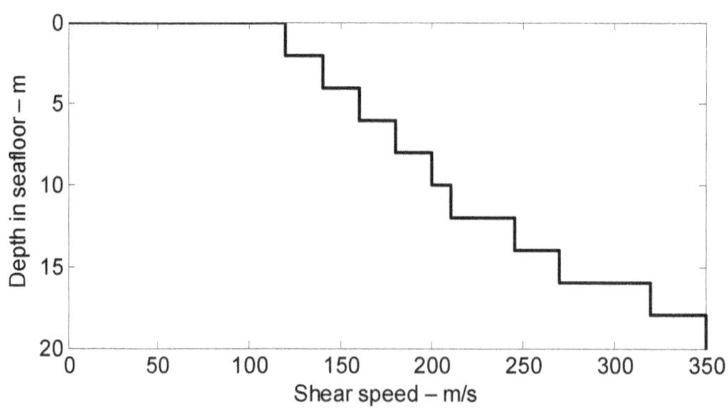


Figure 18.23. Estimate of shear speed as a function of the depth in the seafloor based on inversion of the dispersion data in figure 18.22.

# *References*

Abramowitz, M., and I. A. Stegun. 1965. *Handbook of mathematical functions*. New York: Dover Publications.

Aki, K., and P. G. Richards. 2002. *Quantitative seismology,* 2nd ed. Sausalito, CA: University Science Books.

Albers, V. M. 1972. *Underwater sound: Benchmark papers in acoustics.* Stroudsburg, PA: Dowden, Hutchingson, and Ross.

Allard, J.-F. 1993. Propagation of sound in porous media: Modelling sound absorbing materials. London: Elsevier Applied Science.

Anderson, A. L., and L. D. Hampton. 1980a. Acoustics of gas-bearing sediment: I. Background. *J. Acoust. Soc. Am.* 67, 1865–1889.

Anderson, A. L., and L. D. Hampton. 1980b. Acoustics of gas-bearing sediment: II. Measurements and models. *J. Acoust. Soc. Am.* 67, 1890–1903.

Anderson, V. C. 1950. Sound scattering from a fluid sphere. *J. Acoust. Soc. Am.* 22, 426–431 (1950).

ANSI S1.20. 1988. Procedures for calibration of underwater electro-acoustic transducers. New York: Acoustical Society of America.

Bachmann, W. 1973. A theoretical model for the backscattering strength of a composite-roughness sea surface. *J. Acoust. Soc. Am.* 54, 712–716.

Bartram, J. 1972. A useful analytical model for the parametric array. *J. Acoust. Soc. Am.* 42, 1042–1044.

Beckmann, P., and A. Spizzichino. 1963. *The scattering of electromagnetic waves from rough surfaces.* Oxford, UK: Pergamon Press.

Beranek, L. L., *Acoustics*. 1954. New York: McGraw-Hill; currently available from Acoustical Society of America, 1986.

Bergmann, P. G. 1946. The wave equation in a medium with a variable index of refraction. *J. Acoust. Soc. Am.* 17, 329–333.

Berktay, H. O. 1965. Possible exploitation of nonlinear acoustics in underwater transmitting applications. *J. Sound Vibration* 2, 439–461.

Berktay, H. O., and D. J. Leahy. 1974. Farfield performance of parametric transmitters. *J. Acoust. Soc. Am.* 55, 539–546.

Berryman, J. G. 1980a. Long-wavelength propagation in composite elastic media: I. Spherical inclusions. *J. Acoust. Soc. Am.* 68, 1809–1819.

Berryman, J. G. 1980b. Long-wavelength propagation in composite elastic media: II. Ellipsoidal inclusions. *J. Acoust. Soc. Am.* 68, 1820–1831.

Beyer, R. T. 1960. Parameters of nonlinearity in fluids. *J. Acoust. Soc. Am.* 32, 719–721.

Beyer, R. T. 1974. *Nonlinear acoustics.* Providence, RI: Brown University. Written for Naval Ship Systems Command, U.S. Department of the Navy.

Biot, M. A. 1962. Generalized theory of acoustic propagation in porous dissipative media. *J. Acoust. Soc. Am.* 34, 1254–1264.

Biot, M. A. 1956a. Theory of propagation of elastic waves in a fluid saturated porous solid: I. Low-frequency range. *J. Acoust. Soc. Am.* 28, 168–178.

Biot, M. A. 1956b. Theory of propagation of elastic waves in a fluid saturated porous solid: II. Higher frequency range. *J. Acoust. Soc. Am.* 28, 179–191.

Bjørnø, L. 1977. Finite-amplitude wave propagation through water-saturated marine sediments. *Acustica* 38(4), 195–200.

Blackstock, D. T. 1964. Thermoviscous attenuation of plane, periodic, finite-amplitude sound waves. *J. Acoust. Soc. Am.* 36, 534–542.

Bobber, R. J. 1970. Underwater electroacoustic measurements. Washington, DC: U.S. Government Printing Office, Naval Research Laboratory. Reprinted by Peninsula Publishing, Los Altos Hills, CA, 1988.

Boyle, C. A. 1984. *Acoustic waveguides.* New York: John Wiley & Sons.

Brekhovskikh, L. M. 1980. *Waves in layered media,* 2nd ed. New York: Academic Press.

Brekhovskikh, L. M., and O. A. Godin. 2002. *Acoustics of layered media I. Plane wave and quasi plane waves.* Berlin, Germany: Springer-Verlag.

Brekhovskikh, L. M., and O. A. Godin. 1999. *Acoustics of layered media: II. Point sources and bounded beams,* 2nd ed. Berlin, Germany: Springer-Verlag.

Brekhovskikh, L. M., and Yu. Lysanov. 2003. *Fundamentals of ocean acoustics,* 3rd ed. New York: Springer-Verlag.

Bucker, H. P. 1970. Sound propagation in a channel with lossy boundaries. *J. Acoust. Soc. Am.* 48, 1187–1194.

Burdic, W. S. 1984. *Underwater acoustic system analysis.* Englewood Cliffs, NJ: Prentice Hall. Reprinted by Peninsula Publishing, Los Altos Hills, CA, 2003.

Caiti, A., R. D. Stoll, and T. Akal. 1992. Determination of shear profiles by inversion of interface data. In *Shear waves in marine sediments,* J. M. Hovem, M. D. Richardson, and R. D. Stoll, eds. Dorthrecht, The Netherlands: Kluwer Academic Publishers, 557–565.

Caiti, A., R. D. Stoll, and T. Akal. 1994. Estimation of shear wave velocity in shallow marine sediments. *IEEE J. Ocean. Eng.* 19, 58–72.

Chapman, N. R. 1985. Measurement of the waveform parameters of shallow explosive charges. *J. Acoust. Soc. Am.* 78, 672–681.

Chapman, N. R., L. Jaschke, M. McDonald, H. Schmidt, and M. Johnson. 1997. Low frequency geoacoustic tomography experiments using light bulb sound sources in the Haro Strait sea trial. In *Oceans 97 MTS/IEE Conference Proceedings*, vol. 2, 763–768. Piscataway, NJ: IEEE Oceanic Engineering Society.

Chapman, R. P., and D. H. Scott. 1964. Surface backscattering strengths measured over an extended range of frequencies and grazing angles. *J. Acoust. Soc. Am.* 36, 1735–1737.

Clay, C. S., and H. Medwin. 1977. *Acoustical oceanography: Principles and applications.* New York: John Wiley & Sons.

Colladon, J. D., and J. C. F Sturm. 1887. Memoir on the compression of liquids, *Ann. Chim. Phys.* (2) 36, 225–257.

Commander, K.W., and A. Prosperetti 1989. Linear pressure waves in bubbly liquids: Comparison between theory and experiments. *J. Acoust. Soc. Am.* 85, 732–746.

Crocker, M. J., ed. 1997. *Encyclopedia of acoustics*, vol. 1. New York: John Wiley & Sons.

Del Grosso, V. A. 1974. New equations for the speed of sound in natural waters (with comparison to other equations). *J. Acoust. Soc. Am.* 56, 1084–1091.

Devin, C., Jr. 1959. Survey of thermal, radiation, and viscous damping of pulsating air bubbles in water. *J. Acoust. Soc. Am.* 31, 1654–1667.

DiNapoli, F. R., and R. L. Deavenport. 1980. Theoretical and numerical Green's function of solution in a plane layered medium. *J. Acoust. Soc. Am.* 67, 92–105.

Dobrin, M. B. 1976. *Introduction to geophysical prospecting*, 3rd ed. New York: McGraw-Hill.

Dosso, S. E., and M. Riedel. 2001. Array element localization for towed marine seismic arrays, *J. Acoust. Soc. Am.* 110, 955–966

Dushaw, B. D., P. F. Worcester, B. D. Cornuelle, and B. M. Howe. 1993. On equations for the speed of sound in seawater. *J. Acoust. Soc. Am.* 93, 255–275.

Dutta, N. C., and H. Odé. 1983. Seismic reflections from a gas-water contact. *Geophys.* 48, 148–162.

Dziewonski, S., S. Bloch, and M. A. Landisman. 1969. A technique for the analysis of transient seismic signals. *Bull. Seism. Soc.* 59, 427–444.

Eckart, C. 1953. The scattering of sound from the sea surface. *J. Acoust. Soc. Am.* 25, 566–570.

Engdahl, G., and I. Mayergoyz. 2000. *Handbook of giant magnetostrictive materials.* San Diego, CA: Academic Press.

Essen, H.-H. 1994. Scattering from a rough sedimental seafloor containing shear and layering. *J. Acoust. Soc. Am.* 95, 1299–1310.

Etter, P. C. 1996. *Underwater acoustic modeling: Principles, techniques, and modeling,* 2nd ed. London and New York: Elsevier Applied Science.

Everbach, E. C., 1997. Parameters of nonlinearity of acoustic media. In M. J. Crocker, ed., *Encyclopedia of acoustics*, vol. 1. New York: John Wiley & Sons.

Ewing, W. M., W. S. Jardetzky, and F. Press. 1957. *Elastic waves in layered media.* New York: McGraw Hill.

Faran, James J., Jr. 1951. Sound scattering by solid cylinders and spheres. *J. Acoust. Soc. Am.* 23, 405–418.

Fjær, E., R. M. Holt, P. Horsrud, A. M. Raaen, and R. Risnes. 1992. *Petroleum-related rock mechanics.* Amsterdam: Elsevier.

Foote, K. G. 1987. Fish target strength for use in echo integrator surveys. *J. Acoust. Soc. Am.* 82(3), 981–987.

Fowler, C. M. R. 1990. *The solid earth: An introduction to global geophysics.* New York: Cambridge University Press.

Francois, R. E., and G. R. Garrison. 1982. Sound absorption based on ocean measurements: Part I: Pure water and magnesium sulfate contributions. *J. Acoust. Soc. Am.* 72(3), 896–907.

Francois, R. E., and G. R. Garrison. 1982. Sound absorption based on ocean measurements: Part II. Boric acid contribution and equation for total absorption. *J. Acoust. Soc. Am.* 72(6), 1879–1890.

Freedman, A. 1962a. Mechanism of acoustic echo formation. Acustica 12, 12–21.

Freedman, A. 1962b. The high frequency echo structure of some simple body shapes. *Acustica* 12, 61–70.

Frisk, G. V. 1994. *Ocean and seabed acoustics: A theory of wave propagation*. Englewood Cliffs, NJ: Prentice-Hall.

Frivik, S. A., and J. M. Hovem. 1998. Determination of shear properties in the upper seafloor using seismo-acoustic interface waves. In *IEEE/OES Conference Proceedings of Oceans '98*. Nice, France; 28 September–1 October 1998; 682–686.

Geertsma, J., and D. C. Smith. 1961. Some aspects of elastic wave propagation in fluid saturated porous solids. *Geophys*. 26, 169–181.

Grant, D., J. Andrew, J. Shooter, and K. Focke. 2003. Light bulbs as underwater acoustic sources. Technical Letter ARL-EV-TL-03-21. Applied Research Laboratories, The University of Texas at Austin.

Hamilton, E. L. 1980. Geoacoustic modeling of the sea floor. *J. Acoust. Soc. Am.* 68, 1313–1340.

Hamilton, E. L. 1987. Properties of sediments. In *Acoustics and ocean bottom*, A. Lara-Sáenz, C. Ranz-Guerre, and C. Carbó-Fité, eds. II F. A. S. E. Specialized Conference, Madrid, June.

Hamilton, E. L., and R. T. Bachman. 1982. Sound speeds and related properties of marine sediments. *J. Acoust. Soc. Am.* 72, 1891–1904.

Hamilton, M. F. 1997. Nonlinear effects in sound beams. In *Encyclopedia of acoustics*, vol. 1, M. J. Crocker, ed. New York: John Wiley & Sons, Inc.

Hamilton, M. F., and D. T. Blackstock. 1997. *Nonlinear acoustics*. Hoboken, NJ: John Wiley & Sons.

Harker, A. H., P. Schofield, B. P. Stimson, R. G. Taylor, and J. A. G. Temple. 1991. Ultrasonic propagation in slurries. *Ultrasonic* 29, 427–438 (November).

Haslett, R. W. G. 1970. Acoustic echoes from targets under water. In *Underwater acoustics*, R. W. B. Stephens, ed. New York, Wiley Interscience, 1970, 129–198.

Hobæk, H. 1967. Experimental investigation of an acoustical end-fire array. *J. Sound Vib.* 6, 460–463.

Horne, J. K. 2000. Acoustic approaches to remote species identification: a review. *Fish. Oceanogr.* 9(4), 356–371.

Hovem, J. M. 1995. Acoustic waves in finely layered media. *Geophysics* 40(4), 1217–1221.

Hovem, J. M. 1979. The nonlinearity parameter of saturated marine sediments. *J. Acoust. Soc. Am.* 66, 1463–1467.

Hovem, J. M. 1981. Transmission of sound through a porous disk. *Appl. Phys. Lett.* 39, 590–591.

Hovem, J. M., and G. D. Ingram. 1979. Viscous attenuation of sound in saturated sand. *J. Acoust. Soc. Am.* 66, 1807–1812.

*IEEE Journal of Oceanic Engineering*. 2003. Special issue on geoacoustic inversion in range-dependent shallow water environments, vol. 28, no. 3, July.

Ishimaru, A. 1978a. Wave propagation and scattering in random media, vol. 1: Single scattering and transport theory. New York: Academic Press.

Ishimaru, A. 1978b. Wave propagation and scattering in random media, vol. 2: Multiple scattering, turbulence, rough surfaces, and remote sensing. New York: Academic Press.

Jackson, D. R., D. P. Winebrenner, and A. Ishimaru. 1986. Application of the composite scattering model to high-frequency bottom backscattering. *J. Acoust. Soc. Am.* 79, 1410–1422.

Jebsen, G. M., and H. Medwin. 1982. On the failure of the Kirchhoff assumption in backscatter. *J. Acoust. Soc. Am.* 72, 1607–1611.

Jensen, F. B., and H. Schmidt. 1986. Shear properties of ocean sediments determined from numerical modeling of Scholte wave data. In *Ocean seismo-acoustics: Low frequency underwater acoustics,* T. Akal and J. M. Berkson, eds. New York: Plenum Press, 683–692.

Jensen, F. B., W. A. Kuperman, M. B. Porter, and H. Schmidt. 1994. *Computational ocean acoustics*. New York: AIP Press, American Institute of Physics, Woodbury.

Johnson, D. H. 1982. The application of spectral estimations methods to bearing estimation problems. *Proc. IEEE* 70(9), 1018–1028.

Kinsler, L. E., A. R. Frey, A. B. Coppens, and J. V. Sanders. 2000. *Fundamentals of acoustics,* 4th ed. New York: John Wiley & Sons.

Koch, R. A., C. Penland, P. J. Vidmar, and K. E. Hawker, 1983. On the calculation of normal mode group velocities and attenuation. *J. Acoust. Soc. Am.* 73, 820–825.

Kreyszig, E. 1999. *Advanced engineering mathematics,* 8th ed. New York: John Wiley & Sons, Inc.

Kuster, G. T., and M. N. Toksös. 1974a. Velocity and attenuation of seismic waves in two-phase media: Part I. Theoretical formulations. *Geophysics* 39(5), 597–606.

Kuster, G. T., and M. N. Toksös. 1974b. Velocity and attenuation of seismic waves in two-phase media: Part II. Experimental results. *Geophysics* 39(5), 607–618.

Landow, L. D., and E. M. Lifshitz. 1959. *Fluid mechanics*. Oxford, UK: Pergamon Press.

Langevin, R. A. 1954. The electro-acoustic sensitivity of cylindrical ceramic tubes. *J. Acoust. Soc. Am.* 26(3), 421–427.

Laskey, M. 1977. Review of undersea acoustics to 1950. *J. Acoust. Soc. Am.* 61(2), 283–297.

Leighton, T. G. 1996. *The acoustic bubble*. London: Academic Press.

Levinson, S. J., E. K. Westwood, R. A. Koch, S. K. Mitchell, and C.V. Sheppard. 1994. An efficient and robust method for underwater acoustic normal-mode computations. *J. Acoust. Soc. Am.* 97, 1576–1585.

Lurton, X. 2002. An introduction to underwater acoustics: Principles and applications. Chichester, UK: Springer Praxis Publishing.

Mackenzie, K. V. 1961. Bottom reverberation for 530 and 1030-cps sound in deep water. *J. Acoust. Soc. Am.* 33(11), 1498–1504.

MacLennan, D. N. 1990. Acoustical measurements of fish abundance. *J. Acoust. Soc. Am.* 87(1), 1–15.

Marsh, H. W. 1963. Sound reflection and scattering from the sea surface. *J. Acoust. Soc. Am.* 35, 240–244.

Medwin, H., and S. C. Clay. 1998. *Fundamentals of acoustical oceanography*. Boston: Academic Press.

Medwin, H., E. Childs, and G. M. Jebsen. 1982. Impulse studies of double diffraction: A discrete Huygens interpretation. *J. Acoust. Soc. Am.* 72, 1005–1013.

Mellen, R. 1973. Recent developments in parametric sonar. SACLANT ASW Research Centre Memorandum SR 23, La Spezia, Italy, 15 September.

Mellen, R. H., and Moffett, M. B. 1971. A model for parametric sonar radiator design model. NUSC Technical Memo PA41–229-71, Naval Underwater Systems Center.

Midtun, O. 1962. *Fiskeletning med sonar* (English translation: *Fish detection with sonar*). Oslo: Simonsen Radio A.S.

Minnaert, M. 1933. On musical air-bubbles and the sounds of running water. *The London, Edinburgh and Dublin Philosophical Magazine and Journal of Science*, Series 7 16, 235–248.

Moffett, M. B., and R. H. Mellen. 1977. Model for parametric acoustic sources. *J. Acoust. Soc. Am.* 61, 325–337.

Moffett, M. B., and R. H. Mellen. 1981. Nearfield characteristics of parametric acoustic sources. *J. Acoust. Soc. Am.* 69, 404–409.

Moffett, M. B., and W. L., Konrad. 1997. Nonlinear sources and receivers. In *Encyclopedia of Acoustics*, vol. 1, M. J. Crocker, Ed. New York: John Wiley & Sons, Inc.

Morse, P. M., and H. Feshback. 1953. *Methods of theoretical physics*. New York: McGraw-Hill Book Company.

Morse, P. M., and K. U. Ingard. 1968. *Theoretical acoustics.* New York: McGraw-Hill, Inc..

Muir, T. G. 1974. A survey of several nonlinear acoustics experiments on traveling wavefields. In *Proc. 1973 Symp. Finite Amplitude Wave Effects in Fluids*, L. Bjørnø, ed. Copenhagen: Guildford.

Muir, T. G., and J. E. Blue. 1969. Experiments on the acoustic modulation of large-amplitude waves. *J. Acoust. Soc. Am.* 46, 227–232.

Muir, T. G., and J. G. Willette. 1972. Parametric acoustic transmitting arrays. *J. Acoust. Soc. Am.* 52, 1481–1486.

Munk, W., P. Worcester, and C. Wunsch. 1995. *Ocean acoustic tomography*. New York: Cambridge University Press.

Nielsen, R. O. 1991. *Sonar signal processing.* Norwood, MA: Artech House.

Officer, C. B. 1958. Introduction to the theory of sound transmission. New York: McGraw-Hill.

Ogilvy, J. A. 1991. *Theory of wave scattering from random rough surfaces*. Bristol, UK: Institute of Physics Publishing.

Papadakis, E. P. 2000. *Ultrasonic instruments and devices*. London: Academic Press.

Papoulis, A. 1984. Probability, random variables, and stochastic processes, 2nd ed. New York: McGraw-Hill.

Pekeris, C. L. 1948. Theory of propagation of explosive sound in shallow water. *Geol. Soc. Am., Mem.* 27.

Pettersen, P., J. M. Hovem, A. Løvik, and T. Knudsen. 1977. A new sub-bottom profiling sonar using a nonlinear sound source. *Radio and Electronic Engineer* 47(3), 105–111.

*Physics of sound in the sea,* Parts I–IV, NDRC Summary. 1969. Technical Reports. Washington, DC: U.S. Government Printing Office. Reprinted by Peninsula Publishing, Los Altos Hills, CA, 1996.

Pierce, A. D. 1989. *Acoustics: An introduction to its physical principles and applications.* New York: American Institute of Physics. Reprinted by Peninsula Publishing, Los Altos Hills, CA, 1998.

Porter, M. B. 1991. The Kraken normal mode program. Rep.SM-245. SACLANT Undersea Research Centre (Nato Undersea Research Centre), La Spezia, Italy.

Porter, M. B., and E. L. Reiss. 1984. A numerical method for ocean acoustic normal modes. *J. Acoust. Soc. Am.* 76, 244–252.

Rauch, D. 1980. Experimental and theoretical studies of seismic interface waves in coastal waters. In *Bottom-interacting ocean acoustics,* W. A. Kupermann and F. B. Jensen, eds. New York: Plenum Press, 307–327.

Schmidt, H. 1988. SAFARI: Seismo-acoustic fast field algorithm for range-independent environments: User's guide. SACLANT Undersea Research Centre, SACLANTCEN Report SR-113, La Spezia, Italy.

Sheriff, R. E., and L. P. Geldhart. 1994. *Exploration seismology,* 2nd ed. New York: Cambridge University Press.

Simmonds, J., and D. MacLennan. 2005. *Fishery acoustics: Theory and practice,* 2nd ed. Oxford, UK: Blackwell Publishing.

Spiesberger, J. L., and K. Metzger. 1991. A new algorithm for sound speed in seawater. *J. Acoust. Soc. Am.* 89, 2677–2688.

Stakgold, I. 1979. *Green's functions and boundary value problems.* New York: John Wiley & Sons, Inc.

Stansfield, D. 1991. *Underwater electroacoustic transducers.* Bath, UK: Bath University Press and Bath University Institute of Acoustics. Reprinted by Peninsula Publishing, Los Altos Hills, CA, 2003.

Stephens, R. W. B. 1970. *Underwater acoustics.* New York: Wiley-Interscience.

Stoll, R. D. 1977. Acoustic waves in ocean sediments. *Geophys.* 42, 715–725.

Stoll, R. D. 1979. Experimental studies of attenuation in sediments. *J. Acoust. Soc. Am.* 66, 1152–1160.

Stoll, R. D. 1985. Marine sediment acoustics. *J. Acoust. Soc. Am.* 77, 1789–1799.

Stoll, R. D. 1980. Theoretical aspects of attenuation in sediments. *J. Acoust. Soc. Am.* 68, 1341–1350.

Stoll, R. D. 1989. *Sediment acoustics*, Lecture Notes in Earth Sciences. Berlin: Springer Verlag. Reprinted by Peninsula Publishing, Los Altos Hills, CA, 2006.

Stoll, R. D., and G. M. Bryan. 1970. Wave attenuation in saturated sediments. *J. Acoust. Soc. Am.* 34, 1440–1447.

Stoll, R. D., and T. K. Kan. 1981. Reflection of acoustic waves at a water-sediment interface. *J. Acoust. Soc. Am.* 70, 149–156.

Strutt, John W., Baron (Lord Rayleigh). 1945. *Theory of sound,* 2 vols. New York: Dover Publications Inc. (orig. 1877–78).

Sund, O. 1935. Echo sounding in fishery research. *Nature* 135, 956.

Svensson, P. U., and R. I. Fred. 1999. An analytical secondary source model of edge diffraction impulse response. *J. Acoust. Soc. Am.* 106, 2331–2344.

Thorp, W. H. 1965. Deep-ocean sound attenuation in the sub- and low-kilocycle-per-second region. *J. Acoust. Soc. Am.* 38, 648–654.

Thorsos, E. I., and D. R. Jackson. 1989. The validity of the perturbation approximation for rough surface scattering using a Gaussian roughness spectrum. *J. Acoust. Soc. Am.* 86, 261–277.

Tindle C. T., and G. E. J. Bold. 1981. Improved ray calculations in shallow water. *J. Acoust. Soc. Am.* 70(3), 813–819.

Tindle, C. T. 1983. Ray calculations with beam displacement. *J. Acoust. Soc. Am.* 72, 1581–1586.

Tindle, C. T., and N. R. Chapman. 1994. A phase function for finding normal mode eigenvalues over a layered elastic bottom. *J. Acoust. Soc. Am.* 96, 1777–1782.

Tindle, C. T., K. M Guthrie, G. E. J. Bold, M. D. Jones, D. Jones, K. O. Dixon, and T. G. Birdsall. 1978. Measurements of the frequency dependence of normal modes. *J. Acoust. Soc. Am.* 64(4), 1178–1185.

Tolstoy, I. 1966. *Wave propagation.* New York: McGraw Hill.

Tolstoy, I., and C. S. Clay. 1966. *Ocean acoustics: Theory and experiment in underwater sound.* New York: McGraw-Hill. Reprinted by American Institute of Physics, Woodbury, New York, 1987.

Tucker, D. G., and Gazey, B. K. 1966. *Applied underwater acoustics.* Oxford, UK: Pergamon Press.

Urick, R. J. 1983. *Principles of underwater sound,* 3rd ed. New York: McGraw-Hill. Reprinted by Peninsula Publishing, Los Altos Hills, CA, 1996.

Urick, R. J., and W. S. Ament. 1949. The propagation of sound in composite media. *J. Acoust. Soc. Am.* 21, 115–119.

Van Trees, H. 1968. *Detection, estimation, and modulation theory, Part I.* New York: J. Wiley & Sons.

Van Trees, H. 1971. *Detection, estimation, and modulation theory, Part III: Radar-sonar signal processing and Gaussian signal in noise.* New York: J. Wiley & Sons.

Van Trees, H. 2002. *Detection, estimation, and modulation theory, Part IV: Optimum array processing.* New York: John Wiley & Sons.

Waite, A. D. 2002. *Sonar for practicing engineers,* 3rd ed. Chichester, UK: J. Wiley & Sons.

Wenz, G. M. 1962. Acoustic ambient noise in the ocean: Spectra and sources. *J. Acoust. Soc. Am.* 34, 1936–1956.

Westervelt, P. J. 1963. Parametric acoustic array. *J. Acoust. Soc. Am.* 35, 535–537.

Westervelt, P. J. 1957a. Scattering of sound by sound. *J. Acoust. Soc. Am.* 29, 199–203.

Westervelt, P. J. 1957b. Scattering of sound by sound. *J. Acoust. Soc. Am.* 29, 934–935.

Weston, D. E. 1960. Underwater explosions as acoustic sound sources. *Proc. Phys. Soc., London* 76(B), 223–249.

Westwood, E. K., and C. T. Tindle. 1987. Shallow water time simulation using ray theory. *J. Acoust. Soc. Am.* 81, 1752–1761 (1987).

Westwood, E. K., and P. J. Vidmar. 1987. Eigenray finding and time series simulation in a layered-bottom ocean. *J. Acoust. Soc. Am.* 81, 912 –924.

Westwood, E. K., C. T. Tindle, and N. R. Chapman. 1996. A normal mode model for acousto-elastic ocean environments. *J. Acoust. Soc. Am.* 100, 3631–3645.

Williams, A. O. 1970. Normal-mode methods in propagation of underwater sound. In *Underwater acoustics*, R. W. B. Stephens, ed. New York: Wiley-Interscience

Wilson, O. B. 1988. *An introduction to the theory and design of sonar transducers.* Los Altos, CA: Peninsula Publishing.

Wilson, P. S., J. L. Ellzey, and T. G. Muir. 1994. Experimental investigation of the combustive sound source. *IEEE JOE*, vol. 20, no. 4.

Wood, A. B., and D. E. Weston. 1964. The propagation of sound in mud. *Acustica* 14, 156–162.

# *Glossary*

| | |
|---|---|
| $\varepsilon_{33}^{T}$ | dielectric constant |
| $\tau$ | relaxation time |
| $\varpi$ | spatial angular frequency |
| $\alpha$ | absorption coefficient, N/m |
| $\theta$ | angle |
| $\varphi$ | angle |
| $\omega$ | angular frequency, $\omega=2\pi f$ |
| $\Gamma$ | aperture function |
| $\delta$ | bubble attenuation |
| $\rho$ | density |
| $\Delta$ | dilation |
| $\phi$ | displacement potential |
| $\Phi$ | displacement potential in frequency domain |
| $\Gamma$ | Goldberg number |
| $\lambda$ | Lamé coefficient |
| $\beta$ | nonlinear parameter |
| $\nu$ | Poisson number |
| $\phi$ | porosity |
| $\Re$ | reflection coefficient |
| $\Omega$ | solid angle |
| $\Theta$ | rotation |
| $\sigma$ | scattering cross section |
| $\mu$ | shear modulus |
| $\sigma$ | sonar scattering cross section |
| $\Omega$ | spatial angle |
| $\Lambda$ | spatial wavelength |
| $\tau$ | travel time |
| $\eta$ | viscosity, Pa·s |
| $\beta$ | viscous skin depth |
| $\kappa$ | wave number |
| $\alpha$ | wave number component |
| $\beta$ | wave number component |
| $\gamma$ | wave number component |
| $\Delta\theta$ | beam width |
| $\Re^{(0)}$ | reflection coefficient |
| $\Re^{(1)}$ | reflection coefficient |
| $\Psi(k_x, k_y)$ | two dimensional wave number spectrum |
| $\xi(s)$ | ray parameter |
| $\zeta(s)$ | ray parameter |
| $\delta$, $\delta_{p}$, $\delta_s$ | log decrement, nepers/wavelength |
| $\delta_{\mathrm{dB}}$ | log decrement dB/wavelength |
| $\sigma_0$ | saturation parameter |
| $\omega_{0,m}$ | cut-off frequency, angular |
| $\alpha_{33}$, $\beta_{33}$, $h_{33}$ | piezoelectric constants |
| $\sigma_a$ | surface scattering strength per unit area |
| $\sigma_{bs}$ | backscattering cross section |
| $\gamma_d$ | dipole contribution |
| $\alpha_{dB}$ | absorption coefficient, dB/m |
| $\delta_{dB}$ | log decrement, dB/wavelength |
| $\rho_f$ | fluid density, kg/m$^3$ |
| $\sigma_h$ | standard deviation of surface heights |
| $\Delta L$ | length increment |
| $\gamma_m$ | eigen values of vertical wave numbers |
| $\gamma_m$ | monopole contribution |
| $\psi_m(r)$ | mode function range dependent |
| $\xi_m(z)$ | mode functions, depth dependent |
| $\sigma_n$ | statistical variance |
| $\Lambda_{nm}$ | mode interference period |

| | |
|---|---|
| $\omega_p$ | parallel resonance frequency |
| $\Delta r$ | range resolution $\omega_s$ series resonance frequency |
| $\rho_s$ | solid density, kg/m$^3$ |
| $\sigma_s$ | total scattering cross section |
| $u_x, u_y, u_z$ | particle displacement |
| $\mathbf{u}$ | particle displacement vector |
| $\dot{u}_x, \dot{u}_y, \dot{u}_z$ | particle velocity components |
| $SL_p$ | primary source level |
| $SL_s$ | secondary source level |
| $SL_s^*$ | secondary source level |
| $Q$ | source strength |
| $d_{31}$, $d_{33}$ | piezoelectric constants |
| $Y_{11}^T$, $Y_{33}^E$ | elastic constants |
| $k^2$, $k_{33}^2$, $k_{31}^2$ | electromechanical coupling coefficients |
| $\mathbf{a}$ | acceleration vector |
| $A$ | radiation area |
| $a_p$ | poresize parameter |
| $a_x$, $a_y$, $a_z$ | acceleration components |
| $B$ | band width; permeability |
| $BL$ | bottom loss |
| $\tilde{c}$ | finite amplitude sound speed |
| $C$ | capacitance; elastic constant |
| $c^*$ | effective sound speed |
| $C_0$ | capacitance |
| $c_e$ | effective propagation speed |
| $c_i$ | imaginary component of sound speed |
| $C_p$ | heat capacity at constant pressure; parallel capacitance |
| $c_p$ | pressure wave speed |
| $c_s$ | shear wave speed |
| $c_t$ | time averaged speed |
| $C_v$ | heat capacity at constant volume |
| $d$ | detection index |
| $D$ | directivity factor; electrical displacement |
| $d$ | element separation |
| $B(\theta)$, $B(\theta,\varphi)$ | directivity function |
| $DI$ | directivity index |
| $D_p$ | downward going pressure wave |
| $D_s$ | downward going shear wave |
| $DT$ | detection threshold (dB) |
| $E$ | Young's modulus; energy; electrical field strength |
| $EL$ | echo level |
| $F$ | force |
| $f$ | frequency, Hz |
| $F(\theta)$ | form function |
| $f_0$ | resonance frequency |
| $f_{0,m}$ | cut-off frequency, Hz |
| $f_1, f_2$ | relaxation frequencies |
| $f_D$ | doppler shift |
| $f_{kHz}$ | frequency, kHz |
| $f_{opt}$ | optimal sonar frequency |
| $G$ | parametric gain |
| $G(\omega,k,z)$ | Green's function |
| $G(K)$ | isotropic wave number spectrum |
| $G(t)$ | time dependent gain |
| $g(z)$ | sound speed gradient |
| $h$ | height deviation |
| $H$ | plane wave modulus be |
| $I$ | intensity (watt/m$^2$); current |
| $I_{bs}$ | backscattered intensity |
| $I_i$ | incident intensity |
| $I_r$ | reflected intensity |
| $I_s$ | scattered intensity |
| $I_t$ | transmitted intensity |

| | |
|---|---|
| $I_{tot}$ | total scattered intensity |
| $J$ | reciprocity factor |
| $k$ | horizontal wave number component |
| $K$ | spring stiffness; bulk modulus; spatial wave number |
| $K^*$ | effective bulk modulus |
| $K_0$ | volume stiffness, bulk modulus |
| $K_f$ | fluid bulk modulus |
| $K_r$ | frame bulk modulus |
| $K_s$ | gas bulk modulus |
| $K_s$ | solid bulk modulus, Pa |
| $k_x$ | wave number component |
| $k_y$ | wave number component |
| $L$ | elastic constant; electrical inductance |
| $L_a$ | absorption distance |
| $L_s$ | shock formation distance |
| $M$ | mass; elastic constant; Mach number |
| $m_n$ | mean value |
| $M_v$ | receiver voltage sensitivity |
| $N$ | transduction factor; viscous element |
| $NL$ | noise level |
| $NSL$ | noise spectral level |
| $OL$ | surface loss |
| $p$ | sound pressure |
| $p_0$ | sound pressure amplitude |
| $P_a$ | acoustical power |
| $P_D$ | probability of detection |
| $P_e$ | electrical power |
| $p_{eff}$ | effective sound pressure amplitude |
| $p_f$ | pore pressure |
| $P_F$ | probability of false alarm |
| $P_m$ | mechanical power |
| $p_n(z)$ | probability density function |
| $p_p$ | primary sound pressure |
| $p_s$ | secondary sound pressure |
| $Q$ | source strength, |
| $Q$ | quality factor, $Q$-value |
| $q(x)$ | aperture function |
| $Q_m$ | mechanical quality factor, $Q$-value |
| $\mathbf{Q}_{n-1,n}$ | transfer matrix |
| $\mathbf{r}$ | spatial position vector |
| $R_\varphi$ | reflection coefficient displacement potential |
| $R(s)$ | radius of curvature |
| $r(t,f)$ | ambiguity function |
| $r, \varphi, z$ | rectangular coordinates |
| $r, \varphi, \theta$ | spherical coordinates |
| $R_a$ | acoustic radiation resistance |
| $\gamma$ | ratio of specific heats |
| $R_b$ | bottom reflection coefficient |
| $R_l$ | lower reflection coefficient |
| $RL$ | reflection loss, dB |
| $RL$ | reverberation level |
| $R_m$ | mechanical loss resistance |
| $R_n$ | reflection coefficient |
| $r_{n,\,n-1}$ | reflection coefficient |
| ROC | receiver operating characteristics |
| $R_p$ | electrical leakage resistance; reflection coefficient pressure |
| $r_p$ | pressure wave transmission coefficient |
| $R_R$ | Rayleigh distance |
| $R_{tot}$ | total reflection coefficient |
| $R_u$ | reflection coefficient particle displacement |
| $R_u$ | upper reflection coefficient |

| | |
|---|---|
| $S$ | mechanical strain |
| $s$ | condensation |
| $S$ | surface area; salinity |
| $S(\omega)$ | frequency function |
| $s(t)$ | source signal |
| $s(x, y, z)$ | ray path coordinates |
| $S/N$ | signal-to noise ratio |
| $S_a$ | surface scattering strength per unit area |
| $S_f$ | volume scattering strength |
| $S_i$ | transmitter current sensitivity |
| $SL$ | source level |
| $\mathbf{S}_{n-1,n}$ | scattering matrix |
| $S_v$ | transmitter voltage sensitivity |
| $T^{(0)}$ | transmission coefficient |
| $T^{(1)}$ | transmission coefficient |
| $T$ | mechanical stress |
| $T_{\phi}$ | transmission coefficient displacement potential |
| $T_{em}$, $T_{me}$ | transduction coefficients |
| $TL$ | transmission loss, dB |
| $\mathbf{T}_n$ | transport matrix |
| $t_{n, n-1}$ | transmission coefficient |
| $t_p$ | pressure wave transmission coefficient |
| $T_p$ | transmission coefficient pressure |
| $t_{red}$ | reduced time |
| $t_s$ | shear wave transmission coefficient |
| $TS$ | target strength |
| $T_u$ | transmission coefficient particle displacement |
| $U(z)$, $L(z)$ | upper and lower wave functions |
| $U_n$ | upward going wave in layer $n$ |
| $U_p$ | upward going pressure wave |
| $U_s$ | upward going shear wave |
| $v$ | phase speed, m/s |
| $V$ | voltage |
| $\gamma$ | vertical wave number |
| $v_t$ | target velocity |
| $w$ | group speed |
| $W(k, z)$ | Wronskian |
| $W(x, y, z)$ | wave front coordinates |
| $\mathbf{W}_n$ | stress vector |
| $Y$ | electrical admittance |
| Z | acoustic impedance |
| $Z_a$, $Z_b$, $Z_0$ | acoustic impedances |
| $Z_a$, $Z_b$, $Z_p$, $Z_s$ | electro-mechanical impedances |
| $Z_e$, $Z_{eb}$ | electrical impedances |
| $Z_l$, $Z_m$, $Z_{mo}$ | mechanical impedances |

# *Subject Index*

# *About the Author*

Dr. Hovem has broad experience in most applications of underwater acoustics and geophysics, with main emphasis on wave propagation and signal processing. At the Department of Telecommunication at the Norwegian University of Science and Technology (NTNU), he teaches both introductory and advanced courses in marine acoustics. This book is based on the lecture notes he has collected over many years in the classroom.

After obtaining his doctorate in telecommunication from NTNU in 1966, Dr. Hovem worked as a research scientist at the NATO Undersea Research Center (NURC) in Italy until 1972. In the years 1972 to 1987 Dr. Hovem held several positions at SINTEF in Norway, mainly in the fields of telecommunication and geophysics. SINTEF, headquartered in Trondheim, is the largest independent research organization in Scandinavia.

In 1987, Dr. Hovem returned to NURC as chief scientist and head of the environmental research division responsible for the research programs in acoustic modeling, field surveys, and oceanography. Twice, in 1979 and in 2002, he held one-year appointments as a visiting scientist at the Applied Research Laboratories, the University of Texas at Austin (ARL:UT). Since 1991 and until retirement in 2004 he has been a professor of acoustics at NTNU. Dr. Hovem is currently professor emeritus at NTNU and technical advisor at SINTEF.

www.ingramcontent.com/pod-product-compliance
Lightning Source LLC
Chambersburg PA
CBHW081123180926
51611CB00004B/38
*9780932146656*